Cave and Karst Systems of the World

Series editor

James W. LaMoreaux, Tuscaloosa, USA

More information about this series at http://www.springer.com/series/11987

William B. White
Editor

Caves and Karst of the Greenbrier Valley in West Virginia

Springer

Editor
William B. White
Department of Geosciences
The Pennsylvania State University
University Park, PA
USA

ISSN 2364-4591 ISSN 2364-4605 (electronic)
Cave and Karst Systems of the World
ISBN 978-3-319-65800-1 ISBN 978-3-319-65801-8 (eBook)
https://doi.org/10.1007/978-3-319-65801-8

Library of Congress Control Number: 2017949491

Printed on acid-free paper

This Springer imprint is published by Springer Nature
The registered company is Springer International Publishing AG
The registered company address is: Gewerbestrasse 11, 6330 Cham, Switzerland

Preface

In a series documenting the important cave and karst systems of the world, one could hardly overlook the karst of the Greenbrier Valley in West Virginia, USA. The valley exposes the Mississippian Greenbrier Limestone, and in the Greenbrier Limestone are developed some of the most extensive caves in the USA. Within a three-county area, there is a pattern of karst drainage basins, each with one or more associated long caves. There are more than 2000 caves so far discovered, and of these, there are 24 caves with surveyed lengths greater than 5 km, adding up to 508 km of cave passage. Because of the similarity of geologic setting and the similarity of the processes of cave development, it seemed reasonable to base the book on a series of descriptions of the long caves and to show how they fit into a pattern of cavern development dictated by the local geology.

The structure of the book is as follows: Chaps. 2–4 lay out the geology, hydrology, and geomorphology of the region in broad terms as background for the more detailed cave descriptions and interpretations that follow. Cave exploration in the Greenbrier Valley has a rich history, and this is the subject of Chap. 5. Then, follow Chaps. 6–17 that describe in detail the individual drainage basins and their associated caves beginning with Swago Creek on the north and proceeding southwestward to the Laurel Creek Basin in the southernmost exposure of the cavernous limestone. The final three chapters address the biological and paleontological aspects of the cave systems.

The authors for the chapters were chosen for their detailed knowledge of the cave systems that they were describing. However, quite different approaches were used. Some authors chose to write a scientific treatise on their assigned cave. Others provided highly detailed passage-by-passage descriptions, most profusely illustrated. Both approaches have advantages and disadvantages, so it seemed inappropriate to attempt to force the discussion into some re-determined template. Those readers who want a very close-up-and-personal feel for the inside of a Greenbrier Valley cave are urged to read Chap. 10 with its detailed description of the Culverson Creek System.

In describing the history of cave exploration and in following the exploration history of individual caves, there is a considerable text devoted to who did what, when. The authors and the editor offer no apology. The exploration and survey of these long and sometimes difficult caves require skill, time, and dedication. Those who pursued these tasks deserve at least the small recognition that they receive in this volume.

There is an inconsistency in the units used in the chapters. Some authors converted all measurements into metric units; some did not. However, it must be remembered that some of the cave surveys and cave descriptions date back as much as 60 years to the early 1950s. All such maps and descriptions were in the English units of feet and miles, not meters and kilometers. Much of the surface topography is described from the US Geological Survey's series of 7.5-min quadrangle maps. These are scaled in both miles and kilometers, but the essential parts of the maps, the contour lines that describe elevation, are in intervals of 20 or 40 ft. The editorial decision—which is likely to satisfy nobody—was to allow authors to choose their own units. Those who felt comfortable in metric wrote in metric. Those who felt comfortable in English units used English units.

University Park, USA
June 2017

William B. White

Acknowledgements

The factual information underlying these descriptions, particularly the cave maps, has been obtained by many explorers and surveyors whose work extends back over many decades. Primary explorers of the Greenbrier Valley Caves have agreed to write chapters so that the accounts are based on first-hand knowledge—knowledge often acquired by many lengthy and difficult exploration trips. But behind the few names that appear on chapter headings are many other individuals all of whom have made important contributions. Our present-day knowledge of the caves of the Greenbrier Valley is due to tens of thousands of person-hour spent in exploration and the grueling business of surveying. Thanks are due to them all.

Thanks are due to the West Virginia Speleological Survey for permission to reproduce their maps and to extract much written material from their publications. The West Virginia Speleological Survey Bulletins are the primary repository for the detailed knowledge of West Virginia Caves.

The caves of the Greenbrier Valley are on private land. The landowners are owed an immense debt of gratitude for their allowing access to their land and their caves over many years. As is noted in several of the historical accounts in the chapters that follow, the landowners have not always been well-treated for their generosity and as a result some caves are now closed to all exploration.

A large thank-you is owned to the photographers whose images illustrate the caves. The images are credited individually, but it must be said that without the generosity of the photographers, a book such as this would not be possible.

The editor expresses his personal thanks to the authors for their hard work. He also thanks Elizabeth L. White for her invaluable assistance with proofreading and with improving the appearance of many of the figures.

Contents

1 **The Greenbrier Karst** 1
William B. White

2 **Geology of the Greenbrier Valley** 9
William B. White

3 **Hydrology of the Greenbrier Karst** 21
William K. Jones

4 **Karst Geomorphology** 45
William B. White and Elizabeth L. White

5 **The Exploration History of the Greenbrier Valley Caves** 63
George Dasher

6 **The Swago Creek Basin** 97
William B. White, Gary D. Storrick and Mark J. Johnsson

7 **The Little Levels and the Hills Creek Basin** 119
William B. White

8 **The Friars Hole System** 135
Douglas M. Medville and Stephen R.H. Worthington

9 **Caves, Karst, and Science in the Buckeye Creek Cave Watershed** 153
Gregory S. Springer

10 **The Culverson Creek Cave System** 163
Philip C. Lucas

11 **The Contact Caves of Central Greenbrier County** 207
William M. Balfour

12 **Hydrology of the Sinking Creek/Muddy Creek Karst Basin** 231
William K. Jones and William M. Balfour

13 **Organ Cave** 249
Raymond Cole

14 **Caves of the Dickson Spring Basin, Monroe County** 301
William K. Jones

15 **Windy Mouth Cave** 313
David A. Shank, Megan Curry Fucci and Ira D. Sasowsky

16 **Scott Hollow Cave** 339
Stacey B. Lane, Melisa R. Bishop, Michael J. Dore and Ira D. Sasowsky

17 **The Laurel Creek Basin and Greenville Saltpeter Cave** 359
William B. White

18 **Terrestrial Fauna in the Greenbrier Karst** 371
David C. Culver and Daniel W. Fong

19 **The Subterranean Aquatic Fauna of the Greenbrier Karst**. 385
Daniel W. Fong and David C. Culver

20 **Pleistocene Vertebrates from Greenbrier Valley Caves**. 399
E. Ray Garton and Frederick V. Grady

1 The Greenbrier Karst

William B. White

Abstract

The segment of the Appalachian karst known as the Greenbrier karst is located in the lower valley of the Greenbrier River in southeastern West Virginia. The karst is developed in the Mississippian Greenbrier Limestone which thickens from 100 to 365 m northeast to southwest. The region can be subdivided into drainage basins which drain by subterranean routes to big springs. The Greenbrier karst contains more than 2000 caves of which 24 have surveyed lengths exceeding 5 km. The accumulated length of those 24 caves is 503.7 km.

1.1 Introduction

The Greenbrier Valley in southeastern West Virginia contains some of the longest caves in the USA. The Greenbrier karst is part of the extensive Appalachian karst and is located in Pocahontas, Greenbrier, and Monroe Counties, West Virginia (Fig. 1.1). For an overview showing the relationship of this karst area to other karst regions in the Appalachians and in the USA overall, see Palmer and Palmer (2009). A state-wide description of the karst of West Virginia is given by Dasher (2012). Details of the topography are provided by the US Geological Survey 7.5 min (1:24,000) for which an index is provided in Fig. 1.2.

1.2 Regional Setting

The Appalachian Mountains are usually divided into the Valley and Ridge Province with strongly folded and faulted strata and the Appalachian Plateaus Province where the strata are only slightly deformed. The two provinces are separated by a steep escarpment, called the Allegheny Escarpment (or Allegheny Front) in the north and the Cumberland Escarpment in the south. In eastern West Virginia, there is substantial deformation in the rocks on the eastern margin of the plateau so that the topographic front and the structural front do not coincide. East of the main plateau and west of the mountains that form the topographic front, the Greenbrier Valley has developed in the intermediate geological setting.

The Greenbrier River has its headwaters in the Allegheny Mountains near Spruce Knob, West Virginia, and flows south and southwest following the regional strike for 278 km to its confluence with the New River. The total basin area is 4290 km^2. For much of its length, the Greenbrier Valley is underlain by the Greenbrier Limestone which forms a karstic zone between the Allegheny Mountains to the east and the main Appalachian high plateau to the west (Fig. 1.3). The controlling structural features include the Browns Mountain Anticline and other folds which are formed west of the Allegheny Escarpment. To the west, the limestone dips beneath the younger rocks of the plateau. What is here described as the Greenbrier karst occupies the middle section of the Greenbrier River Valley. Omitted is the headwaters region where the limestone is thinner, and although it contains significant caves, the karst is less well developed. Also omitted is the downstream section where the river is flowing on clastic rocks.

In the northern part of the Greenbrier karst, the river flows along the eastern margin of the valley where it has incised a secondary valley into the clastic rocks that underlie the limestone. Because of the westerly dip, these clastics form a groundwater dam that prevents direct discharge from the karst into the river. Thus, tributary streams entering the river from the east flow entirely on clastic rocks, whereas the

W.B. White (✉)
Department of Geosciences, Deike Building,
University Park, PA 16802, USA
e-mail: wbw2@psu.edu

W.B. White (ed.), *Caves and Karst of the Greenbrier Valley in West Virginia*,
Cave and Karst Systems of the World, https://doi.org/10.1007/978-3-319-65801-8_1

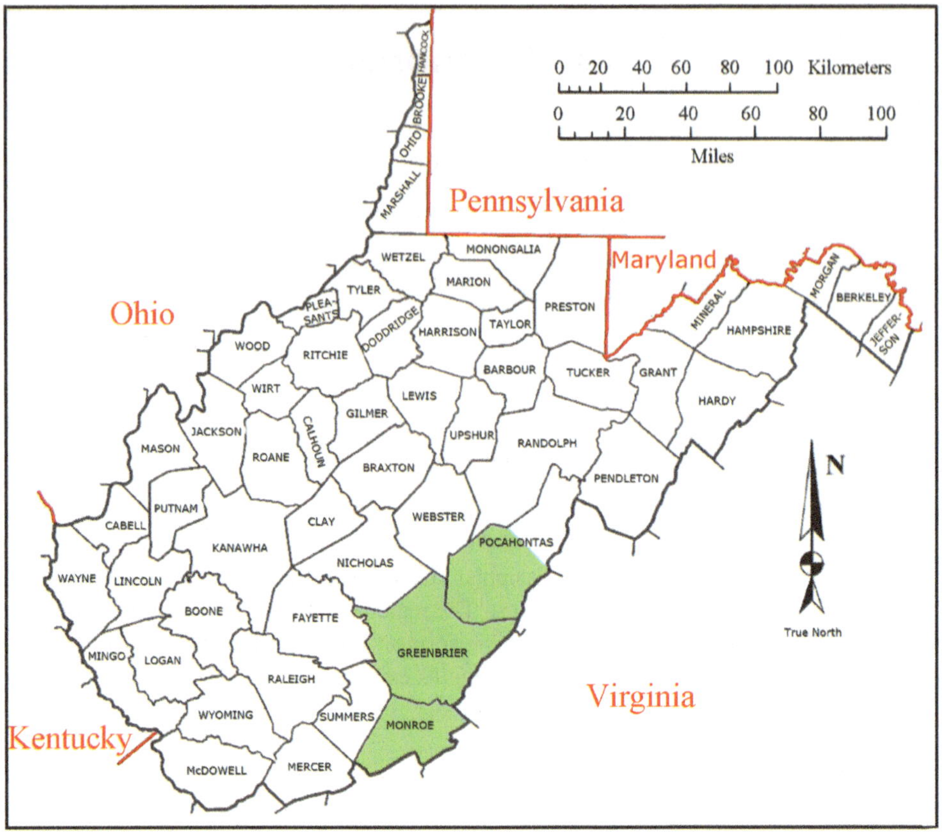

Fig. 1.1 Location map for the Greenbrier karst. Outline map from the US Department of Commerce, 1990

tributaries entering from the west have a pronounced fluviokarstic component.

The limestone thickness varies from about 100 m in the headwaters of the Greenbrier River to 365 m at the southern limit in Monroe County (McCue et al. 1939). The narrow band of Greenbrier Limestone widens to the southwest, and the topography changes from high-gradient fluviokarst basins such as the Swago Creek Valley in central Pocahontas County to the sinkhole plain called the Little Levels in southern Pocahontas County to wide sinkhole plains north and south of the Greenbrier River in Greenbrier and Monroe Counties (Fig. 1.4).

The interbedded shales within the Greenbrier Limestone are important controls on cave development. Near the top of the section, the Greenville Shale is rarely penetrated so that caves in the overlying Alderson Limestone tend to be perched on the shale. The Taggard Formation, a limey shale, is sometimes breached underground and sometimes not. Perched underground drainage is common.

1.3 Karst and Karst Hydrogeology

The Greenbrier karst can be subdivided into a sequence of small drainage basins nearly all of which ultimately drain through springs to creeks tributary to the Greenbrier River. The character of the individual basins changes north to south. The Swago Creek basin, the northernmost basin, is predominantly fluviokarst with primary recharge from mountain streams sinking at the limestone contact. The basins are of more mixed character in the intermediate region of southern Pocahontas and northern Greenbrier Counties. The Locust Spring basin receives surface stream recharge from Millstone Creek and from Hills and Bruffey Creeks and also an extensive recharge from the doline karst of the Little Levels. The Friars Hole System, separated from the Little Levels by Droop Mountain, is recharged primarily by mountain runoff with drainage to the south into Spring Creek. The Culverson Creek basin has a large surface catchment feeding into a very large cave system.

The thickening of the limestone beds and the development of additional anticlinal structures cause the karst region to widen in Greenbrier and Monroe Counties (Fig. 1.4). The result is a broad region of doline karst with mainly internal drainage and most recharge through closed depressions. The close depressions form on a range of size scales and appear to be guided by fractures and faults (Lessing 1979). The southern section of the karst is split into two segments by the Greenbrier River which east of Lewisburg changes from its generally southern course to a generally westward course. The northern segment was subdivided into smaller drainage basins by a series of tracing experiments with fluorescent

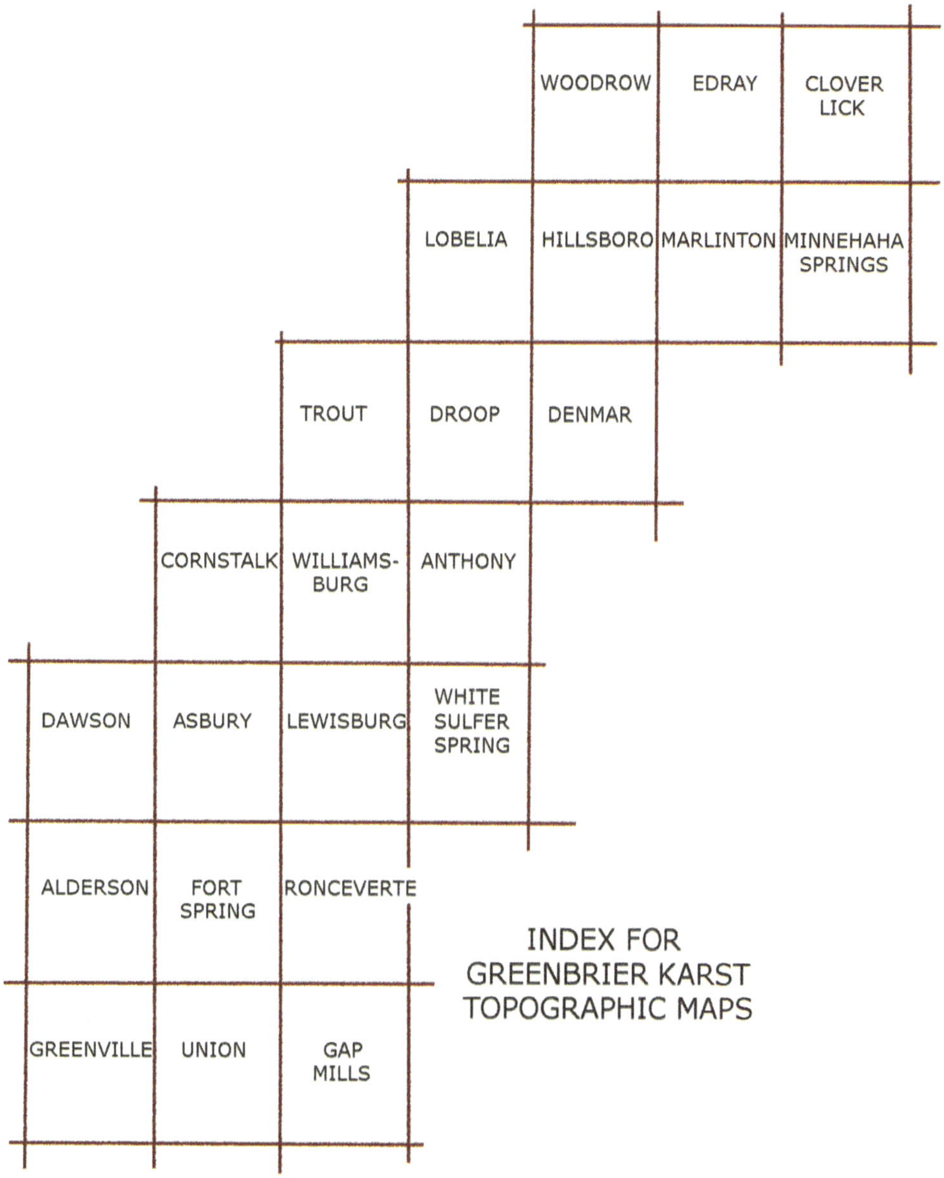

Fig. 1.2 Index map for US Geological Survey 7.5 min quadrangle maps spanning the Greenbrier karst

dyes (Jones 1973). These subbasins have associated large cave systems. The southern segment is also subdivided into distinct groundwater basins (Jones 1997). Further detail is provided by two PhD theses written on the hydrogeology of Greenbrier County (Heller 1980) and Monroe County (Ogden 1976).

1.4 Caves

Many exceptionally large caves occur in the thick and gently dipping limestone in the southern reaches of the Greenbrier Valley along with hundreds of smaller caves. Early cave descriptions for West Virginia were compiled by Davies (1949, 1958). Later, many cave explorers and their organizations contributed to the exploration and mapping of West Virginia caves. The two most prominent organizations were the West Virginia Association for Cave Studies (WVACS) and the West Virginia Speleological Survey (WVASS). The Bulletins of the West Virginia Speleological Survey are the primary documentation for West Virginia caves. The history of exploration in the Greenbrier karst is given in some detail in Chap. 5.

The three counties of the Greenbrier Valley karst are the most cavernous in the State. Dasher (2012) lists the number of caves in each county:

Pocahontas County	621 caves
Greenbrier County	1375 caves
Monroe County	435 caves

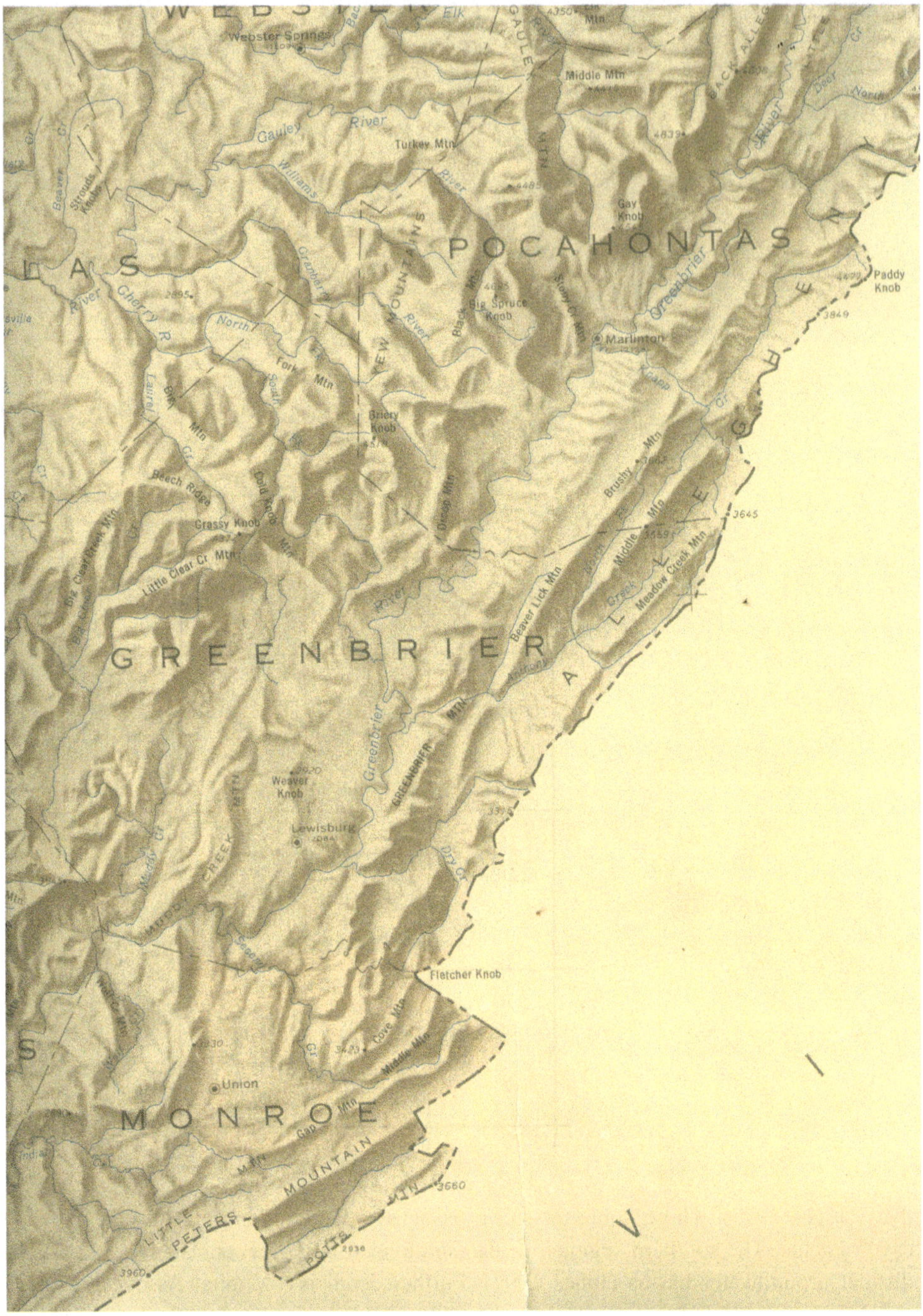

Fig. 1.3 Relief map of Pocahontas, Greenbrier, and Monroe Counties, West Virginia. Extracted from US Geological Survey 1:500,000 shaded relief man of West Virginia, 1968

The total is 2431 caves although some of the Pocahontas County caves are north of the area considered in this volume.

Table 1.1 lists the caves with surveyed lengths exceeding 5 km. The rank on the US long cave list is given for those with lengths greater than 20 km and the world rank for those with lengths greater than 30 km. Adding up the lengths produces a total of 503.7 km (312.9 miles) of mapped cave passage, making the Greenbrier Valley one of the most cavernous regions on the planet. A complete tally would require adding in the accumulated length of more than 2000 smaller caves, many of which are of substantial length. Most, but not all of the listed caves appear somewhere in the description chapters.

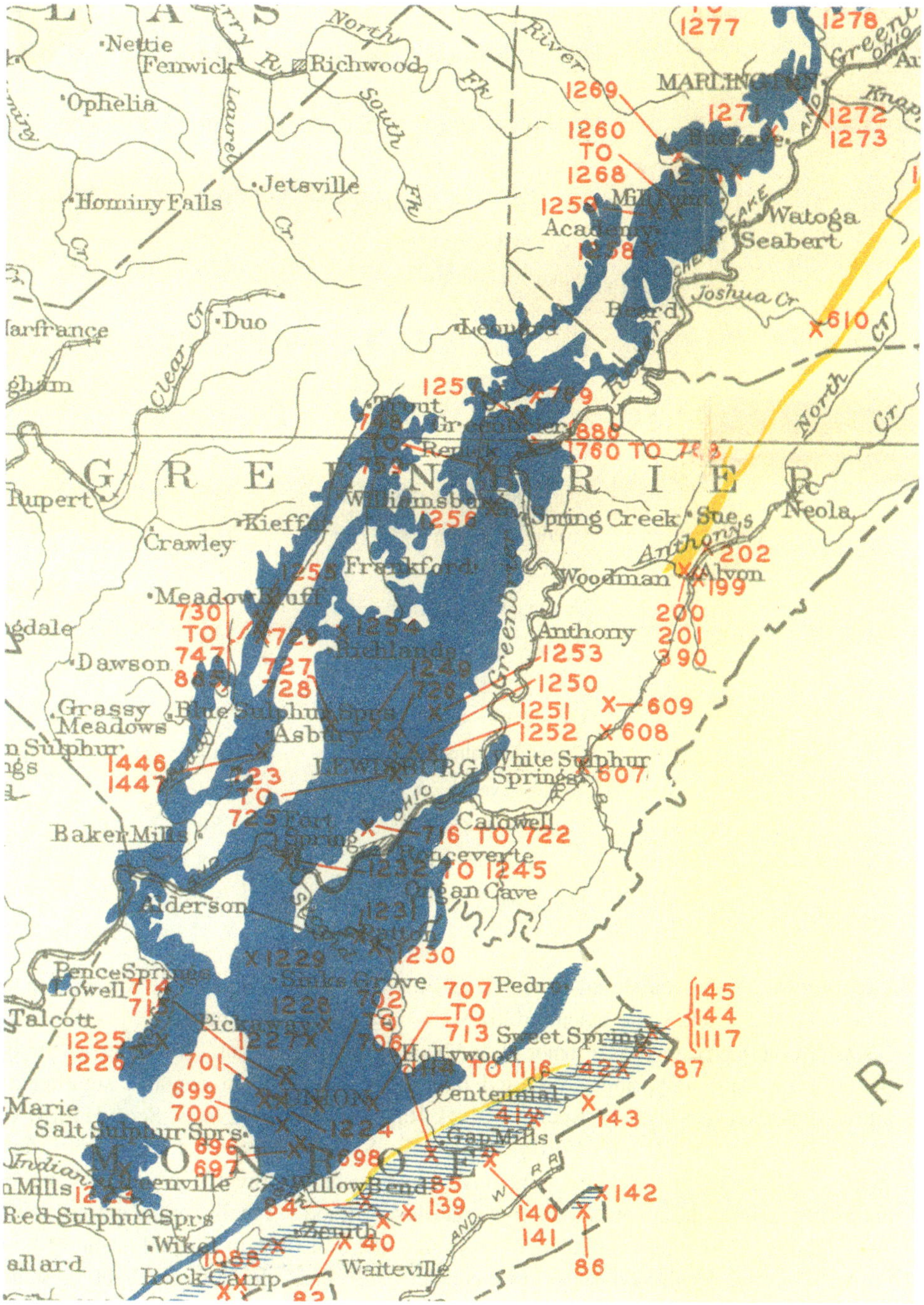

Fig. 1.4 Outcrop area of Greenbrier Limestone in Pocahontas, Greenbrier, and Monroe Counties, West Virginia. *Red numbers* refer to limestone quarries. Extracted from the limestone map of McCue et al. (1939)

1.5 Presentation of the Greenbrier Karst

For purposes of description, the Greenbrier karst was subdivided partly by drainage basins and the large cave systems. These subdivisions become the individual descriptive chapters.

The northernmost basin, Swago Creek (Chap. 6) is an isolated basin connected directly with the Greenbrier River and separated by Rogers Mountain from the next basin to the south. The Little Levels (Chap. 7) is a limestone upland with caves along the perimeter but which does not function as an integrated drainage basin. Separated from the Little Levels by Droop Mountain, but hydrologically interconnected, are the valleys of Hills and Bruffey Creeks. To the south extends the abandoned karst valley of Friars Hole beneath which is the longest cave in West Virginia (Chap. 8).

Moving south into the Big Levels, the wide doline karst on the Greenbrier Limestone, dye-tracing studies have defined drainage basins and their associated cave systems. Some basin boundaries are created by geologic barriers and other by the west-trending Greenbrier River which bisects

Table 1.1 Long caves of the Greenbrier Valley

Cave	Length (km)	US rank	World rank
Friars Hole System	73.4	6	31
Organ Cave	61.9	10	42
Scott Hollow Cave	47.5	17	68
The Hole	37.0	24	104
Culverson Creek System	33.7	30	134
McClung Cave System	29.1	36	
Windy Mouth Cave	29.0	37	
Benedict's Cave	23.9	52	
Bone-Norman System	22.7	55	
Maxwelton Sink Cave	18.8	64	
Portal-Boar Hole System	16.8	71	
Ludington's Cave	14.7	81	
Acme Quarry Cave System	13.6	87	
Overholt Blowing Cave	12.6	98	
Dry Cave	9.2		
Destitute Cave	8.0		
Union Cave	7.4		
Buckeye Creek Cave	7.2		
Carpenter's—Swago System	7.0		
Greenville Saltpetre Cave	6.7		
Wades Cave	6.4		
Laurel Creek Cave System	5.8		
Zicafoose Blowhole	5.8		
Plastic Bag Cave	5.5		

the region. The area is large and complex, and the subdivision into chapters is to some extent arbitrary.

North of the River, there is a major divide between caves that drain eastward to Spring Creek and those that drain southwest to Davis Spring. The Buckeye Creek-Rapps Cave System (Chap. 9) and the Culverson Creek System (Chap. 10) drain to Spring Creek and have large surface catchments making these caves very dynamic during flood flow conditions. Along the eastern edge of the area, the limit of the limestone is formed when the contact with the underlying clastics reaches the land surface. Along this extensive contact zone are formed a sequence of very large caves, the contact caves (Chap. 11). These collect drainage from the clastic terrain between the contact and the Greenbrier River and collectively drain to Davis Spring. To the west, the linear karst valley of Sinking Creek forms an independent basin (Chap. 12).

South of the Greenbrier River, the Organ Cave area (Chap. 13) is an isolated plateau bounded on the north by the river, on the east by the limestone contact, and in the south by the incised valley of Second Creek. Dickson Spring (Chap. 14) is one of the largest karst springs in the area. Its basin contains significant caves but not the exceptionally long ones. Scott Hollow Cave and associated Windy Mouth Cave are in a north-flowing drainage basin that drains directly into the Greenbrier River (Chaps. 15 and 16). Finally, at the extreme southern edge of the Greenbrier karst is an isolated island of limestone surrounded by clastic rock, the Laurel Creek system (Chap. 17). Laurel Creek is a south-flowing tributary of Indian Creek which flows westward into the New River below its confluence with the Greenbrier River and so is part of the Greenbrier Limestone karst but not a tributary of the Greenbrier River.

Not every detail of every cave is described in these chapters, but in broad brush terms, at least, there is a reasonable picture of the type of cave systems and associated karst drainage patterns that developed in this particular geologic and geomorphic setting.

References

Dasher, G.R. 2012. *The Caves and Karst of West Virginia*, vol. 19, 264pp. Barrickville, WV: West Virginia Speleological Survey Bulletin.

Davies, W.E. 1949. *Caverns of West Virginia*, vol. 19, 353pp. Morgantown, WV: West Virginia Geological Survey.

Davies, W.E. 1958. *Caverns of West Virginia*, 2nd ed., vol. 19A, 330pp. West Virginia Geological and Economic Survey.

Heller, S.A. 1980. A hydrogeologic study of the Greenbrier Limestone karst of central Greenbrier County, 167pp. West Virginia. Ph.D. Thesis, West Virginia University.

Jones, W.K. 1973. *Hydrology of limestone karst*, vol. 36, 49pp. Morgantown, WV: West Virginia Geological and Economic Survey Bulletin.

Jones, W.K. 1997. *Karst hydrology atlas of West Virginia*, vol. 4, 111pp. Charles Town, WV: Karst Waters Institute Special Publication.

Lessing, P. 1979. *Karst subsidence and linear features, Greenbrier and Monroe Counties, West Virginia*. Morgantown, WV: West Virginia Geological and Economic Survey Environmental Geology Bulletin EGB-17, one sheet.

McCue, J.B., J.B. Lucke, and H.P. Woodward. 1939. *Limestones of West Virginia*, vol. 12, 560pp. Morgantown: West Virginia Geological Survey.

Ogden, A.E. 1976. The hydrogeology of the central Monroe County karst, West Virginia, 262pp. Ph.D. Thesis, West Virginia University.

Palmer, A.N., and M.V. Palmer. 2009. *Caves and Karst of the USA*, 446. Huntsville, AL: National Speleological Society.

Geology of the Greenbrier Valley

2

William B. White

Abstract

The Greenbrier Karst is located in the Appalachian Highlands in the boundary region between the strongly folded rocks of the valley and ridge province and the gently folded rocks of the Appalachian Plateau. The outcrop of the karstic Greenbrier Limestone occupies portions of Pocahontas, Greenbrier, and Monroe Counties in southeast West Virginia. The Greenbrier Formation is subdivided into the Alderson Limestone, the Greenville Shale, the Union Limestone, the Pickaway Limestone, the Taggard Formation, the Patton Limestone, the Sinks Grove Limestone, and the Hillsdale Limestone. Intermediate shale beds exert an important controlling influence on cave development. Below the limestone is the Maccrady Shale which acts as an aquiclude. A sequence of relatively gentle north–south folds controls the limestone outcrop area at the land surface. Fractures provide further controls over the drainage pathways.

2.1 Introduction

The objective of this chapter is to provide a regional scale description of the geological substrate on which the caves and karst landforms are developed. The information that follows is summarized from existing accounts of West Virginia geology. The West Virginia geological survey was a pioneer in providing detailed investigations of the geological features of the state. The products of these efforts were a series of thick volumes describing the geology, county by county. Of relevance are Pocahontas County (Price and Reger 1929), Greenbrier County (Price and Heck 1939), and Mercer, Monroe, and Summers Counties (Reger and Price 1926). More recent studies of the geology were obtained as part of hydrogeologic investigations for the limestone portions of Monroe County (Ogden 1976) and Greenbrier County (Heller 1980). The present-day geology of the Greenbrier Valley is the end product of a long sequence of sedimentary, and tectonic processes that extend back through the long and complicated history of the Appalachian Mountains. The tectonic events that produce the structural setting are mostly the result of the Allegheny Orogeny (Hatcher et al. 1989).

W.B. White (✉)
Department of Geosciences, Deike Building, University Park, PA 16802, USA
e-mail: wbw2@psu.edu

2.2 Appalachian Geology—The Place of the Greenbrier Valley

2.2.1 Tectonic History

For the north-central Appalachians, Faill (1997–98) has provided a detailed overview of the tectonic history (Fig. 2.1). Events begin in Grenville time, 1200–1000 Ma ago, with the formation of the supercontinent Rodinia. A long period of erosion and then rifting of Rodinia provided an ocean basin for the deposition of an extensive carbonate shelf in Cambrian and Ordovician time. Further orogenic events, the Potomac Orogeny, the Taconic Orogeny, and the Acadian Orogeny, provided the basins for other carbonate deposition in the Silurian/Devonian and in the Mississippian as well as a complex of clastic sediments separating the carbonates.

The last and most important major tectonic event was the Allegheny Orogeny in early Permian time. This was a convergence between the North American plate and the

W.B. White (ed.), *Caves and Karst of the Greenbrier Valley in West Virginia*,
Cave and Karst Systems of the World, https://doi.org/10.1007/978-3-319-65801-8_2

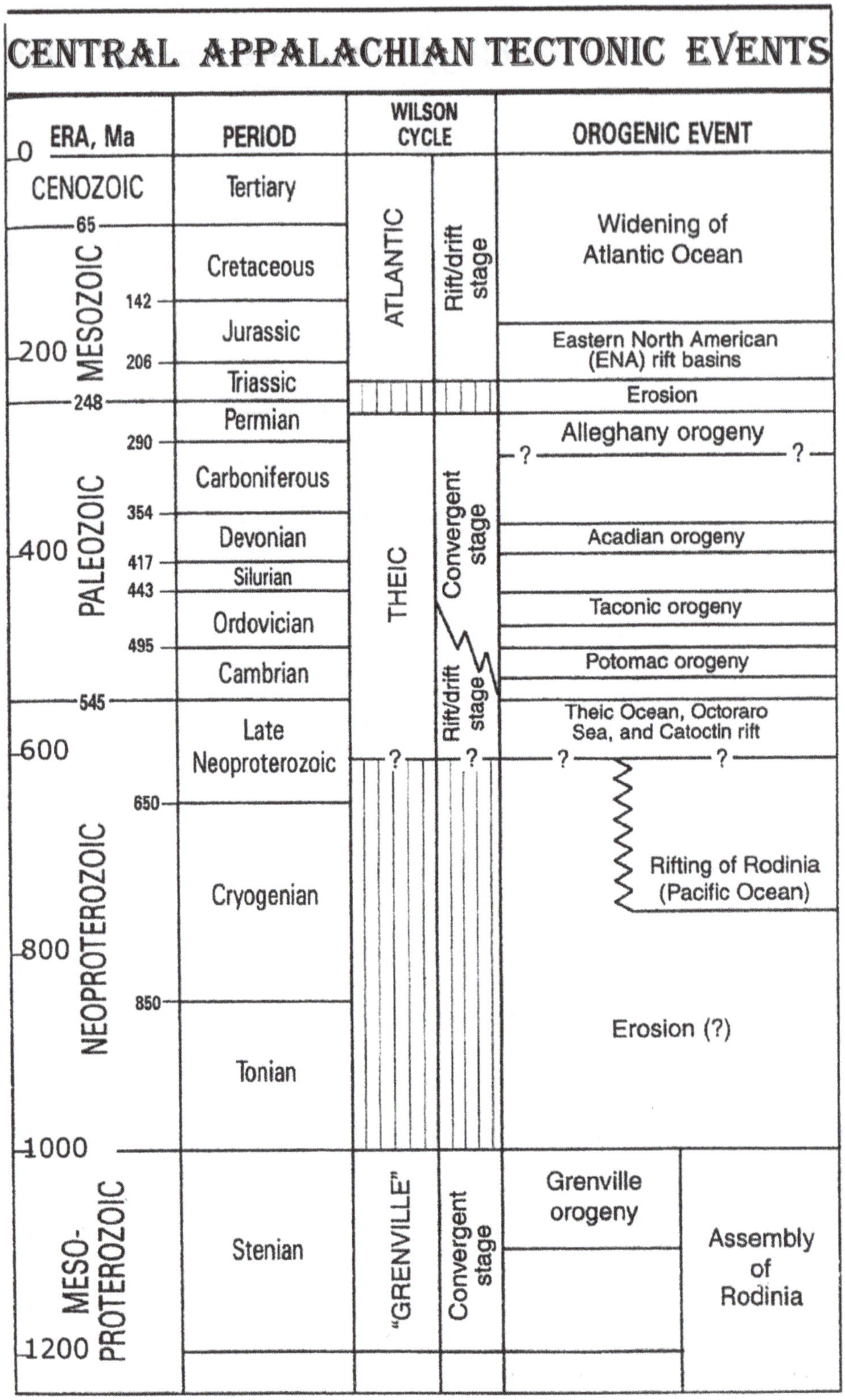

Fig. 2.1 Tectonic events in the Appalachians since Grenville time. Adapted from Fail (1997–1998)

African plate. The southeastern core of the Appalachians was metamorphized, and various igneous bodies were created. Sediments were piled up to form a mountain range estimated to be 3500–4500 m in elevation (Slingerland and Furlong 1989). To the northwest, low angle thrust faults in the middle crust splayed upward, deforming, and transporting Paleozoic sediments and building ramps of sediments one on top of the other (Kulander and Dean 1986). Riding the main thrust faults were sequences of folds. Intense folds later became the Valley and Ridge Provence but less intense folds extended well back into the Allegheny Plateau. Paleomagnetic measurements suggest that the Allegheny Orogeny occurred over a fairly short time period between 275 and 255 Ma ago (Stamatakos et al. 1996).

Rifting of the continental plates in Triassic and Jurassic time produced some extensional normal faulting and infilling of Triassic sediments with the opening of the Atlantic Ocean. Erosion since the Cretaceous has worn down the Appalachians, exposing carbonate rocks of various ages as well as the igneous and metamorphic rocks in the Piedmont and the Blue Ridge, the core of the high Appalachians. Caves and karst landscapes have come and gone as carbonate rocks have been exposed and then eroded away. The present-day karst landscapes date from only the late tertiary.

2.2.2 Physiographic Setting

The usual physiographic provinces and sub-provinces of the Appalachian Mountains are sketched in Fig. 2.2 which also show the position of the Greenbrier Karst.

From central Pennsylvania southward through Maryland and West Virginia, the Appalachian Plateau is separated from the Valley and Ridge by a steep escarpment, the Allegheny Front. The escarpment increases in elevation from 550 m at its northern limit near Williamsport, Pennsylvania and gradually rises to the south. However, the folding induced by the low angle thrust sheets of the Valley and Ridge continues beneath the plateau resulting in a series of anticlines and synclines in the eastern section of the plateau. Erosion has produced a series of ridges known as the Allegheny Mountains that extend from Pennsylvania to east-central West Virginia. The high point of the Alleghenies is Spruce Knob in West Virginia with an elevation of 1480 m. Although the structural deformation is much less severe than in the Valley and Ridge, the Alleghenies are among the most rugged mountains in the Appalachians. West of the Allegheny Mountains, the structural deformation decreases into the central core of the Appalachian Plateaus, a broad synclinal trough with clastic rocks and coal measures but no karst.

Spruce Knob (USGS Spruce Knob Quadrangle) is an important 3-way drainage divide. On the eastern side is the Allegheny Front which drains to the North Fork of the South Branch of the Potomac River. To the northwest, drainage is into tributaries of the Cheat River which flows north into the Monongahela River, a tributary of the Ohio River. To the southwest is the Greenbrier River which joins the New River which flows westward as the Kanawha River to the Ohio River. These interfingering drainage systems have dissected the plateau and exposed the Mississippian limestones. With the high relief, the geologic setting is optimized for the development of large cave systems.

In the northern portions of the Greenbrier Karst, the limestone exposures follow the Greenbrier River along the structural trend of the Allegheny Mountains in the zone between the topographic front and the structural front. The Allegheny Mountains march to an end in Monroe County, and the Greenbrier Limestone becomes part of the dissected edge of the plateau. The narrow band of Greenbrier Limestone widens to the southwest, and the topography changes from high gradient karst drainage basins such as the Swago Creek Valley to the sinkhole plain called the Little Levels in southern Pocahontas County to a wide sinkhole plain north and south of the Greenbrier River in Greenbrier and Monroe Counties called the Big Levels. To the southeast, the Greenbrier Karst is cut off sharply by the line of the structural front where the St. Clair Fault brings up Ordovician carbonates in southeastern Monroe County.

2.3 Stratigraphy

2.3.1 The Greenbrier Limestone

The Mississippian Greenbrier Limestone varies in thickness from about 100 m in the northern limits of the Greenbrier Karst to 365 m at the southern limit in Monroe County (Fig. 2.3). Many descriptions of the Greenbrier Limestone have been reported. The descriptions that follow are a composite but depend largely on the report by McCue et al. (1939) and the geologic map of the limestone exposures in Greenbrier County Heller (1980). The Greenbrier Limestone is equivalent to the Mississippian carbonates of the Cumberland Plateau to the south and the Mammoth Cave area to the southwest but it has retained different stratigraphic names and varies somewhat in lithology. The names assigned to the units of the Greenbrier by Price, Reger, and other early West Virginia geologists were derived from type localities in Monroe County.

Alderson Limestone

The type locality is south of the town of Alderson in Monroe County. It is characteristically a thin-bedded, very variable limy shale to impure shaley or argillaceous limestone that weathers in outcrop into yellow shaley banks. In Greenbrier County, the Alderson is a series of siliceous, coarse-grained, fossiliferous, oolitic beds interspersed with fine-grained argillaceous limestone units. Caves form in the Alderson but tend to be isolated from caves in the underlying limestones.

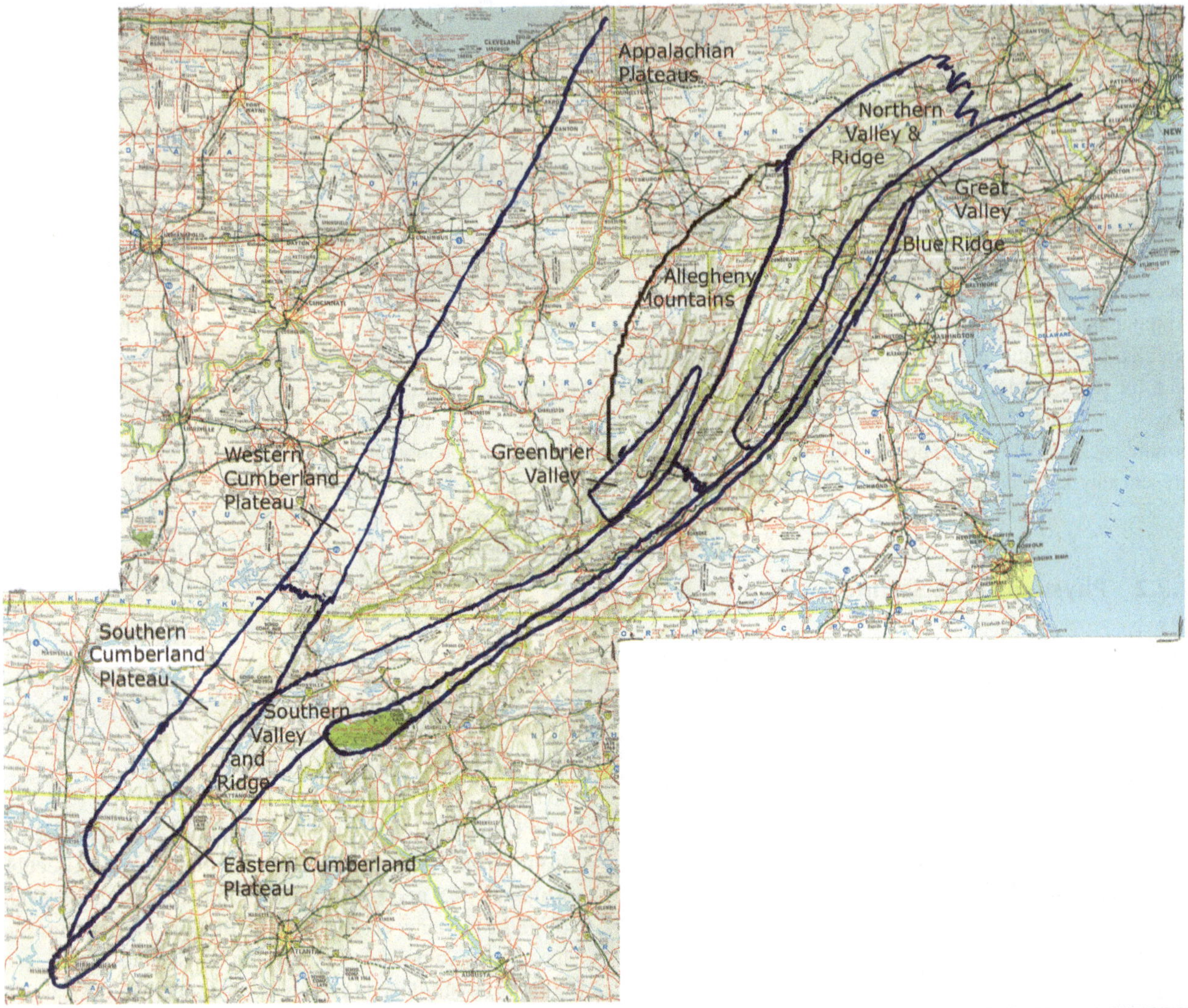

Fig. 2.2 Physiographic provinces of the Appalachians

Greenville Shale

The type locality is near Greenville in Monroe County. The Greenville is a dark fissile shale that acts as a very effective aquiclude. It is a calcarous shale, weathering tan, that contains many marine fossils. The Greenville is rarely (if ever) breached underground.

Union Limestone

The type locality is at the west edge of the town of Union, Monroe County. It is a white to gray, hard limestone, shaley at the top, thick-bedded, and oolitic in part. The Union is one of the most persistent units of the Greenbrier Limestone and is the object of many quarrying operations. The Union is nearly bisected by a shaley or sandy clastic unit identified with the Bethel Sandstone. The Bethel Sandstone is better developed to the north and is represented in the Greenbrier area by a more shaley unit.

Pickaway Limestone

The type locality is the town of Pickaway, Monroe County. The Pickaway is a dark, variegated, silty impure limestone with occasional red streaks and sandy lenses (Fig. 2.4). The unit can be identified by extensive stylolite development. It weathers to a wavy, banded appearance due to the silty and shaley partings. Heller (1980) divides the Pickaway into three members: a lower fossiliferous calcilutite member, a middle superficial oolite member, and an upper laminated calcilutite member. A highly detailed section of the Pickaway was measured in Greenbrier County, 2 km west of Lewisburg (Heller 1980).

Taggard Formation

The type locality for the Taggard Formation is along Taggard Creek in Monroe County. It is a complex formation with a limy red shale on top, a shaly limestone in the middle,

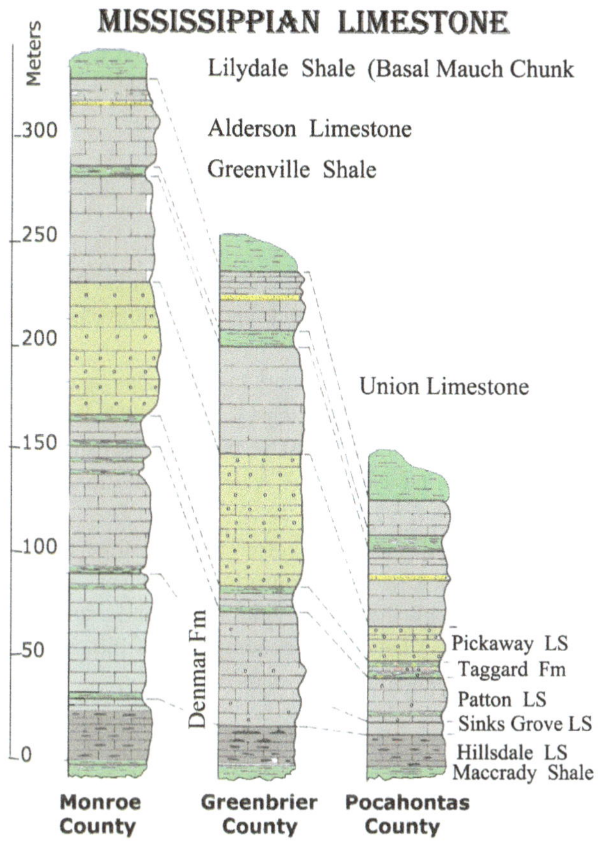

Fig. 2.3 Stratigraphic sections for the Greenbrier Limestone

and a second limy red shale at the bottom. As a shale, the Taggard acts as an aquiclude but because of the high carbonate content, it is frequently breached in the subsurface. It weathers to a red shale which identifies the formation in surface outcrop. These and other details were measured in an exposure on Elk Mountain just north of the Swago Creek Basin (Fig. 2.5). A detailed column of the Taggard was also measured in Greenbrier County (Heller 1980).

Denmar Formation

In some recent literature, the Patton Limestone and the Sinks Grove Limestone are combined into a single unit called the Denmar Formation. We retain the older nomenclature.

Patton Limestone

The type locality for the Patton Limestone is on the south side of Second Creek just south of the village of Patton. It is a hard, gray, pure limestone that usually contains a 2–3 m layer of oolite. There may be a thin shale at the base of the Patton Limestone.

Sinks Grove Limestone

The type locality of the Sinks Grove Limestone is near the town of Sinks Grove, Monroe County. It is a blue, hard, siliceous limestone that often contains nodules of black chert.

Fig. 2.4 Exposure of the Pickaway Limestone at the highway intersection downstream from Fort Spring. Photo by the author

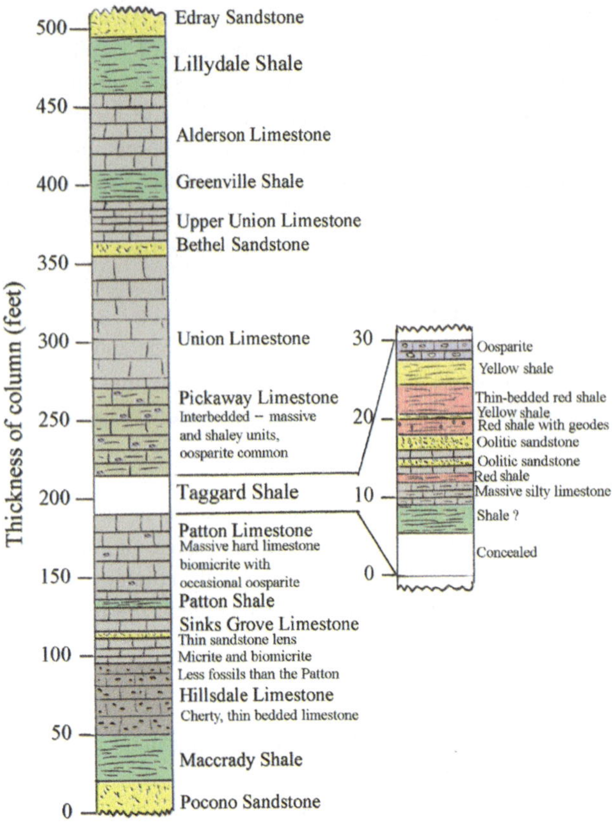

Fig. 2.5 Stratigraphic section of the Taggard Formation measured by the author along US Highway 219 on Elk Mountain north of Marlinton

Hillsdale Limestone

The type locality for the Hillsdale Limestone is just east of the town of Hillsdale, Monroe County. The basal unit of the Greenbrier series is a grayish-blue hard, massive somewhat dolomitic limestone with interbedded chert. The Hillsdale rests unconformably on the underlying Maccrady Shale. The Hilldale Limestone is easy to recognize underground because the insoluble chert layers stand out in relief on cave walls.

2.3.2 Clastic Rocks Above and Below the Karstic Limestone

The highly karstic limestones of the Greenbrier series are bounded above and below by thick sequences of clastic rocks. The overlying clastics provide the catchments for allogenic streams that drain into the karst. The underlying clastics form a lower limit for ground water circulation.

Pottsville Group (Pennsylvanian)

The Pottsville Group is a complex sequence of quartzites, quartz conglomerates, shales, and thin coal beds. The Pottsville is a strongly erosion-resistant formation that provides the resistant caprock for the mountains west of the karst area.

Mauch Chunk Group (Mississippian)

The rocks immediately overlying the Greenbrier are a thick sequence of shales, siltstones, and sandstones and some thin limestones. In sequence, these are:

- Bluestone and Princeton Formations—red, green, and gray shales and sandstones.
- Hinton formation—red, green, and gray shales and sandstones with a few thin limestone beds. The Avis Limestone at the base is thick enough to develop small maze caves.
- Bluefield Formation—red and green shales and sandstones with thin limestone lenses, such as the Reynolds Limestone. The bottom unit is the Lilydale Shale that merges conformably into the Alderson Limestone.

Although the clastic rocks below the limestone are also Mississippian, there is an unconformity between the basal Hillsdale unit and the underlying shale and sandstone. Below these units, the section is composed of thousands of meters of Devonian shale and sandstone.

Maccrady Shale

The Maccrady is a red shale and mudrock with some sandstone. It is present only south of Pendleton County. Although this is a clastic rock, the cave passages of the contact caves are often cut deep into the Maccrady Shale (Fig. 2.6).

Pocono Sandstone

Predominantly hard, massive, but dirty sandstone. The Pocono Sandstone is the resistant rock that forms the eastern boundary of the Greenbrier Karst.

2.3.3 Depositional History

From the broad history of the Appalachians, it is of interest to reconstruct the details of the geologic history that provided the geological substrate on which the Greenbrier Karst was developed. The account that follows was extracted and to some extent paraphrased from a more extensive geologic history of West Virginia limestones (Springer 2000).

The relevant story begins with the Devonian Acadian Orogeny caused by the collision of the North American plate with the Avalon Terrane followed by a collision with the small continent known as Baltica. The Avalon Terrane was crushed into a large mountain chain—the Acadian Mountains—in what is now northeastern North America. Erosion of the Acadian Mountains produced a wedge of Devonian clastics—the Catskill Wedge. The wedge extended into West Virginia which at that time was largely covered by a shallow sea. The clastic sediments deposited in West

Fig. 2.6 Exposure of the Maccrady Shale along highway 219 just south of the Greenbrier River bridge near Ronceverte. Photo by the author

Virginia are known as the Brallier, Chemung, Hampshire, Pocono, and Maccrady formations. The thick Brallier shale was deposited in deep water while the Maccrady was formed by terrestrial rivers.

A period of erosion and folding followed the deposition of the Maccrady Formation. North of Pocahontas County, the Maccrady was completely eroded away. The area appears to have been broadly tilted to the south or southeast with resulting greater erosion in the north. The result was the unconformity between the Maccrady and the basal Hillsdale Limestone. The period of erosion marked by the unconformity was followed by sea level rise as a large portion of the region began to subside in response to the Allegheny Orogeny. Mississippian sea water flowed into the area from the southwest.

The Appalachian Basin has the crude geometry of a trough. The Appalachian Mountains were beginning to rise (again) in the east and the trough paralleled the first range of mountains. Subsidence was greatest east of a hinge line that ran across central West Virginia (Fig. 2.7) with thicker limestones to the south and thinner limestones to the north. The limestones were thin because the northern part of the trough contained deltas created by rivers flowing from the young Appalachian Mountains. These deposits of sandstone, siltstone, and shale are called the Mauch Chunk Group. While the Mauch Chunk was being deposited in the north, the Greenbrier Limestone was being deposited in the south. The sediments inter tongue where they meet. One of these tongues is the Taggard Formation.

The shallow sea produced the Pickaway and Union limestones during a period when sea level and associated tectonics were stable. As the deposition of the Union waned, deltas of the Mauch Chunk began to prograde into the Greenbrier Sea forming the thick red shales and sandstones. The first indication of advancing clastics is the Greenville Shale, followed by the shaley Alderson Limestone, and finally, the sedimentary section was dominated by the shales and sandstones of the Mauch Chunk.

2.4 Structure

2.4.1 Regional Folds

The structural motif of the Valley and Ridge Province extends to the fold structures beneath the plateau but with reduced amplitude. The folds of the Allegheny Mountain sub-province are oriented more or less north to north-northeast, and this orientation is continued into Greenbrier and Monroe Counties. These weak folds are cut off along the southeast boundary of the area by northeast–southwest (52°) trend of the structural front. Identification of the fold structures in the low-dipping rocks requires very careful mapping but they have an important influence on the rocks that crop out at the land surface (Fig. 2.8). The list below contains only a few of the more important structures that have an influence on cave and karst development. These were taken from the geologic maps of Greenbrier County

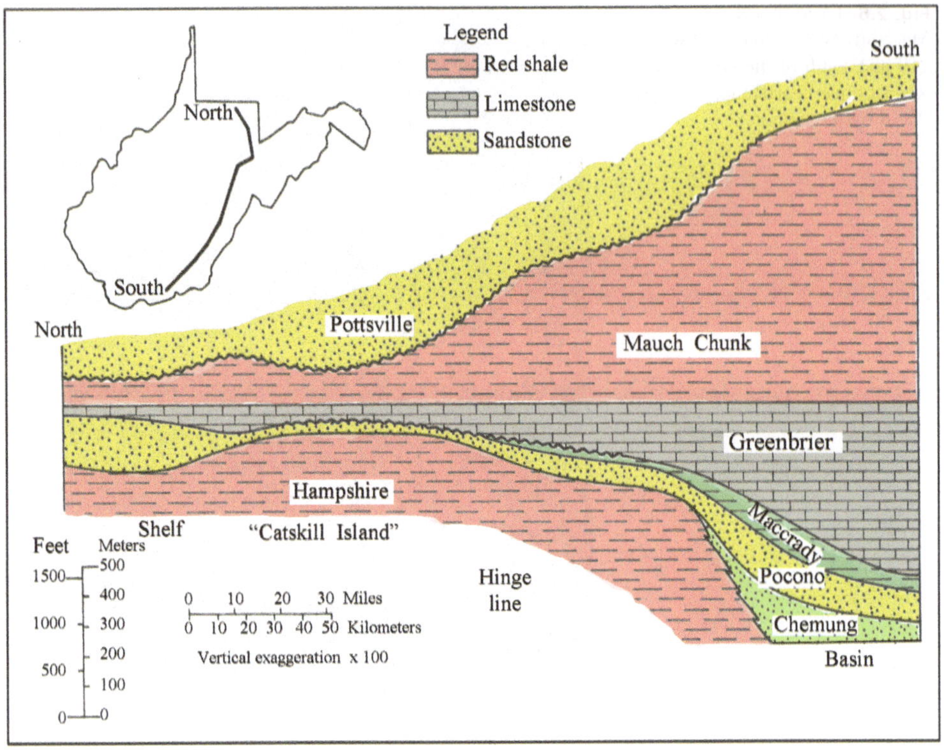

Fig. 2.7 North–south stratigraphic profile through West Virginia showing the hinge line that marks the deposition of thicker limestone units. From Arkle et al. (1979)

(Price and Heck 1939) and Monroe County (Reger and Price 1926). Later, mapping in Monroe County (Ogden 1976) and Greenbrier County (Heller 1980) reveals other fold structures as well as some corrections to the structures previously mapped (Fig. 2.9).

Browns Mountain Anticline

The Browns Mountain Anticline is the largest fold west of the structural front. In character, it is more like the intense folds of the valley and ridge than the much gentler folds of the Allegheny Plateau. It is somewhat domed with the peak of the dome east of Marlinton. The Browns Mountain anticline is strongly folded with dips up to the vertical. The westward dipping limestones of the Swago Creek and Little Levels areas are formed on the west limb of the anticline.

Caldwell (Patton) Syncline

The Caldwell Syncline (Patton Syncline in Monroe County) has its northern limit north of Anthony. The axis follows the course of the Greenbrier River southwestward to Caldwell where the river veers westward, and the syncline axis continues southwestward, passing beneath the Organ Cave Plateau and continuing into Monroe County. The Caldwell Syncline is the structural trough underlying the Organ Cave system. Northeast of Caldwell the surface rocks is the Pocono Sandstones; southwest of Caldwell the Greenbrier Limestone is exposed.

Sinks Grove Anticline

The Sinks Grove Anticline is a major structure of the Greenbrier Karst. The axis trends south-southwest passing 1½ miles east of Maxwellton, just east of Lewisburg, then curves around Ronceverte and continues south to Sinks Grove in Monroe County. The northern segment marks the eastern edge of the Greenbrier Karst where the Maccrady Shale is exposed on the crest of the anticline.

Muddy Creek Mountain Syncline

The Muddy Creek Mountain Syncline is a broad structure with the west limb steeper than the east limb. The syncline axis follows the western edge of Muddy Creek Mountain. The dip in the syncline axis beneath Muddy Creek Mountain carries the Greenbrier Limestone to depth and exposes the basal units of the Mauch Chunk as the surface rocks. The synclinal structure of Muddy Creek Mountain accounts for the steep scarp slopes on the east and west sides.

Williamsburg (Brushy Ridge) Anticline

The Williamsburg Anticline is a narrow but strongly folded structure. The structure begins east of Trout, trends southwest west of sunlight, east of Williamsburg, and follows the crest of Brushy Ridge. Near Asbury, there is an offset to the east. The structure axis then passes through the south end of Muddy Creek Mountain and reaches the Greenbrier River

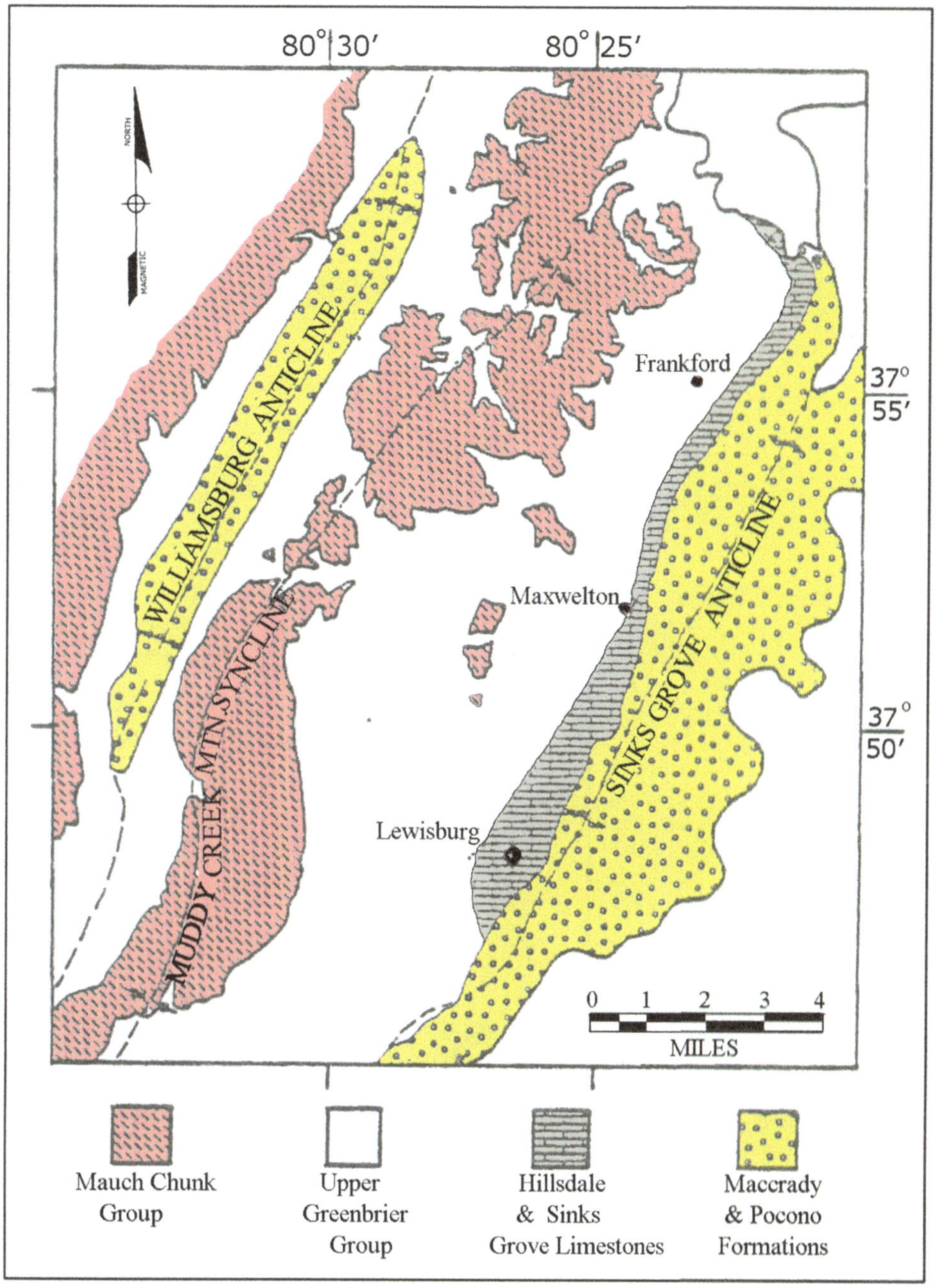

Fig. 2.8 Geologic map of a portion of the Greenbrier and Monroe County karst illustrating the effect of fold structures on the outcrop pattern of the Greenbrier Limestone

east of Alderson. The profile of the structure axis is undulating. From Trout to a point 2 km northeast of Williamsburg, the entire thickness of the Greenbrier Limestone is exposed at the surface. From this point to Asbury, the outcropping rocks belong to the Pocono Series with a thin band of Maccrady at each end. In the structural saddle near Asbury, the basal Greenbrier is exposed. Further south along the axis, the entire thickness of the Greenbrier Series dips below the surface and on the south end of Muddy Creek Mountain, rocks of the Bluefield group form the surface exposures.

Abbs Valley Anticline

The Abbs Valley Anticline is the diminishing northern tail of a major structural feature, the Richlands Fault, in Tazewell County, Virginia. Developing into an anticline, the structure axis crosses Mercer County, a corner of Summers County, and enters Monroe County where it extends through Greenville, veering north and becoming a monocline east of Wolf Creek. The Abbs Valley Anticline is responsible for bringing up the island of Greenbrier Limestone in which the Laurel Creek System is located.

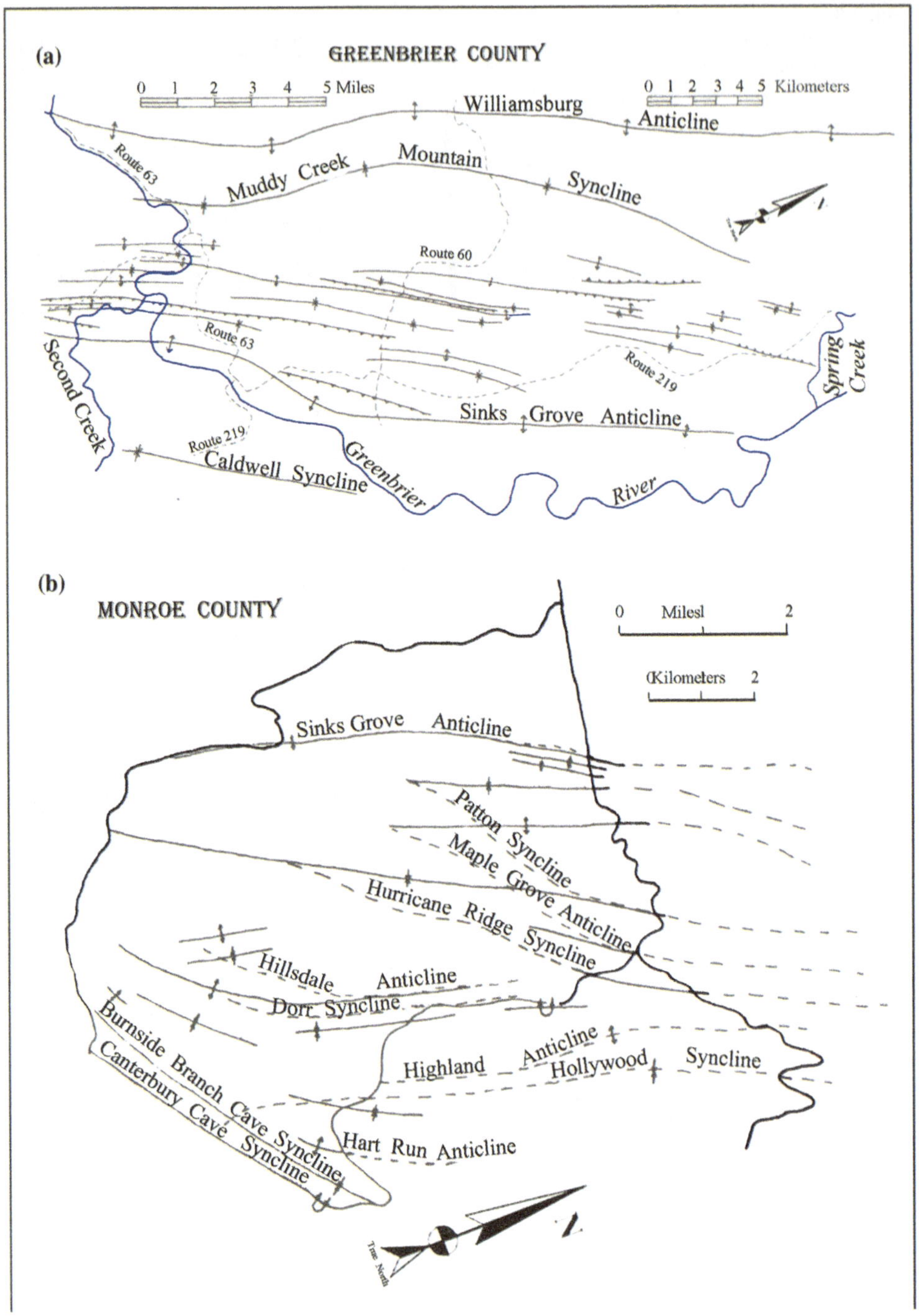

Fig. 2.9 Structure maps for the Greenbrier Karst: **a** Greenbrier County, **b** Monroe County. Adapted from Heller (1980)

2.4.2 Faults, Fractures, and Lineaments

Minor faults occur throughout the Greenbrier Karst. Most are small with displacements in the range of 5–10 m. Some were specifically identified in Greenbrier County (Heller 1980). Their influence on cave and karst development seems to be small.

The Greenbrier Limestone is well jointed, and many cave passages are oriented along joint traces with the guiding joint clearly visible in cave walls or ceiling. However, the overall correlation of cave passages with joint orientations is not good, primarily because in the low-dip limestone, bedding plane partings rather than vertical joints tend to be the initiating pathways. Measured joint orientations in Monroe County (Fig. 2.10a) show a dominant northeast trend which is more nearly parallel to the orientation of the structural front rather than the secondary anticlines and synclines.

Fracture control of surface features takes the form of aligned sinkholes and related features. These can be mapped from aerial photographs. Features visible on aerial

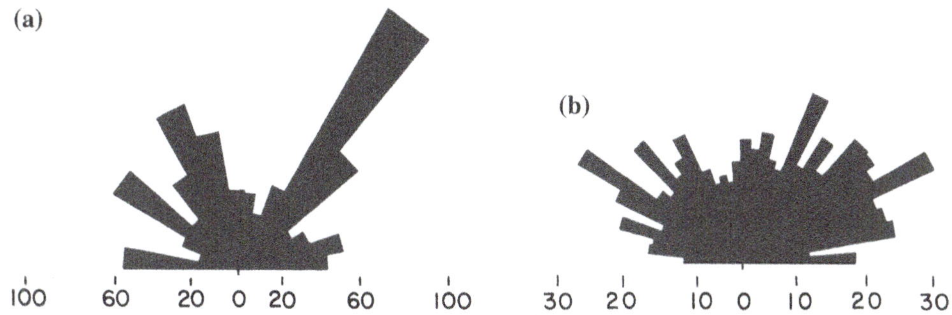

Fig. 2.10 **a** Orientation of 873 joints taken from Greenbrier Limestone, Monroe County. Scale is number of joints sorted by 10° intervals, **b** Accumulated lineament length versus orientation for 749 mapped lineaments. Scale is in thousands of feet using a 5° interval. From Ogden (1974)

Fig. 2.11 Aerial photograph of the monitor lineament, Monroe County. Photo by William K. Jones

photographs are known as fracture traces or photolineaments and have proved useful in locating zones of high permeability for the drilling of water wells. Measurement of photolineaments in Monroe County (Ogden 1976) and Greenbrier County (Heller 1980) does not show any preferred direction (Fig. 2.10b). On the scale of topographic maps, alignments of sinkholes and other karst features can be drawn (Lessing 1979). The major trends parallel the pattern of folding and are simply guided by the local geology. Others, usually strings of sinkholes, cut across the regional structure and appear to be a reflection of the local fracture patterns, which, as seen in the aerial photograph mapping, seem to have no preferred direction.

The term "lineament" is used in several senses. The photolineaments described above are usually spatially limited with lengths of a kilometer or less. But scattered through the Appalachians are major lineaments with lengths measured in tens of kilometers and which appear to be major structural features that usually cut across the characteristic structural grain of the Appalachians. One such is located north of the Greenbrier Karst in Randolph County. The Simmons-Mingo Cave system is developed on the lineament and crosses the spur of a mountain ignoring local geologic structure. In the Greenbrier Karst, the monitor lineament (Fig. 2.11) is identified on the surface by an obvious line of sinkholes.

Acknowledgements I thank William Balfour for his review of this chapter and William K. Jones for permission to use his photograph of the Monitor Lineament.

References

Arkle, T. Jr., D.R. Beissel, R.E. Larese, E.B. Nuhfer, D.G. Patchen, R. A. Smosna, W.H. Gillespie, R. Lund, C.W. Norton, and H.W. Pfefferkorn. 1979. The Mississippian and Pennsylvanian (Carboniferous) systems in the United States—West Virginia and Maryland. *U.S. Geological Survey Professional Paper* 1110-D: 35pp.

Faill, R.T. 1997–98. A geologic history of the north-central Appalachians. Part 1: Orogenesis from the Mesoproterozoic through the Taconic orogeny. *American Journal of Science* 297, 551–619; Part 2: The Appalachian Basin from the Silurian through the Carboniferous. 297: 729–761; Part 3: The Alleghany orogeny. 298: 131–179.

Hatcher, R.D., Jr., W.A. Thomas, P.A. Geiser, A.W. Snoke, S. Mosher, and D.V. Wiltschko. 1989. Alleghanian orogen. In: *The Appalachian-Ouachita Orogen in the United States*, ed. R.D. Hatcher Jr., W.A. Thomas, and G.W. Viele, 233–318. Boulder, CO: Geological Society of America, The Geology of North America, F-2.

Heller, S.A. 1980. A hydrogeologic study of the Greenbrier Limestone karst of central Greenbrier County, West Virginia, 167pp. Ph.D. Thesis, West Virginia University.

Kulander, B.R., and S.L. Dean. 1986. Structure and tectonics of central and southern Appalachian Valley and Ridge and Plateau Provinces, West Virginia and Virginia. *The American Association of Petroleum Geologists Bulletin* 70: 1674–1684.

Lessing, P. 1979. *Karst subsidence and linear features, Greenbrier and Monroe Counties, West Virginia*. Morgantown, WV: West Virginia Geological and Economic Survey Environmental Geology Bulletin EGB-17, one sheet.

McCue, J.B., J.B. Lucke, and H.P. Woodward. 1939. *Limestones of West Virginia*, vol. 12, 560pp. Morgantown, WV: West Virginia Geological Survey.

Ogden, A.E. 1974. The relationship of cave passages to lineaments and stratigraphic strike in central Monroe County, West Virginia. In *Fourth conference on Karst geology and hydrology proceedings*, ed. H.W. Rauch, and E. Werner, 29–32. West Virginia Geological Survey: Morgantown.

Ogden, A.E. 1976. *The hydrogeology of the central Monroe County karst, West Virginia*, 262pp. Ph.D. Thesis, West Virginia University.

Price, P.H., and E.T. Heck. 1939. *Greenbrier county*, 846. Morgantown, WV: West Virginia Geological Survey.

Price, P.H., and D.B. Reger. 1929. *Pocahontas county*, 531. Morgantown, WV: West Virginia Geological Survey.

Reger, D.B., and P.H. Price. 1926. *Mercer, Monroe, and Summers counties*, 963. Morgantown, WV: West Virginia Geological Survey.

Slingerland, R., and K. Furlong. 1989. Geodynamic and geomorphic evolution of the Permo-Triassic Appalachian Mountains. *Geomorphology* 2: 23–37.

Springer, G.S. 2000. The historical geology of West Virginia limestones: an introduction. *West Virginia Speleological Survey Bulletin* 14: 139–151.

Stamatakos, J., A.M. Hirt, and W. Lowrie. 1996. The age and timing of folding in the central Appalachians from paleomagnetic results: *GSA. Bulletin* 108: 815–829.

Hydrology of the Greenbrier Karst

3

William K. Jones

Abstract

Studies of the karst drainage systems of the Greenbrier limestone in southeastern West Virginia began in the early 1960s and were the first to make extensive use of water-tracing techniques and cave mapping in the USA. The carbonate aquifer is about 400 ft (120 m) thick in the Swago Creek area west of Marlinton (Pocahontas County) increasing to 1000 ft (300 m) in southern Monroe County. The basic hydrogeologic setting for the region consists of relatively flat-lying limestones exposed in valleys or plateaus and surrounded by higher elevation clastic units. Recharge to the conduit aquifer system is by capture of surface streams originating on the clastic rocks (allogenic recharge) and water infiltrating through the extensive areas of dolines (autogenic recharge). Only a few surface streams cross the carbonate outcrop, and even, these tend to loose water into the karst drainage systems. Much of the flow through the aquifer is through conduits under open channel conditions much like a surface stream with a roof. Discharge is concentrated at large springs that typically display rapid response to storm events, and the ratio of maximum to minimum discharge exceeds 100:1 for most of the springs. The karst caves and conduits are generally decoupled from surface topographic features, and the patterns of mapped cave passages are influenced by structural and stratigraphic characteristics. Insoluble beds within the Greenbrier Group may perch underground streams well above the apparent base level, and the underlying Maccrady Shale acts as an aquitard with several large caves developed along the contact of the shale and the overlying limestone. Much of this area can be considered a "contact karst" with the clastic rocks delivering concentrated recharge water onto the soluble limestones and the underling shales eventually forcing the return of conduit flow to the surface. The available data on water wells in the limestone suggest that most are actually producing from shaley units with the limestones acting as confining beds. The Greenbrier River and its tributaries represent base level for most of the area, and the relief of several hundred feet provides the hydraulic gradient. The area is underdrained by a well-integrated network of caves.

3.1 Introduction

This chapter presents an overview of studies of the karst hydrology of the Greenbrier limestone in southeastern West Virginia. The area ranges from Swago Creek west of Marlinton in Pocahontas County to Greenville in Monroe County (Fig. 3.1). The Greenbrier Group is exposed in an upland valley or plateau trending northeast/southwest. This is a mature karst aquifer with few surface streams making it across the width of the carbonate outcrop. The Greenbrier limestone outcrop area is less than one-mile wide (1.6 km) in the northern part to about 10 miles (16 km) wide in Monroe County, and the thickness increases from about 400 ft (120 m) in the Swago Creek area to 1000 ft (300 m) in the Greenville area. The sinking streams and caves drain to the

W.K. Jones (✉)
PO Box 356, Warm Spring, VA 24484, USA
e-mail: wkj30@hotmail.com

W.B. White (ed.), *Caves and Karst of the Greenbrier Valley in West Virginia*,
Cave and Karst Systems of the World, https://doi.org/10.1007/978-3-319-65801-8_3

Fig. 3.1 Sketch map showing study area and principal drainage basins

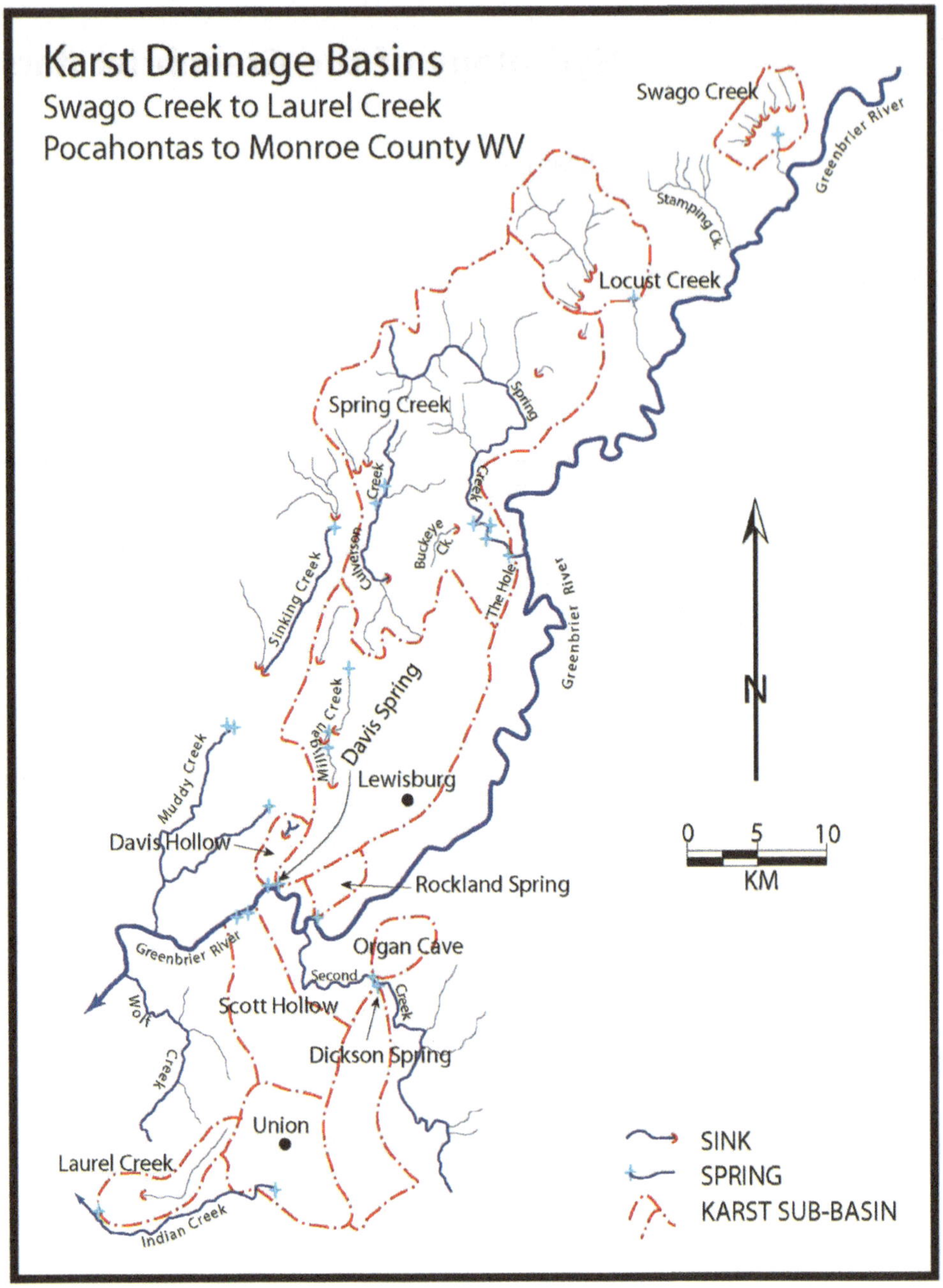

Greenbrier River or its major tributaries of Locust Creek, Spring Creek, and Second Creek for the areas north of Union in Monroe County. Drainage South of Union is to Indian Creek, a tributary to the New River.

The Greenbrier River and Indian Creek represent regional base level for the area. Karst development in the northern part is confined to relatively narrow coves floored by limestone and surrounded by clastic hills. The carbonates thicken, and the width of exposure widens to the south. Karst weathering processes have produced a broad doline plain (Fig. 3.2) with numerous sinking streams, blind valleys, and large springs.

3.2 Climate

All recharge to the aquifer is from precipitation. The humid temperate climate provides relatively uniform rainfall throughout the year. The mean annual temperature at Lewisburg is 50 °F (10 °C) with a mean annual total precipitation of 40 in. (1011 mm) (1981–2010 normals). Average annual snowfall is 32 in. (81 cm). Union is slightly dryer and warmer than Lewisburg while the Swago Creek area is colder and wetter. Marlinton has a mean annual temperature of 47 °F (8.8 °C) and mean annual precipitation of 44 in. (1117 mm) with 40 in. (102 cm) average snowfall.

Fig. 3.2 Aerial photo showing doline plain on mature Greenbrier karst in Monroe County

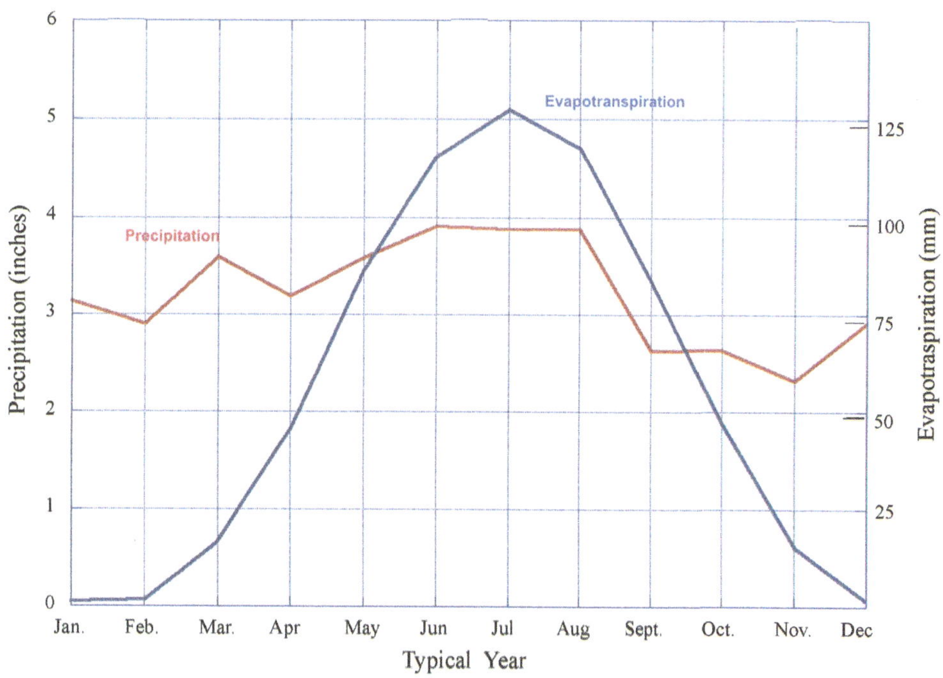

Fig. 3.3 Graph showing monthly precipitation and potential evapotranspiration at Lewisburg

Evapotranspiration exceeds precipitation during the summer months (Fig. 3.3) so stream and spring flows are at a minimum during late summer and early fall. Annual runoff for the Greenbrier River at Buckeye is 22.2 in. (560 mm) and 19.9 in. (505 mm) at Alderson. Roughly, half of the annual precipitation is lost to evapotranspiration.

3.3 Recharge and Yield of the Aquifer

The karst drainage basins are surrounded by clastic rocks but much or all of the flow is subterranean to base-level springs. The discharge at the springs is typically very flashy, and the storm hydrographs are similar to those of nearby surface streams. This is because much of the recharge is from the capture of surface streams at or near the contact with the carbonate outcrop. Recharge is very rapid at the points of capture of surface streams and through the more open drains at the bottom of dolines. Storage within the conduit part of the aquifer is in smaller fractures and the overlying epikarst zone. The spring hydrographs represent a mixture of allogenic recharge from captured surface streams originating on the surrounding clastic units and autogenic recharge captured by the doline plain. A study by Jones and Rauch (1977) showed that most of the springs have a high-flow to low-flow ratio exceeding 100. Discharge from Davis Spring typically ranges from about 7 to over 1000 cfs (0.2–28 cm).

3.4 Water Wells

Water wells that appear to be completed in the limestone may actually be producing from shaley units within the Greenbrier Group or the underlying Maccrady Shale (Ogden 1976; Heller 1980). There are few examples of interchange between water in the conduits and nearby water wells. Water levels in wells are frequently higher than in the conduit part of the aquifer. A study by Demrovsky (2003) examined two cases where water from uncased wells drilled close to cave passages "bursts" into the caves under pressure and created a constant flow or leakage into the cave. Demrovsky termed these "subterranean springs" and found that the constant drainage of the shale aquifer was creating a significant depression in the water table (potentiometric surface) and causing several area wells to go dry. This water was rich in sulfur and iron oxides (ferrihydrite) and of very different chemical quality from streams in the cave. High sulfate concentrations have been reported in scattered water wells in the area (Heller 1980) although these wells appear to be drawing water from sulfur-rich shales (Ogden 1976).

Recharge to the shaley aquifers, especially the Hillsdale limestone/Maccrady Shale, may be from outside of the main areas of the exposed carbonate outcrop (Ogden 1976). The groundwater flow direction for the lower aquifer may not be reflected in the drainage patterns and divides determined using tracer tests in the conduit part of the system. However, a study by Heller (1991) using water levels from wells and MODFLOW to draw the potentiometric surface for the area between Davis Spring and Spring Creek in central Greenbrier County did result in a north–south drainage divide that closely matched the divide predicted by Jones (1973) for the conduit flow part of the aquifer.

3.5 Water Tracing

Some of the earliest water-tracing studies in North America were conducted by Hermine Zotter in the Swago Creek area (Zotter 1965). These tests were planned primarily by cavers from the Pittsburgh Grotto to establish connections between sinking streams and caves and were viewed as an aid to exploring and understanding caves (Fig. 3.4). The tests were qualitative and used packets of activated carbon placed in springs to act as passive collectors for fluorescein sodium dye (CI45350) injected in sinking streams (Fig. 3.5). The use of passive collectors enabled researchers to do weekend trips to the area without the need for constant observation of a number of springs.

The essential procedure was described by Dunn (1957) and used a basic alcohol solution to elute the tracer. The solution was allowed to sit undisturbed for a time period ranging from one-half hour to a couple of days and then visually examined under sunlight or a high-intensity flashlight. A fluorescent-green sheen at the top of the carbon signified a positive test. This system was found to work with other dyes in the Xanthene family (Eosin sodium and Rhodamine WT) so tests starting in the 1980s often used multiple tracers and fluorometers in the analysis to increase the levels of detection and separate the different tracers based on emission wavelength.

The early tests conducted by Zotter in the Swago Creek area typically used 3–8 oz (85–230 g) of fluorescein over distances of one to two miles (1.6–3.2 km). Tests conducted by Jones (1973, 1997) in Greenbrier and Monroe Counties used one to ten pounds of dye over distances of 1–15 miles (1.6–24 km). Many of these tests resulted in visual coloration at the resurgences, and an approximation of the travel time was obtained. Although coloration of springs is discouraged today, one advantage of the high dye concentrations during the early reconnaissance work was that all the springs in a 100-square mile area could be checked from a small plane during a 1-h flight. Visiting all the possible resurgences by road and trail required almost two days of fieldwork.

The early tracer tests in the Swago Creek area were designed primarily to determine the connections between caves, sinking streams, and springs. The tests conducted in the early 70s as part of a US. Geological Survey study (Jones 1973) were an attempt to define drainage divides and determine the catchments for some of the larger springs. The results of the tracer tests were combined with data from cave surveys and plotted on 62,500 scale topographic maps. The

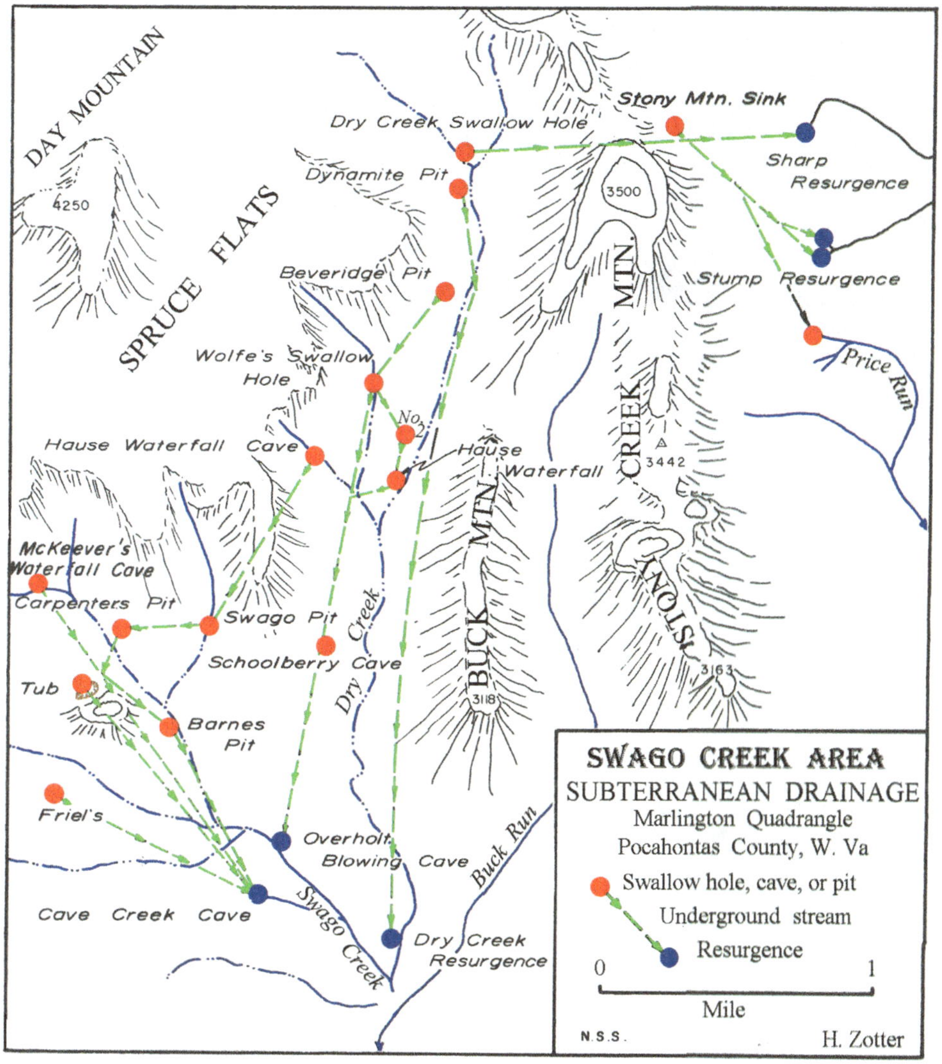

Fig. 3.4 Sketch map showing karst drainage patterns in the Swago Creek area of Pocahontas County (after Zotter 1965)

drainage basins were drawn along the topographic divides on the surrounding clastic rocks and estimated in parts of the limestone outcrop area. Additional tracer tests have been done by different workers over the years, and minor adjustments have been made to the boundaries originally shown in the 1973 publication (Jones 1997; Dasher and Boyer 2000; Tudek 2009).

3.6 Principal Drainage Basins

3.6.1 Swago Creek Basin

The Swago Creek watershed is a karst "cove" of about 4 mi^2 (10 km^2) that contains more than 80 known caves. The rocks are relatively flat lying, but faulting may exert some control over subsurface flow paths. Recharge to the karst aquifer is largely from the capture of surface streams flowing off of the surrounding highlands capped by clastic rocks. Karst groundwater flow is along hydraulic gradients toward the Greenbrier River, but the flow routes are not reflected by the surface topography. This basin has the highest average elevation and greatest relief of the basins described in this volume.

Studies of this area by Zotter (1963, 1965) described the connections between the various stream sinks, caves, and springs (Fig. 3.6). White and Schmidt (1966) used the results of the tracer tests and mapping of caves to define subterranean flow routes and determine drainage divides. They demonstrated that the underground drainage was often unrelated to the surface (or former surface) flow routes. Flow paths that seemed to underdrain the original surface channels were termed "well behaved," and paths that crossed into other basins were "misbehaved." Another concept from this paper was "lost waterfalls" where the cave stream comes to the surface perched on insoluble shale (Fig. 3.7), flows on the surface for some distance, and sinks underground again where it breaches the shale and again reaches the limestone.

Fig. 3.5 Fluorescein sodium dye injected in the stream from Boyds Cave in Monroe County

Streams on the western side of the Swago Creek basin lose water intermittently through gravel-floored stream channels and resurge at Cave Creek Cave spring. This appears to underdrain the dry surface valley and is "well behaved" using the classification of White and Schmidt (1966). Water from Hause Waterfall Cave is pirated from the Dry Creek catchment through Carpenters Cave and is "misbehaved."

Water at the very head of Dry Creek sinks at Dry Creek Swallow Hole and passes eastward under Stony Creek Mountain to resurge at Sharp Resurgence. Dry Creek water sinking slightly to the south at Beveridge Pit more or less underdrains the Dry Creek valley through Wolfs Swallow Hole and resurges at Overholt Blowing Cave.

3.6.2 Upper Spring Creek Locust Creek Cave Basins

The water resurging at Locust Creek Cave (Fig. 3.8) comes from both the Little Levels area west of Hillsboro and a partial contribution from streams sinking on the west side of Droop Mountain. This is a complex basin and was initially studied by Zotter (1965), White and Schmidt (1966). Tracer tests from these early studies showed that water sinking in the bed of Bruffey Creek at the entrance to Bruffey Creek Cave flowed east to Hughes Creek Cave. The water resurges at Upper Hughes Creek where it is perched on the Taggard Shale and flows on the surface for about 600 ft (180 m) until it breaches the shale and sinks again into Lower Hughes Creek Cave. The Bruffey/Hughes water then flows south to resurge at Locust Creek Cave. Also resurging at Locust Creek Cave is water from Grand View Pit and Blue Hole.

The study by Zotter showed that water from Bruffey Creek reached Locust Creek Cave after going through the Hughes Creek Caves, and Hills Creek water was believed to flow under Droop Mountain directly to Locust Creek Cave 1.65 miles to the southeast. Coward (1975) reported tests from Bruffey and Hills Creek Caves with dye recovery in Cutlip and Clyde Cochran Caves with the assumption that all the flow from Hills Creek was to the southwest to Clyde Cochran Cave and then to the east to Locust Creek Cave.

Further studies by Williams and Jones (1983) and Jones (1997) identified a somewhat more complex flow system (Fig. 3.9). Discharge measurements under base flow conditions showed that the flow in Hills Creek alone was greater than the flow in Locust Creek so there were obviously multiple resurgences for this area. Cave exploring and mapping in the Friars Hole system in the 1970s revealed a large master stream passage extending to the southwest toward Spring Creek. A quantitative test (Williams and Jones 1983; Jones 1984) involved the simultaneous injection of an optical brightener in Hills Creek, Rhodamine WT in Cutlip Cave, and fluorescein sodium in Friars Hole cave. The tracer from Hills Creek was recovered at Locust Creek Cave and in Clyde Cochran Cave. However, the tracers injected from Friars Hole and Clyde Cochran Caves were not found at Locust Creek Cave but at JJ Spring on Spring Creek 11 miles (18 km) to the south. Most of the water from Bruffey and Hills Creeks (at least under base-flow conditions) goes through Cutlip, Clyde Cochran, and Friars Hole Caves and ultimately resurges on Spring Creek. Jones (1997) calculated that at low flow only about 5% of the water from Hills Creek went to Locust Creek Cave, and some of the

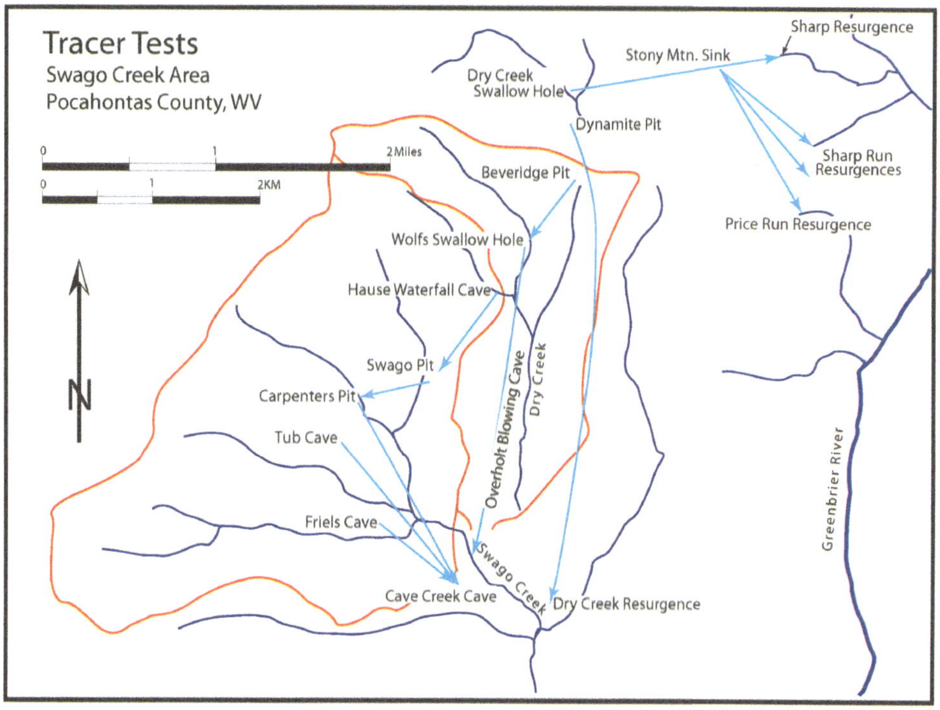

Fig. 3.6 Sketch map showing principle tracer tests in the Swago Creek area (after Zotter 1965)

Fig. 3.7 Lost waterfall on the wall of the Dry Creek Valley, Swago Creek basin. Photo by W. B. White. Used with permission

flow from Bruffey Creek and Rush Run was also diverted south to Spring Creek.

Under low-flow conditions, the bed of Spring Creek is dry for several reaches above the Rt 219 bridge, below the Cannon Hole, and below the upstream entrance to Spur Cave. A 1976 study by Charles and Barbra Williams (personal communication) established the connections between Spring Creek water sinking in the Cannon Hole and the water-filled sinkholes in the Spring Creek floodplain a half mile downstream. These steep-sided sinkholes were termed "floodplain cenotes" by Jones (1997) and are clustered together. The Cannon Hole is an estavelle (Fig. 3.10) that captures the surface flow of Spring Creek during base flow conditions and acts as a spring during periods of high flow. The Spring Creek water reappears in the cenotes and then at JJ Spring on the east side of the surface channel. The water from JJ Spring is on the surface for a couple of hundred feet before entering Spur Cave that acts as a subterranean

Fig. 3.8 Aerial photo of Locust Creek Cave (spring)

meander cutoff and then reappears a mile further downstream. During high-flow conditions, water is flowing in the surface channel, the Cannon Hole is functioning as a spring, and the Circulating Cenote is discharging overflow water as a spring (Figs. 3.11 and 3.12).

3.6.3 Lower Spring Creek Drainage Basin

Karst drainage to Spring Creek from the south (Fig. 3.13) consists primarily of flow from Culverson Creek Cave, Buckeye Creek Cave, and The Hole Cave. Some smaller sinks in the Frankford area also drain northeast to Spring Creek (Jones 1997).

Culverson Creek is fed by streams that flow off of clastic rocks on Cold Knob (Roaring, Little Roaring Creek, and Charley Run), sink near the limestone contact, reappear at springs and flow on the surface for a couple of miles, and sink again at the blind valley entrance to Culverson Creek Cave. The Culverson Creek Water then flows northeast to a series of four springs on the south bank of Spring Creek. The Culverson flow apparently passes underneath the headwaters of Buckeye Creek. A steam sinking near the old Pilgrims Rest Church also resurges with the Culverson Creek water on Spring Creek. The four Culverson resurgences include Matts Black Cave and are clustered along an escarpment on the southwest side of Spring Creek. Tracer tests started from the Culverson and Fullers entrances under low-flow conditions took 14 days to reach Spring Creek (Fig. 3.14).

Wolfe (1973) examined the dry valleys below the sink points and called them "karst sieves" due to the accumulation of clastic sediments left by water percolating through the porous limestone. Rader's Valley is a large dry valley (Fig. 3.15) that is the former surface course of Culverson Creek.

Buckeye Creek flows to the north and sinks at the blind valley entrance to Buckeye Creek Cave. This passage has been surveyed to Spencer Cave on Spring Creek. Under low-flow conditions, the bed of Spring Creek is dry for some distance above Spencer Cave, and the Buckeye Creek water flows on the surface in the Spring Creek channel to sink at the Cannon Hole. A narrow flat-floored valley locally known as the "Race Track" trends south from the Buckeye Creek Cave entrance. This valley is a closed depression and essentially a small polje (Fig. 3.16) about 1 mile (1.6 km) long and averages 300 ft (110 m) in width. Springer (2002, 2004) conducted detailed studies of the sediments and passage sculpturing in Buckeye Creek to characterize paleoflooding in the basin.

The large contact cave system known as "The Hole" drains north to resurge at two springs on the south side of Spring Creek. The eastern most spring is Burns Cave Number 2, and the second spring is the "Blue Hole" (Fig. 3.17) about one-half mile upstream from the Burns Spring. These springs also drain the Frankford area.

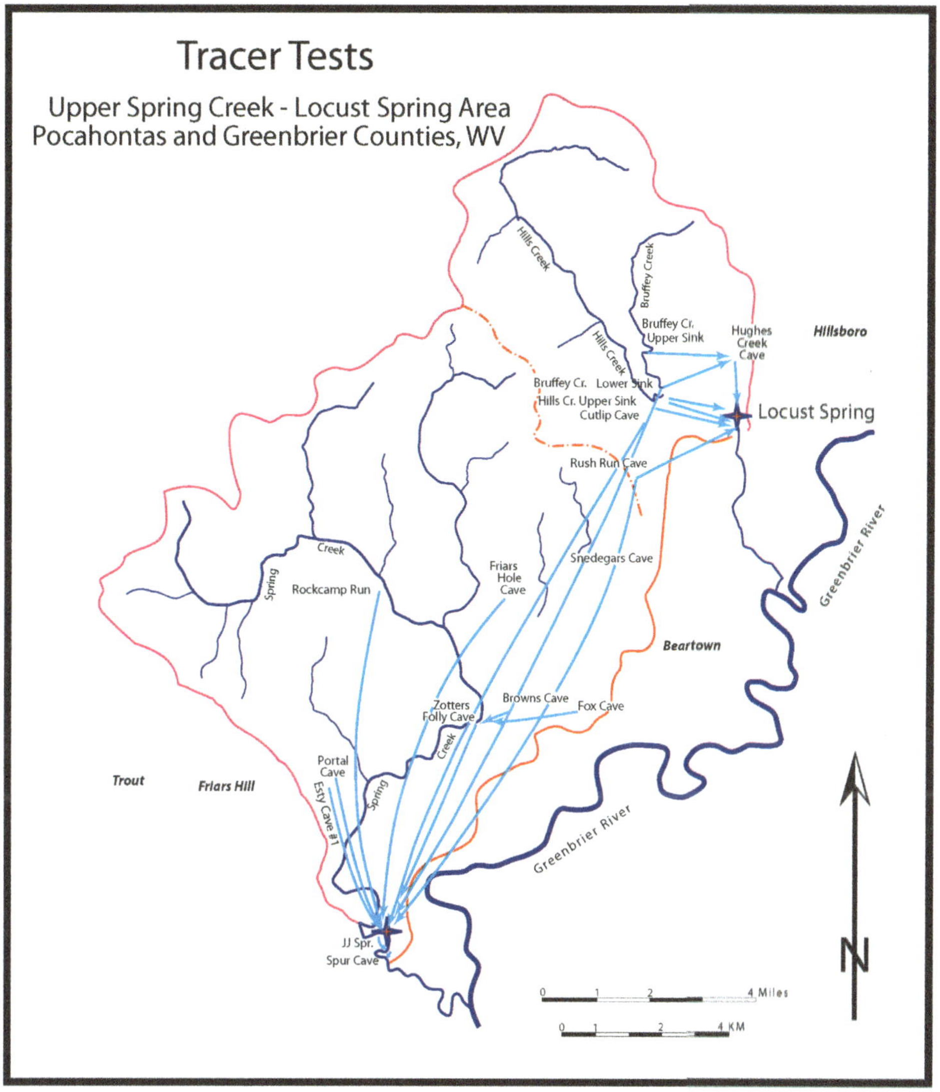

Fig. 3.9 Sketch map showing principle tracer tests in the Upper Spring Creek drainage Basin. Note that flow from Bruffey Creek, Hills Creek, and Rush Run goes to both Spring Creek and Locust Creek. All of the water from Clyde Cochran and Snedegars (Friars Hole) Caves goes to JJ Spring on Spring Creek (after Jones 1997; Dasher and Boyer 2000)

3.6.4 Davis Spring Basin

Davis Spring discharges into the Greenbrier River at Fort Spring southwest of Lewisburg. The spring rises at the base of a cliff (Fig. 3.18) and flows on the surface for about 1000 ft (300 m) to the river. About 80% of the Davis Spring drainage basin (Fig. 3.13) is underlain by carbonates of the Greenbrier Group. Subsurface flow in the basin is to the southwest, and much of it may be along the troughs of several synclines (Tudek 2009). All of the drainage (with the possible exception of extreme flood events) from the basin is through Davis Spring with surface streams flowing off the surrounding clastic highlands and sinking at ponors and blind valleys in the limestone to create a significant volume of allogenic recharge to the spring. Additional recharge to the karst aquifer system occurs through the extensive doline plain developed on the outcrop area of the limestone. Much of the flow through the aquifer occurs through cave passages. Over one hundred caves are known from this basin, and several of these are over 10 miles (16 km) long. Several large contact caves including Ludington, McClung, and Wades caves drain to Davis Spring along with the subsurface runoff from the Lewisburg area. The northern boundary is defined by the Higginbotham Caves west of Frankford. Water from the Higginbotham Number 1 cave reappears in a karst window at Coffman Cave, flows through Savannah Lane Cave, and flows southwest for about 15 miles (24 km) to Davis Spring. Milligan Creek drains the western part of the basin. Milligan Creek is a classic example of an "interrupted" karst stream that sinks and rises at five different points with the final rise at Davis Spring. Most of the tracer tests within this catchment were qualitative, but the straight-line travel velocities for the Lewisburg tests were about 1900 ft (580 m) per day. This basin is described in reports by Jones (1973, 1997) and Tudek (2009).

Fig. 3.10 Barbra Turner Williams adding fluorescein dye to the Cannon Hole estevale at low flow

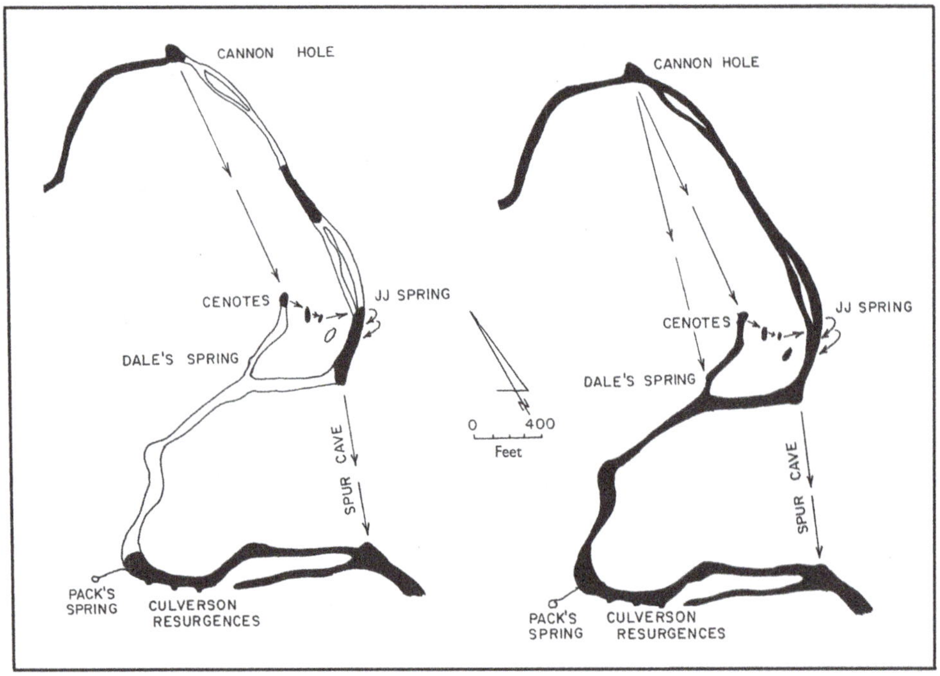

Fig. 3.11 Sketch map of the Spring Creek drainage from the Cannon Hole to JJ Spring (after Williams and Jones 1983; Jones 1997)

The Rockland Indian Spring basin is a small karst catchment that discharges in an alleviated spring alcove directly to the Greenbrier River at Rockland. The basin drains about 3.3 mi^2 (8.5 km^2) and is described by Jones (1973) and Tudek (2009). Watters Cave drains a narrow strip along the southeast edge of the Davis Spring Basin. The cave entrance is just south of the Greenbrier East High School in Fairlea, and the water resurges in a spring above the town of Ronceverte. The catchment is about 1 mi^2 (2.6 km^2).

One full water year (October 1–September 30, 1973) of daily discharge record is available for Davis Spring (Fig. 3.19) and nearby Howards Creek. The record for Davis Spring may slightly underrepresent total discharge because the record is truncated at a maximum flow of 1000 cfs (28.3 cm). The rating curve becomes undefined at high water levels due to backwater from the Greenbrier River at the gauge. Data from the Davis Spring gauging station was used by White (1977) in her study of the effect of karst on flows in Appalachian watersheds.

Fig. 3.12 Aerial photo showing Spring Creek sinking at the Cannon Hole (*far left*), the floodplain cenotes (*center*) and JJ Spring (*far right*)

Paired watershed studies are a common tool in forest and agricultural hydrology for studying the effects of different harvesting or agricultural systems on runoff and water quality from physically similar basins (Woolhiser et al. 1980; Zhang et al. 2001; Schilling 2013). Paired watersheds were also used in a study of the Mammoth Cave area in Kentucky by Hess and White (1989). The Howard Creek catchment is underlain by clastic rocks and is situated just to the east of the Greenbrier River. The Davis Spring catchment lies just to the west of the river. Both Basins are tributary to the Greenbrier River (Fig. 3.20), and the main difference between the basins is bedrock geology. This should appear as differences between the water budgets for the two basins. The basic annual water budget equation is:

$$P = R_s + R_{gw} + E_t \pm \Delta S$$

where P is the annual total precipitation, R_s is surface water runoff, R_{gw} is groundwater runoff, E_t is evapotranspiration, and S is the change in storage between the start and end of the water year. It is generally difficult to separate the surface and groundwater components, and the reported annual runoff for a gauging station is the sum of these components. A rough estimate for gauging stations in West Virginia with significant contributions from karst aquifers is that 85% of the annual runoff is from ground water (Shultz et al. 1995). Note that the water year runs from October 1 to September 30 in an attempt to minimize the changes in storage.

The drainage area for Davis Spring was defined from a series of water-tracing tests (Jones 1997; Tudek 2009). The Davis Spring (USGS Station ID number 03183200) catchment is about 74 mi^2 (192 km^2), and Howard Creek (USGS Station ID Number 03182950) drains an area of 84.5 mi^2 (219 km^2). Precipitation data from the NOAA weather station at Lewisburg, West Virginia, was used to calculate the water budgets for both catchments. The 1973 water year was wetter than the normal precipitation of 40 in. (1030 mm) with 45 in. (1150 mm) of precipitation for the study period. This year had a total of 47 in. (1190 mm) precipitation for the period so runoff from both basins was above the long-term averages. However, this should affect both basins equally so the comparison of water budgets for the two basins (Fig. 3.21 and Table 3.1) should be valid.

The total runoff for the 1973 water year was 21.2 in. (538 mm) from Howard Creek and 25.9 in. (658 mm) from Davis Spring. Runoff at Howard Creek exceeded that from Davis Spring during November (Fig. 3.22) when evapotranspiration was near a minimum but fell below that of the spring during August (Fig. 3.23) when evapotranspiration rates reach a maximum. It is interesting that in terms of area-runoff for November 1972, there was no significant difference between the two basins with Howard Creek having 3.1 in. (78 mm) of runoff and Davis Spring 3.0 in. (76 mm) of runoff. During August 1973, Howard Creek had 0.25 in. (6.3 mm) of runoff compared to 0.47 in. (11.9 mm) at Davis Spring. Adjusted for drainage area, Davis Spring was discharging almost twice as much water as Howard Creek in August.

The paired catchments have the same climate and annual rainfall. The two drainage basins differ fundamentally in only two respects: geology and vegetative cover, so differences in the water budgets should be due to these factors. In terms of the annual water budget, runoff was 22% higher, and evapotranspiration was 18% lower for the karstic catchment. The difference in vegetative cover between the catchment accounts for less than half of the observed differences between the paired basins. The rapid infiltration rates of water from precipitation into the karst aquifer alter the water budget to a certain extent by decreasing evapotranspiration with a corresponding increase in runoff from the karstic catchment.

The analysis of storm hydrographs can also be used to quantify differences between drainage basins (White and White 1974; Mangin 1975; Kresic and Bonacci 2009). A comparison of the storm response for the two basins

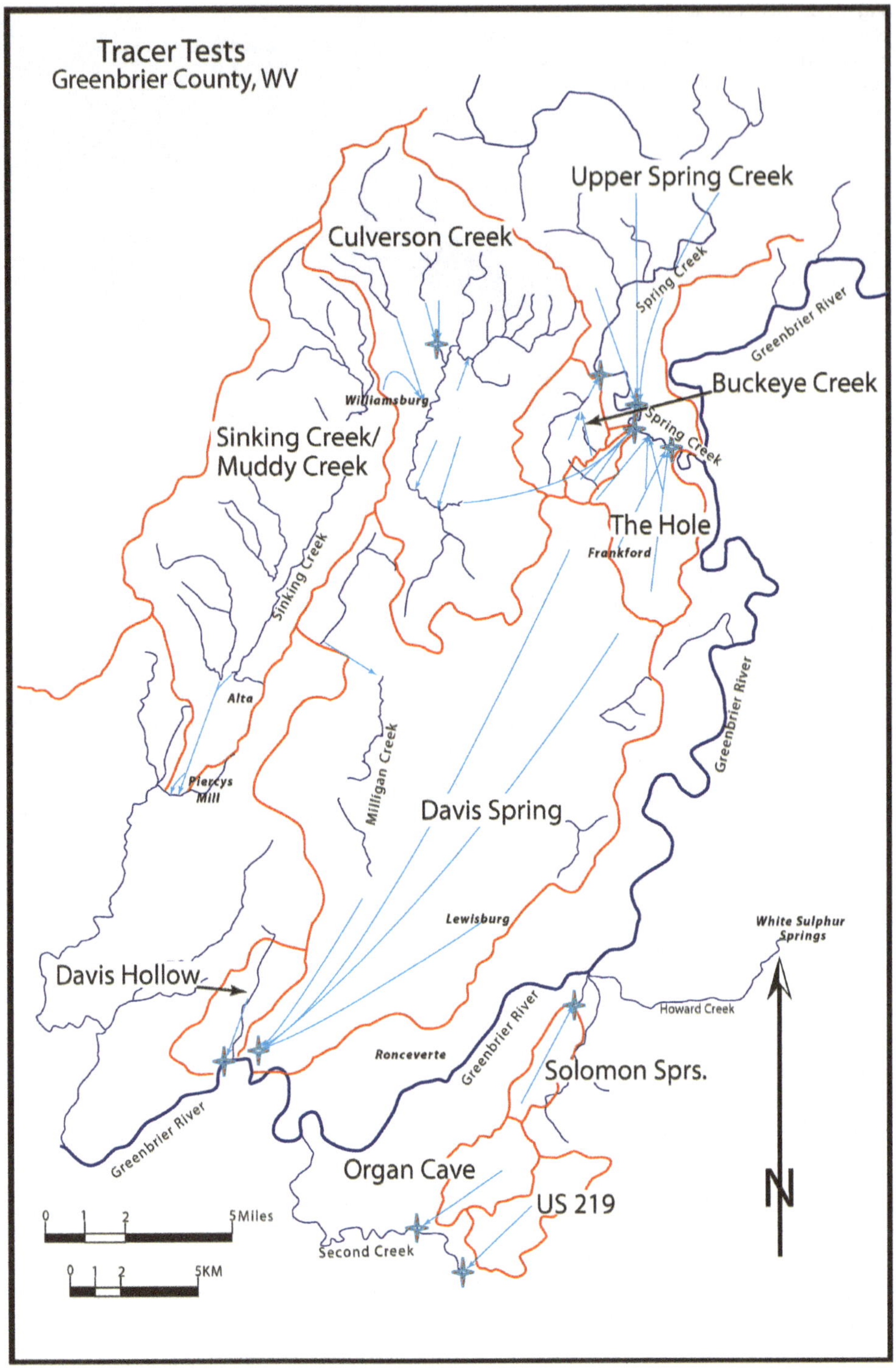

Fig. 3.13 Sketch map of the southern part of the Spring Creek basin (after Jones 1997; Dasher and Boyer 2000), the Davis Spring basin (after Jones 1973; Tudek 2009) and the Organ Cave Basin (after Jones 1997)

shows that Davis Spring typically lags Howard Creek by about a day. The slope of the recession curve following storm events can also show the nature of the release of water from groundwater storage. The recession coefficient (also called the coefficient of discharge) is the slope of the line of the recession hydrograph plotted on semilog paper. Most storm recession hydrographs have distinct periods with different slopes, so this type of analysis is usually conducted on the middle portion of the recession line. To compare recession hydrographs from different catchments, the same time period must be used. A recession analysis from a storm event in August 1973 showed a recession coefficient of 0.0428 for Davis Spring and 0.0625 for Howard Creek (Fig. 3.24). This suggests greater groundwater storage for the Davis Spring Basin or at least a more gradual release of water from storage.

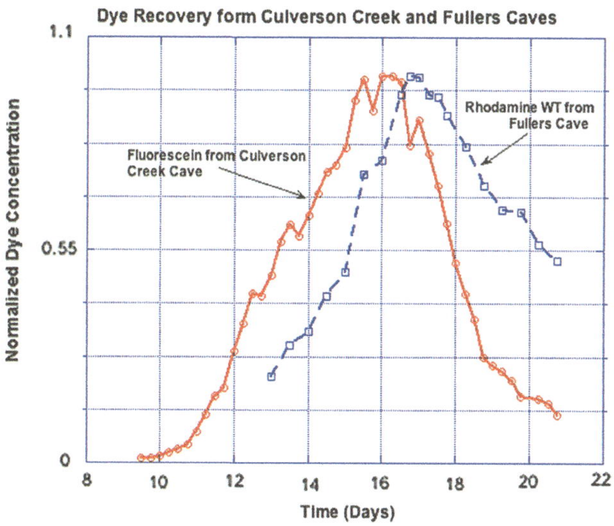

Fig. 3.14 Graph showing recovery of two tracers from the Culverson Creek Cave System at the lower Culverson resurgence on Spring Creek. The trace was started under low-flow conditions on August 24, 2009

3.6.5 Davis Hollow Basin

Davis hollow is underdrained by water sinking at Sinks of the Run Cave, then flowing through General Davis Cave, and resurging in a spring at base level with the Greenbrier River. This 4.4 mi^2 (11.4 km^2) basin is just to the west of Davis Spring at Fort Spring.

3.6.6 Sinking Creek/Muddy Creek Basin

The Sinking Creek/Muddy Creek Basin is northwest of the Davis Spring basin. It is described separately in Chap. 12.

3.6.7 Organ Cave Basin, Soloman Springs, and Route 219 Bridge Basins

These relatively compact catchments are on the southeast side of the Greenbrier River south of Caldwell (Fig. 3.13).

Fig. 3.15 Rader Valley is a large dry valley that was the former course of Culverson Creek when it flowed south to the Greenbrier River near Alderson

Fig. 3.16 Photo of the "Race Track," a polje-like closed depression extending south from the entrance to Buckeye Creek Cave

Fig. 3.17 Blue Hole, one of the two resurgences for the Hole

Water at Soloman Springs flows to the Greenbrier River and is about a mile (1.6 km) from caves.

The headwaters of Second Creek are on Ordovician carbonates on the western base of Peters Mountain in Monroe County. The creek flows to the west across the Greenbrier limestones from about the community of Second Creek to the Greenbrier River, and numerous springs may be found on both banks. A spring on the north bank of Second Creek just upstream of the Route 219 Bridge is fed by water from Helms Cave to the northeast of the spring.

Organ Cave is one of the longest caves in West Virginia and is developed in a synclinal trough at the contact of the Hillsdale limestone and the Maccrady Shale. The catchment for this large cave is only 3.6 mi^2 (9.3 km^2). The cave has a dendritic pattern of passages that drain toward and along the axis of the syncline, and the water resurges at a base-level spring (Fig. 3.25) on Second Creek. The hydrology and ecology of the basin are described in Jones (1988), Culver et al. (1994).

3.6.8 Dickson (Dixon) Spring Basin

Dickson Spring (Fig. 3.26) drains a 27.7 mi^2 (71.7 km^2) catchment in Monroe County on the south side of Second Creek (Fig. 3.27). The water rises at the base of a limestone cliff (Patton limestone) on the western limb of the Hurricane Ridge syncline. During high flow, water also boils up in the spring run about 100 ft (33 m) downstream from the cliff. Flow monitoring reported by Ogden (1976) for the 1974 water year showed a minimum discharge of 6.6 cfs (0.19 cm) and a maximum of 107 cfs (3.0 cm). This represents a high- to low-flow ratio of 16:1 and is much lower than most of the

Fig. 3.18 Aerial photo of Davis Spring

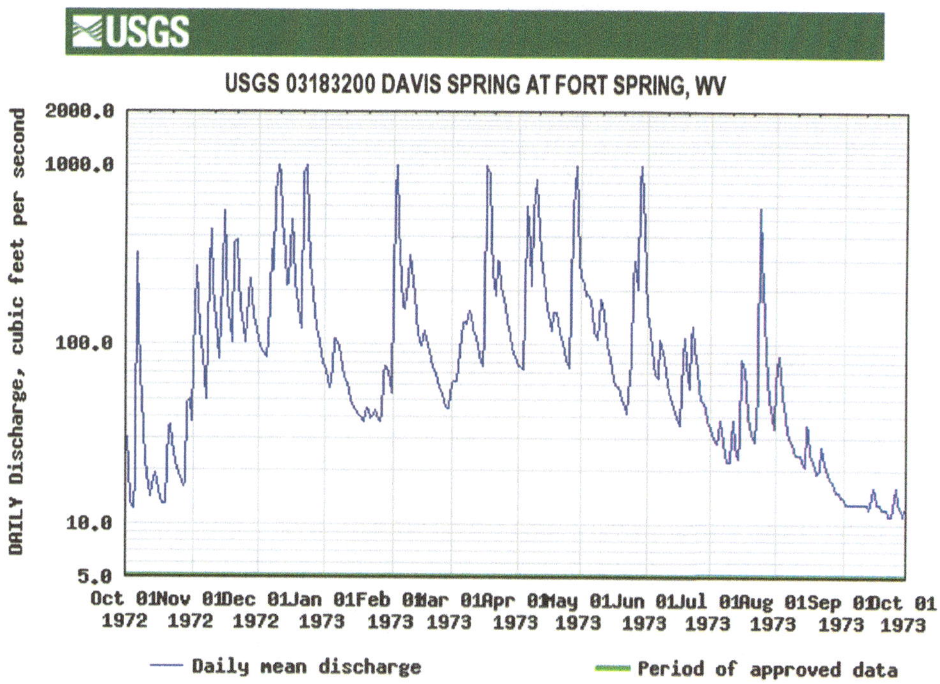

Fig. 3.19 Hydrograph for Davis Spring for the 1973 water year (US Geological Survey)

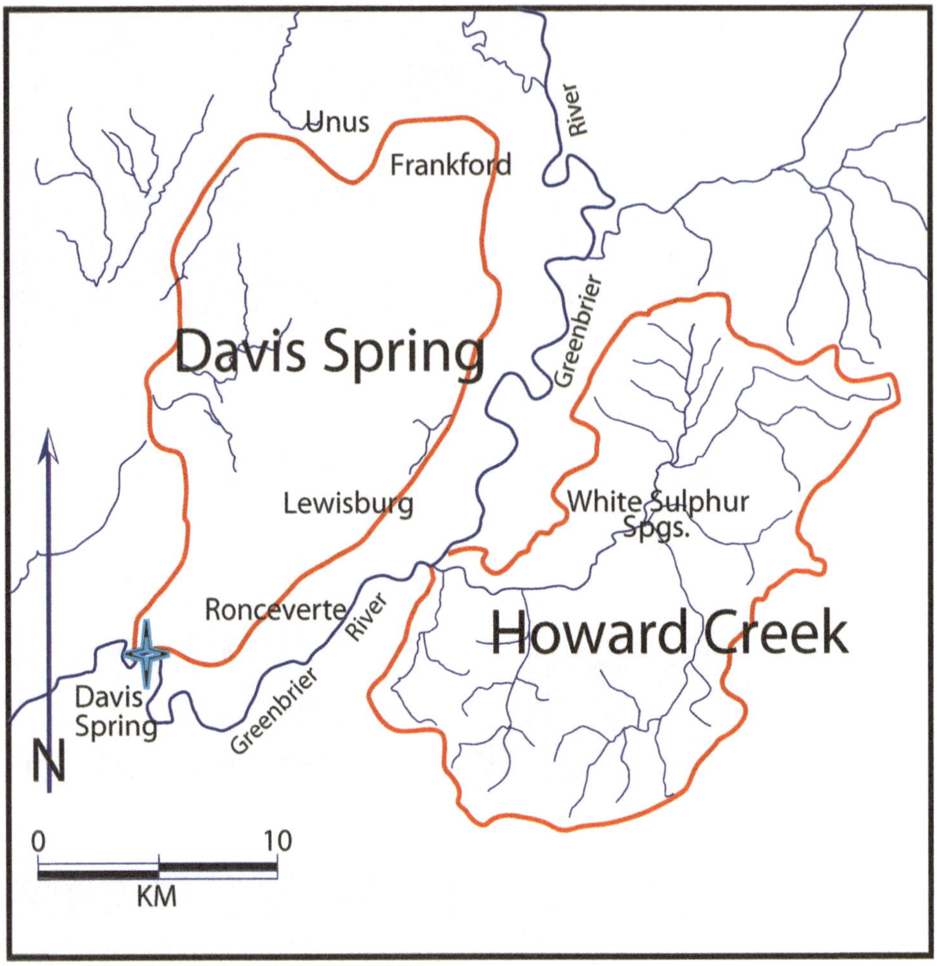

Fig. 3.20 Sketch map showing the Davis Spring and Howard Creek catchments

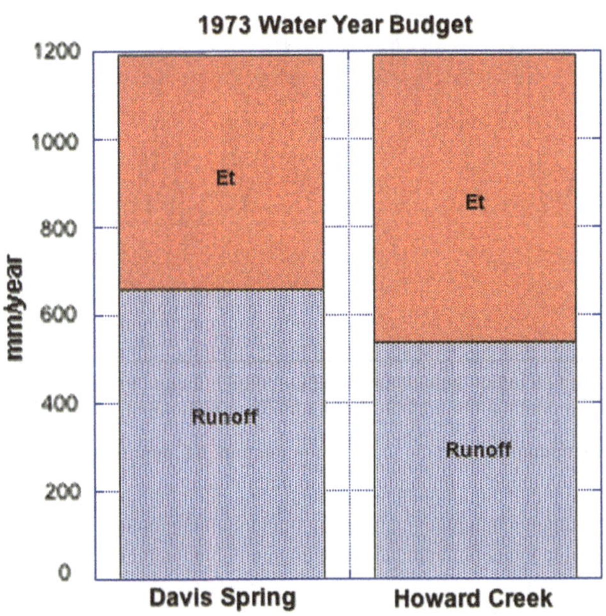

Fig. 3.21 Graph showing water budgets for the 1973 water year

springs draining the Greenbrier karst. The response to individual storm events was also slower than observed for most of the springs in this region. Jones (1997) reported seven tracer tests within this basin. The longest was from Ash Cave about 9 miles (14.5 km) south of the spring. Water sinking in the upper reaches of Burnside Branch goes to Dickson Spring while water from the lower part flows west to Indian Creek through Steel Cave. Straight-line travel velocities for tracer tests in this basin averaged about 2300 ft (700 m) per day.

3.6.9 Scott Hollow Basin

The area around Sinks Grove drains to the north through Scott Hollow Cave and then to several springs in the bed of the Greenbrier River downstream from Fort Spring. Only one small spring (Gloria's Spring) has been observed above normal river pool elevation. The master drain in Scott Hollow Cave is "Mystic River," and most of the underground flow in this basin is channeled through this large passage. The Scott Hollow catchment is about 18.6 mi^2 (48 km^2).

Table 3.1 Comparison of paired karst and non-karst catchments for the 1973 water year

	Davis Spring	Howard Creek
Drainage area	187 km^2	219 km^2
Mean discharge	3.89 cm	3.74 cm
Annual runoff	658 mm	538 mm
Total evapotranspiration	532 mm	652 mm
Q max	28.3 cm[a]	77.0 cm
Q min	0.311 cm	0.207 cm
August runoff	11.9 mm	6.35 mm
Recession coefficient	0.0428	0.0625

[a]Actual maximum exceeds this value

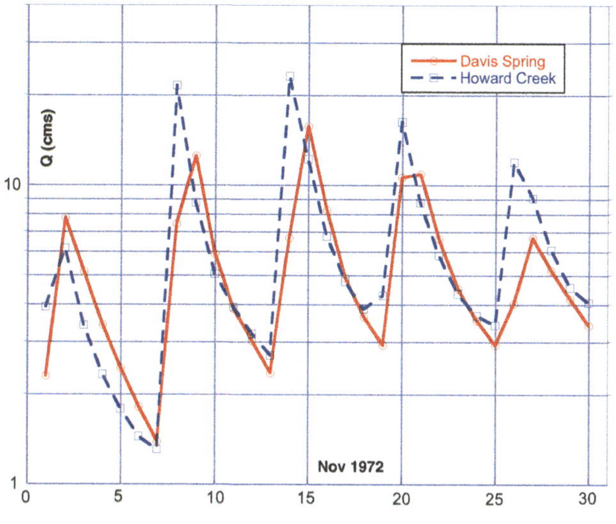

Fig. 3.22 Davis Spring and Howard Creek, November 1972

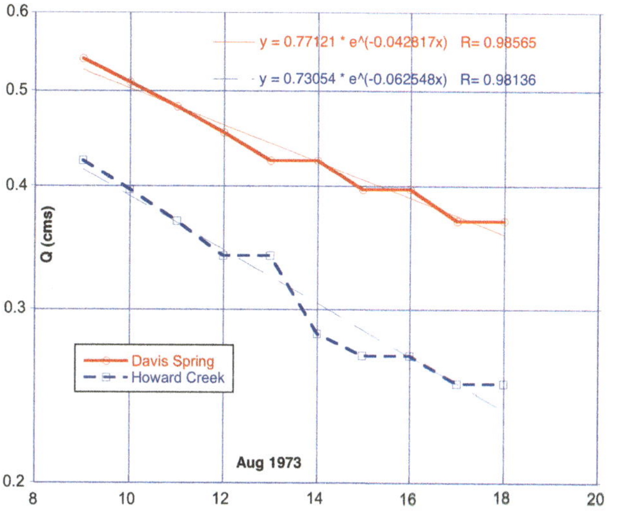

Fig. 3.24 Flow recession curves for Davis Spring and Howard Creek, August, 1973

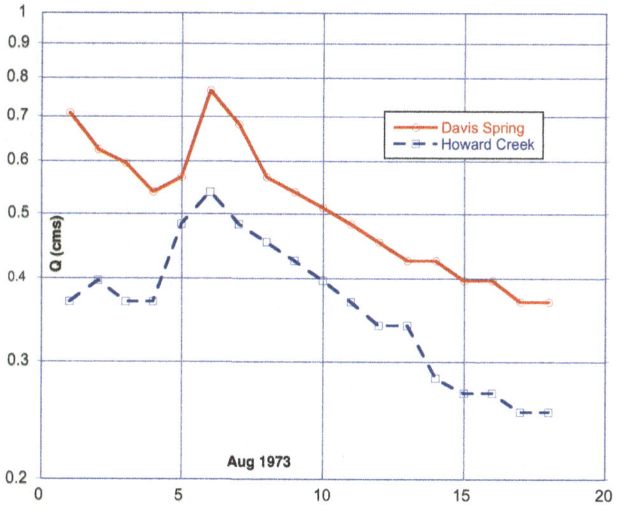

Fig. 3.23 Davis Spring and Howard Creek, August 1973

This basin has been described in reports by Jones (1997), Davis (1999), Bishop et al. (2009), and Demrovsky (2003).

Wolf Creek flows west to the Greenbrier from just to the west of the Scott Hollow basin, but little information is available on the karst of this area. Two large springs are reported near the junction of Broad Run and Wolf Creek at Wolf Creek.

3.6.10 Indian Creek (New River) Basins

Indian Creek rises at a spring (Fig. 3.28) southeast of Union in Monroe County and flows south and then west to join the New River at Indian Mills. Waters sinking in the lower reaches of Burnside Branch and Taggart Branch flow through Steels Cave and then reappear at the head spring of Indian Creek.

Fig. 3.25 Photo showing the Organ Cave resurgence on the north bank of Second Creek

Fig. 3.26 Aerial photo of Dickson Spring

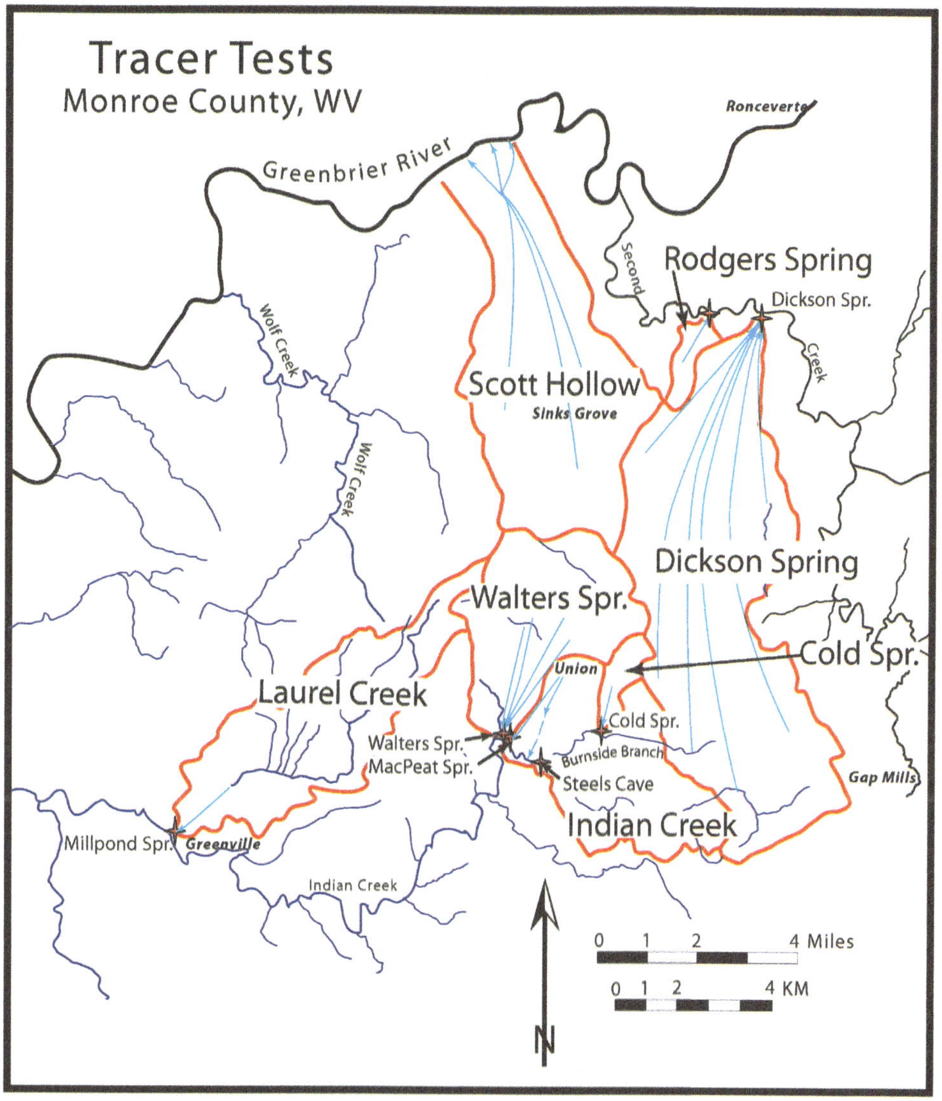

Fig. 3.27 Sketch map showing the Dickson Spring, Scott Hollow, and Indian Creek drainage (after Jones 1997)

MacPeat (McPeak) Spring and Broyles Cave receive water from the town of Union. Water sinking in the town flows partly to Broyles Cave (Broyles Sulfur Spring) and then to a spring on Indian Creek just below the head spring. The rest of the flow from Union reappears at MacPeat Spring on Indian Creek near Salt Sulphur Springs 1.7 miles (2.75 km) southwest of Union (Fig. 3.29). Flow from Union to Broyles probably only occurs during high-flow conditions. This water was contaminated by raw sewage from the town of Union until a treatment plant was constructed 1 mile (1.6 km) south of Union in 1984. The plant discharges into a sinkhole so the (now treated) water still goes to the same springs. The catchment for MacPeat Spring is 2.2 mi^2 (5.7 km^2). Ogden (1976) reported an average flow of 0.2 cfs (5.7 L per second).

Walters Spring is in the next valley about 1000 ft (320 m) to the west of MacPeat Spring but there does not appear to be any connection between the two springs with the MacPeat basin possibly overlying part of the Walters catchment. Both springs (Fig. 3.29) are on the north side of US Route 219. Walters Spring drains an area of 9.6 mi^2 (24.9 km^2) from northwest of Union. Walters Spring is situated at a thrust fault that brings the Greenville Shale to the surface. Ogden (1976) reported discharge ranging from 1 to 125 cfs (0.03–3.54 cm) with an average of about 5–9 cfs (0.14–0.25 cm).

Laurel Creek flows southwest from clastic rocks to the west of the Walters Spring Basin. It sinks at Laurel Creek Cave (Fig. 3.30), flows 0.8 (1.3 km) miles to the southwest to reappear in a karst window at the water entrance to Greenville Saltpeter Cave (Fig. 3.31) and then another 0.5 (0.8 km) miles to reappear at the head of a millpond, and flows to Indian Creek at Greenville. The catchment is 11.9 mi^2 (31 km^2). The Laurel Creek Cave entrance is in a blind valley and is subject to flooding. A study by Groves (1992) traced the changes in the chemistry of the water after the limestone contact was reached and at various locations within the cave and at the spring. Upstream from the limestone contact the Laurel Creek water had a pH of 7.31,

Fig. 3.28 Aerial photo showing the head spring of Indian Creek south of Union

Fig. 3.29 Aerial photo showing Walters and MacPeat Springs near Salt Sulphur Springs

Fig. 3.30 Photo of the blind valley entrance to Laurel Creek Cave

Fig. 3.31 Photo of the water entrance to Greenville Saltpeter Cave

Fig. 3.32 Beginning the tracer test in McClung Cave that established the connection to Sweetwater River, the large stream passage in Maxwelton Cave, May, 2017. Photo by Clifford Lindsay. Used with permission

specific conductance of 62 μS/cm, and hardness of 9 mg/L. Values at the spring were a pH of 7.71, specific conductance of 129 μS/cm, and hardness of 18 mg/L. Discharge was 35 cfs (0.99 cm) at the time of the study.

Indian Draft Cave (Spring) is situated to the northwest of Greenville, and water from Greens Cave has been traced to this tributary of Indian Creek. The catchment is not defined at this time but is probably about 2 mi^2 (5 km^2).

3.7 Closing Comments

In many respects, West Virginia was the caving area of choice for the first caving organization in the Washington DC area. This caving group then became the National Speleological Society, and many of the first scientific investigations of caves and karst features in the USA were conducted in West Virginia during the 1940s. Work on the hydrology of the karst areas began in earnest in the early 1960s with the introduction of passive detectors (carbon traps) to monitor springs for the passage of fluorescent dyes. The area of the Greenbrier limestone karst covered in this book has now been extensively studied by cavers, professional hydrologists, and graduate students from various universities (especially McMaster University in Canada, West Virginia University, and the University of Akron). The area is currently being actively explored by cavers and scientists from many disciplines, and a more detailed picture of the hydrology will continue to evolve. The discovery of a large stream passage in the western end of Maxwelton Cave has once again provided incentive to continue the search for the elusive master drain that must underlie the lower portion of the Davis Spring Basin. A tracer test (Fig. 3.32) from McClung Cave to Maxwelton Cave conducted in May, 2017, established the potential for much additional cave passage and a more detailed picture of the conduit flow system on at least the eastern side of the Davis Spring Basin. If a connection between the western drainage (Higginbotham and Savannah Lane Caves) and the eastern contact caves can be established, a sizable underground river passage should be the reward.

Acknowledgements Many cavers and researchers have contributed to our present understanding of the hydrology of this complex karst area. The references should lead to more detailed information on the tracer tests and hydrogeologic studies summarized in this chapter. Special thanks are due to Roger Baroody for drafting the maps and illustration in this chapter.

References

Bishop, M.R., I.D. Sasowsky, and W.K. Jones. 2009. Preliminary water quality and boundary redefinition of the Scott Hollow drainage basin, Appalachian Plateau. *Proceedings of the 15th International Congress of Speleology* 3: 1422–1427.

Coward, J.H.C. 1975. Paleohydrology and streamflow simulation of three karst basins in southereastern West Virginia, USA, 303pp. PhD Thesis, McMaster University.

Culver, D.C., W.K. Jones, D.W. Fong, and T.C. Kane. 1994. Organ Cave karst basin. *Groundwater Ecology*, 451–473. San Diego: Academic Press.

Dasher, G.R., and D.G. Boyer. 2000. Recent Spring Creek area dye tracings Greenbrier County, West Virginia. *WV Speleological Survey Monograph* 2: 28pp.

Davis, S.B. 1999. Aquifer development in folded and fractured limestone: The Scott Hollow drainage basin, Monroe County, West Virginia, 210pp. MS Thesis, University of Akron.

Demrovsky, N. 2003. The effects of well drilling on karst aquifers in Monroe and Greenbrier County, West Virginia, 198pp. MS Thesis, University of Akron.

Dunn, J.R. 1957. Stream tracing. *National Speleological Society Mid-Appalachian Region Bulletin* 2: 3–9.

Groves, C.G. 1992. Geochemical and kinetic evolution of a karst flow system: Laurel Creek, West Virginia. *Ground Water* 30 (2): 186–191.

Heller, S.A. 1980. A hydrogeologic study of the Greenbrier limestone karst of central Greenbrier County, West Virginia, 421pp. PhD Thesis, West Virginia University.

Heller, S.A. 1991. An attempt to model an Appalachian karst aquifer using MODFLOW. Appalachian Karst. *Proceedings of the Appalachian Karst Symposium*, 177–185. Huntsville: National Speleological Society.

Hess, J.W. and W.B. White. 1989. Water budget and physical hydrology. In *Karst Hydrology*, ed. W.B. White and E.L. White, 105–126. New York: Van Nostrand Reinhold.

Jones, W.K. 1984. Analysis and interpretation of data from tracer tests in karst areas. *National Speleological Society Bulletin* 46 (2): 41–47.

Jones, W.K. 1973. *Hydrology of Limestone Karst in Greenbrier County, West Virginia*, vol. 36, 49pp. West Virginia Geological and Economic Survey Bulletin.

Jones, W.K. 1981. A karst hydrology study in Monroe County, West Virginia. *Eighth International Congress of Speleology Proceedings* 1: 345–347.

Jones, W.K. 1988. Hydrology. In *Caves of the Organ Cave Plateau*, ed. P.J. Stevens, vol. 9, pp. 40–48. Barrickville, WV: West Virginia Speleological Survey Bulletin.

Jones, W.K. 1997. *Karst Hydrology Atlas of West Virginia*, vol. 4, 111pp. Karst Waters Institute, Special Publication.

Jones, W.K., and H.W. Rauch. 1977. Discharge characteristics of karst drainage basins in Greenbrier County, WV. In *Karst hydrogeology*, ed. J.S. Tolson, and F.L. Doyle, 391–398. Huntsville: University of Alabama Press.

Jones, W.K., and J.A. Tudek, Jr. 2012. Hydrology of the Greenbrier Valley karst region. In *Maya Con 2012, Greenbrier Valley, West Virginia*, ed. M. Adler, 94–103. Huntsville: National Speleological Society.

Kresic, N., and O. Bonacci. 2009. Spring discharge hydrograph. In *Groundwater hydrology of Springs*, ed. N. Kresic, and Z. Stevanovic, 129–163. Burlington, MA: Butterworth-Heinemann.

Mangin, A. 1975. Contribution a l'etude hydrodynamique des aquifers karstiques. DES Thesis, University of Dijon, France.

Ogden, A.E. 1976. The hydrogeology of the central Monroe County karst, 263pp. Ph.D. Thesis, West Virginia University.

Schilling, K.E., C.S. Jones, and A.J. Seeman. 2013. How paired is paired? *Journal of Environmental Quality* 42: 1412–1421.

Shultz, R.A., W.A. Hobba, Jr., and M.H. Kozar. 1995. Geohydrology, water availability, and water quality of Berkeley County, WV, with emphasis on the carbonate area. *U.S. Geological Survey, Water Resources Investigations Report* 93: 88pp.

Springer, G.S. 2002. Caves and their potential use in paleoflood studies. In *Ancient Floods, Modern Hazards: Principles and Applications of Paleoflood Hydrology*, ed. K. House, R. Webb, V.R. Baker, D.R. Levish. Water Sciences and Applications, vol. 5, pp. 329–344. Washington, DC. American Geophysical Union.

Springer, G.S. 2004. A pipe-based, first approach to modeling closed conduit flow in caves. *Journal of Hydrology* 289: 178–189.

Tudek, J.K. Jr. 2009. Further hydrogeologic investigations in the Davis Spring drainage basin, Greenbrier County, West Virginia. MS Thesis, West Virginia University.

White, E.L. 1977. Sustained flow in small Appalachian watersheds underlain by carbonate rocks. *Journal of Hydrology* 32: 71–86.

White, W.B., and V.A. Schmidt. 1966. Hydrology of a karst area in east-central West Virginia. *Water Resources Research* 2 (3): 549–560.

White, E.L., and W.B. White. 1974. Analysis of spring hydrographs as a characterization tool for karst aquifers. *Proceedings of the 4th Conference on Karst Geology and Hydrology*, 103–106. Morgantown, WV: West Virginia Geological Survey.

Woolhiser, D.A., D.C. Goodrich, W.E. Emmerich, and T.O. Keefer. 1980. Hydrologic effects of brush to grass conversion. In *Symposium Proceedings of IR Conference Watershed Management*, 293–302. Durango, CO. www.tucson.ars.ag.gov/unit/Publications/PDFfiles/809.pdf.

Williams, C.F., and W.K. Jones. 1983. Hydrology of the upper Spring Creek karst drainage basin, West Virginia (abstract). Huntsville, AL: Proceedings of the 1983 National Speleological Society Convention.

Wolfe, T.E. 1973. Sedimentation in karst drainage basins along the Allegheny escarpment in southeastern West Virginia. PhD Thesis, Geography, McMaster University.

Zhang, L., W.R. Dawes, and G.R. Walker. 2001. Response of mean annual evapotranspiration to vegetation changes at catchment scale. *Water Resources Research* 37 (3): 701–708.

Zotter, H. 1963. Stream tracing techniques and results—Pocahontas and Greenbrier Counties, West Virginia. *NSS News* 21 (10): 136–142.

Zotter, H. 1965. Stream tracing techniques and results—Pocahontas and Greenbrier Counties, West Virginia—Part 2. *NSS News* 23 (12): 169–177.

Karst Geomorphology

4

William B. White and Elizabeth L. White

Abstract

The Greenbrier Karst is formed on a roughly triangular exposure of Mississippian-age Greenbrier Limestone that extends 90 km NNE to SSW in Pocahontas, Greenbrier, and Monroe Counties, West Virginia. The geologic setting is conducive to the development of long caves but not deep caves. Within the Greenbrier Karst are 24 caves with surveyed lengths greater than 5 km of which nine have lengths greater than 20 km for a total of 500 km of cave passage plus many smaller caves. Caves are fragments of conduit system with a fractal length distribution skewed toward long caves. The surface karst consists primarily of dolines and blind valleys. The area is surrounded by clastic rock mountains which provide many surface streams that sink at the limestone contact. The result is a patchwork of large underground drainage systems each discharging through a large spring. Analysis of the profiles of active surface streams and abandoned stream beds gives some insight into the developmental history of the karst.

4.1 Introduction

The term "karst" was used historically to describe the unusual surface landforms that develop in terrains underlain by soluble rock. Caves, historically, were discussed separately. Thus, we have the present situation where books and articles use the term "caves and karst" as if they were different subjects. They are not. Sinkholes, sinking streams, big springs, and bare, intricately sculptured bedrock are the surface manifestation of karst. Caves are the underground manifestation of karst.

There are many types of karst, classified by landforms, by climate, by bedrock, and by ancillary processes that contribute to the development of the landforms (White and White 2013). The Greenbrier Karst is a temperate climate carbonate karst. It is a mix of more doline (or sinkhole) karst in the south trending into more fluviokarst in the north. Description and analysis of the Greenbrier Karst then requires an examination of the characteristics of closed depressions and of the properties of sinking streams along with those properties of the caves which can be analyzed in aggregate.

4.2 Geomorphic Regions

The Greenbrier Karst spans a sufficient area that some subdivision is useful. The areas are underlain by the same carbonate rocks, albeit thickening from north to south, and the areas are drained by the same river. Some of the subdivisions have clear boundaries; others do not. The breakdown is more for convenience of discussion. The sections that follow are an overview. Much more detail is given in later chapters.

4.2.1 The Swago Creek Basin

The Swago Creek Basin (Chap. 6) is the northern-most exposure of the Greenbrier Limestone described in this volume although the karstic limestone continues northeast to the Pennsylvania line. The selection of the Swago Basin as a northern limit of the Greenbrier Karst was to some extent an

W.B. White (✉) · E.L. White
4538 Miller Road, Petersburg, PA 16669, USA
e-mail: wbw2@psu.edu

W.B. White (ed.), *Caves and Karst of the Greenbrier Valley in West Virginia*,
Cave and Karst Systems of the World, https://doi.org/10.1007/978-3-319-65801-8_4

arbitrary choice. The Swago Basin is primarily fluviokarst with only a few remnants of upland surface with expressions of surface karst. Streams crossing the basin are mostly subsurface but with well-maintained surface channels.

4.2.2 The Little Levels

The Little Levels are a remnant of upland surface almost entirely drained in the subsurface except for the northern edge where a dry surface channel crosses the limestone. Caves are located along the northern and southern edges but few caves and no surface streams in the central portion.

4.2.3 The Friars Hole Valley

The Friars Hole Valley is a completely underdrained karst valley bounded by the mountains of the high Allegheny Plateau to the west and by Droop Mountain to the east. Beneath it is the Friars Hole Cave System, the longest in West Virginia. The drainage system is complex with many inputs from sinking streams draining from the western mountain. There is an underground divide of variable position at the northeastern end between drainage eastward to Locust Spring and drainage along the valley axis to Spring Creek in the southwest.

4.2.4 The Big Levels

Largest of the karst areas, the Big Levels, also known as the Great Savannah, is an upland karst surface that extends nearly 30 km northeast to southwest from Spring Creek to the Greenbrier River. It is bounded on the east by the basal contact between the limestone and the underlying shale with the south-flowing Greenbrier River flowing in a deep incised valley still further east. The western boundary is the escarpment of Muddy Creek Mountain and its associated ridges (Fig. 4.1). The northeastern boundary is Spring Creek, a master drain for cave systems both north and south of the stream. The southwestern boundary is the Greenbrier River where it shifts its course westward south of Lewisburg.

The Big Levels are primarily a doline karst with closed depressions on many size scales. Some are visible shallow depressions not sufficiently deep to appear on topographic maps. Others are depressions with depths of more than 30 m.

Most the large cave systems lie within the Big Levels. The Culverson Creek System (Chap. 10) and the Buckeye Creek System (Chap. 9) to the north and the contact caves (Chap. 11) along the eastern margin account for only a fraction of the area. The northern caves drain to Spring Creek, but most of the contact caves contain streamways that ultimately drain to Davis Spring on the Greenbrier River (Fig. 4.2). There are no surface streams crossing the Big Levels.

Fig. 4.1 Limestone surface of the Big Levels with the escarpment of Muddy Creek Mountain in the background. Photograph by the authors

Fig. 4.2 Greenbrier River at Davis Spring. The stream entering the river from the right is the low flow discharge from Davis Spring. Photograph by the authors

4.2.5 The Sinking Creek Valley

The long, narrow valley of Sinking Creek lies west of Muddy Creek Mountain where the limestone is brought to the surface by an anticline. Portions of the Sinking Creek Valley appear to be an unroofed cave passage with a fragment of the cave remaining. The drainage at the southern end is underground (Chap. 12).

4.2.6 The Organ Cave Plateau

The continuation of the Big Levels south of the Greenbrier River is the Organ Cave Plateau. This segment of karstic upland is bounded on the north by the river, on the east by the clastic strata leading up to the White Rock Mountains, and on the southwest by Second Creek, a major tributary of the Greenbrier River which has cut a deep secondary valley into the upland. Beneath is Organ Cave (Chap. 13), the second longest cave in the Greenbrier Karst.

4.2.7 The Dickson Spring Karst

The continuation of the Big Levels south of Second Creek is divided into two parts. The eastern segment is a sinkhole-pocked karst surface with internal drainage that ultimately discharges at Dickson Spring (called Big Spring on the Ronceverte Quadrangle) (Chap. 14).

4.2.8 The Sinks Grove Karst

The largest segment of the karst is near the village of Sinks Grove and extends northward to the river and westward to Flat Top Mountain. Two large systems, Windy Mouth Cave (Chap. 15) and the Scott Hollow System (Chap. 16) drain northward to the river but large areas near Sinks Grove have no known underlying cave.

4.2.9 The Greenville Karst Island

There is an exposed area of limestone bounded on all sides by clastic rocks near the village of Greenville, brought up by an arch in the Abbs Valley Anticline. Known caves in this area drain south to Indian Creek which flows westward to the New River (Chap. 17).

4.3 Surface Landforms

4.3.1 Closed Depressions

Closed depressions—sinkholes—are the best known and most characteristic features of karst terrains. The literature typically distinguishes between solution sinkholes, collapse sinkholes, and soil piping sinkholes. These are better descriptors of the processes involved in sinkhole development than a classification scheme because many closed

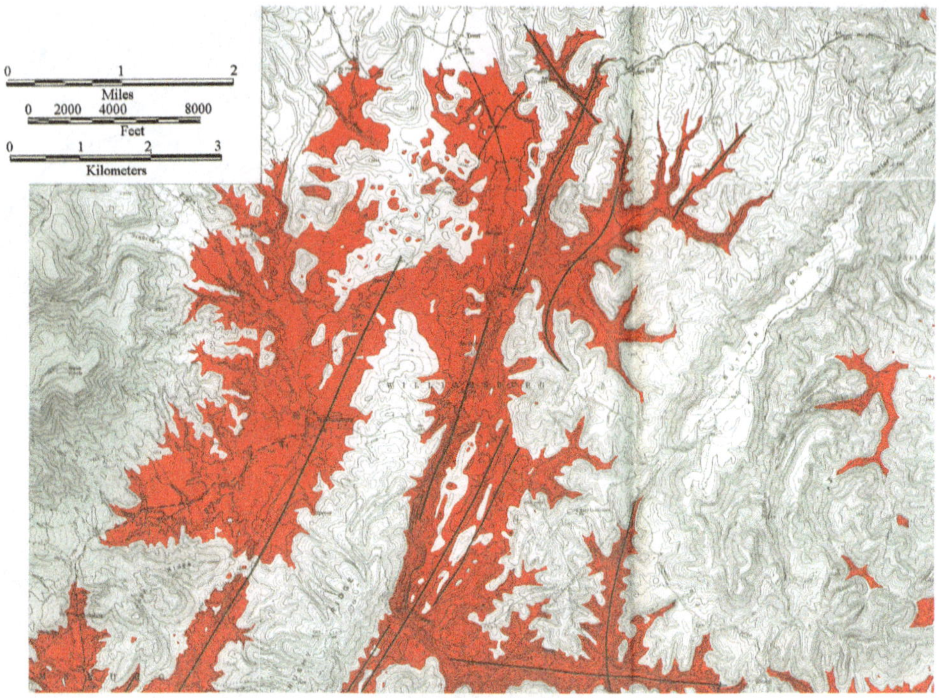

Fig. 4.3 Area of internal drainage northwest of Lewisburg. The *red areas* lie within closed depression contours. This is a segment of a much larger map by Lessing (1979)

depressions involve all three processes. For the Greenbrier Karst, a better scheme is to describe the closed depressions in terms of their size and shape.

On a regional scale, large areas of the Little Levels and the Big Levels are closed depressions (Fig. 4.3). Closed depressions are defined on topographic maps by the highest closed depression contour. There are no continuous surface stream networks in many parts of the Greenbrier Karst. Internal drainage through sinkholes results in an overall lowering of the landscape that leads to broad shallow depressions with no surface outlet.

Dotting the landscape, including the areas bounded by closed depression contours, are many circular sinkholes in a range of diameters and depths. The small sinkholes tend to be roughly circular and are usually soil-mantled (Fig. 4.4). Larger closed depressions tend to be more star-shaped or simply irregular (Fig. 4.5). The irregular shapes may represent the superposition of earlier surface drainage patterns. The blind valleys cut by sinking streams also appear on topographic maps in the form of irregular closed depressions.

Collapses of underlying cave passages to form collapse sinkholes occur but because of the rubble zones and soil mantling, it is difficult to distinguish collapse sinks from solutional sinks. However, there are examples where entire lengths of cave passage have collapsed to form elongated closed depressions. One of the largest is the valley of Sinking Creek. Sinking Creek rises in the mountains north of the karst area and flows into an alluviated blind valley. The stream meanders over the floor of the depression for 7 km to the downstream end of the blind valley and flows into the entrance of a cave known as the "Sinks of Sinking Creek" (Fig. 4.6a). The cave is a 230-m-long fragment of 10–20 m diameter master trunk (Fig. 4.6b). The stream emerges at the upstream end of a narrow, nearly linear, closed valley that extends nearly 10 km to a blind footwall where the stream again goes underground. It reappears as a large spring at Piercys Mill Cave 5 km to the southwest where it forms the headwaters of Muddy Creek. The linear course of the closed valley is dictated by the contact between the limestone and the underlying shale brought up by the Brushy Ridge Anticline to the east. Topographic evidence suggests that this 10-km-long closed valley is an unroofed cave passage with a small fragment of cave surviving at the upstream end and a much large cave segment at the downstream end.

Earlier studies of sinkhole depths in the Appalachians (White and White 1979; Troester et al. 1984) show that sinkhole depths follow an exponential distribution of the form

$$N = N_0\, e^{-Kd} \tag{4.1}$$

N = number of sinkholes of a given depth, d = depth of sinkholes, and N_0 and K are fitting parameters. The parameter K is representative of the sinkhole population and has a value for the Appalachians of $K = 0.22\ \text{m}^{-1}$ independent of rock type (limestone vs. dolomite), structure (flat-lying rocks vs. strongly folded rocks), and age of rocks (Mississippian vs. Devonian vs. Ordovician) based on measurements of more than 5000 sinkholes. A detailed study of the Greenbrier

Fig. 4.4 Small sinkholes northwest of Lewisburg. Larger closed depression in the background. Photograph by the authors

Karst (Soriano and White, unpublished data) gave essentially the same result. Regardless of mechanism and geologic controls, populations of sinkholes have the same distribution of depths. There are many fewer data for the diameters of sinkholes (or even how one defines a diameter for the highly irregular sinkholes) but one limited investigation (White and White 1987) shows that sinkhole diameters are also exponentially distributed. The data gave a good fit to the exponential distribution but not to a power function distribution. Sinkholes are not fractal which distinguishes them from impact crater depth distributions and also cave length distributions which do follow a power function distribution.

4.3.2 The Epikarst

Except for a few towns and villages, most of the Greenbrier Karst is agricultural land with forest on the mountain slopes. Limestone soils make good crop land when they are thick enough to till. Tilled fields occur on the thicker soils, but much of the area is in pasture where the soils are thinner. The limestone bedrock surface is sculptured with deep solutionally widened crevices along joints and intermediate pinnacles that sometime protrude above the land surface (Fig. 4.7). The zone from the base of the solutionally widened crevices, the regolith that fills the crevices, and overlying soil is called the epikarst, and it plays an important role in carrying infiltrating rainfall into the underlying aquifer.

4.3.3 Sinking Streams and Big Springs

The karst drainage basins, for the most part, are bounded on one or more sides by higher terrain underlain by clastic rocks. Surface streams flow down the ridges until they reach the carbonate rock contact. For those basins where the underground drainage system is sufficient to carry all discharges including those of extreme storms, the streams sink and recharge the underground system. Channel erosion ceases downstream from the swallet and relative quickly the channel degrades, sinkholes develop, and the result is a dry valley, often with no trace of steam channel. Upstream from the swallet, the surface stream continues to erode its channel resulting in a small incised valley that ends in blind footwall where the stream is lost. Alternatively, the underground system may not have the carrying capacity to transmit storm flows. In these cases, the surface channel is maintained although it may be used only during periods of high runoff. There are many examples of both types of sinking streams in the Greenbrier Karst, many of which are described in later chapters.

The flow of water in karst drainage basins becomes surface water again at big springs that form the headwaters of substantial streams. Springs are common in most terrains in

Fig. 4.5 Southeast corner of USGS 7.5 min Williamsburg Quadrangle illustrating sizes and shapes of closed depressions. The upper-most closed depression contour is highlighted. The highlighted line in the *upper left corner* is one side of a very large closed depression. Mostly, upper contour is at 2200 ft (671 m). The *red line* is US Highway 219

a great variety of sizes but karst springs tend to be exceptionally large. Jones (1997) notes that in the three-county area of the Greenbrier Karst, there are 68, 78, and 54 listed springs in Pocahontas, Greenbrier, and Monroe Counties, respectively but only 6, 2, and 4 karst springs with discharges greater than 1000 galleons per minute (63.1 L per second). The big karst springs can be used as gauge-points for defining karst groundwater basins.

4.4 Caves

The Greenbrier Karst is one of the most cavernous terrains found anywhere in the world. More than 2400 caves have been cataloged in the three-county area. These, of course, are only the caves that happen to have entrances. It is clear from the distances from sinking streams to their springs that many more caves must exist because known caves account for

(a)

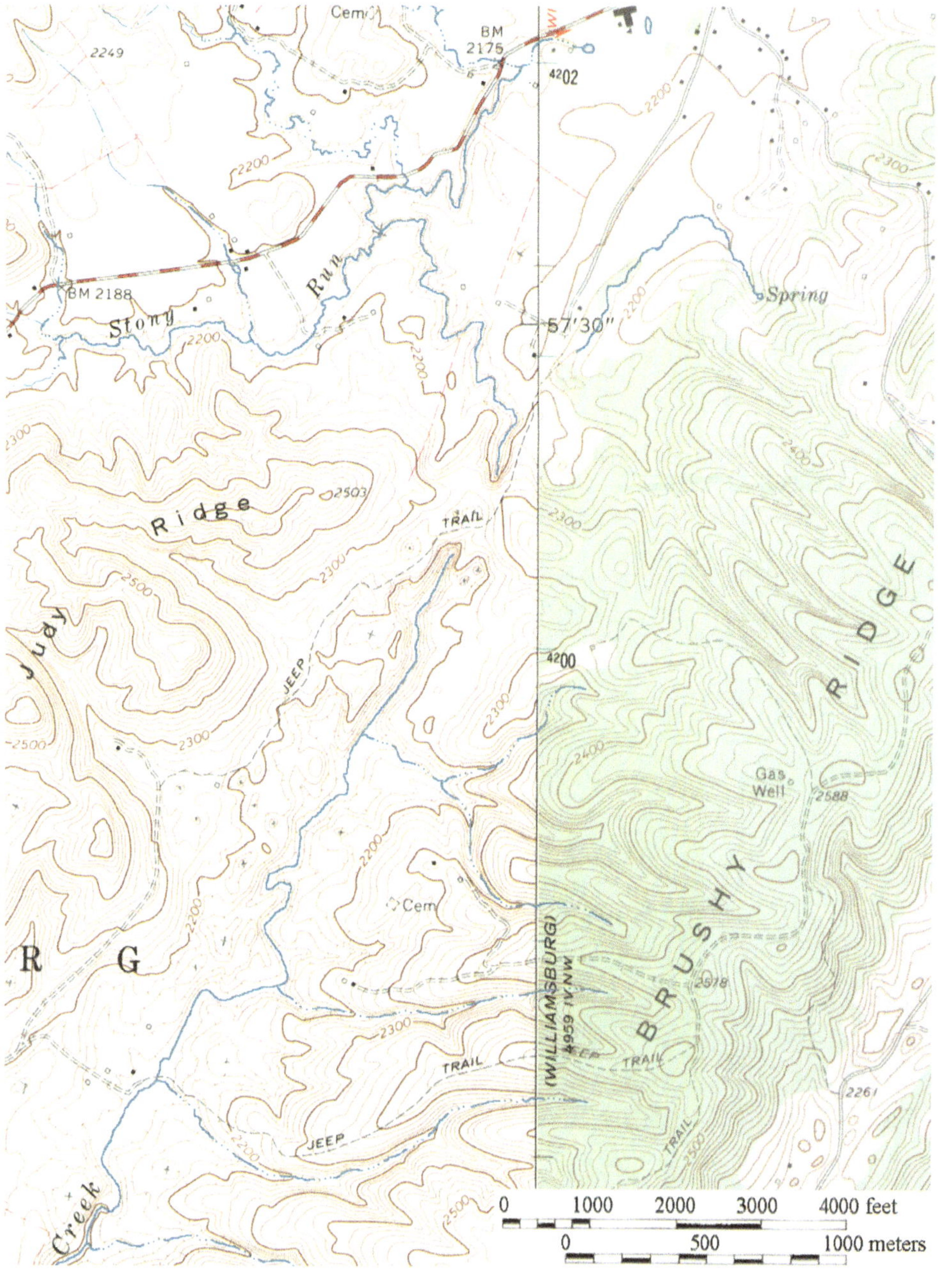

Fig. 4.6 **a** Sections of USGS 7.5 min Cornstalk and Williamsburg Quadrangles showing a portion of the valley of Sinking Creek, its upstream sinking streams, and the underground route of the Sinks of Sinking Creek. **b** Map of the Sinks of Sinking Creek

only a small fraction of the distance over which the streams travel underground.

The caves of the Greenbrier Karst are telogenetic caves, formed at shallow depth by circulating groundwater and sinking streams. The Greenbrier Limestone is dense and well lithified so that the primary pathways initiating cave development are fractures and bedding plane partings. The impermeable Maccrady Shale forms an aquiclude at the base of the limestone so that there was little deep circulation of the groundwater. Some caves in the northern fluviokarst lie at considerable depths below the land surface, but this is more the result of high relief topography rather than intrinsic deep circulation.

The majority of the large Greenbrier caves have branchwork patterns in the sense of Palmer (1991). Sinking streams and the drains of large closed depressions form multiple inputs to the cave system. These inputs form the upstream tributaries to the underground drainage system. The

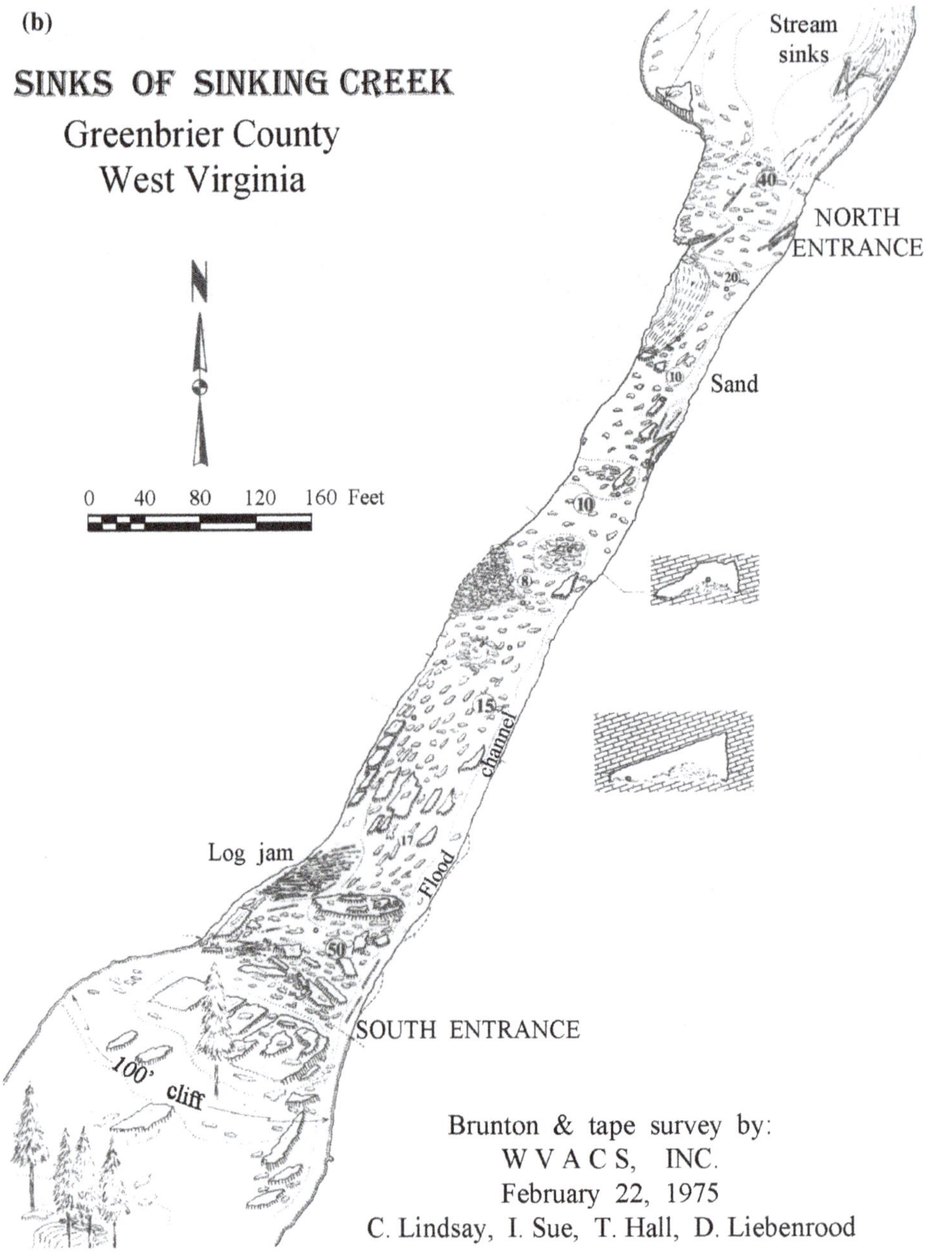

Fig. 4.6 (continued)

branches, however, tend to be elongated along guiding geological controls and thus tend to be parallel to the minor folds that deform the limestone. The various tributaries converge to a single master drain which has its outlet at a big spring. Although some passage cross sections had the tubular shape characteristic of passage enlargement under completely water-filled (phreatic) conditions, most are canyons of various sizes. Many of the canyons have cut deeply into the underlying shale.

Most caves are fragments of conduit systems, truncated by surface erosion, breakdown, flowstone barriers, or sediment plugs. The population of cave lengths in a given region is a natural fractal (Curl 1986). According to Curl, the cave lengths should follow a power law of the form

$$N(\ell) = N(\ell_0)\left(\frac{\ell}{\ell_0}\right)^{\upsilon} \quad (4.2)$$

where $N(\ell)$ is the number of caves longer than ℓ, ℓ_0 is the reference length (m, km, or miles), and υ is the fractal dimension. Figure 4.8 shows length distribution plots for long caves drawn from Robert Gulden's Long Cave Lists (http://www.caverbob.com). The fractal dimension of the US list of long caves is 1.25 and for the world list 1.21, both in

Fig. 4.7 Pasture northwest of Lewisburg showing emergent pinnacles of limestone and thin soils. Photograph by the authors

the range of 1.2–1.6 reported by Curl for other cave populations. The Greenbrier Valley caves with lengths greater than one kilometer, however, have a fractal dimension of 0.86. The population is skewed toward larger caves, a result that is consistent with the hypothesis that the Greenbrier Karst represents an early stage in the dissection of what were originally some extremely large cave systems.

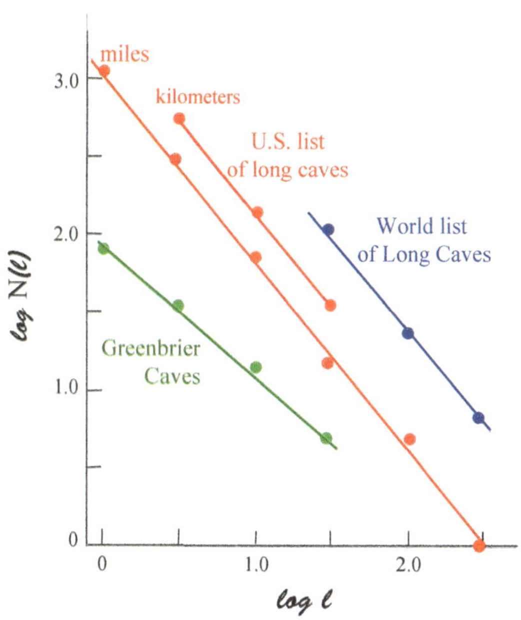

Fig. 4.8 Distribution of cave lengths for the World List, the US List, and for the Greenbrier Karst. All original data taken from R. Gulden's online cave lists (www.caverbob.com). Plots were constructed from lengths given in miles. Lengths in kilometers shift the curves but do not change their slope as shown for the US long cave data

4.5 Stream Channel Profiles

Sinking streams can be investigated by examining their profiles and comparing the segments with perennial flow with losing segments or with segments of dry channels. A selection of profiles was measured for sinking streams in the Greenbrier Karst. The profiles were extracted from topographic maps which have contours in English units. Measurements were converted to metric. Both linear profiles and semilog plots were constructed for comparison.

4.5.1 Background

The long profiles of streams and rivers have been the subject of many investigations. Many of these deal with issues of river hydraulics and sediment size, sorting, and distribution along the channel (e.g., Shulits 1941; Hack 1957, 1973; Howard 1994; Goldrick and Bishop 2007). Other studies use stream profiles to interpret landscape evolution, the effects of changing base levels, tectonic uplift, and migrating knickpoints (Anthony and Granger 2007a; Westaway 2007). A third, less studied, set of stream profiles are those in fluviokarst. Comparisons of channel profiles, especially profiles of dry channels in a set of small basins in the western margin of the Cumberland Plateau, suggested that the underground flow routes maintain grade with the active surface channels and that dry channels can be related to old base levels (White and White 1983).

Many investigators have attempted to fit the long profiles of streams to mathematical functions. Consideration has been given to linear functions, power functions, exponential functions, and logarithmic functions (Strahler 1968). Shepherd (1985) used linear regression methods to compare these functions for 13 Texas streams. Only the exponential and logarithmic forms gave reasonable fits and of these Shepherd proposed that exponential forms work best for streams that drain into aggradational areas and logarithmic forms work best for streams with steep headwalls or scarps in the upstream reaches. In earlier studies, exponential forms seemed to work best for fluviokarst.

4.5.2 Theory

Although it is generally agreed that the plotting variables are elevation of the channel and the horizontal distance along the channel, what has been less frequently discussed is the coordinate system in which these variables are to be plotted. All of the proposed mathematical functions are of indefinite extent while the stream profiles are finite. At the upstream end, the limit is the drainage divide; at the downstream end, the profile is terminated at the confluence with a larger stream or at sea level.

For measurements in the downstream direction, the exponential function has the form

$$y = e^{-x} \tag{4.3}$$

where y and x are dimensionless. The plot remains above the x-axis and only positive values of y appear. The exponential curve extends toward infinity for large negative values of x and crosses the y-axis at the point $y = 1$. This crossing defines the placement of the y-axis and the point on the x-axis where $x = 0$. The exponential curve then approaches the x-axis asymptotically as x takes on large positive values. The resulting curve rises steeply in the northwest quadrant and is much flatter in the northeast quadrant.

To plot a stream profile, it is convenient to draw the y-axis through the drainage divide or near the headwater of the stream and then take this axis as the zero point of the x-axis scale. Distance, d, is then measured downstream from the divide until the channel reaches a confluence with a larger stream or, in the case of karst streams, reaches a swallet. Distance measurements can be scaled from topographic maps and elevations determined from the contours. Rather than plot elevations with respect to sea level, it is convenient to set the elevation to zero at the confluence which will have sea-level elevation E_B. The practical equation for stream profiles is then

$$E - E_B = E_0\, e^{-kd} \tag{4.4}$$

E_0 and k are fitting parameters, necessary because the plots are constructed in real units, not pure numbers. E_0 is the elevation that the channel would have if the exponential curve had continued all the way to the mathematical y-axis. The parameter k, with units of inverse length, is a characterizing parameter for the channel. It is the slope of the straight line that should result from a semilog plot and will have a numerical value dependent on whether the plot is constructed with natural logarithms or base-10 logarithms.

The stream profile is not exponential at the drainage divide. The transition to exponential form takes place at the inflection point when the channel changes from being convex upward near the divide to concave upward in the exponential segment. Mathematically, this is the point where the second derivative goes through zero.

The practical coordinate system does not coincide with the mathematical one. The y-axis should be drawn to cross the exponential curve at the point where $y = 1$, and this position of the axis defines the point $x = 0$. This point is not marked by any physical feature in the channel. To plot a stream profile in dimensionless form, first construct a semilog plot of a segment of the exponential portion and from it extracts the fitting parameters. Then, reconstruct the profile using the dimensionless plotting variables

$$y = \frac{E - E_B}{E_0} \quad \text{and} \quad x = \frac{d}{k} \tag{4.5}$$

Setting $E = 0$ at the confluence cuts off the profile prematurely. The mathematical exponential curve approaches the x-axis asymptotically as x approaches infinity. Thus, the true zero on the E-axis will be below the practical one. When the profile is plotted on a log scale, with the practical coordinate system, the downstream end bends over as the profile approaches the confluence, creating an artifact in the plot.

4.5.3 The Swago Creek Basin

Dry Creek rises in the mountains and descends a steep channel to the Dry Creek Swallow Hole where, during low flow, the stream sinks in its bed beneath a limestone ledge. The dry channel continues down a narrow V-shaped valley for a kilometer and then in a slightly wider valley for roughly 4 km to the eventual confluence of the channel with Swago Creek.

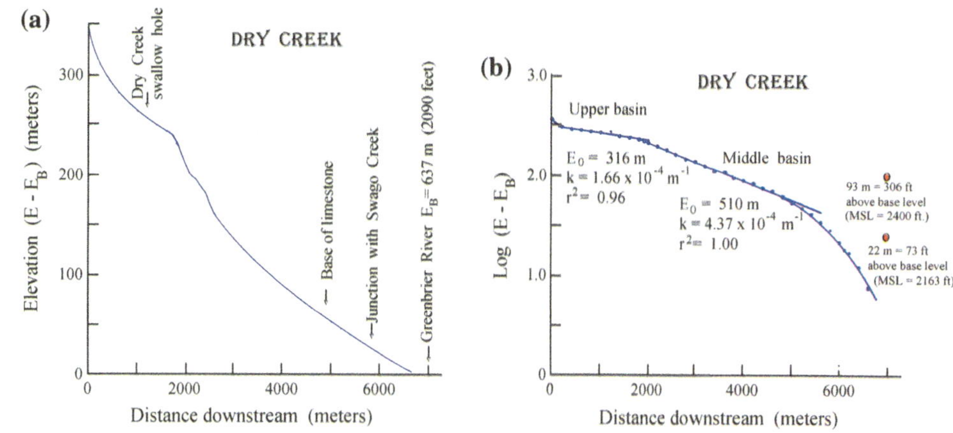

Fig. 4.9 Profile of Dry Creek taken from USGS Hillsboro Quadrangle. **a** Linear profile. **b** Semilog profile

The linear profile (Fig. 4.9a) shows two distinct segments of the basin. The upper basin has the expected exponential form down to the downstream end of the narrow valley. There is no break at the Dry Creek Swallow Hole. The valley steepens below the narrow segment of the valley and then descends smoothly to the confluence. Both segments appear on the semilog plot as straight-line segments with a sharp break between them (Fig. 4.9b). If the upper basin plot is extended, it reaches the confluence at the elevation of the upland karst surface, 93 m (306 ft) above the river channel. The lower, major section of Dry Creek Valley, extrapolates to the river 22 m (73 ft) above the channel. This is a bit above the present-day floodplain of the Greenbrier River, well-developed near the town of Marlinton which is built on the floodplain.

The central branch of Swago Creek is labeled Overholt Run on the topographic map and is the tributary that receives the water from Overholt Blowing Cave. The eastern branch of Overholt Run has been informally called McKeever Run. A profile was constructed from the headwaters above Spruce Flats six kilometers downstream to the confluence with the Greenbrier River.

The linear profile is a smooth exponential (Fig. 4.10a), confirmed by the semilog profile (Fig. 4.10b). Downstream, the channel is located in a wide, alluviated valley. There is a break in the upstream profile where the channel descends from Spruce Flats but otherwise the semilog profile can be fitted with a single straight-line segment. There is no break where the stream from Overholt Blowing Cave empties into the usually dry channel. The linear segment extrapolates to the Greenbrier River only 4 m above the channel. Within the accuracy of the plots, McKeever Branch is at grade with the river in spite of mostly having a dry channel upstream from Overholt Blowing Cave.

4.5.4 The Stamping Creek Basin

Stamping Creek extends for 11.7 km from its headwaters, first down a steep escarpment from the drainage divide on

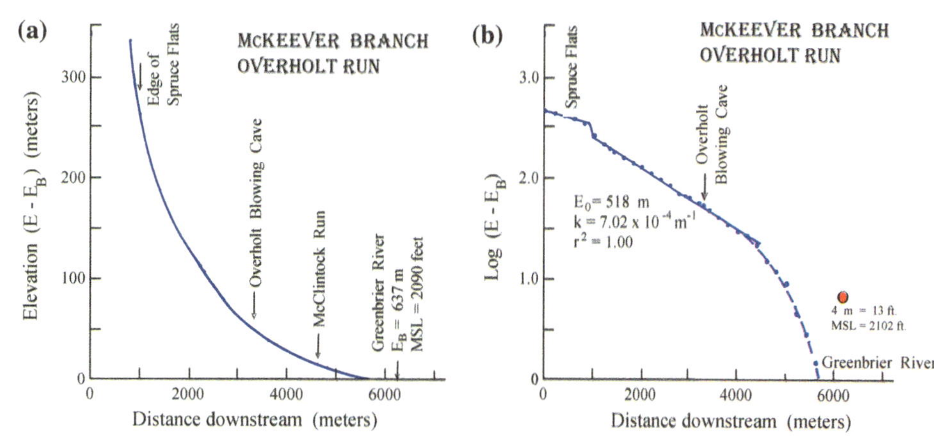

Fig. 4.10 Profile of McKeever Branch of Overholt Run taken from USGS Hillsboro Quadrangle. **a** Linear profile. **b** Semilog profile

Cranberry mountain, then down a long valley extending along the northeastern edge of the Little Levels, and finally in a narrow gorge from the Little Levels to the Greenbrier River. The linear profile of the upper valley has the expected exponential shape, but the stream along the lower valley has an almost flat profile. There is a subtle break at 5 km at the junction with the dry channel of Blue Lick Run which enters from the east (Fig. 4.11a). The semilog profile is nearly flat (Fig. 4.11b) but consists of three distinct linear segments. The subtle break in the linear profile does not appear in the semilog plot. The upstream break, at 1.5 km, is near the base of the mountain and is located near the point where much of the flow of Stamping Creek sinks in its bed. The downstream break, at 6.7 km, is near the Stamping Creek springs just upstream from Mill Point where the channel begins to cut its deep gorge from the Little Levels down to the Greenbrier River.

The long reach of normally dry channel along the northeastern margin of the Little Levels extrapolates to the Greenbrier River at an elevation of 27 m (90 ft) above the present channel. There is no blind valley in Stamping Creek at the upper limestone contact, and the dry tributaries of Blue Lick Run and Tilda Fork enter Stamping Creek at grade. Unlike the drainage on the southwestern side of the Little Levels, the underground routes of Stamping Creek appear to be a fairly recent (mid-Pleistocene) development.

4.5.5 Hills Creek, Bruffey Creek, and Locust Creek

Bruffey Creek and Hills Creek can be considered tributaries of Locust Creek except that the confluence is underground in the cave systems beneath Droop Mountain. Bruffey and Hills Creeks sink on the west side of Droop Mountain in separate cave entrances at about the same elevation of 753 m (2470 ft). The Droop Quadrangle map erroneously shows the two streams converging on the surface at a spring. Locust Creek heads in Locust Creek Spring on the east side of Droop Mountain at the much lower elevation of 637 m (2090 ft). The profile is presented and discussed in Chap. 7.

4.5.6 Rush Run

Rush Run is the next stream southwest of Hills Creek and has a small drainage basin in the dissected hill country on the northwest side of the Friars Hole Valley. Rush Run sinks in a closed depression about 2 km southwest of Hills Creek Cave. Rush Run has the most enigmatic profile of any of the sinking streams examined. The linear profile (Fig. 4.12a) has the same concave upward shape as the other streams with no breaks or other anomalies. The semilog profile (Fig. 4.12b)

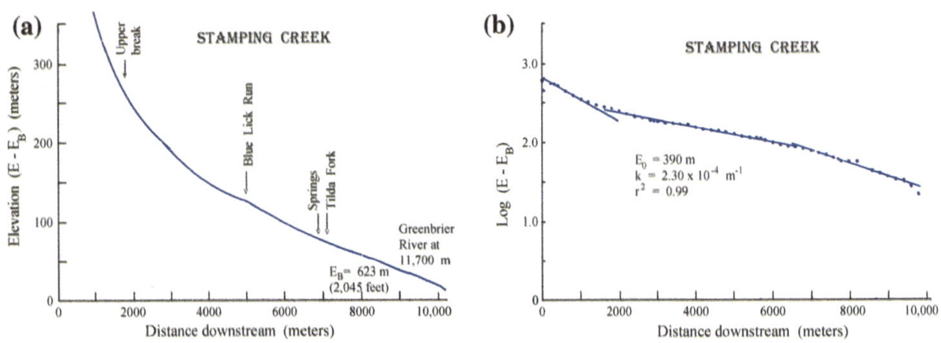

Fig. 4.11 Profile of Stamping Creek taken from USGS Hillsboro Quadrangle. **a** Linear profile. **b** Semilog profile

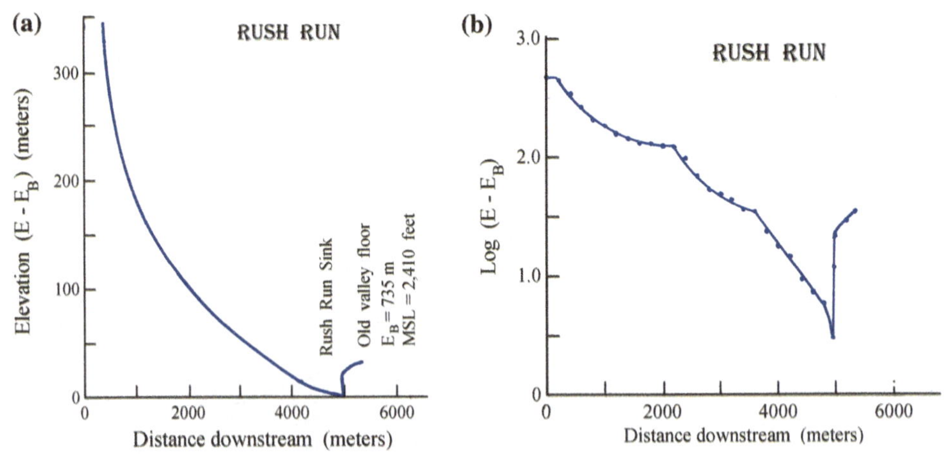

Fig. 4.12 Profile of Rush Run. **a** Linear profile. **b** Semilog profile

is dramatically different. Instead of one or a sequence of straight-line segments, the profile consists of two curved segments in the upstream reach. Only the final segment of about 1 km above the sink point plots as a straight-line segment. The interpretation of this unusual profile is unclear at this time.

4.6 Erosion/Dissolution in the Greenbrier Karst

Given the wealth of information about the caves and drainage systems of the Greenbrier Valley, what can be said about its evolutionary development over geologic time? Unfortunately, the answer is: considerably less than one would like. The landscape, both the karst landscape and the topography on clastic rocks, is constantly evolving. The land surface is lowered, and closed depressions come and go. High-level cave passages are developed and then abandoned as lower-level passages develop in response to lowering base levels. Drainage patterns shift in response to geologic controls. The objective is to draw a timescale for these sequences of events to the extent possible. The discussion will be, of necessity, speculative but, hopefully, not entirely fictitious.

The sections that follow continue earlier discussions on the geomorphic history of the Greenbrier Karst (White and White 1991), other regions of the Appalachians (White 2009), and an interpretation of the karst and caves of Burnsville Cove, Virginia, a karst region in the Valley and Ridge Province (White 2015). The first investigation of the relation of major cave passages to erosion levels is that of Wolfe (1962, 1964).

4.6.1 The Historic View

The sediment pile that accumulated from the Cambrian to the Pennsylvanian and which was folded and distorted during the Permian orogenic episode has been exposed to erosion at least since the Jurassic. There are only rough estimates for the thickness of sediment eroded. Ten km of sediment was thought to be eroded at the border between the Valley and Ridge and the Piedmont, 6 km in the center of the Valley and Ridge, and only one km on the Allegheny Plateau (Judson 1975). MacLachlan (1999) made it 15 km in southeastern Pennsylvania. No estimates have been published for the Greenbrier Valley, but certainly the entire sequence of clastic rocks from the upper Mississippian to the Permian has been removed. This is not a particularly useful guide because the thicknesses of the sedimentary layers were highly variable. Something on the order of 2–3 km seems a reasonable estimate.

Many of the earlier writers argued that most of the erosion took place in the Mesozoic so that the Appalachian landscape had received most of its present form by the beginning of the Tertiary. This left the 65.5 million years of Cenozoic time for the final polish of the landscape into its present form. During this period of time, episodic tectonic uplift provided the hydraulic gradient for rapid downcutting of river valleys. Periods of tectonic quiescence allowed for the lateral erosion needed for the establishment of erosion surfaces (called peneplains in much of the literature). There is a very large and exceedingly confusing literature concerning the number, labeling, location, and age of these features. One of the best discussions is that of Sevon (1985) which, unfortunately, was published only as a field trip guide book. He offers two quotes:

> Little of the Earth's topography is older than the Tertiary and most of it is no older than the Pleistocene—Thornbury, 1969
> .....one can conclude that the great bulk of erosion of the Appalachians took place in the Cretaceous or earlier, and that the present mountains have had a relief not much less than the present since well back in the Tertiary, perhaps since the Eocene.—Rogers, 1967

Melhorn and Edgar (1975) proposed the following sequence for the erosional development of the Appalachians:

> Late Jurassic to early Cretaceous - development of the Fall Zone (Gondwana-Laurasian) surface between 135 and 120 million years ago. Few if any remnants of this surface survive.
> Very late Cretaceous through Paleocene—formation of the Schooley erosion surface during the interval from 85 to about 50 million years ago. The nearly accordant ridge-lines of the mountains in the Valley and Ridge are taken to be remnants of this surface as are portions of the Allegheny Mountains and the Cumberland Plateau.
> Late Eocene through Oligocene - construction of the Harrisburg Surface dating from about 45 to 20 million years ago.
> Upper Miocene to early Pleistocene - possible formation of the Somerville Surface (Coastal Plain) and the Parker Strath during the interval from 12 to 2 million years ago.

Based on sediments deposited by Atlantic slope drainage off the east coast of North America, Poag and Sevon (1989) came up with a similar sequence of erosion surfaces and a similar chronology.

What is now available that was not available to the earlier geomorphologists are some accurate time markers and some reasonably accurate estimates of the rates at which land surface erosion and denudation takes place. With this information in hand, one can calculate backward and determine when the erosion surfaces were created. The results are considerably at variance with the traditional timescales.

4.6.2 The Harrisburg Erosion Surface

The Harrisburg Surface receives the most attention because the doline plain of the Greenbrier Karst appears to be the Harrisburg Surface. The rolling limestone plains of the Big Levels and the Little Levels correlate well with the other valley uplands that have been assigned to the Harrisburg Surface.

The defining locality of the Harrisburg Surface was the Cumberland Valley in the Susquehanna River drainage near Harrisburg, Pennsylvania. A broad rolling upland is at an elevation of 160 m. The Cumberland Valley rises somewhat to the south, reaching 210 m north of Hagerstown, Maryland, and then falls again to elevations near 160 m along the Potomac River. West of Harrisburg, there is a pronounced valley upland formed by accordant hill summits along the incised valley of the Juniata River. The upland surface reaches an elevation of 305 m in Huntingdon County and 365 m at the basin divide in the broad Nittany Valley in central Pennsylvania.

In the Great Valley Province south of the Potomac River, the Harrisburg Surface can be traced along the Shenandoah River as a valley upland that reaches 600 m at the drainage divide. There are further remnants of the surface along the James and New Rivers. The Burnsville Cove Karst is on a tributary of the James River. The Greenbrier River is a tributary of the New River, and a distinct karst surface can be followed irregularly from the New River Valley to the Greenbrier Karst.

4.6.3 Time Markers: What Is the Age of Nothing?

Erosion surfaces develop slowly over long time spans, so the precise age of an erosion surface is difficult to define. A more sharply defined point in time is the onset of dissection of the surface in response to tectonic uplift, lower base levels, and increased hydraulic gradients. Unfortunately, very few high-precision techniques are available for dating what is essentially the absence of rock.

Other features which are essentially the absence of rock are the cave systems. Large abandoned, low gradient trunk passages are formed close to regional base levels. Tiers of master trunks are effective markers for the position of past base levels. Cave development is a very rapid process on a geological timescale. Calculation based on laboratory measurements of dissolution rates indicates that a 10-m diameter conduit can be formed in as little as 50,000–100,000 years (Palmer 1991). Caves, in effect, are a more precise time marker than river terraces or erosion surfaces. The difficulty has been finding a technique for establishing the age of caves.

A major breakthrough was the development of the cosmogenic isotope technique for determining the burial ages of cave sediment (Granger and Muzikar 2001). What follows is a very brief outline of the technique. The Earth is continuously bombarded by high energy particles from space (cosmic rays). Spallation from atoms in the upper atmosphere produces secondary particles, mostly neutrons with some muons, which reach the Earth's surface. There is an interaction with quartz pebbles or quartz sand in which ^{28}Si is converted into ^{26}Al and ^{16}O is converted into ^{10}Be. Both isotopes are radioactive. The half-life of ^{26}Al is 1.02 Ma, and the half-life of ^{10}Be is 2.18 Ma. New isotopes are produced by continuous bombardment and earlier isotopes decay resulting in an equilibrium Al/Be ratio of about 6 for quartz-containing sediment on the surface. If the sediment is washed into a cave as part of the allogenic sediment load, exposure to cosmic rays is shielded, and because of the different half-lives, the isotope ratio begins to decrease. By measuring the $^{26}Al/^{10}Be$ ratio in quartz-containing clastic cave sediments, an age for the deposition of the sediment can be calculated. In the case of abandoned fossil levels, the age of the clastic sediment is the last date at which the cave was part of the active drainage systems and thus for the erosion level to which the cave passage was graded. Unlike shorter half-life dating isotopes such as ^{14}C and $^{234}U/^{230}Th$, the half-lives of ^{26}Al and ^{10}Be are in the proper range for cave passages whose development is expected to have happened from the mid-Pliocene through the Pleistocene.

Unfortunately, the concentrations of cosmogenic isotopes are very small, on the order of millions of atoms. Measurement of these tiny quantities can only be done with an accelerator mass spectrometer, a large, expensive, and uncommon instrument. As a result, relatively few cosmogenic isotope dates are available but those that have been measured place severe constraints on geomorphic interpretation.

Sediment dating on caves in the Cumberland Plateau (Anthony and Granger 2004, 2007b) showed that the dissection of the Highland Rim began only 3.5 million years ago. The Highland Rim is generally considered equivalent to the Harrisburg Surface, so the same date can be applied to the beginning of dissection of the Harrisburg Surface. It appears that valley deepening and dissection of the Harrisburg Surface began only 3.5 million years ago rather than the 12 million years in the traditional chronology. The terrace level called the Parker Strath formed shortly afterward, and its dissection began 2.0 million years ago.

4.6.4 Denudation Rates in the Greenbrier Karst

Rather than assign ages to specific landscape features, an alternative approach is to estimate the rate at which landscapes evolve and then calculate backward. Such estimates have been made for large river basins by measuring the sediment load at the river mouth over a period of time and normalizing to the area of the basin. For the North Atlantic which includes rivers from the Delaware to the Rapidan, the average denudation rate was 22.4 mm/ka (or the numerically equivalent m/Ma) (Judson and Ritter 1964). This measurement includes only the suspended load. Adding the bedload and the dissolved load increases the average denudation rate for the seven rivers draining into the North Atlantic to 46.6 mm/ka.

Karst areas are denuded mainly by the dissolution of the limestone with mass transport out of the basin as dissolved load. There will be a clastic component derived from allogenic sediment injected into the karst aquifer by sinking streams, and there will be a clastic component derived from the insoluble residue from the limestone, but the dissolved load is likely to be dominant. Even in the large river basins, the dissolved load made up 45% of the total.

A variety of methods have been developed for the measurement of the denudation rate in karst (White 2000):

(1) Direct measurement of the dissolution of exposed rock surfaces by means of a micrometer—useful for stream beds, exposed rock surfaces, and the walls of shafts but not representative on a regional scale.
(2) Weight loss measurements on limestone tablets that have been buried in the soil for extended periods of time—gives good estimates of the dissolution in the epikarst.
(3) Basin-scale calculations from the total dissolved load leaving the basin—requires complete hydrographs and chemographs over extended periods of time to account for seasonal variability and for variations in response to storm flow but the best way to obtain regional averages.
(4) Theoretical estimates, based on the assumption that dissolution reactions at the base of the epikarst have reached equilibrium, from the equation

$$D_n = \frac{M_{\mathrm{cal}}}{\rho\sqrt[3]{4}}\left(\frac{K_C K_1 K_{\mathrm{CO_2}}}{K_2\,\gamma_{\mathrm{Ca^{2+}}}\,\gamma^2_{\mathrm{HCO_3}}}\right)^{\frac{1}{3}} P^{\frac{1}{3}}_{\mathrm{CO_2}}\,(P-E) \qquad (4.6)$$

where D_n is the denudation rate in mm/ka. M_{cal} is the molecular weight of calcite (or a weighted mix of calcite and dolomite), and ρ is the rock density in g/cm^3. The K's are the usual equilibrium constants for carbonate reactions, and the γ's are activity coefficients. $P - E$ (precipitation minus evapotranspiration) is the annual runoff in mm/year.

There have been several measurements of denudation rates in the Greenbrier Karst based on discharge and chemical composition of big springs (Table 4.1) (Ogden 1982). Ogden's values seem low compared to other denudation rates measured in the Appalachians which, although they are highly variable from site to site, cluster around 30 mm/ka (White 2007).

In contrast to the limestones, denudation rates on the sandstones and quartzites that cap the Appalachian ridges are about an order of magnitude lower. Savon's (1989) compilation of erosion rates gave 1.5, 2, and 5 mm/ka as the available measurements for sandstones. Anthony and Granger (2004) estimated the denudation rate for the quartzite conglomerate on the Cumberland Plateau as 3–5 mm/ka.

4.6.5 River Downcutting

Tectonic uplift rejuvenates river downcutting, lowers base levels and increases hydraulic gradients within the karst aquifers. The deepening river valleys allow conduit systems to drain, leaving higher cave levels as abandoned dry passages. The few measurements of downcutting rates from Appalachian rivers are given in Table 4.2. They are in the same range as the rate of limestone denudation but much faster than the erosion rates on sandstones. The fact that

Table 4.1 Denudation rates for three spring basins in Monroe County

Spring	Basin area km^2	Area on limestone (km^2)	Denudation rate (mm/ka)
Dickson spring	64	56	22.6
Walters spring	27	24	22.3
Cold spring	2.4	2.4	19.0

From Ogden (1982)

Table 4.2 Rates of river downcutting in the Appalachians

River	Downcutting rate	Method
Cheat River, WV	56–63	Magnetic reversal
East Fork, Obey River, TN	30	Cosmogenic isotopes
Juniata River, Newport, PA	27	Sediment load
New River, Pearisburg, VA	27	Cosmogenic isotopes
South River, Grottoes, VA	23–41	Magnetic reversals

From White (2009)

Savon's estimate of 27 mm/ka for the Juniata River based on sediment load is similar to the others based on cave levels suggests that present-day rates are not dramatically different from what they have been in the past.

The downcutting rate for the New River at Pearisburg was determined by cosmogenic isotope dating of cave sediments (Granger et al. 1997). This value, 27.3 mm/ka, is particularly useful because it is extracted from burial ages from 0.26 to 1.47 million years and thus averages over the climatic fluctuations of the Pleistocene. Pearisburg is only a few tens of kilometers southeast of the Greenbrier Karst.

4.6.6 The Karst Conundrum

With the questionable assumption that karst denudation rates measured under contemporary climatic conditions are reasonable measures of long term averages, one can analyze the landscape backward. The analysis begins with the present-day landscape and then using the measured rates, calculate where the land surface should have been at various times in the past. A karst surface is lowered as a surface, with infiltrating water carrying dissolved limestone into sinkholes and down fractures. The surface need not be dissected because there are no surface streams. Applying this line of argument to karst terrains leads to some conclusions greatly at variance with traditional geomorphology.

If the rate of downcutting of the New River is multiplied by Anthony and Granger's (2007b) date of 3.5 million years for the onset of dissection of the Harrisburg Surface, the result is 90 m of downcutting, a number in good agreement with the present elevation of the Greenbrier River channel compared with the elevation of the karst surface. The onset of dissection of the Harrisburg Surface began much less than 20 or even 12 million years ago.

An even greater discrepancy is found for the Schooley Surface. Taking account of the low denudation rates for the quartzites that make up the ridge tops, and using several arguments including cosmogenetic isotope dated cave deposits, estimation of the date for the onset of dissection of the Schooley surface in the Valley and Ridge and in the Cumberland Plateau produced the same date of 9–12 million years (White 2009). Again, the timescale is greatly shortened from the 50 million year date in the traditional timescale.

4.7 Evolution of the Greenbrier Karst

4.7.1 The Evolution of the Karst Land Surface

Visualize the Greenbrier Karst as it might have appeared in mid-Pliocene time, roughly 3.5 million years ago. The Harrisburg Surface is well established as a low-relief valley floor extending the length of the Greenbrier Karst. The surrounding mountains provide high relief terrain and a continuing source of allogenic recharge to the karst. The Greenbrier River is a low gradient stream meandering along the valley, sometimes on the clastic rocks and sometimes on the limestone. The valley floor has a slight gradient to the southwest. The present-day elevations are—very roughly—730 m at the Swago Creek uplands and the Little Levels, 700 m at the Big Levels north of Lewisburg, and 685 m south of the river near Sinks Grove. Assuming that the river has maintained its present course, the fall would have been only 45 m over the roughly 120 km reach of the river or a gradient of 0.04%. Rejuvenation of the river in the mid-Pliocene produced the present-day deep narrow valley incised into the Harrisburg Surface. The profile (Fig. 4.13) is linear to within the accuracy of elevations scaled from topographic maps and yields a gradient of 0.14%.

The karst surfaces of the Little Levels and the Big Levels have not remained static over the 3.5 million years since rejuvenation began. The limestone is removed by dissolution and limestone surfaces are gradually lowered. There is not necessarily any dissection of the surface during this process because dissolved carbonate is carried into the subsurface through sinkholes and epikarst fractures with little lateral transport. Ogden's (1982) average value of 21.3 mm/ka seems low compared to other denudation rates measured in the Appalachians, but it can be used to calculate the thickness of limestone that has been removed in the past 3.5 million years. The result suggests that there has been 75 m of average lowering of the karst surface (or about 100 m if the more typical value of 30 mm/ka is used).

The Muddy Creek Mountain Syncline preserved the sandstones of the Mauch Chunk Formation which, because of

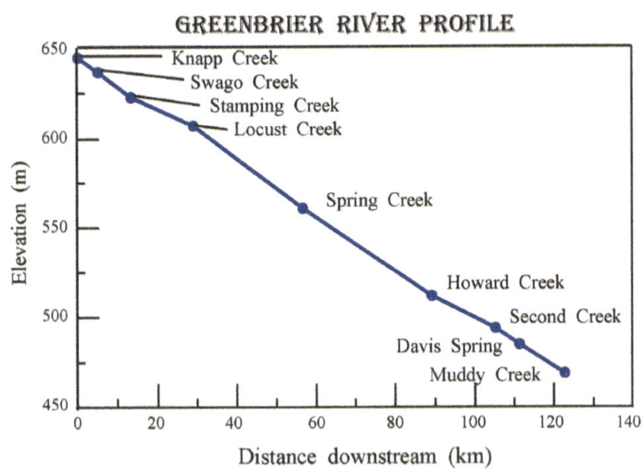

Fig. 4.13 Profile of the Greenbrier River. Distance downstream is measured along the river channel including all of the meander bends. Elevations were taken from topographic maps at the confluence with the specified tributaries

the much higher resistance to erosion, produced the topographic high of Muddy Creek Mountain. The relief of Muddy Creek Mountain and other sandstone-protected topographic highs such as Flattop Mountain is about equal to the calculated denudation of the Big Levels. During the mid-Pliocene, the Harrisburg Surface would have extended across the entire area with all rocks planated to about the same level. Carbonate rock dissolution lowered the limestone surface faster, leaving the sandstones behind as the present-day topographic highs. The elevation of the Harrisburg Surface prior to the onset of its dissection would have been near 800 m.

4.7.2 Time Sequence of Cave Development

One must always begin a discussion of the origin and development of a cave system with this caveat: We never see all of the caves. Passages in explored caves are blocked by breakdown and by silt plugs. There are blank areas between explored caves and their known spring discharge points. Entrances are in short supply and it seems likely that other caves exist, unknown simply because no one has discovered (or dug) an entrance.

Cave development did not begin with the Harrisburg Surface. Adding an additional 75–100 m of rock will expose, in places, the upper strata of the Greenbrier Limestone which have now been mostly eroded away. The low gradient between the water table and the river would have provided less driving force for cave development. Caves that predate the Harrisburg Surface have been destroyed with at most a few surviving fragments. Martens Cave (Chap. 7) is one such possibility. Tub Cave in the Swago Creek Basin (Chap. 6) may be another.

The present-day caves, in spite of their considerable length, seem to be in the decay phase of cave history. Passages have been truncated by deep surface tributaries such as Spring Creek and Second Creek. Deep closed depressions suggest massive collapses in underlying cave passages. Breakdown is very common in many cave passages.

The development of the large cave systems appears to track with the deepening of the secondary valley of the Greenbrier River which has outran the denudation of the karst surface and now provides about 100–150 m of relief between the limestone upland and the river. The time sequence over the past 3.5 million years appears to be one of continuous downcutting of the river and thus of continuous cave development. Wolfe (1964) identified two terraces along the river in addition to the flood plain. At Seebert on the eastern edge of the Little Levels, these terraces are at 625 (flood plain), 650 (second terrace), and 660 (third terrace) meters. The terraces may represent pauses in downcutting which gave the cave passages time to shift into new routes but did not result in neatly stacked tiers of passage. Some passages have the elliptical passage cross sections characteristic of development under water-filled conditions. Many others are canyons cut deeply into the limestone and even into the underlying Maccrady Shale. Many have active streams; the effects of flood flow are expected to be considerable. Greenville Saltpeter Cave, with its smooth rounded passages and anastomotic pattern, clearly formed by slow-moving water below base level. Cave development was interrupted by massive in-filling events of clastic sediment. These events may be related to climatic changes during the Pleistocene ice ages and may be related to the massive boulder fills found in many of the surface valleys cut back into the mountains.

A paleomagnetic reversal found in the sediment of an upper level passage in Windy Mouth Cave by Ira Sasowsky and his students (Chap. 15) suggests a sedimentation age on the order of a million years (older than the 788,000 year reversal date). The passage is 30 m above the present-day river and so would be consistent with the expected 100 m of downcutting over the past 3.5 million years. A magnetic reversal was also found in the sediments of Buckeye Creek Cave (Chap. 9) with a similar result. The conclusion is that the long cave systems of the Greenbrier Karst represent a continuous process of cave development dating at least from Harrisburg time, 3–4 million years. The investigation of the Friars Hole System (Chap. 8) assigned an age of 4 million years to the system. Details of cave patterns were controlled by local structure, particularly the sequence of fold axes, and by lithologic factors, particularly the presence of shaly layers within the limestone. Pauses in river downcutting and cave development were short-lived and likely related to changes in sediment load in response to Pleistocene climatic oscillations.

References

Anthony, D.M., and D.E. Granger. 2004. A late tertiary origin for multilevel caves along the western escarpment of the Cumberland Plateau, Tennessee and Kentucky, established by cosmogenic ^{26}Al and ^{10}Be. *Journal of Cave and Karst Studies* 66 (2): 46–55.

Anthony, D.M., and D.E. Granger. 2007a. An empirical stream power formulation for knickpoint retreat in Appalachian Plateau fluviokarst. *Journal of Hydrology* 343: 117–126.

Anthony, D.M., and D.E. Granger. 2007b. A new chronology for the age of Appalachian erosional surfaces determined by cosmogenic nuclides in cave sediments. *Earth Surface Processes and Landforms* 32: 84–887.

Curl, R.L. 1986. Fractal dimensions and geometries of caves. *Mathematical Geology* 18: 765–783.

Goldrick, G., and P. Bishop. 2007. Regional analysis of bedrock stream long profiles: evaluation of Hack's SL form, and formulation and assessment of an alternative (the DS form). *Earth Surface Processes and Landforms* 32: 649–671.

Granger, D.E., J.W. Kirchner, and R.C. Finkel. 1997. Quaternary downcutting rate of the New River, Virginia, measured from differential decay of cosmogenic ^{26}Al and ^{10} Be in cave-deposited alluvium. *Geology* 25: 107–110.

Granger, D.E., and P.F. Muzikar. 2001. Dating sediment burial with in situ-produced cosmogenic nuclides: Theory, techniques, and limitations. *Earth and Planetary Science Letters* 188: 269–281.

Hack, J.T. 1957. Studies of longitudinal stream profiles in Virginia and Maryland. *U.S. Geological Survey Professional Paper* 294-B: 97p.

Hack, J.T. 1973. Stream-profile analysis and stream-gradient index. *Journal of Research of the U.S. Geological Survey* 1: 421–429.

Howard, A.D. 1994. Modeling fluvial erosion on regional to continental scales. *Journal of Geophysical Research* 99: 13971–13986.

Jones, W.K. 1997. *Karst hydrology Atlas of West Virginia*, vol. 4, 111pp. Leesburg, VA: Karst Waters Institute Special Publication.

Judson, S. 1975. Evolution of Appalachian Topography. In *Theories of landform development*, ed. W.N. Melhorn, and R.C. Flemal, 29–44. Binghamton, NY: State University of New York.

Judson, S., and D.F. Ritter. 1964. Rates of regional denudation in the United States. *Journal of Geophysical Research* 69 (16): 3395–3401.

Lessing, P. 1979. *Karst Subsidence and Linear Features, Greenbrier and Monroe Counties, West Virginia*. Morgantown, WV: West Virginia Geological and Economic Survey, map ERG 17.

MacLachlan, D.B. 1999. Mesozoic. In *The geology of Pennsylvania*, ed. C.H. Shultz, 435–449. The Pennsylvania Geological Survey: Harrisburg, PA.

Melhorn, W.N., and D.E. Edgar. 1975. The case for episodic continental scale erosion surfaces: A tentative geodynamic model. In *Theories of landform development*, ed. W.N. Melhorn, and R.C. Flemal, 243–276. Binghamton, NY: State University of New York.

Ogden, A.E. 1982. Karst denudation rates for selected spring basins in West Virginia. *National Speleological Society Bulletin* 44: 6–10.

Palmer, A.S. 1991. Origin and morphology of limestone caves. *Geological Society of America Bulletin* 103: 1–21.

Poag, C.W., and W.D. Sevon. 1989. A record of Appalachian denudation in postrift Mesozoic and Cenozoic sedimentary deposits of the U.S Middle Atlantic continental margin. *Geomorphology* 2: 119–157.

Sevon, W.D. 1985. Pennsylvania's polygenetic landscape. Guidebook for the 4th annual field trip of the Harrisburg Area Geological Society, 55pp.

Savon, W.D. 1989. Erosion in the Juniata River drainage basin, Pennsylvania. *Geomorphology* 2: 303–318.

Shepherd, R.G. 1985. Regression analysis of river profiles. *Journal of Geology* 93: 377–384.

Shulits, S. 1941. Rational equation of river-bed profile. *Transactions of the American Geophysical Union* 22: 622–630.

Strahler, A.N. 1968. Quantitative geomorphology. In *The Encyclopedia of geomorphology*, ed. F.W. Fairbridge, 898–912. New York: Reinhold.

Troester, J.W., E.L. White, and W.B. White. 1984. A comparison of sinkhole depth-frequency distributions in temperate and tropic karst regions. In *Sinkholes: Their geology, engineering, and environmental impact*, ed. B.F. Beck, 65–73. A.A. Balkema: Rotterdam.

Westaway, R. 2007. Late Cenozoic uplift of the eastern United States revealed by fluvial sequences of the Susquehanna and Ohio systems: Coupling between surface processes and lower-crustal flow. *Quaternary Science Reviews* 26: 2823–2843.

White, W.B. 2000. Dissolution of limestone from field observation. In *Spleogenesis: Evolution of Karst Aquifers*, ed. A.B. Klimchouk, D. C. Ford, A.N. Palmer, and W. Dreybrodt, 149–155. Huntsville, AL: The National Speleological Society.

White, W.B. 2007. Evolution and age relations of karst landscapes. *Acta Carsologica* 36 (1): 45–52.

White, W.B. 2009. The evolution of Appalachian fluviokarst: Competition between stream erosion, cave development, surface denudation, and tectonic uplift. *Journal of Cave and Karst Studies* 71 (3): 159–167.

White, W.B. 2015. Geomorphic evolution of the Burnsville Cove caves. In *The Caves of Burnsville Cove, Virginia*, ed. W.B. White, 443–457. Heidelberg: Springer.

White, E.L., and W.B. White. 1979. Quantitative morphology of landforms in carbonate rock basins in the Appalachian Highlands. *Geological Society of America Bulletin* 90: 385–396.

White, E.L., and W.B. White. 1983. Karst landforms and drainage basin evolution in the Obey River Basin, north-central Tennessee, U.S.A. *Journal of Hydrology* 61: 69–82.

White, W.B., and E.L. White. 1987. Ordered and stochastic arrangements within regional sinkhole populations. In *Karst hydrogeology: Engineering and environmental applications*, ed. B.F. Beck, and W. L. Wilson, 85–90. A.A. Balkema: Rotterdam.

White, W.B., and E.L. White. 1991. Karst erosion surfaces in the Appalachian Highlands. In *Appalachian Karst*, ed. E.H. Kastning, and K.M. Kastning, 1–10. Huntsville, AL: National Speleological Society.

White, W.B., and E.L. White. 2013. Karst landforms: Scope and processes in the early twenty-first century. In *Treatise on Geomorphology*, vol. 6, *Karst Geomorphology*, ed. J.F. Shroder (editor in chief), Frumkin, A. (volume editor), 14–22. San Diego: Academic Press.

Wolfe, T.E. 1962. An investigation of the controls of cavern development in the Greenbrier Limestone of southeastern West Virginia, 66pp. MA thesis, University of Pittsburgh.

Wolfe, T.E. 1964. Cavern development in the Greenbrier series, West Virginia. *National Speleological Society Bulletin* 26: 37–60.

The Exploration History of the Greenbrier Valley Caves

5

George Dasher

Abstract

The earliest accounts of caves in the Greenbrier Valley document mining for saltpeter, which occurred in 23 caves at various periods from the late 1700s, through the War of 1812, and the American Civil War. Technical descriptions of the caves began in the late 1940s with the work of William E. Davies. Exploration by cavers began in the 1950s and expanded rapidly thereafter. Exploration and survey were greatly enhanced by the organization of the National Speleological Society (NSS), West Virginia Association for Cave Studies (WVACS) and the West Virginia Speleological Survey (WVASS). This chapter summarizes the overall exploration history and provides exploration details of 14 selected individual caves.

5.1 Introduction

Caves are very much a part of the landscape in the Greenbrier Valley. Many caves have large open entrances. Streams flow into open cave entrances, and streams emerge as large springs from open cave entrances. Certainly, the more obvious entrances were known to the first settlers to enter the Greenbrier Valley and were likely known to the early Americans who preceded them. Caves were mined for saltpeter in the Nineteenth Century, and there were doubtless many early explorations. However, systematic exploration and documentation of the Greenbrier Valley caves began only in the mid-Twentieth Century and has increased to the present in both the persistence of exploration and the quality of documentation.

The sections that follow summarize the exploration of the major caves. It is not a complete history of the exploration of every cave in the Greenbrier Valley, but it does touch on the large caves that are the primary subject of this book.

G. Dasher (✉)
620 Valley Drive, Elkview, WV 25071, USA
e-mail: wvcaver@juno.com

5.2 Early Cave Exploration to the Mid-Twentieth Century

5.2.1 The Prehistory of the Greenbrier Valley

While numerous Indian burial caves have been discovered in southwestern Virginia, only one such cave has been found within the Greenbrier Valley to date. This is Rapps Cave, located within the Buckeye Creek Basin, where Indian remains and pictographs have been discovered near the entrance. Unfortunately, only the pictographs remain, as the human bones were heavily collected in the mid-twentieth century. This cave is now gated to protect its remaining contents.

5.2.2 The Saltpeter Caves

There are 52 caves within West Virginia where saltpeter mining or prospecting has been documented. Twenty-three of these are within the Greenbrier and New River Valleys of Monroe, Greenbrier, and Pocahontas Counties. Although, in most cases, the miners most likely did not explore beyond the passages and rooms required to collect and process the saltpeter, it is known that extensive mining occurred in at least three caves: Greenville Saltpeter, Haynes, and Organ Caves. Mining took place in Haynes and Greenville

W.B. White (ed.), *Caves and Karst of the Greenbrier Valley in West Virginia*,
Cave and Karst Systems of the World, https://doi.org/10.1007/978-3-319-65801-8_5

Saltpeter Caves prior to the formation of Monroe County in 1799, and then by the Confederacy during the American Civil War. In Organ Cave, it is thought that saltpeter was removed from the 1812 Room during that war, from the Saltpeter Room by the Confederacy during the American Civil War, and from at least two locations along the Organ Mainstream during an unknown period, but probably during the American Civil War.

Those caves known to have been mined or prospected for saltpeter in the Greenbrier and New Rivers Basins are the following:

- Alta Vista Saltpeter Cave, Greenbrier County: American Civil War
- Big Sink Saltpeter Cave, Monroe County: unknown
- Bob Gee Cave, Greenbrier County: probably the American Civil War
- Brady Saltpeter Cave, Monroe County: unknown
- Crowder Cave, Monroe County: American Civil War?
- Dickson's Saltpeter Cave, Monroe County: unknown
- Doen Ballard Cave, Monroe County: probably the American Civil War
- Eagle Rock Saltpeter Cave, Pocahontas County: probably the American Civil War
- Greenville Saltpeter Cave, Monroe County: Antebellum and American Civil War
- Hanna Cave, Greenbrier County: unknown
- Haynes Cave, Monroe County: Antebellum and American Civil War
- Higginbothams Cave #2, Greenbrier County: unknown
- Higginbothams Cave #3, Greenbrier County: unknown
- John Henry Green Saltpeter Cave, Monroe County: unknown
- Judy's Cave, Greenbrier County: unknown
- Knights Saltpeter Cave, Greenbrier County: unknown
- Lobelia Saltpeter Cave, Pocahontas County: unknown
- McFerrin Saltpeter Cave, Greenbrier County: possibly in Lord Dunmore's War
- Lost Cave, Greenbrier County: War of 1812 and Mexican War
- Organ Cave, Greenbrier County: War of 1812 and American Civil War
- Overholt Saltpeter Cave, Pocahontas County: unknown
- Pollock Saltpeter Cave, Greenbrier County: probably the American Civil War
- Snedegars Cave, Pocahontas County: probably the American Civil War; possibly before
- Walnut Grove Cave, Greenbrier County: unknown

5.2.3 West Virginia's First Commercial Cave

Organ Cave is located at the junction of US Route 219 and State Route 63 in southern Greenbrier County. The first recorded owner of the cave was John Gardner, who was granted the land from the Commonwealth of Virginia in 1783. Ownership passed to several other people in the early nineteenth century, by which time the western Virginia resorts and the stagecoach lines connecting them were well established, and the result was that the travelers often stopped at the cave.

A later owner of the cave, James Boone, opened the cave commercially in the early 1900s and provided every third visitor with a candle for illumination. An improved walkway and Edison light bulbs, powered by a Delco generator and 72 storage batteries, were installed in 1914. James Boone sold the cave in 1926 to George Carter and Seth Sively, who opened the cave each summer for tourism. The property then passed to George and Lee Sively, who continued to run the commercial operation until 1997, when the cave was purchased by Sam and Jamie Morgan. Sam and Jamie have made significant improvements to both the outside and underground facilities and continue to operate the commercial section of the cave, as well as offering wild-caving trips for the more adventurous (Fig. 5.1).

5.2.4 Cave Publication by the West Virginia Geological and Economic Survey

The West Virginia Geological and Economic Survey produced geological publications on all of the state's counties early in the Twentieth Century. No caves were described in the Monroe County publication, but the Pocahontas County bulletin—which was printed in 1929 and authored by Paul H. Price—mentioned Overholt Blowing Cave, Overholt Saltpeter Cave, and Snedegars Cave, the last which was described as "an abandoned stream channel, but still active in rainy weather" that contained a great deal of surface debris.

The Greenbrier County bulletin was published in 1939, authored by Paul H. Price and E.T. Heck, and documented fourteen caves. Of these, Organ Cave was given the most text and was described as "one of the few commercial caves in the state," "colored water placed in the cave emerged in Second Creek" (making this the first recorded dye trace in West Virginia), 37 saltpeter hoppers were constructed by the Confederates during the American Civil War, and the bones of the Pleistocene Xenarthra *Megalonyx Jeffersonii* were thought to have been found in this cave. This last item was

Fig. 5.1 Show cave entrance to Organ Cave. Photograph by W.B. White

described in detail, included a complete description of the bones, and stated that they were collected by Colonel John Stewart and were later described by both Thomas Jefferson and Dr. Caspar Wistar.

Of the remaining 13 caves described by Price and Heck, seven are tiny, while the other six were described as follows:

Rapps Cave:	"The floor is very rough with fallen rocks," "no actual stream was present," and "considerable numbers of stalactites and other formations were present."
Arbuckle's Cave:	"a roomy passage, with a smooth floor, that extends 100 yards into a hill."
McClung's Cave:	"its chambers are very roomy," "then contract to a high, narrow cleft with many fallen rocks," "a small stream flowed toward the west, away from the entrance," and "the walking was difficult."
Higginbotham Cave #1:	"a slow-moving stream flows towards the south," and "numerous stalactites are present."
Higghinbotham Cave #2:	"this is a small cave," and "no running water was present during the time it was visited."
Coffman's Cave:	"a good stream flows into the cave," "which has to be waded at places."

5.2.5 The William Davies' Years

Although documented cave exploration began in the Potomac River Valley prior to World War II, serious work did not begin in the Greenbrier River Valley until after the war. Principal among this work was the exploration of the caves with large, obvious entrances, such as Greenville Saltpeter, Laurel Creek, Organ, McClung, Ludington, Culverson Creek, Rapps, and Buckeye Creek Caves. Much of this early exploration was documented in William E. Davies' *Caverns of West Virginia*, which was published by the West Virginia Geological and Economic Survey in 1949.

Davies described 32 caves in Monroe County within the Greenbrier Valley, eight caves in the Greenville area of Monroe County within the New River Basin, 96 caves in Greenbrier County, and 37 caves in Pocahontas County within the Greenbrier Valley. Also included within his publication were 31 maps and 37 photographs. William Davies not only entered, explored, and described many of the caves in his original 1949 publication and in the 1958 edition and 1965 supplement, but also included the contributions of many other individuals and organizations, such as the Potomac Speleological Club and the Pittsburgh, Charleston, and District of Columbia Grottos of the National Speleological Society (Davies 1958, 1965).

Some of the prominent caves documented by William Davies included the following:

Greenbrier County:

- *Bone Cave*, located north of Renick in a quarry above the C&O Railroad tracks, where 2,000 ft of passage was surveyed and the skull of a Pleistocene peccary found.
- *Buckeye Creek Cave*, where almost two miles of cave, the main stream, and several pits were described.
- *Cricket Cave*, where a map was produced and about 4,000 ft of cave was described.
- *Culverson Creek Cave*, where a large stream passage was explored from the main Culverson entrance, 5,000 ft from the Fullers Entrance, and several thousand feet from the SSS Entrance (none of which were known to be a part of a larger cave system at that time).
- *Foxhole #2 Cave*, where almost 2,000 ft in two passages was explored.
- *Grapevine Cave*, where the 120-foot-deep entrance drop was described and a map produced of the large room below. This cave was documented as first descended in July of 1942 by four cavers from Charleston: John Wingfield, Leroy Frazier, George Mann, and John Suter.
- *Higginbothams #1 Cave*, where over 2,000 ft of cave was explored and mapped, and it was noted that the first grist mill west of the Allegheny Mountains had been constructed inside the cave's entrance in 1769.
- *Higginbothams Cave #4*, where more than 3,500 ft of passage was described, a map produced, and the discovery of puma (mountain lion) bones in an upper-level passage was documented.
- *Jewel Cave*, where an extensive description of the cave was produced, as well as a map, and the cave was described as the most visited cave in West Virginia.
- *Ludington Cave*, where about a mile of cave was discovered, as well as a 40-foot-high waterfall, and a map produced.
- *McClung Cave*, where approximately 3,000 ft of cave was explored.
- *Organ Cave*, where the Charleston Grotto explored and described close to a mile of passage. Included were a sketch map of the cave (which was then known to have three entrances), a map of the main commercial section, and a report that the bones of the Pleistocene sloth *Megalonyx Jeffersonii* had been found in the cave.
- *Piercys Cave*, where over a mile of cave passage was described.
- *Piercys Mill Cave*, where about 1,000 ft of cave was described and a map produced.
- *Rapps Cave*, where several thousand feet of passage, as well as several pits, were described.
- *Windy Mouth Cave*, where over a mile of passage was described, an extensive map produced, and the cave was speculated to be almost six miles in length.

Monroe County:

- *Argobrites Cave*, where several thousand feet of cave was documented, a map produced, and the cave was described as having an extensive moonshining operation at one time, as well as the cave stream powering both a grist and sawmill.
- *Chambers Cave*, where a map of the main passage was produced.
- *Crowder Cave*, where a map of the cave was produced.
- *Fletchers Cave*, where over 2,000 ft of passage was described.
- *Greenville Saltpeter Cave*, where almost 2.5 miles of passage was described, a map produced, and the extensive saltpeter mining artifacts were documented.
- *Haynes Cave*, where a map was produced, the cave was described, and the extensive saltpeter mining artifacts were documented.
- *Indian Creek Cave*, where approximately 1,000 ft of passage was described.
- *Laurel Creek Cave*, where a map was produced and several thousand feet of cave passage in a two-level cave was described.
- *Mott Hole*, where a map was produced and the cave's entrance pit was documented as being 203 ft deep.
- *Patton Cave*, where a map was produced and approximately 4,000 ft of passage was described.
- *Steele's Cave*, where a map was produced and several thousand feet of passage was described.

Pocahontas County (Southern Half):

- *Carpenters and Swago Pits*, where a cave in excess of two miles was described, and two maps were produced.
- *Martins Cave*, where two maps were produced and about 1,000 ft of passage was described.
- *Overholt Blowing Cave*, where several miles of passage were described and a small map produced.
- *Overholt Saltpeter Cave*, where a map was produced and the cave was described.
- *Poor Farm Cave*, where a map was produced and about a mile of cave was described.
- *Snedegars Cave*, where a map was produced, the saltpeter works were documented, and the main cave passage was described.

5.3 The Modern Period of Cave Exploration

5.3.1 The Decade of the 1960s

William Davies' publication became the foundation on which much of the future speleological work in West Virginia would be constructed. Exploration continued in many of the major caves in the three counties, but now the emphasis was less on the discovery and exploration of new caves and passages, but rather on surveying—which was first done utilizing lensatic compasses, and then with Brunton compasses and steel tapes. Cave map symbols were standardized, and caves such as Overholt Blowing, Culverson Creek, Bone, and Organ were all increased in length and connected to other nearby caves. Many of the early explorers maintained their interest in the Greenbrier karst for many decades (Fig. 5.2).

Two national publications on caves were produced during this period. The first was a *National Geographic* article, "Exploring America Underground," which was written by Charles Mohr and printed in June 1964. This was 37 pages in length, included 45 color pictures, and described many of the caves then famous within the USA. It did not discuss any of the caves within the Greenbrier Valley, but it did include pictures of one of the saltpeter vats in Haynes Cave, a caver at the bottom of the frozen waterfall in Crookshank, and a two-page photograph of a climber in Cass Cave ascending the Belay Loft Drop, while the nearby Suicide Falls was in flood.

The second publication was *Depths of the Earth*, by William R. Halliday, first printed in 1966, also described a wealth of caves within the USA (Halliday 1976). This publication mentioned the Pleistocene ground sloth, *Megalonyx jeffersonii*, but did not name the cave from which the bones were collected. It also included a 10-page, two-picture description of the exploration in Overholt Blowing Cave, as well as a summary of the early discoveries in Organ Cave and a list of the work in many of the other large caves in Greenbrier County.

Also of importance, the West Virginia Association of Cave Studies (WVACS) was formed in 1961, with the purpose and passion of exploring and surveying the long cave systems of the central Greenbrier Valley. WVACS quickly became the dominant organization doing exploration work, as well as surveys, ridge walking, and dye traces within the valley, and they published several articles on their work. The group still maintains a field station north of Frankford and continues its long tradition of surveying and documenting the caves of the Greenbrier Valley (Fig. 5.3).

In short, thanks in part to Davies' publication, exploration and mapping took off at an incredible pace during the latter part of the 1960s, with work taking place in the following caves:

- An entrance to *Maxwelton Sink Cave* was dug open at the downstream end of the immense blind valley east of the community of Maxwelton. This highly decorated cave was only open for a few years, and saw an extraordinary amount of exploration and survey (coordinated by Chuck Hempel). A tremendous effect was made to stabilize this cave's only entrance, which easily filled with flood debris, only to have a hurricane close the cave for good.
- *Canis Majoris* and *Piddling Pit* by Bob Swensson, the VPI Grotto, and cavers from the National Radio Astronomy Observatory (NRAO). This was followed by a resurvey of the same caves by members of the Monongahela Grotto.
- *Organ*, *Ludington*, *McClung*, *Wades*, *Benedicts*, *The Hole*, *Bone-Norman*, and *Culverson Creek Caves* (which are some of the longest caves in West Virginia) by WVACS. The survey of the Bone-Norman System was originally started by John Davis and Charlie Maus,

Fig. 5.2 Some of the longtime explorers of the Greenbrier caves. *From the left* Joey Fagan, David Newsome, Rocky Ward, Bill Royster, Al Stewart, Doug Medville, Robert Thren, William Balfour, Philip Lucas, Robert Handley, Bill Biggers, Rocky Parsons, Roger Baroody, Bill Jones

Fig. 5.3 WVACS Frankford field station. Photograph by the author

Ludington Cave by Bud Rutherford and Lew Bicking, Wades Cave by Alan Armstrong, Benedicts Cave by Roger Baroody and Charlie Maus, The Hole by Edwin Coffman, Charlie Maus, and Jack Gravemier, and McClung Cave by Roger Stafford, Lyle Conrad, and Walter Lipton.

- *Windy Mouth* by John Davis, Frank Maddrick, Jack Gravemier, and WVACS.
- *Dry Cave*: Phil Lucas and WVACS began a survey.
- *Grapevine*, *Rapps*, and *Buckeye Creek Caves* by the Baltimore Grotto.
- *Cook Pot*, located near Hillsboro, by members of the York and Philly Grottos.
- John Barnes and WVACS completed a series of surface surveys, providing accurate locations of the contact caves located north of Lewisburg. John did this work by measuring an extremely accurate baseline, and then using triangulations (determined with a transit) to calculate the locations of the cave entrances.
- Radio survey work was completed in the Greenbrier Valley by Roy Carlton, John Davis, Charlie Maus, and Lew Bicking to more accurately locate the underground survey stations.
- Dye tracing was completed by a variety of cavers—including Bill Jones, Hermine Zotter, and Doug Medville—in a large number of caves, ranging all up and down the Greenbrier Valley.

5.3.2 The Decade of the 1970s

Advances in technology had a tremendous impact on caving in the 1970s. The use of dynamic laid ropes decreased and was replaced by static kernmantle ropes, which did not spin or stretch, making vertical caving less difficult. Scaling poles, which could bend or drop the climber, were replaced by bolts and aid climbing directly on the cave walls. Synthetic clothing, which did not become saturated with mud or water began appearing, making the caving trips more comfortable, and the Brunton compasses and steel tapes were replaced by Suunto compasses and inclinometers and fiberglass tapes, which made the surveys easier, more enjoyable, and much more productive.

Of particular note was Bill Douty's map of Bone-Norman Cave, which was awarded second place at the first Cartographic Salon at the 1978 Texas NSS Convention. This map not only showed all the passages in this two-entrance, 14-mile-long cave, but included incredible (and beautiful) interior passage detail. Other cavers quickly emulated Bill's work, as well as the other maps entered in subsequent Salons, and these maps became a forerunner of things to come, as now the sketcher on the survey team became the leader, and the entire team endeavored to produce not just a basic map of the main passages in the cave, but rather a highly detailed map showing interior detail, cross sections, ceiling heights, pit depths, passage elevations, and water depths.

One result of this new cartographic standard was that many of the "old" caves were resurveyed, and the resulting new maps contained a much higher level of information.

In addition, while the previous cave data had been processed by hand or by rudimentary computer programs that managed the data one survey at a time and did not close loops, John Davis, Bob Amundson, Bob Thrun, Doug Dotson, and George Dasher all wrote FORTRAN programs during the 1970s that could process the immense amount of data generated by the large cave surveys taking place in the Greenbrier Valley. All of these programs could store the entire data set in one database, build surveys off previous data, close loops, and produce line plots. John Davis and Bob Amundson's program *Cavemap* was the first widely utilized and was used to process the enormous amount of cave data WVACS was producing. *Cavemap* and George Dasher's *Cavesurv* closed loops subsequently and one at a time, but Bob Thrun's *CMAP* and Doug Dotson's *SMAPS* closed multiple loops simultaneously using least square math. Bob Thrun's program, in particular, was used to process the data from Organ and Friars Hole Caves, and his simultaneous loop-closure process allowed the cavers to locate and correct errors within those databases. This was a very time-consuming, tedious process and very much needed with Organ's data, as there were a lot of bad loop closures, as well as old and redundant surveys.

Several other important things occurred during the 1970s. First, in 1973, Bill Jones and the West Virginia Geological and Economic Survey published the 50-page *Hydrology of Limestone Karst in Greenbrier County, West Virginia*. This book, which described the documented dye traces completed within the county, allowed others to build and add to the completed work (Jones 1973).

Second, the West Virginia Speleological Survey (WVASS) was formed, principally by Roger Baroody, and began publishing a long series of bulletins documenting West Virginia's caves and karst. The first of these, edited in 1971 by Doug Medville, documented the known caves in Randolph County, but later bulletins described the *Karst and Karst Hydrology of Monroe County* (Hempel 1975), and *Northern Pocahontas County* (Medville and Medville 1976).

Third, a series of field houses was established along the length of the Greenbrier Valley. These buildings were often nothing more than shacks, camping areas, or old, abandoned houses, but they provided a friendly location at which cavers could sleep, cook meals, and recruit (or be recruited by) other cavers for project or caving trips. These field houses included the Sugar Shack on Swago Creek (Fig. 5.4), the Organ Cave Fieldhouse (Fig. 5.5), two locations at Friars Hole (the Lower and Upper Campgrounds), two Monroe County houses (one north of Union and a later one in Sinks Grove), and four buildings used by WVACS—the first being the Arbuckle house near the airport, the second south of Frankford (Fig. 5.6), a third in a trailer east of the Weaver Knobs, and the fourth on Butler Mountain Road (where WVACS finally purchased the property).

Fourth, WVACS maintained a series of log books, first in the Court Restaurant in Lewisburg and then in the La Strada pizza restaurant in Fairlea. These gave cavers a place to congregate after their caving trips, sign the log books (often with humorous notations), and meet other cavers and describe their caving trips. The comradery established by these focal locations and these log books cannot be over emphasized, as on occasion—at La Strada—there would sometimes be forty or so cavers, all from different organizations and projects, discussing what they had done or planned to do, and enjoying each other's company. In short, these informal meetings were important, as the different cavers and caving groups tended to blend together, often informally assisting in a project, and then moving on to work with a different group of people on other projects.

Finally, WVACS continued efforts to map the caves within the Greenbrier Valley, as well as two projects to map the two longest caves in West Virginia, moved the state to the forefront of caving within the USA. These two projects were Organ Cave and Rubber Chicken Cave.

The Organ Cave Project was begun in the early 1970s by the District of Columbia Grotto, who built on the previous WVACS' work and surveyed not only the large, 36-mile-long Organ Cave, but also the entirety of the caves on the Organ Cave Plateau. Dye traces and geological and historical research were completed, and the end result was a 200-page bulletin, edited by Paul Stevens, produced by WVASS, and which described the caves, the exploration history, and the geology and hydrology, and included quadrangle maps of the Plateau's caves and a lexicon of the caves' place names.

Rubber Chicken Cave was dug open in the Friars Hole area in the mid-1970s, and immense discoveries were made within months. The end result was that, by the middle of the 1980s, the resulting cave, the Friars Hole Cave System, was found to have ten entrances and had been explored to a length that made it one of the deepest caves in West Virginia and the longest cave east of Mammoth Cave (taking that "title" from Organ Cave).

Other surveys that were completed and maps produced during the 1970s included the following:

Bransford, Benedicts, and Bone-Norman Caves by Bill Douty and WVACS. (This effort totaled more than 30 miles of cave surveyed.)

The Hole by WVACS and the Pittsburgh Grotto, chiefly Charlie Williams and Don Schleicher, but also by Charlie Maus and Jack Gravinmire. (This cave is currently 23 miles long.)

Fig. 5.4 Sugar shack on Swago Creek. Photograph by W.B. White

Fig. 5.5 Organ Cave field station. Photograph by the author

Greenville Saltpeter and Rimstone-Crossroads Caves by the Pittsburgh Grotto. The resulting Greenville map was drawn by Don Schleicher and Ray Povirk, while the Rimstone-Crossroads map was drawn by Barbara Turner. Both maps are of such quality that there has never been a need to resurvey the two caves. (These two caves are 4.2 and 0.4 miles long.)

Chambers and Indian Draft Caves, as well as Mott Hole, by the Pittsburgh Grotto. (These caves total about 4 miles of passage.)

Clyde Cochrane Cave was surveyed in 1970 by the Baltimore Grotto as a part of the J.B. Hill Project, who described the cave as "about 1,000 feet of passage and a large lake." Later, in 1980, the cave was resurveyed by the Droop Mountain Cave Club, and the upstream sump in the cave was dove by Gary Storrick, Bill Stone, and Roberta Swicegood. This was found to be 80 ft long, and several thousand feet of passage was then surveyed upstream of the sump and the cave's length extended to 1.2 miles.

Fig. 5.6 Old WVACS Frankford field station. Photograph by the author

Dozens of caves including—*Argobrites*, *Bob Cantabury*, *Boyd Spring Cave*, *Broyles Sulfur Spring*, *Pearson Hollow*, *Fletcher*, *McClung-Zenith*, *Neel Insurgence*, *Odgen's*, *Old Lebanon*, *Patton*, *Rayburn Draft*, *Rehobeth Church*, *Rock Camp Creek*, *Senator*, *Union*, *Walker Farm*, and *Walker Well Caves*—were mapped by WVASS (usually John Hempel and Doug Medville).

The 1.3-mile-long *Watters Cave* was surveyed by Bill Balfour and the Greenbrier Grotto.

The 1.3-mile-long *Herns Mill Cave* by Bill Balfour and WVACS.

Laurel Creek Cave by the VPI Grotto.

The 1-mile-long *Scott Cave* by the University of Virginia Grotto.

The 1-mile-long *Lewis Cave* by Jim Borden and the Organ Cave Project.

Hellems Cave by Paul Stevens and the Organ Cave Project.

Foxhole #2 and *Cricket Caves* were mapped by George Dasher and the Organ Cave Project. (These caves are 2.75 and 0.8 miles long.)

Patton, Laurel Creek, and McClungs-Zenith Caves by George Dasher and the Greenbrier Grotto. (These caves are 3.6, 2.0, and 0.3 miles long. This was the third time that Patton and Laurel Creek Caves were mapped, and the second time McClung-Zenith was mapped.)

Also of note, a second entrance was dug into Grapevine Cave in the 1970s, and the cave was opened as the commercial Lost World Caverns—making it and Organ Cave the only two show caves within the Greenbrier Valley.

WVACS continued to survey the long Big Levels contact caves during this time, such as McClung, Ludington, Wades, and The Hole. Roger Baroody, Phil Lucas, and Bill Biggers connected the two entrances of Buckeye Creek Cave, and WVACS connected all ten entrances of Culverson Creek Cave. In addition, Jim Hixson began a survey of the Acme-area caves, which are located on the south side of the Greenbrier River below the mouth of Second Creek. These caves include Acme, Windy Mouth, Jewel, Frazier, Upper Frazier, and Doodle, and total about 25 miles of mapped passages.

Important paleontological discoveries were also made in Organ Cave, most of which were donated to the Carnegie Museum of Natural History. These were all of Pleistocene or Recent age, and included dire wolf, saber-tooth cat, brown bear, black bear, beautiful armadillo, porcupine, complicated tooth horse, caribou, long-nosed peccary, and American mastodon.

5.3.3 The Decades of the 1980s and 1990s

In 1984, Mike Dore dug open a cave entrance in a blind valley in northern Monroe County and broke into the upper end of a cave passage that contained juvenile mastodon bones and which led to a large stream passage that was named Mystic River. Additional exploration discovered more passages, as well as more mastodon bones and teeth, with the result that the newly named Scott Hollow Cave became, at more than 20 miles, the third-longest cave in West Virginia.

The West Virginia Speleological Survey continued to document caves in the Greenbrier Valley during these two decades, on this occasion publishing bulletins on *The Organ Cave Plateau* (Stevens 1988), *Southern Pocahontas County*

(Storrick 1992), and the *Buckeye Creek Basin* (Dasher and Balfour 1994), as well as a monograph on the caves of the Richland area of Milligan Creek (Ashbrook 1995). These last two publications were co-produced by WVASS and WVACS. In addition, in 1997, Bill Jones and the Karst Waters Institute published the 110-page *Karst Atlas of West Virginia*, which described various karst processes and documented most of the known dye traces in West Virginia, including Pocahontas, Greenbrier, and Monroe Counties (Jones 1997).

The West Virginia Cave Conservancy was also formed during this period by Bob Handley, Tim Brown, Ed Swepston, Cliff Lindsay, and others, and began to purchase caves to keep them open to the caving community. Most notable among their efforts was to gain access to Rapps Cave and a property over the physically closed Maxwelton Sink Cave. Then, using a large trackhoe, the Conservancy managed to dig into a known passage in Maxwelton, install a vertical culvert, and allow WVACS access to begin surveying this extensive cave system.

Surveys and maps completed during the 1980s and 1990s included the following:

- *Carpenters-Swago*, *Tub*, *Roadside*, and *Haynes Cave* were surveyed by Mark Johnsson and friends. (These caves together are about 5.5 miles long.)
- *Lost World*, the three *Piercys Caves*, *Sinks of Sinking Creek*, the *Pollock Caves*, *Rimstone Falls Cave*, *Reynolds Sink Cave*, *Lost Cave*, *Lost World Caverns*, the commercial section of *Organ Cave*, and the *Plastic Bag Cave System* were mapped by Bill Balfour and WVACS. (These caves total about 10 miles of surveyed passage.)
- *Rapps Cave* and *Sinks of the Run* by Randy Rumer and friends. (These two caves are 1.1 and 0.6 miles long.)
- The 2.7-mile-long *General Davis Cave* by Scott Robertson and friends.
- The 1.4-mile-long *Rehobeth Church Cave* by Bob Frostick and the Charleston Grotto.
- The 6-mile-long *Portal* by Doug Medville, Bill Storage, and friends.
- The 4-mile-long *Boarhole* by Dave Cowan and friends.
- *Handline* and *Dollar Cave* were mapped by Charlie Lucas and WVACS.
- *Stove* and *Upper Buckeye Caves* were surveyed by Greg Springer and WVACS.
- *Piddling Pit* and *Canis Majoris* were mapped by Greg Springer and friends. These two caves are together about 2.2 miles long.
- *Helix Cave* was by Aaron Bird, Mark Passerby, Greg Springer, and WVACS.
- The 1.6-mile-long *Hide-a-Bed Cave* by Mark Passerby and WVACS.
- *Poorfarm*, *Greenbrier*, *Bob Gee*, and *McFerrin Saltpeter Cave* were mapped by Tom Spina and WVACS. (These caves total about 2 miles of surveyed passage.)
- The Richlands Project on Milligan Creek was completed by Bert Ashbrook and WVACS.
- *Destitute Cave*, the third-longest cave in Monroe County, was mapped by Mike Futrell and friends. This cave is 5 miles long.
- *Steeles* and *Hunt Caves* by Joseph Caldwell and WVACS. (These caves together are about 4 miles long.)
- *Hurricane Ridge*, *Union*, and *Dickson Saltpeter Cave* were surveyed by the Monroe County Survey, who were mostly West Virginia University graduates. Hurricane Ridge and Union Caves are long water caves that drain north to Dickson Spring. Together, these caves are currently about 7 miles long.
- *Zicafoose Blowhole* and *Middle Earth* were surveyed by Mark Passerby, Bob Kirk, and friends. These two caves, which are located in Raders Valley, total about 4.6 miles of surveyed passage and provide an important link to the hydrology of the valley.
- *Buckeye Creek*, *Casteret*, *Spout*, *Fletcher*, and *Dogwood Sink Caves* were mapped by George Dasher and WVACS. (These caves total about 7 miles of surveyed passage.) This was part of two projects to completely ridge walk the Buckeye and Culverson Creek Basins and locate and document all the caves within these areas.
- Bill Jones and Charlie Williams, along with Doug Medville and Steve Worthington, continued to conduct dye tracing in the *Friars Hole Cave* area.
- George Dasher and Doug Boyer completed a 42-trace dye-tracing project within the Buckeye Creek, Spring Creek, and Culverson Creek Basins.

Projects that continued included the following:

- The survey of *Friars Hole Cave* by Doug Medville, Bob Gulden, and friends.
- Gary Storrick and the Pittsburgh Grotto restarted the survey of *Overholt Blowing Cave*, located in the Swago Creek Basin.
- Phil Lucas and WVACS continued to map *Culverson Creek Cave*.
- Bill Balfour, followed by Joseph Caldwell, and WVACS continued to map *Ludington* and *McClung* Caves.
- Charlie and Phil Lucas began a new WVACS survey of *Dry Cave*, which is a Tonoloway Limestone stream cave located in eastern Greenbrier County. This project was later also taken over by Joe Caldwell.

In addition, Randy Rumer and Tom Hay, during their survey of Rapps Cave, found two mastodon teeth. Marshall

Homes discovered the skull of a long-nose peccary in Poorfarm Pocahontas Cave, Aaron Bird and Greg Springer found a musk ox tooth and caribou tooth in Helix Cave, and Mark Botkin and Greg Springer discovered a mastodon tooth in Cass Cave. Additional mastodon teeth and bones (including a humerus) were found in Scott Hollow Cave; and a group from the D.C. and Greenbrier Grottos collected more than a dozen individuals of the Pleistocene flat-head peccary from Patton Cave. These last were donated to the Smithsonian Museum of Natural History in Washington, D.C.

5.3.4 Exploration in the Twenty-First Century

The pace of surveying continued into the new century, with the West Virginia Speleological Survey publishing three monographs, two of them just north of the area covered in this volume: *Recent Spring Creek Dye Traces* (Dasher and Boyer 2000), *The Survey of Cassell Cave* (Bob Zimmerman 2009), and *The Survey of Cass Cave* (Bob Zimmerman 2011). The first monograph was co-produced by WVASS and WVACS, and the last two included descriptions of Cass and Cassell Caves, as well as clear, high-quality quadrangle maps of two intensively complicated cave systems. Also completed during this time was the bulletin, *The Caves and Karst of West Virginia* (Dasher 2012), which contained not only a description of all of West Virginia karst areas and many of its caves, but also lists of long caves, deep caves, deep pits, saltpeter caves, karst springs, dye traces, Virginia Region accidents, and reprints of the 2000 and 2012 NSS convention geological field trips.

These decades also saw the development of two survey-reduction computer programs, *Compass* and *Walls*, both of which could not only process the immense data generated by large caves, but could also close loops, plot the station coordinates, and provide a host of other attributes—such as the ability to plot the passage dimensions, produce profile views, blend different sketches together, integrate the line plots with the sketches using modern drawing software, and incorporate the cave map and surface maps. These two programs, which were developed in Colorado and Texas, use modern software, allow data sharing between the survey-project members, and are offered to cave surveyors free of charge. They have become the favorite data-reduction software throughout the USA.

The use of laser range finders also became popular during the 2010s. These devices allowed a fast and easy measurement of the slope distance of the cave survey and provided a quick method of determining the dimensions in large, sensitive, or inaccessible passages. A compass and inclinometer were then added to the laser range finder (for a price), and the resulting instrument allowed the cave surveyors to have a single device that could quickly and easily measure slope distance, azimuth, and inclination.

Surveys and maps that were completed within the Greenbrier and New River Valleys during this period include the following:

Ludington and McClung Caves by Joseph Caldwell, which terminated a 50-year effort by WVACS to survey these caves and produce a map. These two caves, which are located only a few hundred feet from each other, are 16.4 and 9.9 miles in length.

Wolf Creek River Cave by Doug Medville, Bill Balfour, Rocky Parsons, and Ed Saugstad. This is a significant half-mile-long water cave located within the Wolf Creek Basin.

Indian Draft Cave by Doug Medville, Dick Graham, Ron Simmons, Mike and Andrea Futrell, and Bill Balfour. This is another significant water cave and is located just north of the community of Greenville. It is the longest cave in West Virginia formed in the Alderson Limestone and is 1.5 miles long and 328 ft deep.

Hunt Cave was mapped by Joe Caldwell and WVACS. This cave, which is located between Windy Mouth and Scott Hollow Cave, is 2.4 miles long and 209 ft deep.

Savannah Cave was dug open and surveyed by Bruce Fries, John Tudek, and WVACS. This stream cave, which is located north of Lewisburg, is formed in the Pickaway Limestone. As such, WVACS conducted a major dig to get into the cave and down into the better-cave-forming Patton Limestone, with the hope that the cave and its water would lead to the long-sought cave system that drains the entire western part of Big Levels. The cave, once entered, did penetrate both the Pickaway and Taggard formations at an 85-foot-deep drop, but the downstream passage was blocked by an immense breakdown collapse. The overall cave is now 0.6 miles long and 185 ft deep.

Burntwood Cave was dug open and mapped by Bruce Fries, Greg Springer, and WVACS. This cave drains a large, blind valley between the Spring Creek Cenotes and the Friars Hole Cave System. Several groups have dug at the cave over the years, but a breakthrough did not occur until Bruce Fries, Carroll Bassett, and others opened up the Worm Way—a 300-foot-long, 1-foot-high, flood-prone crawlway. The crawlway ended at a series of pits in a major dome and canyon complex, beyond which over a mile of crawlway and trunk passages. Many good leads remain, but floods refilled Worm Way with mud and debris and stopped the exploration.

Smoke But No Flaim was discovered and mapped by Jeremy Browning, Adam Byrd, and WVACS. This is a stream cave located north of Williamsburg that is formed not in the Greenbrier limestone, but rather in the Glenray Limestone, a

member of the Mauch Chunk Group, which is late Mississippian in age. The cave is 1.2 miles long and 182 ft deep.
Rainbow Run Cave was surveyed by George Dasher and WVACS. This cave, formed in the Union Limestone, has an incredible potential, as it drains to Walters Spring, one of the largest springs in Monroe County. The cave, unfortunately, ended after about 3,000 ft.
Forest Cave was dug open and surveyed by Larry Lilly and friends.

In addition, Mike Dore continued to work in the Scott Hollow Basin, WVACS, and a group of the original explorers of the Friars Hole Cave continued to work, bringing the length of that cave system up to 47 miles, and WVACS continued survey work in Dry Cave, Maxwelton Sink Cave, The Hole, and Culverson Creek Cave. WVACS also attempted several major digs—including ones in Savannah, McClung, Robbins Run, and Burntwood Caves—and they connected the Portal with the Boarhole in December of 2012, resulting in a cave system that was named the Boartal.

Lastly, a tapir tooth was found in Forest Cave by Clayton Lilly. Fred Grady purchased Haynes Cave in northern Monroe County. Grady, who was a paleontologist with the Smithsonian Institute, then discovered the bones of a Pleistocene ground sloth in Haynes. These bones, when compared to the original specimens that were supposedly collected from Organ Cave in the late Eighteenth Century, were found to be from the same individual, proving that the type specimen for Thomas Jefferson's *Megalonyx jeffersonii* had come from Haynes Cave and not Organ.

5.4 Individual Cave Exploration Histories

Exploration of the Greenbrier Caves has been a long and arduous activity. The examples given below are illustrative but incomplete. They do illustrate the time commitment and the persistence required of the cavers who have explored and surveyed the caves of the Greenbrier Valley. Some of the caves mentioned below are described in later chapters, but some are not. Some additional historical detail is provided in later chapters as part of the cave description.

5.4.1 The Boartal Cave System

The entrance to the Portal is located in an obvious sinkhole near Spring Creek and was discovered in 1982 by Ann Harman and Dave Goldman. The two cavers found a 35-foot-deep pit inside the sinkhole and—after returning with rope and vertical gear—Dave moved a rock and Ann squeezed through a tiny opening. She dug through some gravel and found a large passage.

Randy Rumer, Mike Dyas, and Roy Jameson then entered the cave and mapped over 3,000 ft from the entrance to complex of canyons and infeeders. They encountered a 15-foot-high nuisance drop and stopped at a flat-out, sump-prone, 80-foot-long belly crawl where one of the surface streams entered the cave. A subsequent trip surveyed this passage and found a wet 30-foot-high drop. The project was then put on hold, as there were too many ongoing survey projects in the area and not enough cavers to go around.

The next trip was in 1988, when Bill Storage and John Ganter found a dry, fossil passage that led north for 2,000 ft to a climbable pit with a large passage below with the rumble of a large stream. Serious surveying began, and the cave began taking shape. An entrance infeeder complex was discovered, as well as a dry fossil passage going north for "3/4" of a mile, followed by a stream passage that was another ¾-mile long. Additional surveying took place over the next 2 years. The wet 30-foot-high drop was descended and 1,600 ft surveyed and a dry, upper level found. The complicated entrance area was surveyed, its abundant leads pushed, and another 7,700 ft was added to the cave's length. Mike and Andrea Futrell discovered the Zombie Zoo, where they mapped 4,000 ft. More than 2,000 ft of cave was surveyed off Harman Creek, the cave's downstream water passage. By October 1990, with all the leads ending in mud chokes and sumps, over six miles had been surveyed in the cave (Medville 2003).

Dave Cowan purchased the land containing the Portal and then discovered a sinking stream nearby. Subsequent digging in July of 1994 opened a cave that would ultimately be named the Boarhole. The entrance led to a 15-foot-high chimney (the Sausage Stuffer), a small room, a short belly crawl, and then a double dome (the Sower Doughms). This in turn led to a 3-foot-wide, 30-foot-high canyon (the Piggy Wiggly Canyon), another room (the Boar Dello), and a huge, steep passage (the Hog Wild). Surveying began, and substantial discoveries were made over the next year. These included the phreatic Hedgehog High Way; the tight Chitlins Squeeze; the vertical Swine Flue; the huge room Boarache Down Palace; another room with anastomoses, Sowmost Heaven; a southwestern-trending passage, the Southwest Passage; and the impressive, almost 1,300-foot-long Boaring Boulevard.

Carroll Bassett then purchased the property over the Boarhole. Harry Brandle, a caver from Long Island, asked Dave Cowan to take a walkie-talkie into a high, wet area off the Hog Wash. Dave thought this effort would end as a joke, but communication with the outside was established immediately. Brian Pease next drove down from Connecticut with a sophisticated cave radio, and—once he had established radio contact with an antenna inside the cave—marked stations on the steep hillside above the cave. Unfortunately, efforts to dig at these locations encountered rocks that could just not be circumvented. Finally, a breakthrough was made into the cave—only to have the new entrance slump closed. A year later, cavers from the Tri-State Grotto dug the entrance back open and installed a 17-foot-long, 36-inches-diameter, vertical culvert.

In August of 2000, a group from WVACS entered the Boarhole to survey a lead off the New York Section. With more people than instruments, Larry Fisher and Gordon Cole checked out a high lead that appeared to be a fossil stream passage. The passage was nearly filled with sand, but contained a good deal of air. After plotting their survey, it was discovered that the high lead in the Boarhole was above Harman Creek in the Portal. Larry and Gordon continued to dig at their find once or twice a year, and then—on the day of the 2002 WVACS Christmas Party—four cavers (Larry, Yvonne Droms, and Tom and Pam Malabad) entered the Boarhole and again attacked the dig, which was now 80 ft long and 6 in. high. They moved it forward 30 or so feet, and Tom broke into larger cave. The team left the cave to attend the party, but they returned the next day, surveyed their find, and discovered footprints and a survey station on the other side. The 4.5-mile-long Boarhole had been connected to the 6-mile-long Portal, resulting in a 10.5-mile-long cave that was named The Boartal (Cowan 2003; Droms 2003).

5.4.2 Buckeye Creek–Rapps Cave System

Buckeye Creek Cave is located north of the Big Levels plateau along the southeastern edge of the Knob County. It, like Culverson Creek Cave, located just to the west, is a water cave, with many long, large passages, where Buckeye Creek flows into the main entrance, under a tributary ridge of Butler Mountain, and reappears at Spencer Cave to flow into Spring Creek.

Buckeye Creek Cave, with its obvious large entrance, has been known since the first white settlement began in the area in the late 1700s. In fact, the flat valley bottom upstream of the cave was called the Racetrack and was used for horse racing prior to the American Civil War. This "racetrack" extends a full mile to the south, is floored with thick alluvium, and typical of such a karst valley, contains no surface stream.

Buckeye Creek Cave contains a basal stream level and four higher paleostream levels that are now dry. Rapps Cave, located above the eastern part of Buckeye Creek Cave, is geologically and hydrologically related to the longer, more complex Buckeye Creek Cave. At present, the two caves have been connected via a voice connection, and—while the connection passage has not been dug open or surveyed—the Buckeye Creek Cave System and Rapps Cave are assumed to be physically one cave system.

The earliest reference to Buckeye Creek, Spencer, and Rapps Caves is in Paul H. Price and E.T. Heck's *Greenbrier County*, published by the West Virginia Geological Survey in 1939. Not only did the two authors theorize that the probable resurgences of both Buckeye Creek and Culverson Creek Caves were on Spring Creek, but Rapps Cave was one of thirteen Greenbrier County caves entered and partially explored. A partial account of their visit in October 1932 follows:

> ...the cave is about 200 yards behind and below the house, opening by a fairly large hole into the side of a steep hill. The rooms are fairly large but do not extend very far, perhaps 200 yards. The floor is very rough with fallen rocks; it was damp but no actual stream was present at time of visit. Considerable numbers of stalactites and other formations were present. No rats were found but three bats were collected, two *Pipistrellus subflavus* subflavus, and one brown, *Myotis lucifugus*. The only other animals found were a few crickets, *H. subterraneous* Scudder.

The first exploration of the three caves began in the early 1950s, when the Charleston Grotto entered the caves and theorized that Spencer Cave was the resurgence of Buckeye Creek. An early date, August 8, 1953, is found in three places in Buckeye Creek Cave, one of which is adjacent to the names of Dave Bowen and Bob Handley.

William Davies' descriptions of Rapps and Buckeye Creek Caves in his 1949 *Caverns of West Virginia* are both very short. He reported that Buckeye Creek Cave "*has not been completely explored but 4,000 ft of stream passage has been followed.*" Rapps was the most famous of the three caves, and Davies described Rapps' large entrance room and went on to narrate:

> On the east side of the room a number of formations, mainly columns, are developed. The cave continues N-30-E as a passage floored with huge slabs of breakdown covered by formations and it is with considerable difficulty that the cave is traversed for 500 feet. The passage may continue but no way through the maze of rocks could be found. A large cave entrance, directly opposite Rapps Cave on Spring Creek, may be a continuation of the passage.

In April 1957, John Glazer (or Glaser) authored an account of a trip into Rapps Cave. This included the following:

> near the farthest point reached, a tiny crevice in the east wall close to the floor gives one a restricted view of a roomy lower passage running parallel to the upper one, a flowstone cascade pouring from a broad ledge, a rather extensive secondary passage opening above this ledge, and one of the most frustrating and monotonously regular crawlways I have ever had the dubious privilege of traversing.

In 1960, two WVACS members, Jim Berry and Gordon Rutzen, were the first to launch a serious exploration in Buckeye Creek Cave. The Baltimore Grotto also worked in both caves in the early 1960s and started a map under the direction of Harvey Duchene. As a part of this effort, John Cooper and others explored both Rapps and Buckeye Creek Caves, discovered a pool of 6-foot-deep water below a 25-foot-deep drop in Rapps, surveyed the main passages of both caves, and found extensive life in the main Buckeye Creek streamway.

The Baltimore Grotto cavers also tried unsuccessfully to dive the Spencer Sump, which—at the time—contained no air space. In addition, several incidents were reported of cavers being burned by ignited methane gas as they navigated the Near-Siphon in Buckeye Creek Cave.

In 1965, William Davies published a supplement to his *Caverns of West Virginia* and, thanks to the efforts of the Baltimore Grotto, the description of both Buckeye Creek and Rapps Caves was far more extensive. Formations, formation rooms, side passages, the Near-Siphon, and the large passage beyond were described in Buckeye. A length of 4,880 ft was given for the main stream passage, and the stream passage was reported as "*terminating in a siphon.*"

Additional passages were reported in Rapps Cave, and a 25-foot-deep pit and a 40-foot-deep pit with a waterfall were also reported. Further descriptions included "*prismatic jointing in the ceiling,*" several short, lower-level passages extending from under the breakdown in the entrance room, and a domepit "*35 feet in diameter and 100 feet high with water pouring down it.*" The length of the upper Rapps Cave passage was given as 1,375 ft.

WVACS added to the Baltimore Grotto survey in the late 1960s and published the first map of the Buckeye Creek Cave. Roger Baroody, Phil Lucas, and Charlie Maus were a part of this effort, which was completed with a Brunton compass and steel tape. Their map included both the Buckeye Creek and Spencer Entrances and the three lower levels of the cave, but had very little detail and no elevations. The original survey data have been lost, but a few blue-line copies of the old WVACS map have survived to the present. Roger Baroody, during this resurvey, collected bones that were identified by the Carnegie Museum of Natural History as a partial lower jaw of a tapir and part of a skeleton of a flat-headed peccary.

While mapping the downstream portion of Buckeye Creek Cave near the Spencer Sump, Roger tried to convince Phil Lucas that Phil could not swim through the deep water and into Spencer Cave. At the time, the Spencer Sump had a small amount of airspace and little wind. Phil was unconvinced and swam off downstream in search of adventure and daylight. He soon disappeared from sight. When he returned, Roger asked if he had gotten into Spencer. Phil did not answer, but when he reached Roger and Bill Biggers (who was also on the trip), Phil pulled his hands from beneath the water and presented Roger with a bouquet of freshly picked wild flowers. Phil had made the first recorded connection between Buckeye Creek and Spencer Caves.

In 1972, Bob Handley returned to Buckeye Creek Cave with Charlie Maus and Bud Rutherford. These three cavers searched unsuccessfully for a connection between Buckeye Creek and Rapps Caves. Bob climbed Prism Canyon just below the Bowen Waterfall, was able to traverse upstream above the falls, and discovered the upper, eastern end of Prism Canyon.

Then, with the aid of a piton-secured cable ladder, Bob, his son Todd, Charlie, Bud, and Jan Rutherford were able to explore the upper end of Prism Canyon and penetrate the massive breakdown. In the process, they discovered a remnant room above the breakdown and the highest level in the cave. In addition, Bud negotiated a breakdown crawl and found the last long passage discovered in Buckeye Creek Cave. He named this passage the Attic, but it has since become known as the Too-Good-To-Last Passage.

During the early 1980s, Randy Rumer secured temporary permission to enter Rapps Cave. Over a period of 1 year, with the help of Leslie Becker, Tom Hay, John Fredericks, and John Robinson, Randy mapped 1.1 miles and produced a high-quality map that shows a great deal of passage detail.

In June of 1985, between aborted Cricket Cave survey trips, Bill Dorsey, Bill Balfour, and George Dasher initiated the third and final survey of Buckeye Creek Cave, mapping from the Buckeye Entrance to a large room located a few hundred feet downstream of the Canyon. The mapping project did not properly begin until the survey of Cricket was completed. Then, on October 19, three WVACS mapping crews, composed of nineteen cavers, entered Buckeye Creek Cave and carried the survey downstream, through the Near-Siphon, and into the central portion of the cave system.

Enthusiasm in the project immediately waned, and George had to make do with two and then one survey parties. In addition, there were horrible loop closures between the various levels in the cave, and much passage had to be resurveyed.

In July of 1986, Bob Handley and Paul and Jeremy Pittman dug a trench in the Buckeye Creek stream below the south end of McClungs Avenue, lowering the water level in the Near-Siphon by half a foot. On this same trip, the three also cut a narrow, airy ledge into several mud banks near the top of Prism Canyon. This trail, the Skyline Drive, not only offered breath-taking views into Prism Canyon (the bottom of which is located 90 ft below), but provided a safer shortcut to the passages of the cave above the Bowen Waterfall.

A month later, as the final survey of the cave was nearing completion, Bob, Todd, and Scott Handley investigated the log jam located just inside the Spencer Entrance, where

Buckeye Creek was dammed. Then, in October, the three Handleys, Bud Rutherford, Dave Seslar, and Sonja Ostrander labored in a low, wet area just inside the entrance, moving logs, boards, and sticks in a successful effort to break the dam. Later that day, after the water in the sump had receded, Sonja and Bob made the first entrance to entrance traverse of the stream passage in nearly 20 years. Several weeks later, Phil Lucas' old route through the log jam was discovered, and Bill Boehle, Sonja Ostrander, and George Dasher surveyed the connection between the two caves.

The final WVACS survey of Buckeye Creek Cave required 15 survey trips and was finished on May 9, 1987. All told, 39 persons, mostly WVACS members, surveyed 4.47 miles in and around the cave in 125 surveys and 81 person trips. However, only four people went on more than three trips into the cave, and—ultimately—George and Sonja Ostrander surveyed most of the cave. George drafted both the working and final maps, and the map won the highest award possible in the Cartographic Salon at the annual NSS Convention in Sault Saint Marie, Michigan—thus continuing the high cartographic excellence begun by Bill Douty and Bill Balfour in the previous decade and which is now being continued by younger cavers such as Greg Springer.

Work continued in the Buckeye Creek Basin after the surveys of the caves were finished. First, in November of 1992, Bob Handley and Walt Pirie in Rapps Cave established a voice connection with Liz McGowan and Robert Stepp in Buckeye Creek Cave.

Next, in the summer of 1994, George Dasher and Bill Balfour published a WVASS-WVACS bulletin on the Buckeye Creek–Rapps Cave System (Dasher and Balfour 1994). This book included not only descriptions of all the known caves within the Buckeye Basin, but also maps of most of the caves; descriptions of the history, geology, geomorphology, paleontology, and biology of the basin; and a fold-out, composite topographic map of the cave system produced by Joe Caldwell. And lastly, as the century was ending, George Dasher and Doug Boyer completed 42 successful dye traces in the Buckeye, Spring, and Culverson Creeks Basins. This too was documented in a subsequent WVASS publication (Dasher and Boyer 2000).

5.4.3 Culverson Creek Cave

Culverson Creek Cave is West Virginia's premier water cave. It is located in the Knob Country of northern Greenbrier County, west of Buckeye Creek, and is a large cave with passages that are commonly described as the size of those in Mexico. The cave contains 20 miles of surveyed passages and is the seventh-longest cave in West Virginia. It has approximately 60 square miles of drainage area above its main entrance and drains to a series of large springs on Spring Creek. The cave has eight entrances, and the main entrance, into which Culverson Creek flows, is 40 ft wide, 20 ft high, and formed in a limestone headwall 100 ft high.

Exploration of the cave began in June of 1953 when Bob Handley, Earl Thierry, and Rod Scheer found a way over the large log jam at the Culverson Creek Entrance and discovered a deep lake on the other side. To negotiate the lake, they tied two 55-gallon oil drums together that had been washed into the cave and made a makeshift raft. After crossing the lake, they followed a wide stream channel for nearly a mile until stopping at a 10-foot-high waterfall now called the Hairy Place. They entered the cave a second time in October 1953 and climbed past the Hairy Place and continued another half mile downstream. Thierry (1954) published two reports in the 1954 *NSS News* that described these two epic trips, down the huge stream passage with the roar of a large cave stream, negotiating obstacles along the way until finally, after two miles, reaching a huge log jam that blocked their path.

Two more hard-charging trips also took place in July of 1953. Tony Solar and Ray Moore entered Fuller Cave, located up Thorny Hollow about three miles east from the Culverson Entrance, and chased a small stream downstream. They followed a twisting, wonderfully carved stream canyon passage for a mile until they came to the top of a waterfall. They returned the following weekend with a rope and descended three waterfalls and then went down a cascading stream canyon for about a mile before turning around. Their exploration was a remarkable achievement, and it would have been a great surprise had they penetrated the breakdown and made their way down to the huge intersection where the main flow of Culverson Creek is encountered.

Next, in December of 1963, Mason Sproul, Jim Scrookie, and Stuart Sprague entered the SSS Entrance (named after the first letters of their last names) and explored a descending narrow, twisting canyon passage that intersected the Fuller Passage. Then, in June of 1965, 12 years after the original exploration from the Fuller entrance, Rocky Ward, Bill Mauck, Gene Rawlins, and Marty Haas reached the Lower Culverson Trunk, discovered Dream Lake (the cave's terminal sump), and travelled upstream for over a mile in a huge passage. The story of that trip is truly a spine-tingling tale, as Rocky had a dream the night before that fully described Dream Lake and because the group heard unexplained noises and rock fall during their most remote exploration.

The McLaughlin Entrance to the cave system had been shown to Bill Biggers and Mike Hamilton, who entered the cave in July of 1965, explored one mile downstream, and reached the point where the Culverson Creek stream enters the McLaughlin Trunk Passage. Then, in 1966, members of the VPI Cave Club dug open the Wild Cat Entrance connection to the McLaughlin Cave. Two years later, during a two-day effort in September, Mike Hamilton, Paul Davis, Bill Turner, Doug Yeatts, and Tom Vigour opened a connection from the Wild Cat Entrance to the Balcony Passage in Culverson. Thus, two large caves containing the large Culverson Creek Cave stream were connected.

The connection between Culverson Creek and Fuller Caves took place on May 10, 1970, by Roger Baroody and Phil Lucas during a surveying trip beyond the Hurricane Lake Passage in Culverson Creek Cave. This was a surprising connection, because the Hurricane Lake Passage starts as a small tributary infeeder and was thought unlikely to drop the 180 vertical feet to the main stream passage a mile distant.

After the connection of Culverson Creek and Fuller Caves, WVACS made an assessment of the early surveys and determined that most of these contained little detail. It was therefore obvious—because of the connections and other discoveries within the cave system—that a new map was needed. Thus, the agonizing decision was made to resurvey the entire cave system. Phil Lucas accepted the role as coordinator for this decades-long project. Bill Royster became a co-conspirator, along with Bill Balfour and Sandy Van Luik, all of who went on many long trips. Bert Ashbrook joined in on the exploration and processing of survey data in the 1990s, as the project was nearing its completion. Of course, there were many others who have braved the cold water and voluminous mud to survey this cave system.

Later connections included the Hinkle-Unus Entrance and the LL Entrance, the last of which was named for the original explorers, Cliff Lindsay and Phil Lucas. The Oh-No, No Truck Entrance was also discovered from inside the cave by Phil, Cliff, and Ed Swepston. Cliff attempted to exit this tiny entrance, but became stuck, so Phil and Ed left for the nearby Fullers Entrance to pull him out. Cliff, meanwhile, was able to work his way out of the cave, and his next action was to move (and hide) Phil's truck. Phil, when he climbed out of the steep, funnel-like sinkhole that is the Fuller Entrance, thought that his truck had been stolen, and yelled, "Oh no! No truck!" And the entrance was so named.

To state that the cave project is complete is misleading because there are many leads left in this cave system. For instance, the main cave stream, Culverson Creek, has only been followed half the distance to its resurgence at Spring Creek.

5.4.4 Destitute Cave

Destitute Cave is located on the southern part of the Organ Cave Plateau, and is almost five miles long and the second-longest cave in Monroe County. Its entrance is in the bottom of a large sinkhole and takes a considerable stream (Fig. 5.7). This reappears on the surface at Destitute (US Route 219) Spring, which is located just upstream of the US Route 219 bridge over Second Creek. The cave was first described in the 1975 *Caves of Monroe County*, and a superb map showing about 600 ft of passage was prepared by Pat Moretti of the D.C. Grotto. A second map was included Stevens' (1988) *Caves of the Organ Cave Plateau*, but this was just a redrawn version of the Moretti map.

Fig. 5.7 Entrance to Destitute Cave. Photograph by Ron Simmons

In 1990, seeing the potential for a large cave system, Andrea and Mike Futrell and Jack Kehoe began a series of digs, following strong air at the end of the known cave. They effort slowed after several trips, and—although the lead was not forgotten—their focus shifted to other caves.

Andy Murphy, a caver, purchased the property containing the cave in the summer of 1994. The dig was restarted, and Andrea, Mike, and Jack were joined at the dig by Dick Graham, Ron Simmons, Doug and Hazel Medville, and Miles Drake. The dig was soon large enough for Miles to slip past a boulder into going passage. However, just as he did so, the boulder shifted, trapping Miles in virgin cave.

The others offered to pass in the survey gear so Miles could get started on mapping the cave while they worked on the boulder, but he instead spent 5 hours chiseling the rock until he could slip back through. With the "rescue" complete, the entire group set to enlarging the squeeze.

A final dig trip was made by Mike Futrell and Ben Schwartz a few weeks later, and the cave was now open for mapping. The next month, Mike, Andrea, Miles, and Paul Gillis started the survey at the dig. By the end of the day, 4,600 ft of nice parallel upper- and lower-level trunk had been mapped. Mike and Andrea returned the following weekend and mapped from the entrance to the dig.

The cave was pushed to nearly three miles in the weeks that followed. The key was the Eveready Bunny Tube, which is a 2,000-foot-long, belly and hands-and-knees crawl heading north, with mud, water, and Rimstone squeezes (Fig. 5.8). The surveyors thought it would never end, but it finally intersected the main trunk. This passage continues to the north for about 1,400 ft to an upstream sump. Not far along the right wall is an unexplored crawl that trends toward Pond Cave.

Downstream, the main passage trends south-southwest for over 4,000 ft to a sump. The Hurricane Passage is located at the southwest end of the initial trunk passages This is a small wet, sloppy, windy, muddy crawl that trends south along the same lineament as the Eveready Bunny Tube for about 1,850 ft, but is going upstream. This lead also remains to be completed.

Most of the work was finished by the fall of 1994. A second effort was made 1 year later, and then additional trips in each October from 1996 to 1998, for a total of 14 survey trips. The cave stands at 3.76 miles long and 228 ft deep, and awaits further exploration and a final map.

5.4.5 Dry Cave

Dry Cave is located in eastern Greenbrier County on Anthony Creek, and is formed in the Silurian-age Tonoloway Limestone within the eastern limb of the Browns Mountain Anticline. Dry is famous for its long, linear water passage that follows the strike of the anticline, its vertical maze, its numerous and incredible speleothems, and its long, wet, windy trips, where cavers have to hold their heads sideways to accommodate the dip of the limestone.

Dry Cave was described by William Davies as a small cave "300 feet long, 8 feet high, and 6 feet wide." The cave contained no water and ended in a collapse, and was located above an immense spring that supplies water for the Greenbrier resort in White Sulphur Springs. Because it was the dry karst feature, as opposed to the spring, the cave was named Dry Cave. Locals said that, before the collapse, it was possible to follow a stream upstream for "a long ways."

Roger Baroody and Jack Gravemier investigated the cave in 1968 and found that they could crawl through the collapse, where they came to a room full of shattered rock. They found a window in the rock, and followed a tight belly crawl for about 20 ft, where they encountered a walking passage with many speleothems. Following this passage led the two cavers to a mud bank that sloped down to the cave stream. They then exited the cave, as they had other plans for the day.

Fig. 5.8 Eveready Bunny Tube, Destitute Cave. Photograph by Ron Simmons

Roger returned to the cave on a solo trip, and made his way upstream in a nice walking passage for about 1,000 ft. He turned around at a water crawl, but told Jack of his discovery. Phil Lucas joined Jack for a later trip, and the two pushed through the water crawl and discovered that a major survey project was required. This started in May of 1971 when Phil, Charlie Williams, and Ray Povirk mapped 1,268 ft in the cave. A 50-pound rock fell from the ceiling in the entrance crawl during this trip and onto Charlie's back. He could not dislodge the rock, but eventually slithered far enough along the passage that his two companions could remove it. The crawlway was then named "William's Wiggle."

More WVACS survey trips took place over the next 2 years, with Phil as the project leader. The upstream and upper levels of the cave were explored, and 2.5 miles of cave was mapped. There were then no more survey trips into the cave until the 1980s, when WVACS again began to map the cave. This project was first led by Charlie Lucas (Phil's son), and then by Joseph Caldwell. This too was a multi-year effort, and more of the upper levels were mapped and unsuccessful efforts were made to penetrate the terminal collapse at the upstream end of the cave. In addition, ridge walking on the surface failed to locate the cave's water source.

The survey project again stalled, although some passage was mapped by Gordon Cole, Larry Fisher, and Ed Saugstad in 2008. The prospect of reactivating the survey was discussed at WVACS membership meetings throughout 2010 and Greg Springer volunteered to lead the effort. Amid these discussions, Bill Balfour acted as an intermediary between Greg and Phil Lucas, the cave's longtime explorer and cartographer. Phil very graciously provided Greg with the Dry Cave data files, survey notes, and Corel Draw map files. Most importantly, Phil helped Greg introduce himself to Johnny Lynch, the land owner, whose permission was needed to enter the cave.

The surveys of the late 1960s and 1980s left many upper-level leads throughout the cave. In fact, the 1980s crew estimated they had surveyed well less than half of the cave. With so many leads, Greg and his collaborators decided to begin by tackling major leads in order of their distance from the entrance. The first of these leads was in the 1960s-era D survey, which is about 1.5 hours from the cave's entrance. After surveying this, the project was reactivated in earnest on New Year's Day 2011 when Gordon Cole, Larry Fisher, and Greg Springer surveyed 505 ft of breakdown trunk and assorted crawls in the MD Area. Two more trips were taken to that area in 2011, with each netting over 500 ft of new survey, but no other trips took place that year.

The pace quickened somewhat in 2012 after the MD Area was completed. Although complex and poorly decorated by Dry Cave standards, the area yielded 2,200 ft of survey and the cave's unique character intrigued everyone involved with the first four trips.

The next major lead was atop the aptly named Crappy Layer Climb about 800 ft upstream of the MD area. The area surprised its surveyors by containing some nice, easy passages heavily decorated with aragonite bushes (Fig. 5.9). Over the course of multiple trips, a total of 15 WVACS members helped survey the Aragonite Attic, and this helped increase attendance at WVACS project weekends (which was one goal of the renewed project). All told, the Aragonite Attic yielded 2,100 ft of new survey and set the stage for a very productive 2013.

Completion of the Aragonite Attic was followed by a systematic search for leads between the Crappy Layer Climb and the Blowhole. None were found, so Nick Socky, Greg Springer, and Rebecca Stewart checked a lead upstream of the Blowhole near Station 5077. This area is a little less than 3 h from the cave's entrance and the trio was surprised to find large walking passages heading both upstream and downstream. The Strolling Passage yielded nearly 1,000 ft on that first trip and 2,800 ft would be surveyed in the area in 2013.

The final survey of 2013 took place in September and culminated in the surveyors climbing upward into a tall, rift-like passage moving large quantities of air. This tantalizing lead had to be set aside until the next year because the landowner had asked that the survey pause during hunting season, but two teams returned in March and were rewarded with 1,000 ft of heavily decorated and mostly walking passages. Many leads were generated and another 1,800 ft would be added that year. The year 2014 also saw leads around Chris' Cantilever tackled and once again many leads were generated. Over one mile of survey was added in 2014 alone.

The area beyond Chris' Cantilever continued to yield good footage in 2015 (3,600 ft), but attempts to find a way beyond The Bitter End failed to turn up leads. Thus it became clear in 2015 that the rest of the survey would consist of mopping up upper levels and no major extension of the cave is likely. Nonetheless, 4,200 ft were surveyed in 2015 and there appears to be at least another year of work remaining. The cave is currently 6.24 miles long, and is documented by a beautiful map drawn by Greg.

5.4.6 Friars Hole Cave System

The Friars Hole Cave System is currently about 47 miles long, and straddles the Greenbrier and Pocahontas County line, with five entrances in each county. The Snedegars (Saltpeter) Entrance of what became the cave system has been known since pioneer days, and the large passage inside that entrance was mined for saltpeter by the Confederacy during the American Civil War.

Fig. 5.9 Spray of aragonite crystals, Dry Cave. Photograph by Dave Socky

Active exploration of the cave system began with the descent of the 99-foot-deep "Crookshank Hole" by Huntley Ingalls in 1951 and the discovery of a 500-foot-long crawlway that required some digging. A subsequent trip by Huntley and George Moore on New Year's Day 1956 found that the crawlway had washed open and was now a stoopway. The party explored all the way to the Snedegars-Crookshank Sump, and George estimated the length of the passages they had explored and suggested that the connection to Snedegars Cave should be sought.

Lew Bicking took an interest in the cave when he visited Crookshank in November of 1962 with George Titcomb, John Cooper, and John Holsinger. John and Lew descended the pit and found the cave flooded, and Lew was only able to prusik out while John held the rope away from the waterfall. Other cavers had to be summoned from a Lewisburg restaurant to haul John out. Lew returned to the cave a year later and began a serious exploration of the system. In December of 1963, he entered the Snedegars Entrance and pushed through the sump, but found the passage leading to Crookshank blocked by fill. He discovered the Snedegars Staircase Entrance the following day, and the first connection between the various entrances (that would become the Friars Hole Cave System) occurred in May 1964 from the Staircase Entrance to the Saltpeter Entrance. A second connection occurred the following month between the Saltpeter and Crookshank Entrances.

In August of that year, George Titcomb and Charlie Schwab dug open the entrance to the Friars Hole Cave. A month later, Lew entered this cave and found two short drops and a beautiful trunk passage. Lew made seven more trips into the area in 1964, followed by an additional ten trips in 1965. He had, on these trips, a variety of companions, as few people could endure more than one trip. He also loaned his friends dry suits, as much of the cave was wet. On one occasion, Lew, Eddie Baurer, Robert Whittemore, and Dixon Hoyle were trapped behind the Snedegars sump for three days, an event that had the cavers on the surface scrambling to purchase all the garden hoses in Marlinton to drain the sump.

Lew came very close on several occasions on making what later turned out to be critical connections, and there was a lot of enthusiasm to continue to explore the Friars Hole Cave after Lew's untimely 1966 death in a motorcycle accident in Washington D.C. However, there were not many volunteers to do it. At that time, 4.5 miles had been surveyed in the area's caves.

Another close look at the area began in the mid-1970s, when the Met Grotto began mapping in Crookshank and WVACS (Charlie Williams and Bill Balfour) started surveying in Friars Hole Cave. A sinking stream in the upper Friars Hole Valley began to wash open a cave in March of 1976, and Barry Baumgardner, Gordon Mothes, and Doug Medville began digging at this location. They uncovered a cave entrance and found a small, yellow rubber chicken—for which the entrance was named. Subsequent caving parties followed a narrow tube downstream to a 12-foot-high waterfall drop, a 72-foot-high drop, and two other 12-foot-high drops. These led to a low, phreatic tube that, after 300 ft, opened into a high canyon, a stoopway, and then an overhung climbdown. Beyond this was the main cave stream, which was followed upstream for about 1,000 ft. Then, in April of 1976, ten cavers surveyed 9,600 ft of passage, and, by May, had mapped 4.5 miles in the cave. The cave data, as quickly as it was produced, were key-punched into Bob Thrun's data-reduction program and was used to provide plots of the system that were overlaid on aerial photographs and topographic maps.

Two trips then connected Rubber Chicken to Friars Hole Cave. The first, occurred prior to the first major survey trip into Rubber Chicken, when Canadian cavers from McMaster University unknowingly found a connection between the two caves. The second was in May, when Ed Strausser, Chuck Hempel, and Charles and Barbara Williams mapped into a narrow, southwest-trending passage in upstream Rubber Chicken, crossed a chest-deep pool of water, and began to go down another passage. They then descended a 10-foot drop and discovered one of Lew Bickings' old Friars Hole survey stations.

Landowner relations then took a turn for the worse, as the owner of the Rubber Chicken Entrance was not happy with the number of cavers visiting his property. Because of this, attention now turned to Crookshank-Snedegars Cave. A party entered Crookshank in late May, and Mike Dyas pushed a low crawl south to a series of crawls and stoopways. Once surveyed, it was discovered that this passage was only 35 ft from The Highway in Rubber Chicken. Parties were sent into both Snedegars and Rubber Chicken with a plan to try to communicate. The Snedegars crew (Doug Medville and friends) arrived first and—while talking—heard a voice answering on the other side of some breakdown. The two groups moved some rocks and could soon see each other's lights. After some additional rock removal, the Rubber Chicken party (Tom Hay, Tom Curtis, and Ralph Kennedy) squeezed through the newly made connection. The cave now had five entrances and was 12.5 miles long.

Rubber Chicken was next connected to Canadian Hole (Ro's Pit), which was then a short, multi-pitch cave located north of the Snedegar Entrance. This connection occurred in the summer of 1976 in a trip to the end of the Rubber Chicken Highway by Bru Randall, Walter Plunkett, and Ed Strausser. They crossed from downstream Rubber Chicken and into Canadian Hole, but didn't realize what they had done. Then, on the 1977 Labor Day Weekend, several Canadians surveyed at the south end of Canadian Hole and, after negotiating several wet crawls, found footprints (from the 1976 trip) in what turned out to be the upper end of the Rubber Chicken Highway.

The final connection of the caves that would ultimately form the Friars Hole Cave System was made during the Veterans Day weekend in 1978, when Bob Anderson, Linda Baker, and Patty Mothes pushed the low Toothpick Cave stream downstream into large passage and surveyed 2,500 ft in 70 stations. They then explored for several more thousand feet, and found footprints at the far upper end of Rubber Chicken Cave. They next went downstream in Rubber Chicken, doing a lot of route finding since none of them had been to that part of the cave before, and finally exited the Snedegars Entrance after a 16-h trip.

5.4.7 Greenville Area Caves

The extensive plateau formed by the Greenbrier carbonates within the Greenbrier Valley also extends south in Monroe County and into the upper New River Valley. Of the caves located here, the largest and best known are Greenville Saltpeter and Laurel Creek Caves, which are located in the Greenville area in an outlier of the Union Limestone.

Each cave contains four entrances, several levels, and a number of large rooms and passages. The area's surface stream, Laurel Creek, first flows through Laurel Creek Cave, then Rimstone-Crossroads Cave, and finally through Greenville Saltpeter Cave. It exits near the Water Entrance of that cave and empties into Indian Creek. The three caves are together over eight miles long and, if they were not broken by surface erosion, would be a single, large cave system. Greenville is now closed because of endangered bats, and Laurel Creek was closed after the rescue of three cavers who were trapped by high water in 1982.

Greenville Saltpeter Cave was mined extensively for saltpeter prior to the American Civil War and by the Confederacy during the war, and William Davies noted in 1958 that the cave contained many "cart ruts, burro tracks, mattock marks, and other relics of saltpeter operations…" The cave was visited in August 1787 by the Reverend John Smith, who described the trip thus: "We took lights in our hands and traviled thro' many delightful apartments Some Places Apparently 30 or 40 ft high and as many yards broad….".

Reverend Smith also noted that, after leaving the cave, "A mile from where we entered in ransacking this Dark Vault I saw the manner of making a Salt Peter which is made here in great quantities." Greenville was then known as "Singing Cave," as Sunday school classes and other groups would enter the cave to sing and listen to the sound of their voices being amplified by the cave walls. In addition, "Singing Cave" was one of three natural wonders (in what is now West Virginia) described in print by Thomas Jefferson in the

1790s, and the cave was also the most popular of the caves visited during the 1948 NSS Convention in Charleston (which was the first convention to offer led caving trips).

Greenville Saltpeter Cave was first mapped by the Charleston Grotto in the late 1940s and a map produced for Davies (1958) *Caverns of West Virginia*. The cave was also extensively described for that publication. Laurel Creek Cave was mapped by William Davies and a map and description produced. Crossroads Cave was described by Davies, but Rimstone was not documented. Both Greenville Saltpeter and Rimstone-Crossroads Caves were then mapped in the early 1970s by the Pittsburgh Grotto and high-quality maps produced. These were included in the 1975 WVASS publication *Caves of Monroe County*. Also included in that publication was a map of Laurel Creek Cave drawn by Roger Baroody. This map showed the connection, for the first time, between the main Laurel Creek Entrance and Saddle Cave.

Laurel Creek was mapped for the third and last time in the late 1970s by George Dasher and the Greenbrier Grotto. This resurvey pushed the cave very close to the 4.2-mile length of Greenville Saltpeter Cave, completely explored the lake area in the back of the cave, and added the Roles and Hilltop Entrances to the cave system. The resulting map was produced just in time for the 1982 rescue, when three Maryland cavers entered the main entrance of Laurel Creek Cave on a Saturday morning in June and were trapped by high water.

The rescuers found, when they arrived on Sunday morning, that the main entrance to Laurel Creek Cave was completely underwater, and that Laurel Creek, upstream of the entrance, was now a lake 100 yards wide and about a half mile long. Fortunately, the rain had stopped and the sun had come out, and the Hilltop Entrance, a 1- by 2-foot fissure crawlway over broken glass and trash, was not flooded and could be used. However, once the rescuers reached the underground stream, they discovered there was nothing they could do but wait, as the downstream passage was flooded to the ceiling. They exited the cave, and the entirety of the rescue personnel—state troopers, sheriffs, conservation officers, and the cavers—were invited to and attended a dinner in Greenville for the 50th wedding anniversary of a local husband and wife.

Late on Sunday afternoon, the water level finally deceased to a level when it was thought another rescue attempt might be successful. A large number of cavers entered the Hilltop Entrance with ropes, floatation equipment, life jackets, and scuba fins. Shouting brought a response downstream, so it was known that the trapped cavers were alive and a way was open to reach them. Jim Hixson, Bill Balfour, and Joe Musser, the only cavers with wetsuits, were lined more than 1,000 ft downstream to the large Theater Room, where it was discovered that the trapped cavers were waiting. Bill Balfour returned upstream to report on the three trapped caver's condition and that they could swim out. Then he, Don Walton, and George Dasher each took an extra lifejacket and swam downstream to the Theater Room. After that, it was a simple matter to swim everyone back to the dry part of the cave, retrieve all the equipment, and exit through the Hilltop Entrance.

5.4.8 Haynes Cave

Haynes Cave is located on the plateau just south of Second Creek. The cave has a long history of human use, as it was mined for saltpeter both before and during the American Civil War. Haynes is famous for its extensive saltpeter works that still exist and which were constructed by the Confederate saltpeter miners.

The cave was first mapped in 1946 by William Davies and a quality map produced of the main passages. Davies also included an extensive description and three pictures in *Caverns of West Virginia*. Chuck Hempel included a much shorter description of the cave in his 1975 *Caves of Monroe County*, and he reprinted Davies' map. Mark Johnsson next surveyed the cave in a 6-year effort, beginning in 1977, and produced an extremely high-quality map that showed all of the cave passages, as well an incredible amount of passage detail. Mark found the length and depth of the cave to be 4,237 and 147 ft.

Haynes Cave is famous as the site where the bones of the Pleistocene sloth *Megalonyx Jeffersonii* were discovered in 1796 while Monroe County was still a part of Greenbrier County. The bones were sent to Thomas Jefferson by the father of Greenbrier County, John Stuart, and Jefferson then gave a presentation on the fossils the following year to the American Philosophical Society in what is generally agreed to be the beginning of vertebrate paleontology in North America (Fig. 5.10). This discovery was long attributed to have occurred in Organ Cave; however, Fred Grady, a paleontologist with the Smithsonian Institute, found bones from the same individual in Haynes Cave in 1990s, and thus proved that the original sloth remains came from Haynes, and not Organ.

5.4.9 Maxwelton Sink Cave

Maxwelton Sink Cave is one of the Greenbrier Valley's premier contract caves. The cave was opened in the mid-1960s, and then explored and surveyed by the Pittsburgh, Boston, and Philly Grottos. Entry was via the main Cove Creek Entrance, which is located in an immense blind valley east of the community of Maxwelton, and which gave access to the extreme northeast end of the cave. This entrance was notorious for flooding and silting shut, but ten

Fig. 5.10 Jefferson ground sloth. Photograph by the author

miles of cave was surveyed (with Chuck Hempel coordinating the effort) before the entrance was totally silted shut by Hurricane Camille in 1969. The survey efforts did not go to waste; however, because a good-quality map (for the times) of Maxwelton Sink Cave was produced in 1971.

Thirty years later, the West Virginia Cave Conservancy attempted to purchase the cave and the closed Cove Creek Entrance. The attempt failed, so several Lewisburg-area cavers (the Monroe County Cave Survey and WVACS) considered digging into the cave. Dave Scott owned land that was over the cave to the south of the Cove Creek Entrance, and several digs were attempted between the years 2000 and 2002. All of these failed.

In early 2002, Dave Scott, Jeff Bray, Tom Malabad, Bill Liebman, and Ed Swepston conducted three microgravity traverses in order to find the location on the surface that was over a section of the cave called Heaven. These surveys found anomalies at the expected location of the cave, so a drill was used to sink an exploratory well. A camera was lowered into the hole and confirmed that there was cave passage at 105 ft below the surface.

Another microgravity traverse was performed lower down in a sinkhole, and confirmed the location of the upstream end of the Heaven Passage. In August of 2002, a D-8 bulldozer and a large excavator were used to dig in the sinkhole. Eventually, the dig was 40 ft deep and 30 ft in diameter and had exposed blowholes indicating that cave passage was close at hand. The dig had to continue by hand since it was now too deep for heavy equipment. By late September, cavers broke into the Heaven section. But the work wasn't done—the new entrance had to be enlarged, stabilized, and a 40-foot-long, vertical culvert installed. A lid and lock was designed and put on top of the culvert and the raw dig was filled in and the area landscaped (Fig. 5.11). In addition, Dave Scott donated his property to the West Virginia Cave Conservancy.

Because of all of this work and other ongoing caving projects, the new survey of Maxwelton Sink Cave didn't

Fig. 5.11 Entrance pipe to Maxwellton Sink Cave. Photograph by David Socky

start until February of 2004. But when it did start, it attracted a lot of cavers. From February to December, a total of 23 cavers participated in 23 separate survey trips and mapped a total of three miles of passage. The project, of course, started at the end of the Heaven passage and included most of the passages up to the main Cove Creek trunk. Once the two main parallel trunks were reached, the survey turned northeast east, toward the Airport Room and the sealed shut Cove Creek Entrance.

The next year, 2005, saw even more activity. There were a total of 27 survey trips with another 3.8 miles of cave surveyed. The project also picked up a few more cavers after word got out that there was large cave to be explored and surveyed—and a total of 32 cavers helped. Most of the northeast section of the cave was now completed, along with the beginning of the "H Canyon" section to the east, where the sealed Airport Entrance is located. The survey downstream in Cove Creek in the Hall of Kings was also started, and—by the end of 2005—a total of 6.8 miles had been mapped in the cave.

As time progressed, the survey was getting deeper into the cave and the trips were becoming longer and more difficult and more complicated. In addition, the glamour of a new cave to survey began wearing off and even worse, White-Nose Syndrome (WNS) had become a real threat to caving in West Virginia. Because of all this, the pace of the survey slowed down for 2006. Only 15 cavers actively participated in the 14 survey trips for this year, and only a total of 1.6 miles was mapped. The work was now taking place in the downstream Cove Creek, as well as in the upper-level maze section beyond the Hall of the Kings Waterfall, where a lot more cave was discovered than was shown on the old map. A start was made on Spaghetti Junction, which is an area where multiple passages converged on the same spot, but are on many different levels. The lowest level, or course, was Cove Creek. In addition, the parallel passage that led to Thunder Hall was surveyed.

WNS had a more direct impact on activities in West Virginia in 2007. Surveys that year were down to only 8 trips with 11 cavers. In spite of this, 1.4 miles of cave were surveyed. Most of the work for 2007 was in the middle section of the cave, including a start on the downstream Cove Creek passage and some of the complex to the south of Thunder Hall.

Only five trips occurred in 2008, and only two in 2009, for a total of only 4,900 ft surveyed. The work continued in the downstream Cove Creek passage, plus a couple of surveys took place to clean up the complex beyond the "H Canyon" in the far eastern part of the cave. Only nine cavers took part in surveys for the years 2008 and 2009.

By the end of 2009, most of Maxwelton Sink Cave had been surveyed with a little over 10.5 miles logged into *Compass*. There were just a few major sections that needed to be completed, mainly in the Airport Entrance Passage past the "H Canyon" and in the downstream Cove Creek passage, which was known to end in a sump. Interest in the survey had really died down, plus WNS was keeping a lot of cavers home. For these reasons, there were no survey trips at all for both 2010 and 2011, and there were only three trips in 2012, with only 955 ft surveyed. But the maze complex to the east of "H Canyon" was finished, which left only the narrow, twisting Airport Entrance Passage to be completed for the eastern section to be totally done. Also, the very wet and cold section of Cove Creek below Spaghetti Junction was finished in December of 2012—a trip that required full wetsuits.

The year 2013 saw a resurgence in work on the survey of Maxwelton Sink Cave. This was partly due to the fact that WNS was now prevalent throughout all of Virginia and West Virginia and cavers were now no longer concerned about spreading it, and were thus becoming active again. In addition, Greg Springer, who had agreed to do the massive amount of cartography, found that the original notes for a number of surveys were lost, which required some resurvey work. Dave Socky spearheaded work to get these sections remapped, and 2013 saw a total of 21 cavers participating on 16 survey trips with around 4,200 ft surveyed. Five of these trips were resurveys, allowing Greg to continue working on the map. One trip started the survey of the Airport Entrance Passage, which turned out to be fairly tight—the original survey was stopped because not all the cavers could fit down the passage. All the other trips for 2013 were cleanup surveys and checking leads for the map.

Downstream Cove Creek, called Gasoline Alley, was finally completed in 2014 with one survey trip that netted 1,100 ft. The extreme, southwest, end of the cave did in fact end with a sump, in a room that was very muddy and a deep, dark pool of water. Besides completing downstream Cove Creek, 2014 was a year of more resurvey. The notes for a large section around the Airplane Room in the northeast end of the cave had been lost, and so, for the first few months of the year, cavers flooded the cave with six separate survey trips to recover a sketch for this critical part of the cave. All told, a total of 13 cavers participated on seven survey trips to map 3,700 ft of cave. This year also saw an influx of new cavers, a lot of them from the VPI Cave Club in Blacksburg, Virginia.

For the year 2015, 27 cavers helped with 13 separate survey trips that netted a total of 4,300 ft. The numbers, of course, indicate more and more cleanup of the survey, although this year did see the completion of the Airport Entrance passage—which was finished by three skinny cavers! The large side lead off the north side of the Cascade Causeway (downstream Cove Creek) was also completed. So, with a length of 12.3 miles, it was thought that all the major parts of the cave had been surveyed and that there was nothing left but mop-up and lead checking, plus some rework as demanded by the cartographer. But lead checking and resurvey can, and does, result in finding more cave—some of it virgin.

In one case, Greg Springer requested a resurvey of Thunder Hall, because the original survey did not include enough detail. After the resurvey of Thunder Hall was completed, some other leads in the area were checked, resulting in close to 500 ft of nicely decorated virgin passage being discovered. It was decided, after seeing Thunder Hall, that the dome at the end of this passage should be climbed. The bolt climb was started in November of 2015, and possible going passage was found at the top.

There were 11 survey trips into the cave in 2016. Many trips were cleanup, but two were bolting and surveying trips in Thunder Dome. The top of the dome was reached in April and 370 ft of large, virgin passage was surveyed. Later trips added over 1000 ft of new survey, but the two remaining leads do not appear promising.

Some of the surveys conducted after the new entrance was dug open suffered from poor sketches, so 2017 began with two teams re-resurveying the LOVE, LOV, LVS, and LOVEU surveys. Dave Socky, Larry Fisher, and Gordy Cole resurveyed the 2000s-era LOVEU and LVS Surveys, while Greg Springer, Errol Glidden, and Tom Feeney resurveyed in the old LOVE and LOV Surveys. Together, the two teams mapped almost 2,900 in over 100 stations, and substantially eased Greg's job as cartographer.

However, a major breakthrough occurred in the following month when Chris Coates, Nikki Fox, Carl Amundson, and Terry McClanathan discovered what is now known as Sweetwater River. The river is reached by the Pie Passage, which was dug open by Nikki and Chris in 2016. Initially, the Pie Passage yielded 3000 ft of crawling to walking-height passages, but on 11 February the four surveyors crawled through deep water to find river Borehole extending to the NE and SW. Subsequent trips yielded two miles of large stream passage and the cave continues with good air in both the upstream and downstream directions. The size of the stream and its possible hydrological connections to many of the contact caves to the north and south gives the discovery enormous potential and it may be the most significant discovery in the Greenbrier Valley since the initial exploration of the Friars Hole and Scott Hollow Cave Systems.

With this, it is obvious that the survey and exploration of Maxwelton Sink Cave will never be complete, and will continue even after Greg Springer's excellent map is finished. All told, over the past 13 years, more than 100 cavers have assisted in the WVACS survey of Maxwelton Sink Cave, which has a steadily growing surveyed length of 15.27 miles.

5.4.10 Overholt Blowing Cave

The entrance of Overholt Blowing Cave is a spring with an immense discharge located next to the Swago Creek Road. The cave has been known since the American Civil War, was mentioned in Paul Price's 1929 publication on Pocahontas County, was described by William Davies as having several miles of passage, and its early exploration was documented in great detail in William R. Halliday's *Depths of the Earth*.

In 1956, three members of the Pittsburgh Grotto—Ralph Doerzbacher, Rita Battistoli, and Will White—pushed a windy crawl named Davies Didn't Crawl (where William Davies had turned around years before), waded through the chest-deep Lake Rita, and began what was 4 years of extensive exploration (Fig. 5.12). A low-water crawl was discovered a month later, which was named the Dardanelles. Subsequent discoveries led to the Tom-Tom passage, Neptune's Avenue, the Roux Room, Anne's Avenue, the Gypsum Trail, Alcoa Avenue, the Crystal Pool Pit, and The Attic, giving the cave a length of about one mile. Unfortunately, every attempt to circumvent the Dardanelles and make the cave trips less wet and miserable was not successful.

The exploration of Overholt Blowing has always had one overriding goal: to reach the upstream end of the cave. Trips soon required 19 h to reach this part of the cave, where the exploration teams were stopped at the Fourth Waterfall. Finally, on the New Year's Weekend 1959, a full-scale expedition camped for three days in the cave, climbed the Fourth Waterfall, and immediately encountered the Fifth Waterfall. The group was then trapped by high water at the Dardanelles. Baltimore Grotto members attempted a rescue using an improvised diving rig assembled from 225 ft of garden hose and bicycle air pumps, but once underground, encountered the trapped cavers who had swum through the receding flood waters.

Supplies were carried into the cave throughout the summer of 1960, and the Pittsburgh Grotto then spent four days camping in the cave over the Thanksgiving Weekend. They extended the survey to the base of the Fourth Waterfall, giving the cave a length of almost three miles, and used a scaling pole to climb both the Fourth and Fifth Waterfalls. They found 650 ft of walking passage that led to a low room, with the Doubtful Crawl being the only way on. Bob Dunn and Jerry Frederick entered this passage, and after an hour of difficult crawling, reached the Rat Room, where they found a cave rat's nest, a chunk of asphalt, and a woman's garter. Following this big push, the work in the cave came to a stop, and it would be 18 years before anyone returned.

A Canadian effort to survey the cave began in the 1970s, but left few detailed records of their work. According to William Halliday in his book *Depths of the Earth*, the Canadian interest in Overholt began when George Tracey mapped 1,500 ft on a 29-h-long trip in the Penn State Section. Cavers from the McMaster University Caving and Climbing Club soon became involved, and six people entered the cave intending to camp for four days and climb the First Waterfall. Their scaling pole was unfortunately too short, and—after hours of work—they discovered they were at the wrong waterfall.

Fig. 5.12 Water crawl leading new discoveries in Overholt Blowing Cave. Photograph by Victor Schmidt

Cavers from the Societe Quebecoise de Speleologie also became interested in the cave, and Paul-André Auclair, Daniel Caron, and Jean Lamarre found the Doubtful Crawl and surveyed to the Rat Room in 1978. The Canadians continued to make trips into the early 1980s, concentrating on the Sixth Waterfall. This was climbed, but the upstream passage led to a succession of domes where no way on was evident. Green leaves were found in this area, so the cavers knew that they were near the surface, but their work ended without finding a back entrance to the cave. In all, the Canadians mapped over 3.7 miles, including passages beyond the Pittsburgh Grotto surveys, but their map was never published and the notes have been lost.

In July of 1992, one of the Canadian cavers, Mike Palethorpe, took Gary Storrick, Barb Schomer, Bru Randall, and Brian Preaux on a mapping trip to the Disappointment Dome. This proved to be the first in a long series of survey trips, and other Pittsburgh Grotto members soon joined the effort. Surveying continued in 1993 with trips out Anne's Avenue to the Crystal Pool Pit. On one trip, Brian Preaux dislocated his shoulder. This was a common, repetitive event for Brian, so Gary reset it. Barb, Brian, and Gary returned on a later trip to work in the Crystal Pool Pit area. They were able to reach the bottom of the pit, and they surveyed into Willy's Pinchout, where Brian again dislocated his shoulder in a low stream crawl. This time Barb reset the shoulder, and their survey continued.

As the year progressed, Barb, Bru, Walt Hamm, and others returned to map Anne's Avenue and the parallel Gypsum trail. They also surveyed the breakdown north of Anne's Avenue. Doug and Hazel Medville joined the team to help finish the Neptune's Avenue survey and to map the low crawl at the south end of the Roux Room. Later, on New Year's Day, Brian and Gary tied the Anne's Avenue and the Gypsum Trail survey to Neptune's Avenue by mapping the Roux Room. Dwight Livingston helped Brian and Gary start the Cathedral Passage survey. On another trip, Angus McDonald, Bru, and Gary hauled Ian Drummond's cave radio into the cave to obtain surface fixes at the Roux Room, Mountain Room, Empire Room, and Alcoa Avenue junction. These fixes showed that there were no significant cumulative compass errors in the survey.

There were only a few trips in 1994, but these proved quite productive. Brian, Gary, and Chris Werner mapped the Cathedral Passage, and discovered a large, parallel virgin trunk passage running back from the start of Alcoa Avenue. They did not, however, find a lost waterfall discovered long ago by Roswell Jones. The first survey trip in 1995 was in April, when Walt and Bru mapped across the Dam Pit Traverse, while Barb and Gary crossed the pit and proceeded out the Coral Canyon. They mapped past several junctions and discovered several domes and a short climb down that led to stream crawls in two directions. Barb and Gary surveyed both downstream and upstream in passages that became too unpleasant to continue. On the way, Gary discovered a climb to an upper-level canyon that ended at the brink of a short drop into a huge room, which he thought might be the Good Friday Room.

Barb, Bru, Perk, Walt, Gary, and John Murray returned in October to find out. While Barb, Bru, and Walt tied up some loose ends in Anne's Avenue and the Gypsum Trail, John, Ray "Perk" Hilco, and Gary rigged a rappel into what they thought was the Good Friday Room. They then surveyed the room, but left three leads unchecked. When the two teams met, it was discovered that Barb's team found a large room with a new rope hanging from an upper lead. This sounded distressingly familiar, and sure enough, Barb's team had entered the Good Friday Room through one of the unchecked leads.

Chris and Gary also visited the Cathedral Room that year, taking a steel-rung etrier to Jones' Squeeze. Gary wrote afterward:

> Sometimes human inertia amazes me. Forty years ago, Pittsburgh Cavers first got to this spot by free-climbing the slot. For the next forty years, a fixed ladder or line hung in the slot, and people struggled to fit through the squeeze while dragging packs, etc. The slot rigging was the worst I've ever seen: one bolt in the tightest part of the slot, where it prevented moving ascenders up, tied off with not-a-knot, backed up with another bolt whose entire side was exposed where the rock had broken, two bolt hangers I could bend with my fingers, and a carabiner that was so corroded that neither Chris nor I could tell which side was the gate. This thing was dangerous to the max, and I was glad that someone expendable (like Chris) had climbed it instead of me. Yet here I was, on a trivial 20-foot against-the-wall climb that ended on a 2-foot ledge—and it was shown on the map thirty-five years ago.

1996 was a very wet year in West Virginia, and it wasn't until Memorial Day weekend that cavers were able to enter the cave. The Dardanelles were impassable, so they mapped in the Penn State Section. They returned to the area in July, and while Gary sketched, Chris checked out a small side lead that trended west—instead of the usual north-south orientation of the cave. This was a major discovery, and Perk, Gary, and Chris surveyed upstream on a subsequent trip, and ultimately found a downstream station that Barb and Gary had set over a year before. The three cavers had done it—bypassed the Dardanelles, closed the largest loop in the cave, and explained how the water from Schoolberry Cave could get into Overholt without flowing through the dry Penn State Section. The cave was, at that time, 7.8 miles long and 659 ft deep, and was, at the time, the second-deepest cave in West Virginia.

And then, just when everything was looking good, cavers from Connecticut upset the owners by trying to bluff their way into the cave. That incident, plus concerns about liability, caused the owners to close the cave to everyone. This put an end to the project, and underscored how thoughtless acts can ruin otherwise good landowner relations. Because of his disillusionment, Gary did not start drawing the map until 2009. After 700 h of work that year and another 400 h in 2015, he finished a working draft and entered it into the 2015 NSS Convention Map Salon, where it won the Caver's Choice award and a Merit Ribbon.

5.4.11 Patton Cave

Patton Cave is a two-entrance cave located in eastern Monroe County. The cave is formed, not in the Greenbrier limestones, but rather in the Ordovician Big Valley Formation (of the Black River Group) at the eastern base of Peters Mountain. The cave receives its water from Chambers Cave, located to the south, and a spring near its entrance forms the headwaters of Second Creek.

Although the main entrance of Patton Cave is a deep sinkhole that has long been known to locals, there is no documented history of the cave prior to the publication of William Davies' *Caverns of West Virginia*. A map produced by Davies was included in that publication, as well as a description and picture of a large flowstone mound. The map, which only showed the main passage back to the last room in the cave, the Football Field, was an excellent presentation of the much of the cave. Chuck Hempel, in his *Caves and Karst of Monroe County*, published in 1975, also included a description and map. Chuck's map, although it showed both the major passages in the cave, was much more stylized and included very little passage detail.

Patton—because of its large, easy passages, many speleothems, and relatively relaxed access—had become a favorite cave among cavers and spelunkers. This, unfortunately, led to some areas of the cave being vandalized with graffiti. It became obvious, during a Greenbrier Grotto trip to clean the worst of this, that a modern map was required, and the Grotto thus began to resurvey the cave in January 1980. This project, which was coordinated by George Dasher, lasted until August 1981 and mapped the entire cave, brought the cave length up to over two miles, produced a high-quality map, and confirmed that a pit entrance existed off the stream passage in the back of the cave.

Marshall Homes and Dennis King, two of the cavers who had participated in the survey, then dug in a passage north of the Football Field. They pushed through a tight crawlway and into a tiny room, and then they dug through a short, sloping passage and to the top of a pit. This was descended on a subsequent trip, found to be about 30 ft deep, and a large number of peccary bones were discovered at its bottom. Fred Grady, a D.C. Grotto member and a paleontologist with the Smithsonian Institute, was contacted, and the

Fig. 5.13 Flat-head peccary from Patton Cave. Photograph by the author

Greenbrier and D.C. Grottos completed five trips and removed approximately one dozen specimens of Pleistocene flat-head peccaries, *Platygonus compressus* (Fig. 5.13). These were not placed on display in the Smithsonian (they already had a specimen on display); however, a composite skeleton was donated to the West Virginia Geological and Economic Survey and is on display there, and casts were constructed that are currently displayed at the Cincinnati Museum of Natural History.

5.4.12 Plastic Bag Cave

The two entrances to the Plastic Bag Cave System are both located within the Culverson Creek Basin. Both entrances receive a stream, and this water flows south, through Handline and Picnic Caves, and finally emerges on the surface near Culverson Creek.

The northern Plastic Bag Entrance was discovered in 1977 when Dave Goldman and Pete Williams were able to move some rocks and obtain entry into a small entrance room. They found a water-filled tube heading downstream with low airspace that was blasting air. On their next trip, they removed their clothes and packed them into plastic bags. These were floated through the tube, and the cave was thus named. This near sump lasted for about 50 ft before it opened back up into a small dome complex. Leading south out of the Changing Room was a stoopway that contained the stream.

Walking passage was soon encountered. The cave next took several small jags and descended a couple of downclimbs. It then opened into a big canyon several feet wide and up to 30 ft high. Several more scoop trips were taken over the next couple of months before WVACS—with Bill Balfour coordinating the survey—undertook what would be an 8-year project to map the cave. Wetsuits quickly gained favor as the means to negotiate the entrance near sump, and a second sump was reached approximately 8,000 ft from the entrance, ending the scooping phase of the project. When a mapping crew reached this point, Bill Balfour, who had a full wetsuit on, got down into the sump and felt a strong breeze coming through about 2 inches of airspace.

Pete Williams, the other person on the trip, said the sump looked hopeless—and that was the dare Bill needed. Bill continued on and after about 40 ft of nose-to-the-ceiling passage, popped out into big walking canyon again, this time with lots of nice flowstone all around. Bill strolled down this stream passage for about 1,000 ft before turning around and heading back to his now-chilled companion. The following weekend, seven people pushed beyond this near sump and mapped over 2,500 ft before reaching a third sump. This sump was indeed terminal.

As exploration and mapping continued, a small infeeder stream canyon was pushed upstream to a waterfall climb. This climb was ascended, and led into an area of breakdown that was associated with a fault. A low stream crawl was pushed upstream from this area to a point where surface debris and warm air was encountered. Digging in the belly crawl soon revealed light coming in around the corner, and after several hours of work, the explorers were able to squeeze out into the cool night air.

The weather then turned nasty, with lots of rain and then heavy snow. Spring 1978 brought more floods and, when the cavers finally attempted to get back into the cave during the summer, they found that many changes had taken place. The old entrance near sump was now completely shut, and the new entrance was completely clogged with sediment. The cave stood at a survey length of just less than two miles, but Mother Nature had decided that no one was going to see any virgin passage in the immediate future.

Needless to say, the lack of an entrance stopped the project dead in its tracks. Pete Williams' untimely death also dampened enthusiasm. Over the next couple of years the entrance situation didn't get any better either, as more floods and harsh winters put the cave on the back burner. Finally, in 1980, the weather began to cooperate and some folks who were not acquainted with the wet nature of the cave were recruited to try and finish the mapping (a classic suck-in, so-to-speak).

The main Plastic Bag Entrance was still sumped and occupied by a very unfriendly beaver, so the new entrance, named the Paper Bag Entrance, was selected as the entrance of choice. A couple of weekends worth of digging opened the entrance back up and made it tolerable. It actually turned out to be a much easier task than anticipated; however, the belly crawl in the stream still was there.

Mapping resumed, and the Paper Bag Entrance and the infeeder canyon were tied into the system. Rumored upper levels were investigated and surveyed, and proved to be quite extensive. Gypsum and Sparkling Avenues were entered and pushed to their limits. They both contained extensive deposits of gypsum crusts and needles. Quartz crystals were discovered embedded in the ceiling of Sparkling Avenue, and an area of 12-inches-long gypsum needles proved to be the highlight of the project. A huge chamber was found above where Sparkling Avenue and Gypsum Avenue intersect. This, the Hall of Silence, was 75 ft wide, over 200 ft long, and up to 80 ft high. At the end of 1981, the cave had grown to over three miles and all but one lead had been exhausted. Then the weather decided to close the cave once more.

Again the cave was put on the back burner for a period of time. There was not the urgency to open it back up for just one more trip, and consideration was given to drawing the finished map. However, this was put off too and finally another trip was planned in 1985 to finish that last lead. Again the Paper Bag Entrance was excavated and a cast of thousands (actually two mapping teams) entered to wrap up the survey. Surprisingly, the lead ended after a few hundred feet and the cave was finally declared finished. It had actually only taken only ten survey trips to map the cave, but those ten trips had been spread out over an 8-year period. The final numbers for the cave were 3.4 miles surveyed and a depth of 135 ft reached.

5.4.13 Scott Hollow Cave

Scott Hollow Cave was opened in the fall of 1984 when a backhoe was used to excavate down through 20 ft of soil and broken rock to a small hole blasting cool air into the warmer outside air. The hole was enlarged horizontally for 10 ft to a tight, 14-foot-deep drop into large cave passage. Just inside, perched on top of boulders, was the lower jawbone of a baby mastodon which would prove to be the first of many prehistoric bones found in Scott Hollow. Mastodon Avenue trended down dip along the bottom of the Greenbrier limestone. It was also thought that the passages would be similar to Scott Cave, large enough to stand up in most places, but interrupted by occasional tight constrictions. This prediction was wrong in a big way. There were no constrictions and the passages were unusually large. Within 6 months, over eight miles of big trunk passages had been surveyed, and now—after 20 years of work—the cave has been mapped to a length of 28 miles. Scott Hollow Cave has become one of West Virginia's favorites and by 2011, when the cave was closed, over ten thousand individuals had visited the cave.

Many caving groups explored and mapped Scott Hollow Cave. There has always been a local exploration group, in addition to the Baltimore Grotto, The Parkersburg Area Grotto, Virginia Tech (VPI) cavers, Ohio Cavers and Climbers, and a strong independent group from Holmes County, Ohio. Recent explorers include those from northern West Virginia and Maryland. The cave was closed in 2011 to reassess access policies and a year later was slowly reopened under a more restrictive access policy. This is the present status as of this writing in 2016.

Perhaps the most significant discovery in Scott Hollow Cave was the main drain, Mystic River. Mystic River is an amazing passage, a black void stretching over five miles, always large, and in many places either 100 ft high or 100 ft wide. It was obvious from the beginning that Mystic River is draining a much larger area than just Scott Hollow. The North-South Passage, an infeeder to Mystic River, was assumed to be most of the Scott Hollow proper drainage, and Craig's Creek, on the west side of Mystic River, was assumed to be Flat Top Mountain drainage. The very confusing question at the time was the origin of such a large stream passage. Dye tracing would quickly give the answer.

When Scott Hollow was first discovered in 1984, the boundaries of the Scott Hollow drainage basin were not known, and some of those boundaries remain elusive to this day. East of Scott Hollow and east of the Sinks Grove Anticline lies the small community of Sinks Grove, which is sited on a rather large, isolated sinkhole plain. Prior to the cave's discovery, attempts to dye trace this sinkhole plain failed. The discovery of Mystic River provided a closer, interim location to capture the dye. Well-known hydrologist Bill Jones conducted two traces in 1985 connecting the Sinks Grove sinkhole plain to Scott Hollow Cave. Underground drainage near Sinks Grove flows south towards Swopes Knobs, where it is speculated it crosses the plunging Sinks Grove Anticline, then does a 180° turn to flow north through Scott Hollow Cave. This discovery doubled the known size of the Scott Hollow drainage basin.

During the late eighties, Cave diver John Schweyen pushed the downstream Mystic River sump and emerged into several miles of large cave, coming within two thousand feet of nearby Windy Mouth Cave. Many efforts to bypass this sump, including bolting and passage enlargement have yet to succeed. No caver has been in this section of the cave for more than 25 years and the only idea left for getting there is to dig from the surface. This section of cave extends to the northern end of Scott Hollow proper and is thought to be near the location where Mystic River begins its journey through the stratigraphically higher Greenbrier limestone units, ultimately resurging in Gloria's Springs on the Greenbrier River, two miles downstream from Fort Spring, near the top of the Greenbrier series.

Beginning in 1996, University of Akron graduate students, advised by Ira Sasowsky, began a series of master's thesis research in Scott Hollow Cave. The tenth of this series, a water-quality study completed in 2010 by Melisa Bishop, revealed new additions to the Scott Hollow drainage basin. The boundary of the Sinks Grove sinkhole plain was extended further to the north with additional dye tracing, and a smaller basin parallel to Sinks Grove, previously thought to flow to Second Creek, was traced to the Little Mystic stream, an infeeder to Mystic River at the southern portion of the cave system. Another area west of Mystic River, near the Broad Run drainage to the west, was also traced to Scott Hollow. The current Scott Hollow drainage basin is estimated to be at least fourteen square miles.

The last major, multi-mile discovery in Scott Hollow Cave occurred in 2001 when the upstream Mystic River Sump was bypassed and a major extension of Mystic River was discovered. Surprisingly, these passages are the largest, and easiest to traverse in the entire known cave system. The largest room in the cave, the Super Bowl, which is 200 ft in diameter and 70 ft high, is located near the far upstream terminus of Mystic Avenue. Other notable discoveries in the new extension include Corey Hackley's and Keely Owens' far southern extension of the cave, which nearly reaches the Sinks Grove Anticline, and Adam Lake's bolt climb near Super Bowl, revealing theorized major faulting that brings surface water rapidly into Mystic River in that area. The new extension is also the location of the largest prehistoric bone found, a mastodon humerus approximately 3 ft long and 1 foot in diameter, which has been carbon dated to 22,000 years before present. Considerable surface digging efforts have yet to yield an entrance into this area of the cave.

Current exploration efforts include several major surface digs on the east side of the Sinks Grove anticline. It is believed via dye tracing and geology that the Sinks Grove sinkhole plain contains many miles of passage, upstream of known Scott Hollow passages. Within Scott Hollow, the Scoop City level has potential to be extensive. That part of the cave is rarely visited. Most of the current exploration occurs in the upstream Mystic River area, partially due to a fine camp area, established almost 30 years ago, and now contains multiple amenities, including twenty individual camp spots, a living room, a kitchen, and running water. An entrance into the area beyond camp would no doubt be of great assistance in extending the known limits of Scott Hollow Cave.

5.4.14 Spring Creek Cenotes

The Spring Creek Cenotes are located on Spring Creek between Frankford and Renick. Spring Creek is one of the two largest karst basins in the Greenbrier Valley (and in West Virginia), with Davis Spring being the other. Spring Creek drains the south side of Droop Mountain, and thus collects all the water from Friars Hole Cave that flows south, as well as the water from Fox Cave, Buckeye Creek Cave, and Rubble Spring.

The Cannon Hole is located about a half mile north of the cenotes and about a mile below the US Route 219 Bridge over Spring Creek. The Cannon Hole is an estavelle—which means that the surface Spring Creek sinks into the hole during low-water conditions and that subterranean waters resurge out of the hole in high-water conditions. The Cannon Hole was dove by Ron Simmons in the early 1980s. He reported that it is less than 15 ft deep, with walls of bedrock and a bottom that was gravel and full of rocks, logs, and other debris. The Cannon Hole was named because Confederate soldiers supposedly threw a cannon into it to prevent its capture after the Battle of Droop Mountain. However, subsequent research by Terry Lowry has found that the victorious Union troops, when they returned north after the battle, took with them a captured bronze howitzer with a broken carriage.

Spring Creek, during low or normal water conditions, will be dry downstream from the Cannon Hole for about a quarter mile, until JJ Spring is reached. During these conditions, the underground water first appears in the Circulating Cenote, which is the largest and most-obvious of the Spring Creek Cenotes (cenote: a Spanish word meaning limestone well) (Fig. 5.14). Circulating Cenote is approximately 100 ft in diameter and is partially divided by a limestone peninsula. Its water, during normal water levels, flows underground to the Grapevine Cenote. During times of high water, its water will flow in a surface channel 50 ft wide and 2 ft deep to the surface Spring Creek.

The water emerging at Circulating Cenote has been dye traced from the Friars Hole Cave System on separate occasions by Charlie Williams and Bill Jones. The cenote itself has been dove by at least six people. Jim Brown connected the two parts of the cenote through the peninsula, and Jack Lake and Ron Simmons both dove downstream in the Circulating Cenote and resurfaced in the Grapevine Cenote. Both Jack and Ron attempted to follow the water upstream in Circulating Cenote. Jack had to abort his dive early on, but Ron found a way upstream in 1983. This was one of Ron's first cave dives, and he reported:

> I dove between the two cenotes along Spring Creek and connected them. Then I tried to go upstream out of the largest cenote. I was at home in the cave and was used to pushing hard. The passage I headed into was a bedding plane about 4 or 5 feet high and about 20 feet wide. The only problem was that there were anastomoses on the ceiling and floor. They almost touched except for about five inches in the middle. So to go forward I had to keep flipping onto my side and squeezing between the projections. I kept doing this for about 40 to 50 feet, then noticed there was a lot of silt everywhere and decided to look behind me. This was definitely one of those "Oh shit" moments, as there was absolutely no visibility behind me.
>
> So I turned around after some effort and headed out. One slight problem: I had not belayed the line at all on the way in and it was swinging freely back and forth in the passage between the ceiling and floor anastomoses. So I had no way of knowing which projections I had squeezed between on the way in. The one thing that helped me keep it together was that I had plenty of air with me. It took something like 45 minutes to get out, never knowing if I was going the right way or squeezing into a dead end. After that dive I decided that maybe I should get a little more training and experience before returning to sump diving.

That ended Ron's work in the cenotes; however, John Schweyen explored Circulating Cenote in 1988. He found an underwater cave passage about 50 ft deep that was about 2 ft high and 3–4 ft wide. This passage opened up until it was between 5 and 15 ft in height, and at least 20 ft wide. John followed the passage upstream by bouncing from one wall to the other. He could not see both walls at the same time. There was a great deal of scalloping on the passage walls, and not a whole lot of silt. The passage led approximately 1,100 ft to a large room. The room was at least 30 ft in diameter, 12 ft below the surface, and contained large blocks of limestone. John's dives were terminated not because he encountered blockage or a passage ending, but because he had lost the passage walls and could not follow the water flow.

The Spring Creek water, once it leaves Circulating and Grapevine Cenotes, flows underground to JJ Spring, where it rejoins the surface flow (if any) downstream of the Cannon Hole. From JJ Spring, the water flows downstream to the

Fig. 5.14 Circulating Cenote at high flow. Photograph by the author

upstream entrance of Spur Cave, where it again goes underground–on dry days. This water can, however, overflow on the surface and join the overflow water from the Circulating Cenote. Dye tracing completed in 1976 by Charlie and Barbara Williams defined the flow routes for this small, but complex area. They also completed the first tracer test from Friars Hole Cave to the Spring Creek Cenotes in the same year.

The combined overflow water then travels downstream (from the upstream entrance to Spur Cave), where it joins the water emerging from the five Culverson Creek Cave resurgences. These resurgences were dye traced from the Culverson Creek Cave by Hermine Zotter, Bill Jones, and Charlie Maus in the late 1960s and early 1970s. Additional dye tracing was completed in the early 1990s by George Dasher and Doug Boyer, who discovered that the water in the entirety of the upper Buckeye Creek Basin has been pirated to these springs.

Of the five Culverson resurgences, the largest cave is Matts Black Cave, which was originally surveyed in the early 1970s by Phil Lucas and Roger Baroody. The origin of the cave name comes from the black, ubiquitous veneer which coats most of the cave's floor. This substance was analyzed by George Moore, using x-ray diffraction, and found to be birnessite, a manganese oxide that is the most-common manganese mineral found in caves in the USA—and which is found in Culverson Creek Cave. Matts Black Cave has two passages, the Water Passage, which contains the water, and the Relict Passage, which is a paleowater passage. The sump in the Water Passage was dove by Ron Simmons in the 1980s, and Ron found that the underwater passage ended after 800 ft in an enormous collapse that was composed of small breakdown blocks and through which there was no route.

The water inside Spur Cave joins the surface Spring Creek upstream from the lower-most Culverson resurgence (and downstream of the other four). An attempt to dive Spur Cave was attempted in the early 1980s by Jack Lake, but he had to abort his dive because of the large amount of flood debris that had washed into the cave. The cave was later mapped by George Dasher, Bill Balfour, and WVACS.

From Spur Cave, Spring Creek continues flowing south to the Greenbrier River, picking up additional water from Burns Cave #2 and Leggs Springs en route. Burns Cave #2 was also dove by Ron Simmons in the 1990s, and a large cave was discovered beyond the entrance sump. The cave is the resurgence for The Hole, which is the fifth-longest cave in West Virginia. The Spring Creek Cenotes and JJ Spring may be the third-largest spring in the state, while the Culverson resurgences may be the fourth largest. Davis Spring is the largest spring in West Virginia, followed by the Cowgar Mill Spring, which is located on the upper Elk River several miles below the community of Slaty Fork.

5.4.15 Windy Mouth Cave

Windy Mouth Cave is an extensive, 18-mile-long cave located south of the Greenbrier River in western Greenbrier County. Although the cave has been mapped on several occasions, its full extent is not known. This is due to the inaccessibility of the cave, which is located downstream of tall, riverside cliffs, and to the difficulties involved in exploration of a long, complex, single-entrance cave system with a 900-foot-long entrance crawl.

Mapping was started in Windy Mouth Cave in mid- to late-1950s by three cavers from the Charleston Grotto: John Davis, Jack Gravenmier, and Frank Matterick. These cavers later became founding members of WVACS, and—because of this—they ultimately moved on to work in other caves in the Greenbrier Valley. Nevertheless, the map of their efforts is included in the 1965 addendum to William Davies' *Caverns of West Virginia*, and shows an incredibly extensive cave system.

Sometime in the early 1960s, Jim Hixson became interested in Windy Mouth—possibly after a trip from the 1963 NSS Convention at Mountain Lake, Virginia, to see the cave's famous bear track. Jim took over the survey, and because the previous surveys had no vertical control, began a new survey at the cave at the entrance. At the time, Jim lived in eastern Tennessee and, after finding it hard to recruit cavers there, began drafting cavers from the VPI Student Grotto by offering free rides. These were very long trips for Jim, going to the cave and back to Tennessee, and he was only able to get into the cave once a month. Two of the VPI students who assisted Jim were Gary Moss and Bob Amundson, and Jim would mail Bob the survey notes so that Bob could process the data using his computer program before the next trip. Bob assisted Jim until the mid-1970s when he was drawn into his own survey project in McClung Cave.

Many stories about Jim's trips have been told and retold. But, in general, Windy Mouth trips were long, grueling trips that started on Saturday and did not end until Sunday morning. Gary, in particular, remembered coming out in daylight the next day on many trips, and cavers often fell asleep while exiting the entrance crawl. All the surveyors moved on auto pilot while exiting the cave, and they occasionally crawled over the sleeping caver without realizing it. Karl Berge, a new VPI caver, was asked about all the gear on the ground at the Windy Mouth parking lot, and said, "Most cavers who get out of Windy Mouth are usually so tired that they don't give a damn about their equipment and just throw it down on the ground, stomp on it, and grind it into the dirt." Another of Jim's stories involved finding fresh bear tracks on top of the cavers' crawl marks while exiting the cave. Fortunately, because he was recruiting surveyors from a student grotto, Jim had a renewing source of young cavers to bring into the project.

Trips needed to start from Lewisburg early, and getting Jim away from his coffee was a daunting task. Caving with Hixson also meant caving with Linda, his dog. Linda was allowed to run free once underground, and sometimes even brought the surveyors the tape. Linda would direct cavers into the correct route on the way out of the cave, saving them a great deal of crawling. She would also run to the entrance and then return to see how the exiting cavers were doing.

Jim, in early 2001, before his health deteriorated, allowed Dr. Ira Sasowsky (University of Akron) to make copies of all of his survey notes. They also converted data from old disks to a readable format. Graduate student David Shank then produced a line map of the cave using *Compass*. No one seriously worked in the cave until the late 1990s and early 2000s, when Mike Dore led a multi-year effort to connect Windy Mouth to Scott Hollow Cave. Mike and his companions were extremely hard-charging cavers and explored much of the southern, upstream part of Windy Mouth, but they were not successful in connecting the two caves. His parties did no surveying.

WVACS then began to map the cave in April of 2015. This project, which is being coordinated by Nick Socky and Tommy Polson, began at the entrance and is a complete resurvey of the cave. It is a DistoX2 and Suunto survey, with the intentions of generating a modern map, as well as finding and pushing leads in the hopes of discovering new cave. In 2015, a total of 12 trips mapped 9,041.5 ft, or 1.71 miles.

In 2016 there were a total of 20 trips into Windy Mouth Cave. Nearly all of the trips were resurvey except for a total of four bolting trips. In the E-survey, bolting by Tommy Polson and Jon Lillestolen up a 50-foot dome led to 335 ft of virgin cave. Then in 2017, Tommy, Jon and Jenn McGuire returned to the E-survey and found another 665 ft of virgin cave that still goes with plenty of air flow to follow. Nick Socky and Joe Calderone began bolting up two domes in the L-survey in 2017 The first 20-foot dome was completed and about 40 ft of going passage was found. There have been five trips into Windy Mouth as of April, 2017. Several trips to the upper levels show the cave to be breaking out in at least 20 different directions. Many more trips will be needed to complete this section of the cave.

In 2016 at total of 22,414.8 ft in 970 shots have been surveyed in 39 different surveys involving 47 surveyors. In 2017, 8,835.9 ft have been surveyed in 470 stations with 36 participants. The total current surveyed length of the cave is 9.1 miles with a depth of 307.3 ft.

Acknowledgements This historical account is based on the author's personal experience, on published documents as cited, and on personal communications from many cavers. Thanks are given to Dave Cowan (Boarhole), Mike Dore (Scott Hollow Cave), Mike and Andrea Futrell (Destitute Cave), Phil Lucas (Culverson Creek Cave), Doug Medville (Portal Cave and many others), Gary Moss and Nick Socky (Windy Mouth Cave), Dave Socky (Maxwelton Sink Cave), Greg Springer (Dry Cave), and Bob Handley, Bill Jones, Charlie Maus for various caves.

References

Ashbrook, B. 1995. *The Caves of the Richlands Area of Greenbrier County, West Virginia*, vol. 1, 37pp. Barrackville, WV: West Virginia Speleological Survey Monograph Series.

Cowan, D. 2003. The exploration and mapping of the Boar Hole. *NSS News* 61 (9): 238–243.

Dasher, G.R. *The Caves of East-Central West Virginia*, the guidebook of the National Speleological Society's 2000 convention, the National Speleological Society and the West Virginia Speleological Survey, Bulletin #14, May 2000, 294 pages, includes an interview with Ron Simmons about his almost-fatal dive in the Circulating Cenote.

Dasher, G.R. 2012. *The Caves and Karst of West Virginia*, vol. 19, 264pp. Barrackville, WV: West Virginia Speleological Survey Bulletin.

Dasher, G., and W.M. Balfour. 1994. *The caves and karst of the Buckeye Creek Basin, Greenbrier County, West Virginia*, vol. 12, 238pp. Barrackville, WV: West Virginia Speleological Survey Bulletin.

Dasher, G.R., and D.G. Boyer. 2000. *The Recent Spring Creek Area Dye Tracings, Greenbrier County, West Virginia*, vol. 2, 28pp. Barrackville, WV: West Virginia Speleological Survey Monograph.

Davies, W.E. 1958. *The Caverns of West Virginia*, 2nd ed., vol. 19A, 330pp Morgantown, WV: West Virginia Geological and Economic Survey.

Davies, W.E. 1965. *The Caverns of West Virginia, Supplement*, vol. 19A, 72pp. Morgantown, WV: West Virginia Geological and Economic Survey.

Droms, Y. 2003. Connection. *National Speleological Society News* 61 (9): 244–248.

Halliday, W.R. 1976. *Depths of the earth*, 432. New York: Harper & Row.

Hempel, J. 1975. *The Caves of Monroe County*, vol. 4, 145pp. Barrickville, WV: West Virginia Speleological Survey Bulletin.

Jones, W.K. 1973. *The Hydrology of Limestone Karst*, 36, 49pp. Morgantown, WV: West Virginia Geological and Economic Survey Bulletin.

Jones, W.K. 1997. *Karst hydrology Atlas of West Virginia*, vol. 4, 111pp. Leesburg, VA: Karst Waters Institute Special Publication.

Medville, D. 2003. The exploration of the Portal. *National Speleological Society News* 61 (9): 236–238.

Medville, D., and H. Medville. 1976. *The Caves and Karst hydrology of Northern Pocahontas County*, vol. 6, 160pp. Barrickville, WV: West Virginia Speleological Survey Bulletin.

Stevens, P. 1988. *The Caves of the organ Cave plateau*, vol. 9, 200pp. Barrickville, WV: West Virginia Speleological Survey Bulletin.

Storrick, G.D. 1992. *The Caves and Karst hydrology of Southern Pocahontas County and the Upper Spring Creek Valley*, vol. 10, 215pp. Barrickville, WV: West Virginia Speleological Survey Bulletin.

Thierry, E. 1954. Charleston Grotto explores Culverson Creek Cave. *National Speleological Society News* 12: 4.

Zimmerman, R. 2009. *The Cassell Cave Survey Project*, vol. 3, 60pp. Barrickville, WV: West Virginia Speleological Survey Monograph.

Zimmerman, R. 2011. *The Survey of Cass Cave*, vol. **4**, 64pp. Barrickville, WV: West Virginia Speleological Survey Monograph.

The Swago Creek Basin

6

William B. White, Gary D. Storrick and Mark J. Johnsson

Abstract

The Swago Creek Basin is the northernmost drainage in the defined area of the Greenbrier Karst. It is a fluviokarst basin with extensive underground drainage but relatively little expression of surface karst. Swago Creek heads in two large cave-entrance springs: Overholt Blowing Cave and Cave Creek Cave. The Overholt Blowing Cave stream can be explored nearly to the head of the Dry Creek Valley but the upstream segment of the valley is pirated eastward to springs on the opposite side of the mountain. Little of the Cave Creek drainage is accessible to direct exploration but some fragments are represented by Barnes Pit, Tub Cave, and the Carpenters-Swago System. The Carpenter-Swago System has 3.1 miles of surveyed passage, and it is strongly controlled by a N60°E fracture system. Extensive stream tracing in the basin reveals details of drainage pattern which is strongly controlled by the shaly Taggard Formation.

6.1 Introduction

The Swago Creek Basin is the northernmost of the local drainage areas that together make up the Greenbrier Karst. It is a multi-sectioned cove cut into the high Allegheny Plateau to the northwest. The dissected plateau remnants are labeled as mountains on the topographic maps: Rogers Mountain to the west, Swago Mountain to the northwest, Spruce Flats and Day Mountain to the north, and Stony Creek Mountain and Bridger Mountain to the east (Fig. 6.1). Day Mountain at 4255 ft is the highest point of elevation in the basin. Details are shown on the USGS Hillsboro and Marlinton Quadrangles.

Electronic supplementary material
The online version of this chapter (doi:10.1007/978-3-319-65801-8_6) contains supplementary material, which is available to authorized users.

W. B. White (✉)
Department of Geosciences, Deike Building, University Park, PA 16802, USA
e-mail: wbw2@psu.edu

G. D. Storrick
139 Spruce Lane, Poca, WV 35159, USA

M. J. Johnsson
45 Fremont St, Ste 2000, San Francisco, CA 94105, USA

6.2 History

The very early status of knowledge of the caves of the Swago Creek Basin is given by a single quotation from Price and Reger's (1929) volume on Pocahontas County geology. "Known caverns in Pocahontas County are, Saltpeter Cave at the head of Swago Creek, Overholt Blowing Cave near McClintock's Mill, and Sneadegar Cave west of Droop Mountain near the Greenbrier County Line." Davies (1949) offers a description and map of Overholt Saltpeter Cave that has not been modified, a description of Tub Cave, a description of the first 150 ft of Overholt Blowing Cave, and a description of 1500 ft of Cave Creek Cave. Exploration by cavers from Charleston, West Virginia in 1953 added Swago Pit, Carpenters Pit, and Roadside Pit to the cave list.

The decade from 1956 to 1966 might be termed the "Pittsburgh Period" of exploration. Many trips by cavers from the Pittsburgh Grotto of the National Speleological Society and the Pittsburgh Explorers Club pushed the Swago-Carpenters System to about 2.3 miles and Overholt Blowing Cave to 2.9 miles. Extensive ridge walking revealed many pits and small caves so that by 1963, there were 69 caves listed, most of them very small. Most of these explorations and discoveries were written up in the

W.B. White (ed.), *Caves and Karst of the Greenbrier Valley in West Virginia*,
Cave and Karst Systems of the World, https://doi.org/10.1007/978-3-319-65801-8_6

Fig. 6.1 Topographic map of the Swago Creek Basin and surrounding area. Excerpt from U.S. Geological Survey 1:100,000 series Marlinton Sheet. Metric map with 50 m contour interval

newsletters of the two organizations with a later published summary (White 1960, 1964). The Pittsburgh data were made available to W.E. Davies so that the second edition of his book contains many more caves in the Swago Valley (Davies 1958). After 1960, more emphasis was placed on deciphering the drainage system with a series of tracer experiments (Zotter 1963, 1965) which in turn led to the use of the Swago Basin as a demonstration of the importance of conduit flow in karst aquifers (White and Schmidt 1965).

During the 1970s, the Swago Creek Caves came to the attention of cavers from McMaster University in Canada. They extended the exploration of Overholt Blowing Cave and other of the smaller caves in the basin. Most of these explorations appear as field trip reports in the *Canadian Caver*. One of the Canadian cavers, George Tracey, was killed in Barnes Pit on July 1, 1975 when a loose boulder shifted and crushed his chest (Merrin 1975). Much of the Canadian activity shifted south to the rapidly growing Friars Hole System but there were still occasional visits to the Swago Basin, especially Overholt Blowing Cave. A Quebec Group reached the end of the cave in 1978 followed by another group reaching the end in 1991 (Zabrok 1993). There was a final trip to the end by Myrna Diaz-Mundo and Marc Legault in 1995. The Swago Creek Basin was used as a test case for simulation of conduit hydrology (Coward 1975).

The third major effort on the Swago Valley was in the 1980s and 1990s when groups led by Mark Johnsson extended the Swago Carpenters System and undertook a careful resurvey of the cave along with new surveys of Tub Cave, Roadside Pit and others. Many of these maps were published as part of a guidebook (Medville et al. 1983). These maps form the basis for the present document. Also undertaken in the period 1991–1996 was Gary Storrick's resurvey and extended survey of Overholt Blowing Cave. The survey was truncated in 1996 when the landowners closed the cave to all visitors. Pittsburgh cavers returned to the Swago Valley in the early 2000s to prepare detailed maps of many of the smaller caves (Hamm 2002). Storrick (1992) compiled a volume of cave descriptions for Southern Pocahontas County and a bit of northern Greenbrier County. Portions of the cave descriptions that follow are drawn or paraphrased from this report or from Johnsson's guidebook.

6.3 Basin Geomorphology: Surface Karst

6.3.1 Geologic Setting

The Swago Creek Basin is formed on the west limb of the Browns Mountain Anticline with the limestone dipping very gently to the west. There is a minor anticlinal structure oriented roughly north-south and which parallels the Dry Creek Valley. This structure is plunging to the south.

The Greenbrier Limestone is about 400 ft thick in this area. A detailed stratigraphic column measured a few miles to the north is shown in Chap. 2 (Fig. 2.5). The surface outcrop of the Greenbrier Limestone occupies the middle slopes of the basin. Higher on the mountains are the sandstones, siltstones, and shales of the Mauch Chunk Formation. The lower portion of the valley, below the springs, is floored with the Maccrady Shale and the underlying Pocono Sandstone. The Taggard Formation, a complex sequence of shales and shaly limestones, is an important controlling factor in the development of cave systems in the Swago Creek Basin.

6.3.2 The Drainage Pattern

There are three named tributaries to Swago Creek on Hillsboro 7.5 min quadrangle: Dry Creek, Overholt Run, and McClintock Run. Overholt Run, the central tributary, is again split into three branches, for convenience named on Fig. 6.2 as McKeever Run, Friel Run, and McNiel Run. McClintock Run also has two branches, giving overall six main tributaries that head in the mountains, flowing on the shales and sandstones of the Mauch Chunk Formation and overlying clastics. Without exception, these streams sink into their beds or against rock ledges near or just below the 3000-foot contour. With the exception of a small tributary of McKeever Run that drains into Swago Pit, the other streams disappear into coarse cobble and gravel fills or very minor cave entrances. The Swago Creek Basin has topography of moderate relief. The tributaries of Swago Creek form a radiating pattern where they have cut deep valleys back into the plateau. The tributaries have well-preserved stream channels but these are typically dry except after periods of excessive precipitation. The insurgences have limited capacity, forcing high flows from storms or snow melt onto the surface channels, keeping them well-scoured (Fig. 6.3). Some of the underground flow paths roughly parallel the surface drainage; others do not. The underground flow paths are discussed with the individual cave systems. Representative views of the topography are shown in Figs. 6.4, 6.5 and 6.6.

Swago Creek heads in two large springs: Overholt Blowing Cave and Cave Creek Cave both located near the base of the limestone at 2230 ft elevation. The Overholt Blowing Cave spring is marked as the head of Swago Creek on the USGS Hillsboro 7.5 min quadrangle but Cave Creek Spring is not. Although it is the larger of the two tributaries,

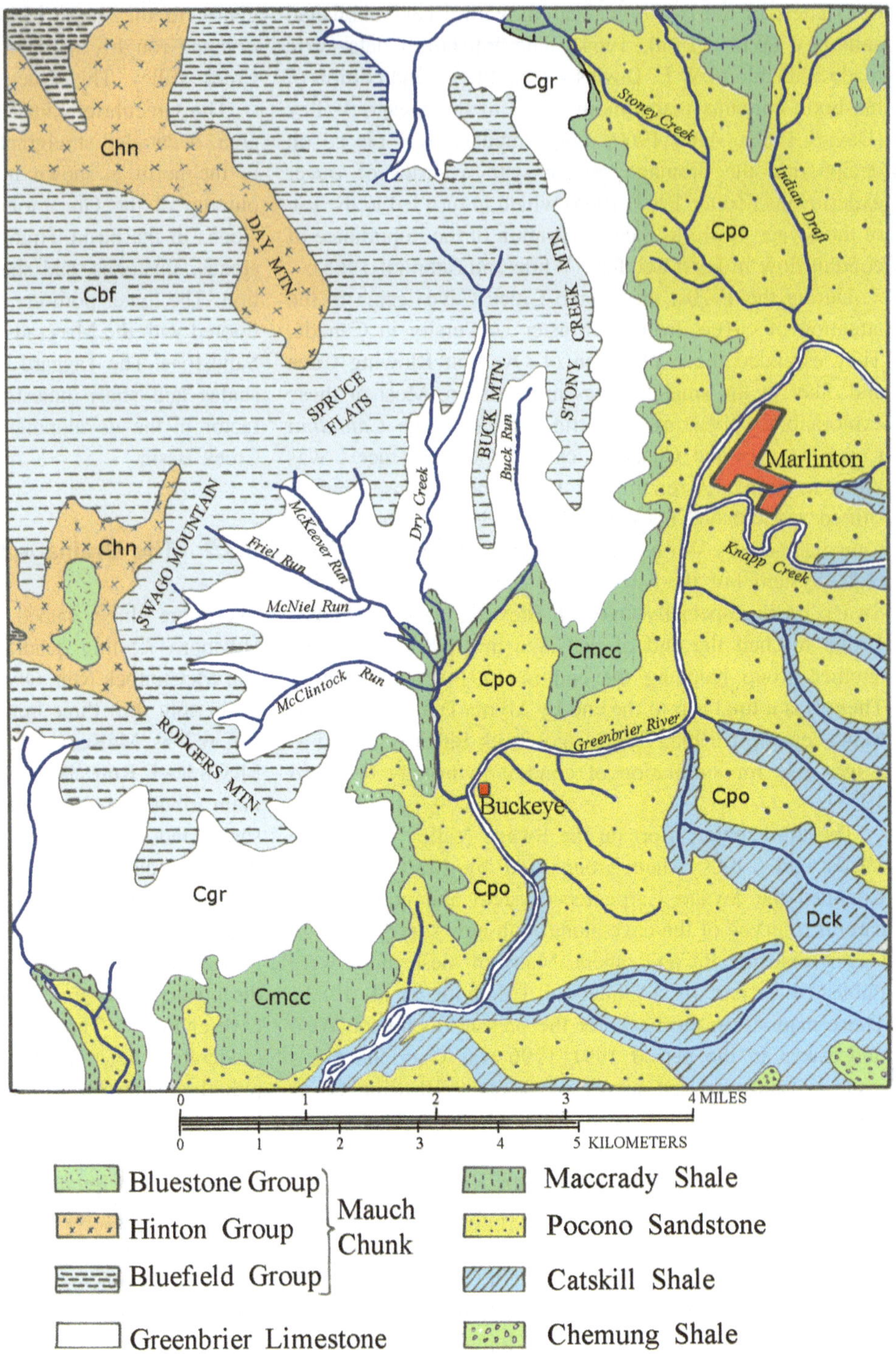

Fig. 6.2 Geologic map of Swago Creek Basin taken from the county geologic map of Price and Reger (1929)

its location at the head of a deep, wooded ravine apparently escaped the attention of the mapmakers. The downstream reach of Swago Creek flows on the clastic rocks of the Maccrady Shale and the Pocono Sandstone and reaches the Greenbrier River at the village of Buckeye at an elevation of 2100 ft.

6.3.3 Lost Waterfalls

The Taggard Formation is a hydrologic confining layer but only a partially effective one. Underground streams perched on the Taggard Formation sometimes breach the confining layer underground but there are examples where the stream

Fig. 6.3 Bed of dry Creek after storm flow has washed out the road. Note cobble and boulder fill in the stream bed. The bedload moves during extreme flood flow. Photo by W.B. White

Fig. 6.4 View to the east with Overholt Blowing Cave quarry in the foreground and Bridger Mountain in the background. Photo by W.B. White

emerges from the hillside as a spring, flows across the confining layer as a short segment of surface stream, and then sinks underground again in the limestone below the confining layer. If the hillside is steep, the segment of surface stream can appear as a waterfall. There are two lost waterfalls on the wall of the Dry Creek Valley, known as the Hause #1 and #2 Waterfalls (Fig. 6.7).

6.3.4 Closed Depressions

There is little surface expression of karst in the Swago Valley except for the sinking streams and dry stream beds. Hillsides are steep and there are few closed depressions. There is a deep depression on the ridge above the entrance to Overholt Blowing Cave, there is the Tub Cave sink, and

Fig. 6.5 View to the west. Overholt Run in the immediate foreground, Barnes Pit in the center of the image, the McClintock run drainage beyond, and Rogers Mountain in the background. Photo by W.B. White

Fig. 6.6 View to the north along the upper end of the Dry Creek Valley. Note dry channel of Dry Creek. The Dry Creek swallet is just around the corner at the beginning of the trees. Photo by W.B. White

there are a few other closed depressions in the uplands but closed depressions are a minor landform in the Swago Valley. What do occur are scattered pits open to the surface. Mostly, they are simple open shafts with no connection to underlying cave passages.

6.4 The Caves of the Swago Valley: Overview

The Swago Valley has been very intensively explored by cavers for more than 50 years, and as a result, there is a very long catalog of caves and pits. The locations are shown in

Fig. 6.7 Hause waterfall #1. **a** Where the stream emerges from the hillside. **b** The waterfall into a closed depression where the water goes back underground. Photos by W.B. White

Fig. 6.8. Most of these are very small. Those that enter the discussion that follows are:

- 2 Barnes Pit
- 15 Carpenters Pit
- 16 Cave Creek Cave
- 25 Dry Creek Swallow Hole
- 43 House Waterfall No. 1
- 44 House Waterfall No. 2
- 45 House Waterfall Cave
- 64 Overholt Blowing Cave
- 71 Roadside Pit
- 84 Swago Pit
- 85 Tub Cave

6.5 Overholt Blowing Cave

6.5.1 Cave Description

Overholt Blowing Cave is a rare example where one can enter at the spring mouth and examine the cave all the way to the headwaters. The main entrance, the spring mouth, is at the head of Swago Creek, in a small abandoned quarry, immediately beside the Spruce Flats Road (Fig. 6.9). A more regional view is shown in Fig. 6.4. There was a second entrance to the cave in the upper corner of the quarry but in the 1990s, it had filled in.

The overall layout of the cave is shown on a stick map (Fig. 6.10). The cave was mapped from the entrance to Disappointment Dome by Pittsburgh Grotto cavers in 1957–

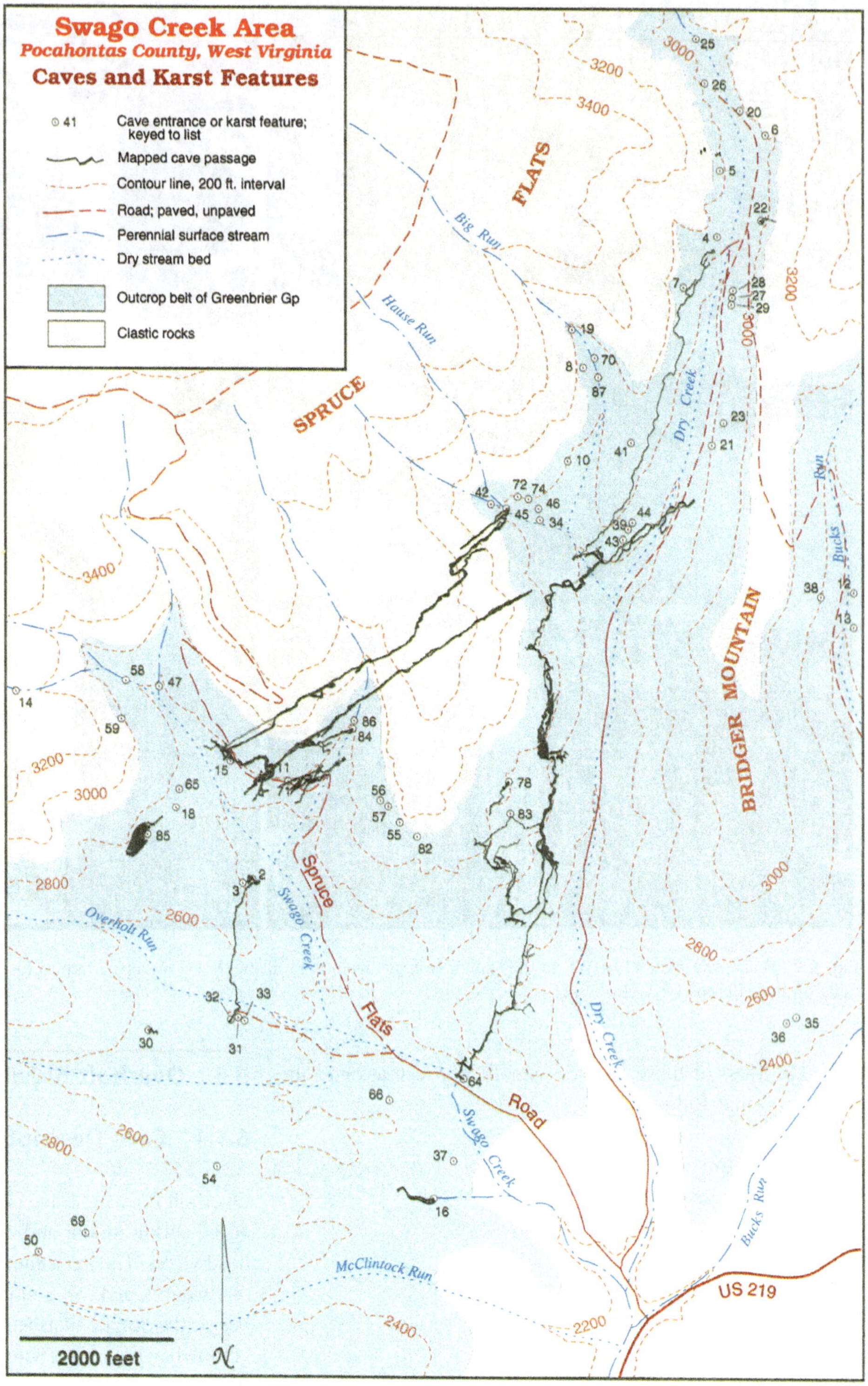

Fig. 6.8 Overview map of the caves and pits in the Swago Valley. Map by Mark J. Johnsson

1959. The extreme back end of the cave was mapped by Canadian cavers in the 1970s. The historical maps are shown on electronic maps M-6.1 to M-6.3. A comprehensive and detailed mapping was undertaken by Gary Storrick and his colleagues in the early 1990s (electronic maps M-6.4 to M-6.13) but the effort was terminated prematurely due to the closing of the cave in 1996. The substantial portion of the cave that was remapped has greater accuracy and far greater detail than the earlier maps.

The spring entrance is an 8–10 foot wide, 4-foot high stream passage which can be traversed for 100 ft to where the passage and stream continue north as a low crawlway.

Fig. 6.9 Entrance to Overholt Blowing Cave. Photo by W.B. White

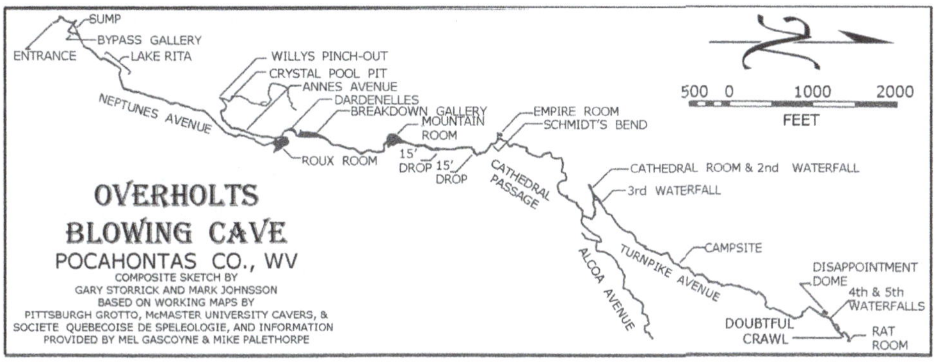

Fig. 6.10 Line map of Overholt Blowing Cave. Map by G.D. Storrick and M.J. Johnsson

A small opening at this point gives access to an upper room floored with rock debris. The second entrance used to connect to a small hole at the top of the slope.

The passage trends N20°E to N30°E as a crawlway in the stream, 18 in. high and 8-foot wide, for 100 ft to where it opens into a room. Upstream, the stream emerges from a sump and forks, one fork flowing into the crawlway and the other west into a mud choke from which it emerges on the surface as a spring in the base of the quarry. The sump is bypassed by an 8-foot high, 15-foot wide dry passage which rejoins the stream after 400 ft. The main stream passage, Neptune's Avenue, is 5–20 ft high and 6–10 ft wide and 3000 ft long. The water is mostly a shallow stream on the level, gravel-covered, passage floor but reaches a depth of 4 ft at Lake Rita. Small infeeders bring in small streams at multiple locations along the passage. These are short, low, stream crawls that end in vertical shafts and serve as drains for water from the hillside above.

Neptune's Avenue ends where the stream passage becomes a stoopway under breakdown in 1–2 ft of water. Fifteen feet above the stream at this point is a fragment of upper level passage called the Roux Room. The Roux Room passage extends south above the steam passage for 300 ft as a low crawl over sediment infilling. At the north end of the room, a small opening over breakdown connects to another upper level passage, Anne's Avenue which trends south parallel and above Neptune's Avenue for 1500 ft. Another smaller passage below Anne's Avenue is called the Gypsum Crawl. Anne's Avenue has a rectangular cross section 3–15 ft high and 5–6 ft wide which merges into a scalloped elliptical tube with a narrow canyon cut in the floor (Fig. 6.11). The southern trend of the passage is blocked by a gravel and cobble fill (Fig. 6.12) beyond which a low crawl leads to a complex of small passages and shafts. The main passage doubles back at the sediment choke and after 200 ft is cut off by the 25-foot deep Crystal Pool Pit. Traversing the pit and continuing another 200 ft brings one to a large breakdown room. Beyond the breakdown rock, a passage leads to a higher level Attic Room and to an area of clear pools, pits, and a lower level canyon. The lower level

Fig. 6.11 Anne's avenue. Photo by W.B. White

Fig. 6.12 Cobble fill at Willie's Pinchout in downstream end of Anne's Avenue. Photo by W.B. White

canyon connects to a complex of passages at stream level which eventually connect back to the main stream passage in the Breakdown Gallery beyond the Dardanelles.

In the main stream passage at the Roux Room, one can continue upstream by crawling under the breakdown in a very low stream crawl for 500 ft. This crawlway, The Dardanelles, has been known to flood, trapping cavers until the water levels recede. Beyond the stream crawl, the passage opens into the 1500-foot long Breakdown Gallery, a 30-foot high, 25-foot wide trunk with the stream flowing around and under a floor of large jumbled breakdown blocks. Breakdown chokes off the stream passage but a small hole leads up into the Mountain Room, a roughly circular chamber more than 100 ft in diameter floored with breakdown. There is a large stalagmite in the room that has grown up into a shaft in the ceiling. A passage at the southeast corner of the Mountain Room leads to the Ivory Palace, a series of small rooms decorated with pure white speleothems. A second passage continues northward for 1000 ft as the 25-foot wide, 15-foot high Helictite Gallery.

Reconnection with the stream passage is made through a 20-foot deep crevice at the end of the Helictite Gallery or through a hidden passage beneath a high boulder in the middle of the Helictite Gallery. This is followed by 600 ft of stream passage of irregular shape. Five hundred feet along the stream passage is a fork in the passage. The left fork leads to the Empire Room, a 100-foot high cylindrical chamber. The right fork becomes the Cathedral Passage leading 2000 ft to the Cathedral Room. At 1600 ft along the Cathedral Passage, the right fork is Alcoa Avenue which extends 2000 ft ending in a low crawl that passes under the Dry Creek Valley Road.

The 50-foot Second Waterfall enters the Cathedral Room from the ceiling, and a second smaller waterfall enters from a hole in the wall. A 30-foot high pile of breakdown allows explorers to reach the top of the smaller waterfall. The cave continues from the top of the Second Waterfall as a narrow twisting passage leading back sub-parallel to the Cathedral Room and then along a narrow ceiling crack to a wider passage. Third Waterfall is in a shaft adjacent to the ceiling crack. Beyond this is a long stream passage leading to the Turnpike, a walkway with a 300-foot long dry, gypsum sand floor. The passage continues 2000 ft to Disappointment Dome via low steam crawls and narrow walkways. Thirty-foot high Fourth Waterfall drops into 50-foot Disappointment Dome. Almost immediately above Fourth Waterfall is 50-foot Fifth Waterfall. At the top, the passage extends 600 ft to a large canyon with breakdown and ends in a small room. Halfway along this passage, a large passage to the left leads to the high point in the cave. A climb on the right wall leads to a narrow passage ending in domes where green leaves have been found on the floor. Back in the main passage, several short crawls lead from the small room, and a 500-foot crawl connects with the Rat Room, a small room marking the presently known end of the cave.

6.5.2 Geology and Drainage Pattern

The Overholt Blowing Cave spring is in the Sinks Grove Limestone near the contact with the underlying Hillsdale Limestone. The stream maintains this stratigraphic position the entire distance to the Alcoa Avenue Junction with the Mountain Room and the Empire Room developed in the overlying Patton Limestone. The Turnpike and upstream passage are in the Pickaway Limestone so that at the sequence of shafts and waterfalls at the Cathedral Room take the stream through the Taggard Formation and Patton Limestone. At the upstream end of the system, the high passage extending from the top of the Fifth Waterfall to the Rat Room must be very close to the top of the Union Limestone and possibly as high as the Alderson Limestone.

The stream passage roughly parallels the dry surface channel of Dry Creek. It is a few hundred feet west of the surface stream and at relatively shallow depth. The Breakdown Gallery passes beneath a pronounced surface gulley, and the decreased depth and leakage from surface flow in the gully may be responsible for the extensive breakdown in the cave. The Cathedral Room is beneath the side valley that contains Hause Run and Hause Waterfall Cave and appears from the surveys to be very close to the surface.

Hause #2 Waterfall on the hillside is only about 100 ft from the Turnpike inside the cave. A dye trace (Zotter 1963, 1965) showed that water that sinks at Wolfe Swallow Hole (the sink point of Big Run, the northwestern branch of Hause Run) emerges at Hause #2 Waterfall. It remains possible that the water emerging from a bedding plane at the top of #2 waterfall is derived from the cave stream because to get from Wolfe Swallow Hole to the waterfall, the stream would have to pass over (or under) the cave stream if it is not a tributary. The dye ultimately appeared at the Overholt Blowing Cave spring. The emergence of water at the top of Hause #1 Waterfall is 52 ft lower than the emergence of water at #2 Waterfall, and a dye trace shows that the water that sinks at the base of #2 Waterfall re-emerges at the top of #1 Waterfall. An altimeter survey in 1961 gave an elevation of the #1 emergence as 2627 ft, and an overland survey gave an elevation of the #2 emergence of 2679 ft.

The surface stream called Dry Creek heads in a deep valley between Spruce Flats and Stony Creek Mountain. Under low flow conditions, the stream sinks in its bed and at the Dry Creek Swallow Hole (Fig. 6.13). There is a well-developed channel downstream from the swallow hole so that storm flow and snow melt that exceeds the capacity of the swallow hole continues as a surface stream. The furthest point of exploration in Overholt Blowing Cave is only half a mile downstream from the Dry Creek Swallow Hole. It would seem obvious that the Dry Creek Swallow Hole is the source of the stream in the cave but a dye test was conducted in the early 1960s to check it. The dye did not appear at Overholt Blowing Cave. Thinking that insufficient dye had been used, the test was repeated and was found to emerge at the Sharp Farm Spring on the east side of Stony Creek Mountain. Unfortunately, the Sharp Farm Spring is the water supply for the village of Campbelltown and the dye trace turned it green.

The upper Dry Creek Valley was mainly sheep pasture at the time of these observations. The sheep found the limestone overhang at the swallow hole to be a cool resting place, so the sink of Dry Creek is through a thick mixture of stream cobbles and sheep dung, not the best source for a water supply. The elevation difference between the swallow hole just below the 3000-foot contour and the Sharp Farm Spring at 2450 ft means a roughly 500-foot drop in elevation with deep flow beneath Stony Creek Mountain. The flow line is across the dip from near the top of the limestone at the swallow hole to near the bottom at the spring. Several attempts have been made to dig open the hypothetical cave at the swallow hole but to date with no success.

Fig. 6.13 The Dry Creek swallow hole. Photo by W.B. White

6.6 The Carpenter-Swago System

6.6.1 Description of the Cave System

The Carpenter-Swago System has two entrances, both pits. Swago Pit is 500 ft north of a sharp bend in the Spruce Flats Road in the bottom of a steep, narrow gully. The gully carries a tributary of McKeever Run which flows off the clastics, across the Alderson Limestone and the Greenville Shale, through a small cave in the upper portions of the Union Limestone, and finally plunges into Swago Pit (Fig. 6.14) at an elevation of 2760 ft. Carpenters Pit is an open shaft on the hillside below the Spruce Flats Road also at an elevation 2760 ft. (Fig. 6.15).

The overall pattern of the cave (Fig. 6.16) displays two long, straight, parallel passages connected at the southwest end by cross passages. The vertical development is complex in the southwest section with multiple overlapping passages. The cave contains 3.1 miles of mapped passage with an estimated 1.5 miles un-surveyed. The total vertical extent is

Fig. 6.14 The Swago Pit entrance with waterfall. Photo by W.B. White

Fig. 6.15 The Carpenters Pit entrance. Photo by W.B. White

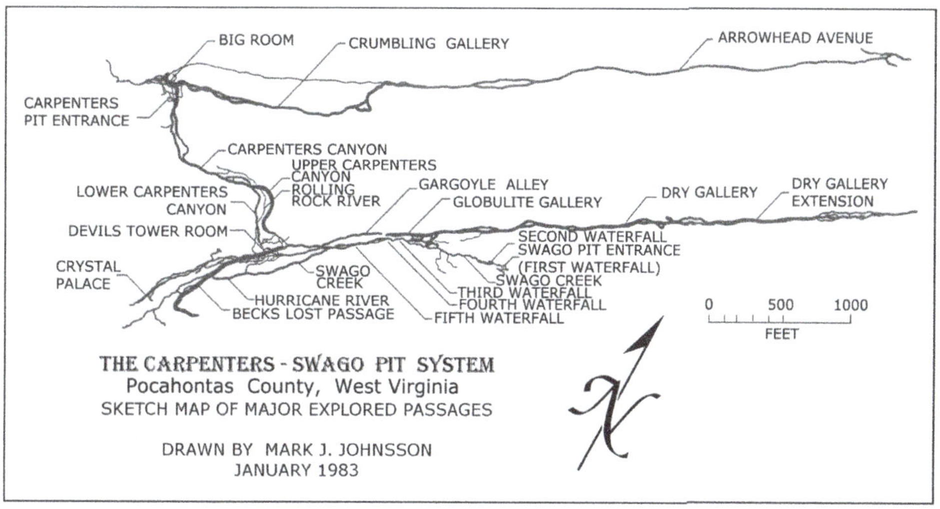

Fig. 6.16 Line map of the Carpenter-Swago System. Map by M.J. Johnsson

239 ft. The survey is presented in the electronic map file as a series of sectional maps with a master map showing the relationships (M-6.14 to M-6.20).

A low stream passage at the base of Swago Pit takes the stream 120 ft to the Second Waterfall where it plunges 25 ft into a deep pool. A fissure passage crosses the stream passage and can be followed to several shafts which allow access to the second level of the stream passage downstream from the Second Waterfall (Fig. 6.17). The stream passage continues southwest for 180 ft with a 5-foot by 5-foot cross section typically with several feet of water (Fig. 6.18), then opens into a 4–7-foot high, 10–30-foot wide passage which continues 320 ft to the Third Waterfall. Beyond the 33 foot drop is 50 ft of additional passage to the Fourth Waterfall. The base of this 24-foot drop in considered the connection with Carpenters Pit.

From the wide section of the second level, a number of small crawlways on the northwest side of the passage lead, via a complex breakdown area, to the eastern of the two long straight passages. To the southwest, the passage is called the Globulite Gallery, 20–30 ft high and wide, that extends 300 ft to a blockage by a massive flowstone slope. To the northeast, it is called the Dry Gallery that extends 1900 ft to an apparent end in breakdown. At the end of the Dry Gallery, one can crawl beneath the breakdown at a number of points and enter a low stream passage that parallels the Dry Gallery beneath the breakdown. After 300 ft, the passage opens into an extension of the Dry Gallery which continues

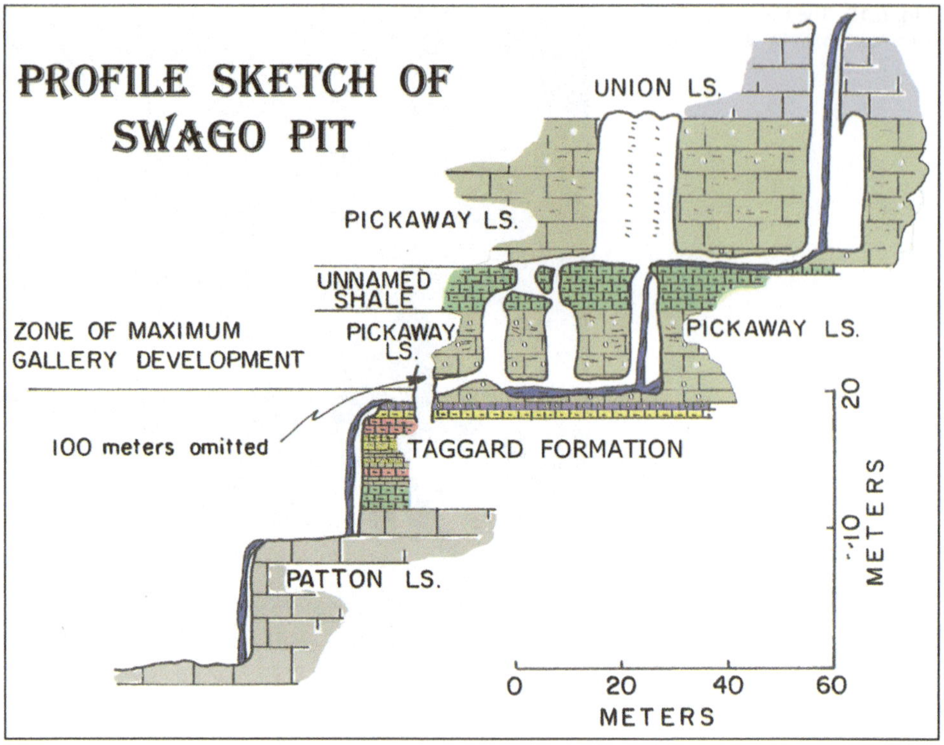

Fig. 6.17 Profile of the Swago Pit entrance area

for and additional 1600 ft. The passage ends in a 124-foot high shaft nearly 4000 ft from Swago Pit and only a few hundred feet from passages in Overholt Blowing Cave. The shaft was scaled in 1979 but no passage was found at the top.

Carpenters Pit is 4–6 ft wide and 40 ft long at the top and 76 ft deep. An additional 16-foot offset pit opens into a room with a canyon cut in the floor. This room or ledge is Upper Carpenter's Canyon. An additional 29-foot drop takes one to the floor of Carpenter's Canyon.

To the northwest from the floor of Carpenter's Canyon, the canyon opens into a large passage 50–100 ft high and 34–40 ft wide that extends north, then west for 300 ft, the Big Room. A stream enters the corner of the Big Room as a 20-foot waterfall emerging from a narrow fissure passage. A steep breakdown slope, the Foggy Breakdown Mountain, descends to the floor of the Big Room. To the north from the entrance pit along Upper Carpenter Canyon, one can cross the canyon and enter the Crumbling Gallery, the southwestern end of the western long straight passage. About 300 ft along Crumbling Gallery, openings in the breakdown floor connect with a narrow canyon below which doubles back and connects with the Big Room. Crumbling Galley continues with occasional pits in the floor for 1500 ft when it reaches a flowing stream. The stream passage is incised below the level of Crumbling Gallery and apparently crosses under it multiple times. This is the stream that emerges from the fissure passage to form the waterfall in the Big Room but has not been surveyed. Upstream 200 ft, the stream emerges from a breakdown that blocks the main passage. A small passage, Paul's Crawl, bypasses the breakdown and connects with the continuation of the main passage, now called Arrowhead Avenue, with the stream flowing on its floor. Arrowhead Avenue is the western of the two long straight passages and extends as an 8–10 foot high, 4–6 foot wide roughly rectangular stream passage for 2000 ft to the northeast, crossing under Spruce Flats, and ending at the base of several shafts on the side of Dry Creek Valley.

Southwest from Carpenters Pit, Carpenter's Canyon averages 30–50 ft high and 10–20 ft wide and continues 350 ft to an obscure climb on the western wall. This climb leads upward 15 ft to the Upper carpenter's Canyon. Lower Carpenter's Canyon continues southeast for 300 ft to end in a series of breakdown rooms beneath the Devil's Tower Room. Rolling Rock River is encountered by descending a 10-foot pit in Lower Carpenter's Canyon. This stream passage may be followed 350 ft and ranges from 1 to 10 ft high and 4 to 10 ft wide. Rolling Rock River is the downstream continuation of the stream in Arrowhead Avenue and from the waterfall in the Big Room.

At the top of the obscure 15-foot climb, Upper Carpenter's Canyon can be followed as a high ledge back to the bottom of the entrance pit. To the southeast, the upper canyon extends 300 ft to the Devil's Tower Room (named for a

Fig. 6.18 Stream passage below Second Waterfall in Swago pit. Photo by W.B. White

similar-shaped stalagmite in the center of the room). This room is one of the major junctions in the cave. A series of passages extending 500 ft southwest comprise the Crystal Palace Section. Gargoyle Alley extends northwest for 200 ft as a 6–10 foot high and wide passage. A steep flowstone slope on the northeast side of the Devil's Tower Room gives access to a lower level. To the southwest is Beck's Lost Passage, a 30–50-foot high, 20–30-foot wide elliptical trunk passage that extends 1000 ft to the southwest. The passage is choked with fill at the end, and it has no obvious hydrologic link to other passages in the system. To the northeast of the Devil's Tower Room is a 20–50-foot high, 4–8-foot wide canyon leading 400 ft to the base of the Fourth Waterfall and the connection to Swago Pit.

Midway along Beck's Lost Passage is the Hurricane River, a low, wet stream passage that continues upstream and downstream. Upstream 100 ft is the confluence of the Rolling Rock River and the Swago Pit stream. Downstream 300 ft is the final sump of the Carpenters-Swago drainage at an elevation of 185 ft below the entrance.

6.6.2 Hause Waterfall Cave

The entrance to Hause Waterfall Cave is at the base of a waterfall on a tributary of Dry Creek. A stream, informally called Hause Run, rises on Spruce Flats and flows down into the Dry Creek Valley where it sinks into Hause Waterfall Cave. A complex entrance series (Fig. 6.19) leads to 500 ft of nearly linear passage that begins as a small phreatic tube and later develops as a deep canyon in the floor. Hause Waterfall Cave is the source of the stream in Arrowhead Avenue but, although the passages line up properly, no physical connection was found during the early exploration. Sometime later, a collapse opened a connection and Hause Waterfall Cave is now a back entrance to the Swago-Carpenter System.

6.6.3 Roadside Pit

Roadside Pit may be considered a satellite cave to the Swago-Carpenter System but has no known connection to it. The entrance is a small hole in the ditch of the Spruce Flats Road about halfway between Swago Pit and Carpenters Pit at an elevation of 2800 ft (Fig. 6.20). The entrance is a 55-foot shaft. The entrance pit is barely large enough at the top to admit an explorer but which bells out 15 ft below and descends through the ceiling of a room.

The main passage is 300 ft long with an irregular breakdown floor. At the end of the main channel, a 6-foot waterfall drains though the breakdown. Beyond is a series of shafts elongated along the guiding fracture. To the right of the main channel about 100 ft from the waterfall, a crawlway leads to the 12-foot high, 70-foot wide, 180-foot long Big Room. Prior to the 1970s, this was the known extent of the cave. A 10-foot pit in the southeast corner of the Big Room gave access to a crawlway which opened into a set of northeast-southwest trending passages on several levels as shown on the electronic map. The present length of the cave is 4457 ft and the vertical extent is 166 ft (electronic map M-6.21).

Water enters the cave through shafts and drips and multiple points. These aggregate into two streams, the Corkscrew River and the Rimstone River, which meet at the Junction Room and then flow through a low stream passage to the southwest. The stream passage can be followed for 500 ft to where the stream is lost in breakdown 166 ft below the entrance.

6.6.4 Geologic and Hydrologic Relationships

The tops of both Swago Pit and Carpenters Pit are in the upper part of the Union Limestone. The Swago Pit stream crosses the

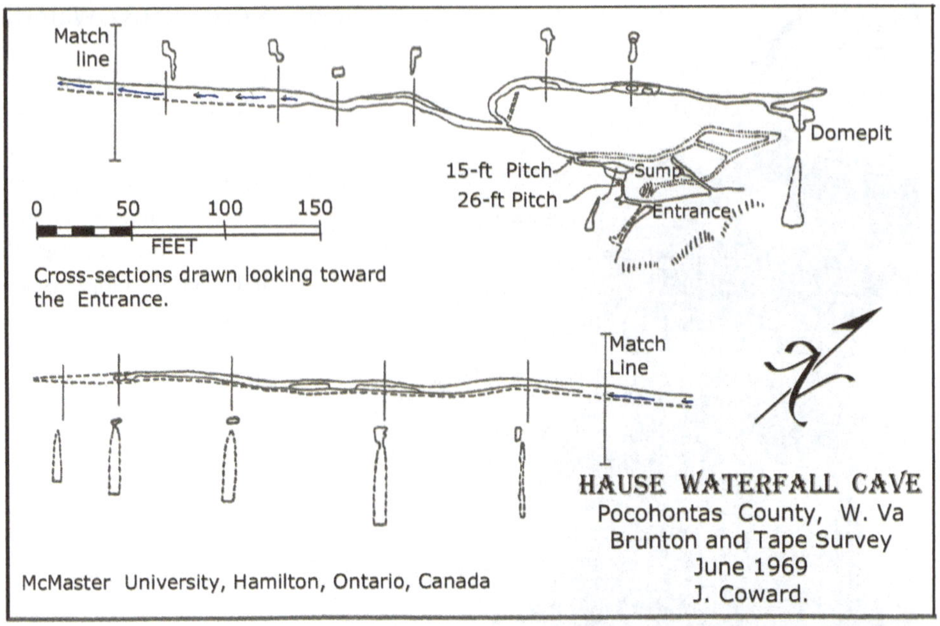

Fig. 6.19 Map of Hause Waterfall Cave. Adapted from Coward (1975)

Fig. 6.20 Entrance to Roadside Pit. Photo by W.B. White

lower units of the Greenbrier Limestone as a sequence of step-like waterfalls (Fig. 6.17). Carpenters Pit passes through the Union and Pickaway Limestones and reaches the first cave passages at the contact between the Pickaway and the Taggard Formation. The Taggard Formation as a confining layer is a dominant control in the Swago-Carpenter System. Both the Dry Gallery and Arrowhead Avenue, the two long straight parallel passages, appear to have developed as perched drainage on top of the Taggard Shale. The floor of the Dry Gallery is somewhat above the Taggard and is now abandoned except for a small stream in the extension. Arrowhead Avenue carries an active stream that derives its flow from the sink of Hause Run at Hause Waterfall Cave. Only at the southwest end, in Crumbling Galley and Carpenter's Canyon, has the stream cut its way through the Taggard into the underlying Patton Limestone.

Although many of the Greenbrier Limestone caves develop along bedding plane in response to various

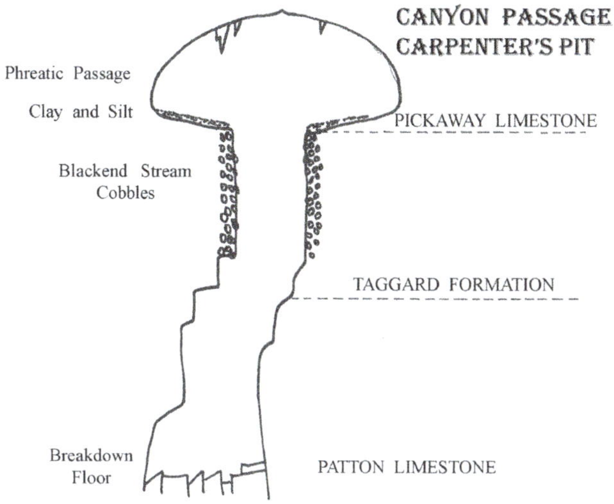

Fig. 6.21 Cross section of Carpenter's Canyon

lithologic controls and the hydraulic gradient, the Carpenter-Swago System displays very strong structural control. The Arrowhead Avenue and Dry Gallery passages are nearly linear, ignore surface topography and drainage, and are clearly guided by N60°E master fractures.

Carpenters Pit made an interesting demonstration of the quite different characteristics of karst aquifers (White and Schmidt 1965). In classical porous media aquifers, the water table is a subdued image of the overlying topography—high beneath the hills and low under the valleys. Hause Run continues serenely down Arrowhead Avenue as a low gradient, free-surface stream in an air-filled cave passage. Where it passes beneath a spur of Spruce Flats, the cave passage is 700 ft below the land surface. What water table?

The time sequence for passage development is unknown. The cross section of Carpenter's Canyon is instructive (Fig. 6.21). After the upper passage had formed at the Pickaway–Taggard contact, there was a massive flux of clastic sediments that deposited a thick layer of stream cobbles in the passage. When renewed downcutting resumed, the stream that formed Carpenter's Canyon cut through the sediment layer but left a thick sequence of cobbles on the walls of the canyon. These deposits must be very old, predating the sections of cave below the Taggard.

There are many great vertical shafts, some in the caves and others as isolated shafts with openings on the hillsides. These form near the edge of the caprock—either the Mauch Chunk Formation or the Greenville Shale—where water flowing off the caprock reaches the limestone and developed open pathways to the subsurface. They are not usually found in cave passages below the caprock. The large, roughly circular, shafts are formed in the vadose zone by fast-moving films of unsaturated water. The mechanism is similar to that developed for the large vertical shafts of the Mammoth Cave area (Brucker et al. 1972).

There remain some interesting inconsistencies. Zotter (1963, 1965) reported a Canadian dye trace that showed the combined Carpenters streams—Hurricane River—resurged at the top of the Barnes Pit Waterfall. The final observation of Hurricane River in the lowest stream passage is 185 ft below the entrance, an elevation of 2575 ft. The Barnes Pit Spring is at 2520 so the elevation difference of 55 ft would be reasonable. However, for the stream to get from the lowest level of Carpenters Pit to the Barnes Pit Spring would require a channel under the valley fill in McKeever Run. The flow path would have to go deep and then come up again.

There is a further inconsistency. The final disappearance of Hurricane River is 50–60 ft below the shaly beds of the Taggard Formation. The Barnes Pit Spring is on top of the Taggard Formation. The Taggard/Pickaway contact at the base of Carpenters Pit is at an elevation of 2660 ft. The top of the Taggard at Barnes Pit Spring is at 2520 ft, resulting in a 140-foot discrepancy in the position of the Taggard/Pickaway contact. Either there is a massive error in the elevation data, a massive error in the identification of the strata, or there is an unknown structural feature along the valley of McKeever Run.

6.7 The Infeeders to the Cave Creek Spring

Cave Creek Spring drains roughly two-thirds of the Swago Basin including the tributaries of McClintock Run and Overholt Run. These tributaries sink at the upper limestone contact but few accessible cave entrances have been discovered. Based on discharge characteristics, Cave Creek Spring should be the outlet for a large and complex cave system but only a few fragments have been discovered.

6.7.1 Cave Creek Cave

The entrance to Cave Creek Cave is at the head of a ravine, a rectangular opening 70 ft wide and 10 ft high. The cave entrance is also the spring mouth with the stream flowing over a picturesque 6-foot waterfall, where the Hillsdale Limestone contacts the underlying Maccrady Shale. The passage divides just inside the entrance. The dry branch is 15 ft wide and 2–4 ft high developed along a bedding plane in the cherty limestone. The other branch is of similar dimensions and carries the stream. The branches rejoin after 350 ft to form a 3-foot high stream crawl that ends 125 ft upstream where massive slabs of breakdown block the passage (Fig. 6.22).

There is anecdotal evidence that the breakdown blockage occurred between 1948 and 1956. Davies (1949) reported 1500 ft of passage beyond the passage junction. According to a personal communication, a local resident traversed a long distance in the cave mostly along a dry passage of walking height. The slabs blocking the passage hinged downward from the ceiling and are likely to now be supported in in-washed stream gravel. An attempt in the 1960s to blast open the blockage was aborted because of lack of any obvious place to plant the charge. Modern techniques of microblasting might be successful.

6.7.2 Tub Cave

The entrance to Tub Cave is in a large collapse sink on the crest of a spur of ridge that extends southward from Swago Mountain (Fig. 6.23). The entrance, a 15-foot by 4-foot slot, is against the wall of a slump area at the bottom of the sink. It descends to the top of a breakdown slope at the ceiling of the main chamber of the cave.

Tub Cave consists mainly of a single chamber, 420 ft long, 200 ft wide and 40 ft high (electronic map M-6.22). The floor of the chamber is a mud flat that appears to be the bed of an intermittent lake. The western wall of the chamber is bedrock but the eastern side is a massive breakdown slope

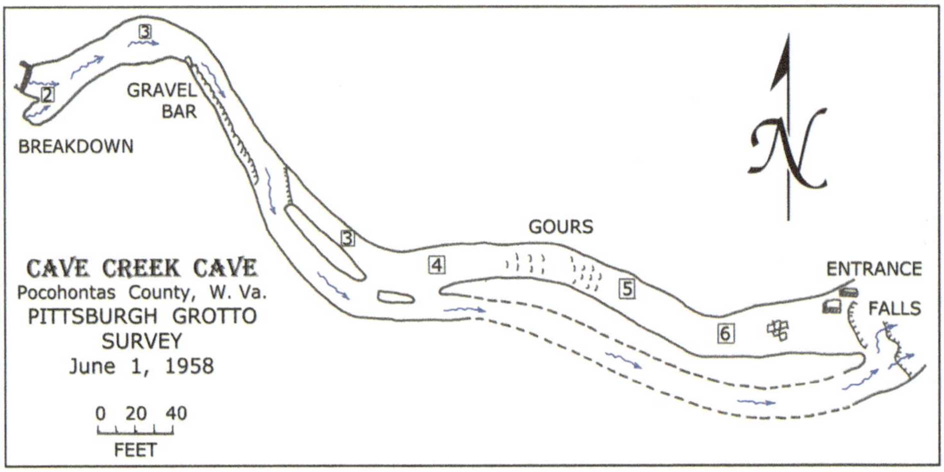

Fig. 6.22 Map of Cave Creek Cave

Fig. 6.23 Sinkhole that contains the Tub Cave entrance. Photo by W.B. White

that extends up to the ceiling. A small waterfall enters the chamber through a shaft in the ceiling and has built up a flowstone mound over the breakdown. The southwestern end of the chamber appears to be blocked by breakdown but it can be circumvented to give access to an additional smaller room that marks the southwestern limit of the cave.

A moderate-size stream enters the chamber from a passage at the northeastern end. It is joined by several other small streams, in addition to the waterfall, that enter the chamber. The combined streams flow in complicated patterns across the mud floor and exit through a small hole beneath the breakdown. The passage carrying the main stream can be followed upstream for some hundreds of feet —the exact distance varying between reports. The stream passage has a left wall (looking upstream) mostly of solid rock but a right wall of breakdown with many small openings extending upward through the breakdown. The passage is clearly formed around the edge of a massive collapse, possibly the entrance sink.

A large boulder on the edge of the road across from the McKeever farmhouse was taken as the benchmark for a number of altimeter and overland surveys conducted in the early 1960s. The best estimate from the topographic map gave an elevation of 2445 ft for the benchmark. One of these surveys was an altimeter traverse to establish elevations of Barnes Pit and Tub Cave with the following result. Elevations are calculated with respect to the benchmark.

Benchmark (P-13)	2445 ft
Lip of Barnes Pit	2493
Top of Barnes Pit waterfall	2520
Saddle at the edge of Tub Cave sink	2703
Tub Cave entrance	2590
Floor of Tub Cave	2534

The mud floor of Tub Cave is only 14 ft above the elevation of the Barnes Pit Spring. It seems likely that the floor of Tub Cave is perched on the Taggard Formation. This would place the cave in the Pickaway Limestone. In spite of the short horizontal distance and the small elevation difference, a dye trace from Tub Cave to the Barnes Pit Spring was negative. The test was positive to Cave Creek Spring.

Tub Cave is a somewhat enigmatic feature. The large cross-sectional chamber has the appearance of being a fragment of very large master trunk conduit blocked by collapse at both ends. Its elevation and stratigraphic position are close to those in the Carpenters-Swago Cave System on the opposite site of the valley but there is no comparable passage in the Carpenters-Swago System. Tub Cave is a hint of a previous large and complex cave system that has been destroyed by erosion except for a few fragments.

6.7.3 Barnes Pit

The entrance to Barnes Pit is a 20-foot pit at the edge of the McKeever Branch of the Overholt Run valley. The very dangerous descent is through the loose boulders that make up the valley fill material. The Barnes Pit stream rises from a broad platform of red shale that marks the top of the Taggard Formation, crosses the shale as a waterfall, and flows into the pit (Fig. 6.24). Continuous stream action contributes to the unstable entrance. Few explorers have felt any urge to push Barnes Pit, but Pittsburgh and McMaster explorers have provided a short description and a map (electronic map M-6.23). The cave is in the Patton Limestone.

The entrance pit leads to 50 ft of passage followed by a low wet crawlway opening into a small room with an 8-foot waterfall and two small wet leads. The right passage continues to a sharp 140-degree bend, then walking-height passage leading to a 22-foot pit. At the base of the pit is an extremely unstable breakdown room. This feature is marked on the McMaster map as a 60-foot high boulder choke. Its location would place it under the valley, and it may represent the site of a previous entrance. From the bottom of the breakdown room, a stream passage continues downstream as a 6- to 8-foot high, 2–3-foot wide passage which continues uniformly for 1800 ft to an 18-foot climbable drop.

Beyond the top of the 18-foot drop, a passage continues for 200 ft to an overhanging drop into a room with no passage leading off. At the base of the 18-foot drop, the passage is nearly blocked by a large rock after which is a 12-foot drop to a place where the stream disappears into a 6-foot wide, 3-in. high crack which marks the point of furthest exploration in Barnes Pit. The stream has been traced to Cave Creek Spring but Barnes Pit does not appear to be an access to the system.

6.8 Mineralogy

The caves of the Swago Creek Basin offer little in the way of exotic mineralogy. Calcite speleothems are found in modest quantities in those portions of the caves that are not protected by a shale or sandstone caprock. Caves beneath the caprock do not receive seepage water from overlying soils and thus tend to be devoid of speleothems. Special attention can be called to a few places. Roadside Pit is one of the better decorated caves. There is the flowstone mound

Fig. 6.24 Entrance to Barnes Pit. The stream emerges from a spring at the top of the Taggard Formation, flows across the shale and into the pit entrance. Photo by W.B. White

beneath the waterfall in Tub Cave. There is the Mountain Room and the Ivory Palace in Overholt Blowing Cave. And, there is the Globulite Gallery in the Carpenters-Swago System.

The Globulite Gallery contains a profusion of the botryoidal or nodular speleothems variously known as globulites, cave coral, cave popcorn, and cave grape (Fig. 6.25). Globulites are layered structures built up slowly as the calcite is deposited. The point of origin for the growth is a point of the cave wall or ceiling. The exact mechanism of their growth is not completely understood but when conditions are right, they usually appear in great numbers.

Gypsum appears as wall crusts and as sand on the passage floor in a few exceptionally dry passages, specifically Anne's Avenue, the Gypsum Crawl, and the Turnpike Passage in Overholt Blowing Cave, and in the Dry Gallery in the Carpenter-Swago System. The origin of the gypsum has not been established. Cave gypsum is commonly derived from the oxidation of pyrite in the limestone. There is an alternate source in the Swago Creek Caves. Nodules of anhydrite occur in some of the shaly limestones and the hydration of anhydrite, $CaSO_4$, to gypsum could be the source. Measurement of the sulfur isotope ratios in the gypsum could determine the source.

(a) (b)

Fig. 6.25 Globulites in the Globulite Gallery, Carpenter-Swago System. **a** On walls. **b** On stalactites. Photos by W.B. White

6.9 Concluding Comments on the Swago Creek Basin

Swago Creek, as the northernmost basin selected as part of the Greenbrier karst, is primarily fluviokarst. Runoff from much of the roughly 7 mi^2 drainage basin is by way of surface streams on clastic rocks. These feed a large and complex underground drainage system the empties into two large springs that provide the headwaters for Swago Creek. Inputs to the underground system are not efficient so that high flows from storms and snow melt override the insurgences and cross the karst in surface channels. As a result, the karst erosion surface is represented only by a few remaining fragments of upland. Surface flow has produced a dendritic pattern of stream channels in deep valleys which have removed large parts of the original erosion surface and also fragmented the cave system.

Acknowledgements This chapter is dedicated to the memory of G. Dallas McKeever, farmer and beekeeper, who owned important cave entrances and whose welcome to the cave explorers who arrived in the 1950s and 1960s allowed the explorations to get underway. Thanks are extended to all landowners in the Swago Valley who have tolerated cavers tramping across their property for more than half a century. A document such as this can only be written because of the exploration and mapping conducted by dozens of individual cavers. Only a few names appear on the maps and documents, but thanks to them all.

References

Brucker, R.W., J.W. Hess, and W.B. White. 1972. Role of vertical shafts in movement of ground water in carbonate aquifers. *Ground Water* 10 (6): 5–13.

Coward, J.M.H. 1975. Paleohydrology and streamflow simulation of karst basins in southeastern West Virginia, 394pp. PhD thesis, McMaster University.

Davies, W.E. 1949. *Caverns of West Virginia*, vol. 19, 353pp. Morgantown, WV: West Virginia Geological Survey.

Davies, W.E. 1958. *Caverns of West Virginia*, 2nd ed., vol. 19A, 330pp. Morgantown, WV: West Virginia Geological and Economic Survey.

Hamm, W. 2002. Working the Swago Creek area. *Pittsburgh Grotto Netherworld News* 51, Part I (1), 3–10; Part II (4), 14–20; Part III, (5), 14–20; Part IV (6), 8–13.

Medville, D.M., G.R. Dasher, and E. Werner. 1983. *An Introduction to the Caves of East-Central West Virginia: Guidebook to the 1983 NSS Convention*, 113–117. Barrackville, WV: West Virginia Speleological Survey.

Merrin, S. 1975. Accident report. *The Canadian Caver* 7 (2): 3–7.

Price, P.H., and D.B. Reger. 1929. *Pocahontas County*, 531. Morgantown, WV: West Virginia Geological Survey.

Storrick, G.D. 1992. *The Caves and Karst Hydrology of Southern Pocahontas County and the Upper Spring Creek Valley*, vol. 10, 215pp. Barrackville, WV: West Virginia Speleological Survey Bulletin.

White, W.B. 1960. Exploration and discovery at Swago Creek. *Pittsburgh Grotto Netherworld News* 8: 198–206.

White, W.B. 1964. Cavern and karst development in the Swago Creek Area, West Virginia, USA. In *Proceedings of the 3rd international congress of speleology*, Vienna, Austria, vol. II, 279–285.

White, W.B., and V.A. Schmidt. 1965. Hydrology of a karst area in east-central West Virginia. *Water Resources Research* 2: 549–560.

Zabrok, P. 1993. Sons of Overholts. *Canadian Caver* 25 (2): 20–28.

Zotter, H. 1963. Stream tracing techniques and results: Pocahontas and Greenbrier Counties, West Virginia. *National Speleological Society News* 21: 136–142.

Zotter, H. 1965. Stream tracing techniques and results: Pocahontas and Greenbrier Counties, West Virginia—Part II. *National Speleological Society News* 23: 169–177.

7 The Little Levels and the Hills Creek Basin

William B. White

Abstract

The Little Levels are a segment of the Greenbrier karst that may also be a segment of the Harrisburg erosion surface. The karst surface truncates the westward dipping Greenbrier Limestone. The underlying Maccrady Shale, an effective aquiclude is exposed at the land surface along the southeastern edge, forming a groundwater dam between the karst and the deeply incised Greenbrier River. All drainage in the central portion of the Little Levels is subsurface either east to Stamping Creek springs or southwest to Locust Spring. However, few parts of the underground are accessible to observation. The caves of the Little Levels are of intermediate size, one to two miles, and many carry active streams which allow some interpretation of the eastern and western edges, but not the central portion.

7.1 Introduction

The "Little Levels" is a name given to a low relief upland plateau with the town of Hillsboro roughly at its center (Fig. 7.1). US Highway 219 crosses the area from northeast to southwest. The upland is underlain by the lower units of the Greenbrier Limestone which is dissected into broad, shallow closed depressions. Surface drainage crossing the Little Levels consists of Stamping Creek to the northeast and Millstone Creek to the southwest. There are no surface streams crossing the central portion of the Little Levels (Fig. 7.2). Millstone Creek ends in the closed depression of Beards Blue Hole so that the western portion of the Little Levels drains only by underground routes. The topography of the Little Levels and its immediate surroundings are shown on USGS Hillsboro, Lobelia, Droop, and Denmar Quadrangles.

Electronic supplementary material
The online version of this chapter (doi:10.1007/978-3-319-65801-8_7) contains supplementary material, which is available to authorized users.

W.B. White (✉)
Department of Geosciences, Deike Building, University Park, PA 16802, USA
e-mail: wbw2@psu.edu

The Little Levels are separated by Droop Mountain from the Hills Creek Valley to the west. The Hills Creek drainage is complex with a portion of the flow connected with the Little Levels while another portion drains through the Friars Hole cave system to the southwest (Chap. 8). Hills Creek and Bruffey Creek come into the discussion of both karst areas.

The caves of the Little Levels for the most part have been known for a long time. Many have large, obvious entrances that would be well-known to local residents. Davies (1949) reports at least partial descriptions of Beards Blue Hole, Blue Springs, Bruffey Creek Cave, Hills Creek Cave, the Hughes Creek Caves, Locust Creek Spring, Martens Cave, Martha's Cave, and Poor Farm Cave. Exploration by cavers from the 1950s onward extended many of these caves and added others. A listing and description including many of the smaller caves were given by Storrick (1992).

The cave descriptions that follow draw heavily on Storrick's account and on various unpublished documents. A caveat: much of the geomorphic interpretation depends on relative elevations of various springs, cave passages, and other features. Many of the available elevations were scaled from topographic maps which in this area have a 40-foot contour interval. Elevations in caves are given as depth below the cave entrance, but vertical controls on cave surveys often have a great deal of uncertainty.

W.B. White (ed.), *Caves and Karst of the Greenbrier Valley in West Virginia*,
Cave and Karst Systems of the World, https://doi.org/10.1007/978-3-319-65801-8_7

Fig. 7.1 Google Earth image of the Little Levels

7.2 Geomorphic and Geologic Setting

The Little Levels is a doline karst at an elevation of 2300–2400 ft. The area was interpreted as a remnant of the Harrisburg erosion surface by White and White (1991). The limestone dips to the west so that the eastern edge of the surface is the contact between the limestone and the underlying Maccrady Shale. On the eastern edge of the Little Levels, the land drops off sharply into the incised valley of the Greenbrier River at an elevation of 2050 ft. The shale and the underlying Pocono Sandstone form a groundwater dam that prevents underground drainage from reaching the river by a direct route. Stamping Creek has cut a deep gorge through the clastic rocks between Mill Point and the river as has Locust Creek on the western side.

The high plateau to the west is collectively known as the Yew Mountains which reach elevations of 4000–4500 ft. The sandstones, conglomerates, siltstones, and shales are only slightly deformed and have a generally westward dip. The plateau has been dissected by the westward flowing Williams River, Cranberry River, and Cherry River into fragments which are also called mountains and have individual names. Immediately to the north of the Little Levels is Little Mountain with a high point of 3400 ft. To the northeast, Rogers Mountain separates the Little Levels from the Swago Creek Basin. The western boundary is Droop Mountain, a topographic and structural anomaly because the Greenbrier Limestone also crops out on the west side to produce the highly karstic Friars Hole Valley. Droop Mountain has a narrow flat top at 3000 ft elevation because of the erosional resistance of the Droop Sandstone, a unit of the Mauch Chunk Formation. There is evidence for structural complexity, likely one or more faults, within Droop Mountain that accounts for its survival as a north–south barrier across the karst.

7.3 The Stamping Creek Drainage

Stamping Creek originates on a saddle of Cranberry Mountain at an elevation of 4100 ft. It flows down a long valley cut deeply into the Yew Mountains (Fig. 7.3). Stamping Creek sinks in its bed near the limestone contact but maintains a surface channel which is dry under low-flow conditions (Fig. 7.4). The resurgence of Stamping Creek is at a group of springs 1200 ft upstream from the highway

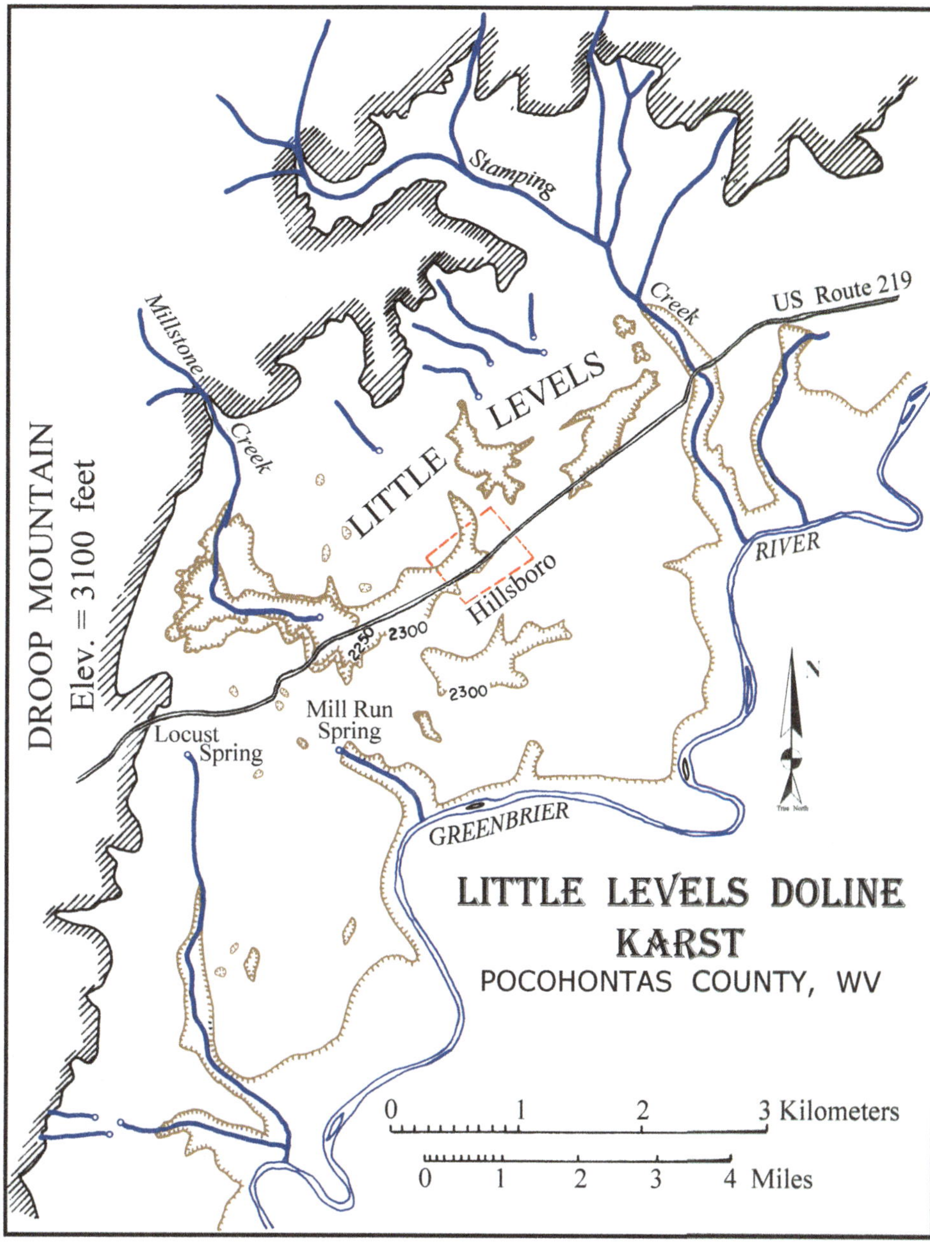

Fig. 7.2 Map of the Little Levels showing the largest closed depression

intersection at Mill Point. The spring southwest of the surface channel is the entrance to Blue Spring Cave. To the northeast of the surface channel, close to the highway are a group of springs called the Roadside Resurgences. A dye trace from a sink point in the stream channel about 3 miles upstream appeared at the Roadside Resurgences but not at Blue Spring Cave. Both springs are in the Hillsdale Limestone at an elevation of 2196 ft. Curiously, Blue Spring is shown on the USGS Marlinton 15-min quadrangle (1923) but not on the Hillsboro 7.5 min quadrangle (1977).

7.3.1 Blue Spring Cave and Spring

The cave entrance is a 20-foot wide by 6-foot high opening at the head of wooded alcove. Inside, one can follow a stream passage for 1500 ft to a breakdown room (electronic map M-7.1). Straight ahead, dry passages can be followed to the south for about 500 ft. The stream in the entrance passage rises from the breakdown. At the breakdown room, the main passage turns sharply to the north and can be followed as a walking height passage and then a stoopway for an

Fig. 7.3 The Stamping Creek Valley seen from route 150 on top of Cranberry Mountain. Photo by W.B. White

Fig. 7.4 The dry channel of Stamping Creek upstream from the springs. Photo by W.B. White

additional 1500 ft. There are three sumps in three short side passages along the west wall of the passage at 600, 850, and 1200 ft from the sharp turn. At the end of the passage is another sump. Continuing ahead is a narrow tube that ends after 150 ft in yet another sump. The north passage appears to be an overflow route with the main discharge in a lower inaccessible passage.

7.3.2 Steam Cave

Relatively little is known about the internal drainage of the Little Levels. Steam Cave is a half-mile fragment of stream passage with the stream flowing east toward Blue Spring to which it has been dye-traced (Fig. 7.5).

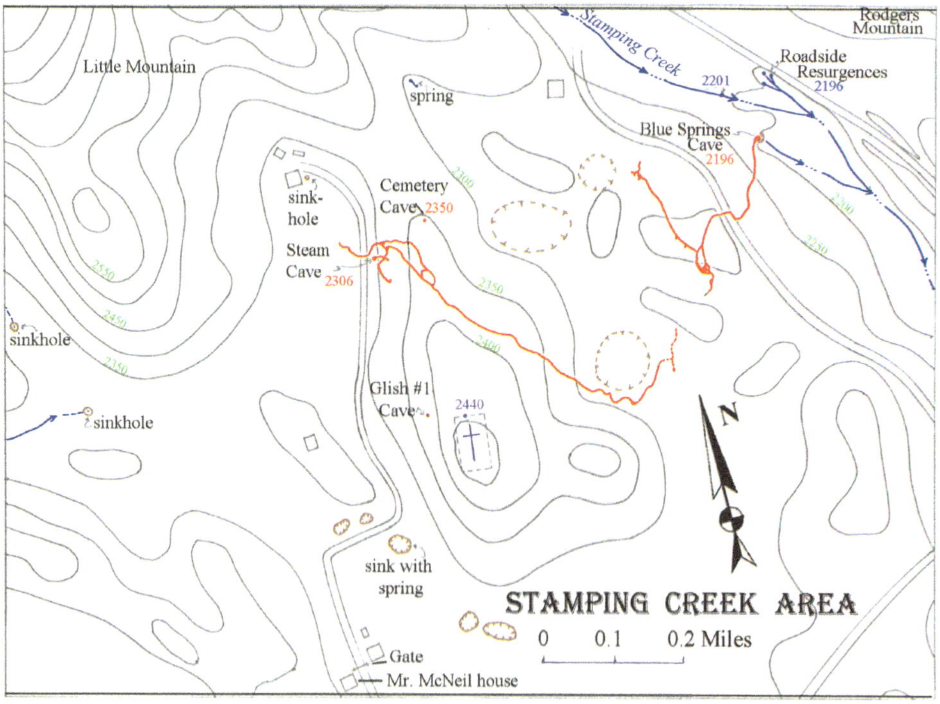

Fig. 7.5 The eastern edge of the Little Levels showing the relation of Steam Cave, Blue Springs cave and the resurgence of Stamping creek. Redrawn from an original map by F. Potter in 1965

The entrance is an obscure 10-foot pit in an open field at an elevation of 2306 ft. At the bottom, a canyon passage leads 50 ft to a junction. Straight ahead, the passage descends a flowstone slope and ends in a flowstone choke. To the right, the passage can be followed for about 100 ft, becoming a low crawl and ending in breakdown at the edge of the hillside. To the left, and beneath a shelf, is a crawlway leading deeper into the cave, reaching the stream passage after 50 ft (electronic map M-7.2).

Upstream, the passage can be followed for 300 ft to a low airspace that continues with some air flow. Downstream, the cave stream and two parallel dry crawlways can be followed to the southeast for 150 ft where the crawls merge and open to pleasant, dry, walking passage 10 ft high and 10–15 ft wide. The stream, an incised canyon, occasionally intersects the dry passage. The dry passage also gradually becomes more canyon-like and trends southeast for 350 ft to a junction with the stream. Beyond, the cave is a wet walkway for 600 ft to a junction with an upper level passage entering from the east. This passage ends in mud fill after a short distance.

Beyond the junction, the cave trends more to the east and drops through a series of rapids and pools in a passage up to 30 ft high. After several short downclimbs in the water, the gradient decreases and the passage turns to the northeast with two natural bridges across it. The passage becomes lower and finally ends in a sump. The sump is about 500 ft from the nearest point in Blue Spring Cave. The sump is 91 ft below the entrance, at an elevation of 2215 ft, only 19 ft above the elevation of Blue Spring. The stream flow in Steam Cave is reported to be significantly less than the flow in Blue Spring, indicating other unobserved tributaries.

7.3.3 Cook Pot

The top of the Greenbrier Limestone lies on the slopes of the surrounding mountains several hundred feet above the Little Levels. Tributary streams flowing from the sandstones and shales of the upper slopes sink near the limestone contact, but the slopes seem to be poorly explored and there are few reports of inlet caves. An exception is Cook Pot. The entrance pit is on a tributary of Stamping Creek named Tilda Fork on the topographic map at an elevation of 2750 ft. Cook Pot, with 1.42 miles of survey, is an excellent example of a high relief inlet cave and is described in detail. The original description was published by Medville (1971). The description was reproduced by Storrick (1992) and is reproduced again below (electronic map M-7.3)

The entrance is an unobtrusive 30-foot pit with a second 12-foot drop at the bottom. At the base of the pit, a passage leads downward for 10 ft and intersects a shaft, the floor of which is 20 ft below the passage. Four passages lead off from the base of the shaft. Two end in breakdown, third becomes too low to follow, and the fourth leads to the continuation of the cave. The continuation passage is an obvious belly crawl from the bottom of the shaft which can be followed for 30 ft to the top of another drop. Rather than descending the drop at the end of the crawlway, one continues out over the drop by traversing over an open-bottomed canyon for another 30 ft to a large chockstone from which one can descend a free 43-foot drop into the room below.

At the base of the 43-foot drop, the cave continues in two directions. To the northeast, the passage can be followed upstream for about 1000 ft through a series of dry crawls, narrow canyons, and high domes. These domes are 40–60 ft in height and are about 20 ft in diameter. At the extreme upstream end of this section, the stream emerges from a sump but the passage continues indefinitely as a crawl over gypsum sand.

To the southwest of the 43-foot drop, the passage can be followed as a stoopway and crawlway for 250 ft to a major passage junction—Fork falls. This waterfall is 15 ft high and can be down-climbed on large breakdown blocks. The stream at the falls can be followed downstream for 350 ft through a low wet passage to another junction. Continuing 15 ft beyond Fork Falls, one comes to a second passage junction. Here the cave can be followed right (north) or left (south) in large passages. A second stream flows in this passage and parallels the Fork Falls stream. Following this stream upstream to the north, one traverses a long ledge to the top of another waterfall, Spoon Falls, and can follow the stream upstream for more than 1500 ft through a series of large domes. These are the largest in the cave. The dome-stream complex in the northwest end of the cave has not been completely explored.

Following the Spoon Falls stream downstream to the south at the passage junction just beyond Fork Falls, the cave continues for 200 ft as a 20-foot wide, 30–50-foot high passage. A small stream enters this passage from the west. It is possible to follow this stream for about 250 ft to a breakdown choke. After the large passage, the Spoon falls stream can be followed through a low crawl for 100 ft to a small done and a passage junction. Going left (east), one climbs up out of the stream passage and into dry canyon. This can be followed for 50 ft to the Fork Falls stream. Continuing straight ahead in the Spoon Falls stream (south), the cave continues as hands and knees crawl for 50 ft to a second junction. Again, by going left (east) and following the stream, one arrives at the Fork Falls stream. The combined stream then continues to the south. If, instead, one continues to the south at the second junction, dry walking passage is entered. This can be followed for 1500 ft to a series of domes. This section of the cave is dry, pleasant, and contains some nice flowstone and rimstone pools.

At the junction of the Fork Falls and Spoon Falls streams, the main part of the cave continues to the south. Beginning as walking passage, the ceiling soon drops to within one foot of the floor and with minor variations, remains there for the next 500 ft. This section of the cave is a long, wet, rocky crawl which can be unpleasant in wet weather and tiring as one leaves the cave. Beyond this crawl, the floor starts to drop and one follows a canyon passage 10–15 ft high for another 300 ft. One then comes to the last drop in the cave, this one being 32 ft down the side of a canyon. It is also possible to descend via a 30 foot waterfall just upstream from this drop. Below the 32-foot drop, one continues to the south for 100 ft to a 10-foot drop into a deep pool. This can be bypassed by traversing a ledge to the left (east) wall of the passage and chimneying down to the stream.

The cave continues for about 1000 ft as a 30-foot high and 6-foot wide passage to junction. To the left at this junction, one can follow large dry passage over breakdown for 200 ft to a terminal breakdown choke. A small stream entering this passage from an adjacent dome can be followed through a pit in the floor to the main cave stream. Continuing straight ahead at the last junction, a large dome, 50 ft high and 30 ft wide, is crossed. The main cave stream enters this dome from a low passage at its base, and a side stream enters via a 20-foot high overhung waterfall. A passage on the right (north) side of this dome climbs steeply until it is 30 ft above the dome. This passage can be followed for several hundred feet until it rejoins the main cave stream upstream from the dome. One can chimney up in this passage for another 20 ft through loose boulders to enter a large upper room. The upper room is about 100 ft in diameter, up to 20 ft high, and has several passages leading off from it.

From the upper room, one can gain access to the stream passage above the overhung waterfall that enters the dome below. This stream rapidly becomes too low to follow. Other passages leading from the upper room become quite small, although exploration here is not complete. Below the room, the main cave stream continues downstream for another 200 ft to the south as a low, wide passage. This passage ends in massive breakdown from above. Apparently at this point, the three passage levels in the cave, which are above one another, have joined to produce the breakdown. The cave terminates beneath a large surface sinkhole, 60 ft across and 30 ft deep.

The final sump of the cave stream is 223 ft below the entrance, placing it at an elevation of 2527 ft. This is about the elevation of the Little Levels surface and more than 300 ft above the elevation of the Blue Springs where the water is most likely to emerge. The entrance is near the top of the Union Limestone. The cave crosses the entire thickness of the Union and Pickaway Limestones with steep gradients, multiple drops, and many vertical shafts.

7.4 Locust Spring

Locust Creek rises at Locust Spring at an elevation of 2090 ft (Fig. 7.6). The stream flows southward for several miles in an incised valley with a small floodplain to its confluence with the Greenbrier River. Locust Spring emerges from the Hillsdale Limestone at the base of Droop Mountain. Locust

Fig. 7.6 Locust Spring. Photo by W.B. White

Spring is the drainage outlet for the western side of the Little Levels and also for a portion of the Hills Creek Basin on the other side of Droop Mountain.

The stream emerging from Locust Spring is sumped a short distance inside except during exceptionally dry weather. In dry times, it is sometimes possible to pass this section (Sump I) without diving equipment and enter a short section of walking passage. About 200 ft inside, a small waterfall enters the right side of the passage while the stream flows under ledges and breakdown. In a short distance, the passage has wall-to-wall water and the ceiling lowers to form Sump II. There is a passage under the left wall that goes upstream as a sump (Sump IIA) for about 250 ft leading to the main part of the cave (electronic map M-7.4). If the stream passage is followed, Sump IIB is reached. On the upstream side of this sump, the passage splits but both passages quickly end in breakdown with water coming out of the breakdown.

Upstream from Sump IIA is a walking passage that leads to a breakdown slope and a segment of large breakdown-floored trunk. This passage leads to a junction of two upstream passages. The right-hand passage carries most of the water that drains to the spring. The water has not been traced as an individual inlet. This stream can be followed for about 800 ft to a breakdown collapse. The stream emerges from the breakdown pile. Above the collapse is a large chamber with a passage heading southwest.

Back at the passage junction, the left-hand passage trends more or less southwest under the flank of Droop Mountain. This is the largest continuous passage in the cave although it doesn't carry much water. About 1400 ft upstream, a large passage comes in from the right. This is the other end of the southwest passage at the breakdown chamber. After another 1200 ft, the main passage intersects a fault. The passage becomes filled with breakdown with the stream underneath and can be followed until completely blocked by breakdown.

The main passages in the cave trend northeast–southwest and are thus more or less parallel with the trend of Droop Mountain well under its southeastern flank. At least three individual infeeders have been identified and there may be more under the breakdown. None of these has been individually connected to any of the sinking streams that have been dye-traced to Locust Spring. The stream at Sump IIB might be the infeeder from Beard's Blue Hole, and the large stream emerging from the Breakdown Chamber might be Hills Creek but this is purely conjecture.

7.5 The Millstone Creek Drainage

Millstone Creek rises on Viney Mountain at an elevation near 3400 ft. It flows off the clastics and sinks at the limestone contact where it reaches the northwest corner of the Little Levels. Although shown as a blue-line stream on the topographic map, in reality Millstone Creek is a grass-grown channel where it winds around the western edge of the Little Levels. The channel ends at Beards Blue Hole. Storm waters overflow the sink points on the mountain and go underground at Beards Blue Hole. There is no continuation of the channel downstream. Millstone Creek and its associated caves and surface karst are shown on the intersecting corners of Lobelia, Hillsboro, Droop, and Denmar quadrangles.

7.5.1 The Hughes Creek Caves

There are two Hughes Creek Caves, one above the Taggard Shale and one below. The cave entrances are in an alcove cut into the east flank of Droop Mountain. Under moderate flow conditions, a stream emerges from the upper cave, flows about 400 ft in a surface channel and then sinks into the lower cave as a waterfall over a lip of Taggard Shale (Fig. 7.7). The upper cave is formed in the Pickaway Limestone; the lower cave in the Patton Limestone.

The entrance to the upper cave is at the base of a 30-foot wide, 20-foot high limestone headwall. The entrance is a low opening which enlarges to a trunk 30 ft wide and 6–8 ft high (Fig. 7.8). At low flow, the cave stream sinks in a narrow slot in the floor and the outside channel is dry. At moderate to high flows, the stream emerges from the entrance. The cave passage is tubular, with large clay banks along the stream (electronic map M-7.5). It ends after 300 ft in a room at the edge of a large breakdown. Here there are three choices. One can climb up the breakdown, eventually reaching nearly 100 ft above the entrance. One can stay low, crawl in the stream, and follow the stream circumnavigating the breakdown pile for 500 ft until the passage chokes out in breakdown. The final choice is to traverse a ledge above the stream passage and reach an upper level tube, much the same size as the stream passage, and which doubles back, parallel and above the stream passage. This passage ends in breakdown against the hillside near the cave entrance.

The entrance to the lower cave is a hole in the stream bed where the Taggard Shale has collapsed into the underlying cave passage. Lower Hughes Creek Cave is an excellent example of what Palmer (1975) has called a floodwater maze (Fig. 7.9). Passage walls are covered with small scallops indicating high flow velocities. Passages tend to be washed clean. The cave has a maze pattern with every available joint opened by dissolution. The passages tend to be 6–10 ft high and 3–6 ft wide. Passages tend to follow a N20°E joint orientation.

By following the passage on the west side of the cave for about 500 ft to the south, one reaches the stream flowing from the upper cave. It can be followed downstream for another 150 ft to a junction with another stream coming from the north. The combined streams flow into a terminal pool 3–4 ft deep, 50 ft across and with about a foot of airspace. On the east side of this pool is a breakdown area with muddy passage above. This is the downstream end of the cave stream and is beneath the west side of the next valley to the east of the cave entrance.

The surface stream channel slopes downward from the entrance of the upper cave to the entrance of the lower cave. A surface channel continues beyond the lower cave entrance, but it slopes upward in the downstream direction to eventually intersect the channel of Millstone Creek. During high flows, the lower cave completely floods, finally forming a lake over the entrance. The lake rises, spreading along the channel until it finally spills over into Millstone Creek.

7.5.2 Martha's Caves

There are two Martha's (or Martha Clark's) Caves, an upper and a lower. Both entrances are in a 50-foot high cliff in a

Fig. 7.7 Surface segment of stream connecting Upper and Lower Hughes Creek caves and the entrance to Lower Hughes Creek cave. Photo by W.B. White

Fig. 7.8 Water crawl into the entrance of Upper Hughes Creek cave. Photo by W.B. White

low hill on the south side of Millstone Creek with the lower cave entrance at an elevation of 2250 ft and the upper cave 2277 ft. The entrances are located very close to the southwest corner of the Hillsboro Quadrangle. The exact stratigraphic relations are not known but probably the caves are in the Patton (possibly Sinks Grove) Limestone. Martha's Caves were explored and surveyed by members of the Pittsburgh Grotto in the early 1960s, and they produced the map included in this volume (electronic map M-7.6).

The lower entrance passage is 2 ft high and 30 ft wide, but increases to 8–10 ft high 40 ft inside. Breakdown nearly blocks the passage just beyond this point, but one can continue and enter a large passage with a ceiling height of 15–35 ft. Six hundred feet from the entrance, a side passage on the right leads northeast for 200 ft to a fork. The right fork leads upstream for 170 ft and then ends in a breakdown choke. The left fork heads downstream and becomes too tight to follow after 350 ft.

The main passage continues past the side lead another 500 ft in 15–25 foot high, 40 foot wide trunk passage to a second junction. One hundred and fifty feet up this passage is a third stream than can be followed downstream through Katherine's Ramble for an addition 450 ft to a breakdown slope.

The main passage continues for 400 ft past the second junction, through a breakdown walled area (the Chaos) to where it reaches the first of a series of pools known as the Great Wet Way. The pool's area is 300 ft long and intersects a large upper level trunk passage. One can follow the trunk northeast for 500 ft, there it overlies the Great Wet Way, or continue 1500 ft southeast in an increasingly large passage known as Alice's Gallery. Passage dimensions increase to 50 ft high and 120 ft wide before ending in a large breakdown room (the Rubble Room). A left side passage, 500 ft from the end (Dallas' Alley), terminates in a breakdown choke after 200 ft.

The entrance to the upper cave is to the left of the lower entrance and 27 ft above it. A 10-foot high, 10–15-foot wide passage heads southwest for 600 ft to where it intersects a stream passage. To the left, one can follow the stream passage upstream for 350 ft to a breakdown choke. The main passage, downstream, continues 250 ft to where a second stream joins it from the left. This stream can be followed upstream 300 ft to a 45-foot shaft with a waterfall. The downstream passage continues 1500 ft through silt banks until the cave finally ends in a mud fill and sump. Along the way one passes a 40-foot waterfall in a shaft on the right.

Although the individual flow paths have not been sorted out, it appears that the Martha's Caves are the convergence of many flow paths from the Little Levels draining toward Locust Spring still 4000 ft to the southwest. The stream gradients in the caves are relatively low with the lowest point in the lower cave only about 40 ft below the entrance. This would place the elevation at the final sump at about 2200 ft, still about 100 ft above the spring. The elevation difference suggests that both upper and lower caves remain in the Patton Limestone.

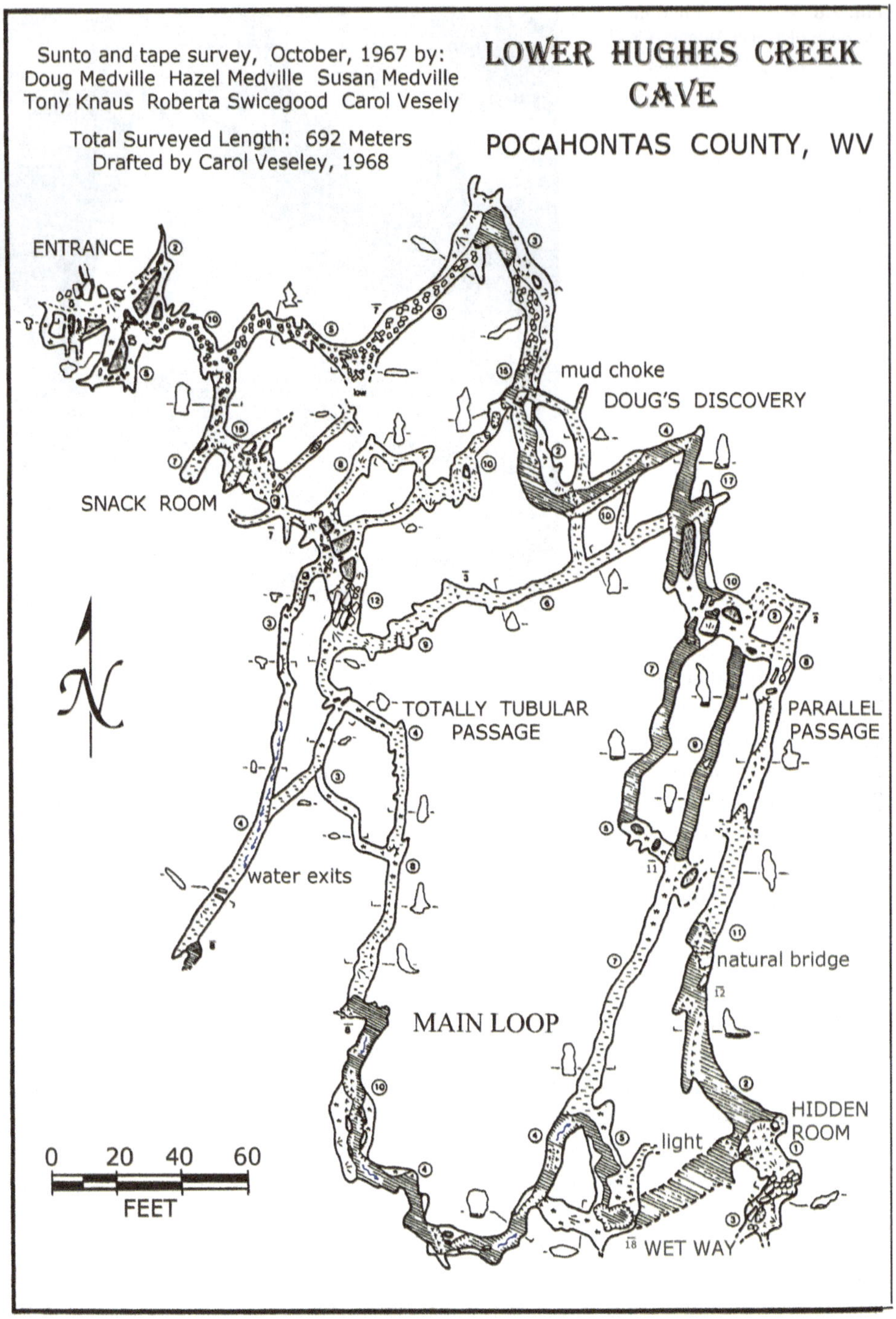

Fig. 7.9 Map of Lower Hughes Creek Cave. From Storrick (1992)

7.5.3 Beards Blue Hole

Beards Blue Hole is the downstream terminus of Millstone Creek. It is a large vertical-walled sinkhole at the bottom of which is a deep pool which is the sump for several infeeders that drain into the sink. On the southwest side of the sink, 2–4-foot high, 6-foot wide stream passage can be followed upstream 550 ft through two pools to the mid-west entrance. This small hole was dug open and emerges into the bed of Millstone Creek. Beyond this entrance, a small crawl leads 200 ft to the west entrance. A second stream enters the pit from the north. This passage, which can be followed for 300 ft to a sump, has been dye-traced from Grand View pit, half a mile to the northeast.

Located in the southwestern corner of the Little Levels in the same large closed depression as Martha's caves, the rising dip of the bedding suggests that Beards Blue Hole is in the Hillsdale Limestone. The pool has been traced to

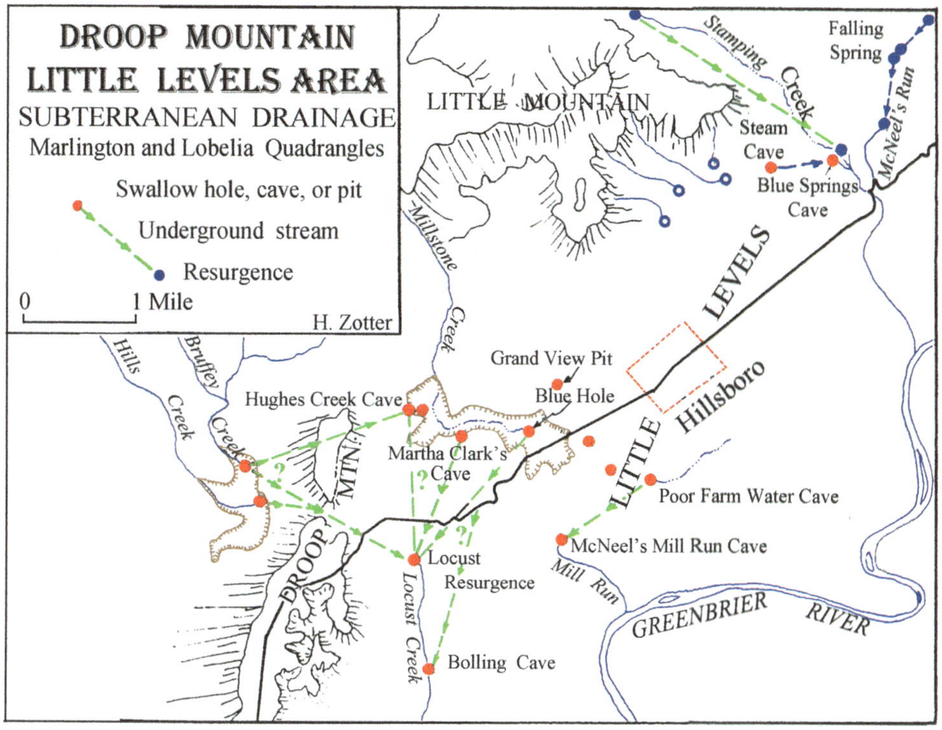

Fig. 7.10 Dye-tracing connections established for the Little Levels. Adapted from Zotter (1965)

Locust Spring. This area of the Greenbrier karst was the subject of Zotter's (1965) pioneering dye-tracing studies and most of what is known about the hydrologic interconnections of the underground drainage in the Little Levels is due to her investigations (Fig. 7.10).

7.6 Poorfarm Cave

Poorfarm Cave is one of the best known caves in the Little Levels. It was described in detail by Davies (1949) and was intensively investigated by Wolfe (1973) as part of his PhD dissertation on cave sediments. The Wolfe map was reproduced along with a detailed description by Storrick (1992) and is reproduced again as electronic map M-7.7.

The entrance to Poorfarm Cave is in an obscure sink in open woodland on the northeast side of a low hill at an elevation of 2360 ft. The cave consists mainly of two master trunks, strike-oriented, and roughly parallel. The entrance is a collapse into the northwest end of the lower passage. It is a smooth-walled passage, 10 ft high and 15–20 ft wide with a smooth clay floor. Six hundred feet from the entrance, an obscure side passage on the left connects with the Canyon, the other master trunk, near its northeast terminus in a large breakdown. The side passage continues and leads to a lower passage with an active stream, fed by several vertical shafts. The entrance passage continues for 800 ft as a 6–10-foot high, 30-foot wide trunk with alternating smooth floors and breakdown. It is interrupted by 500 ft of breakdown maze and then opens to another walkway. After another 600 ft the passage is again interrupted by breakdown where the canyon passage crosses above. After a final 800 ft, the entrance passage terminates in a breakdown plug.

The canyon passage heads southwest about 30 ft higher than the entrance passage. The first 1000 ft average is 50 ft wide and 20 ft high with some areas reaching ceiling heights of 60 ft. The floors alternate from smooth clay to breakdown. There is a 40-foot crawl where sediment has filled the passage almost to the ceiling. The canyon passage crosses the entrance passage with no interconnection. The passage ends in a tangle of small passages with active speleothem deposition.

The two passages that make up Poorfarm Cave are remnants of a paleodrainage system. At elevations above 2300 ft, the passages are well above the active drainage focused on Locust Spring. There are massive infills of clay, silt, and cobbles that demonstrate an active drainage from Droop Mountain in the past. Details of the sediments are given by Wolfe (1973). There are places in the cave where more recent processes have removed earlier sediment (Fig. 7.11). Speleothems occur in a few places, often interlayered with clastic sediments, with many now dry and inactive (Fig. 7.12).

Fig. 7.11 Travertine shelf overlying a fill bank that is now eroded away. Poorfarm Cave. Photo by W.B. White

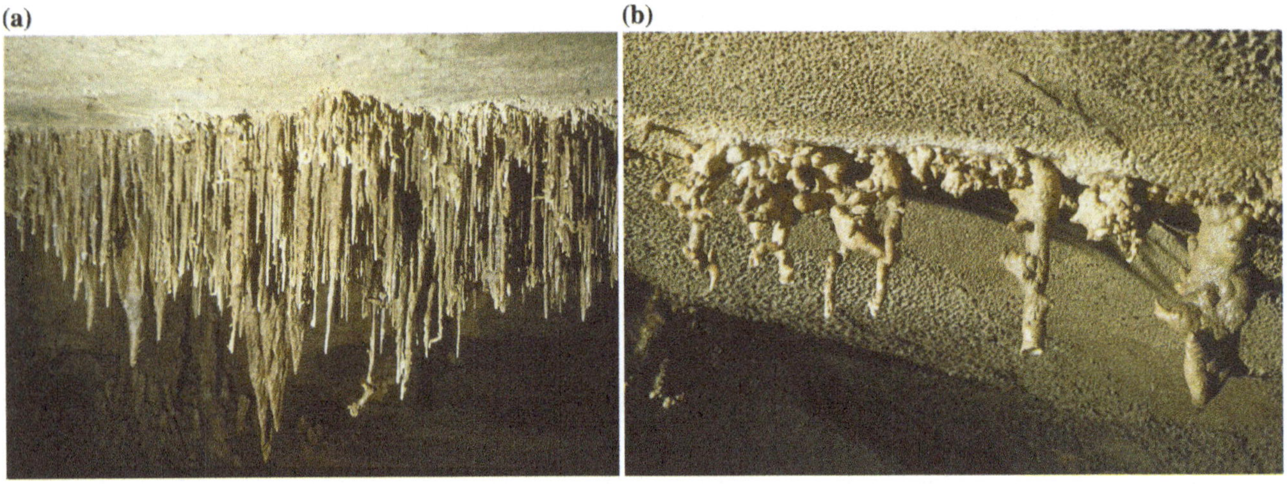

Fig. 7.12 Speleothems in Poorfarm Cave. **a** Erratic stalactites. **b** Helictites. Photos by W.B. White

7.7 The Hills Creek Drainage

The headwaters of Hills Creek are at the northeast end of a high valley in the Yew Mountains confusingly also called the Little Levels. The stream flows west with a low gradient for 2.5 miles parallel to highway 39. The stream then makes a turn to the southeast, drops precipitously over Hills Creek Falls and then follows a steep narrow valley until the gradient begins to flatten into a floodplain below Lobelia (Fig. 7.13). Hills Creek goes underground at the entrance to Hills Creek Cave on the west base of Droop Mountain at an elevation of 2450 ft. The lower reach of the Hills Creek basin is a blind valley requiring water depths of more than 100 ft in order to spill over into the irregular valley southwest along Droop Mountain. The usual interpretation is that the low gradient segment in the high valley was once a tributary of the Cherry River, pirated by Hills Creek when the higher gradient stream cut back into the mountains. A curious feature of Hills Creek is that it is essentially linear over the roughly six miles from the Hills Creek Falls to the cave entrance suggesting that the stream is following a lineament.

Bruffey Creek is a more conventional stream with multiple tributaries that head on the mountain flanks and converge into a single stream that sinks at the entrance to

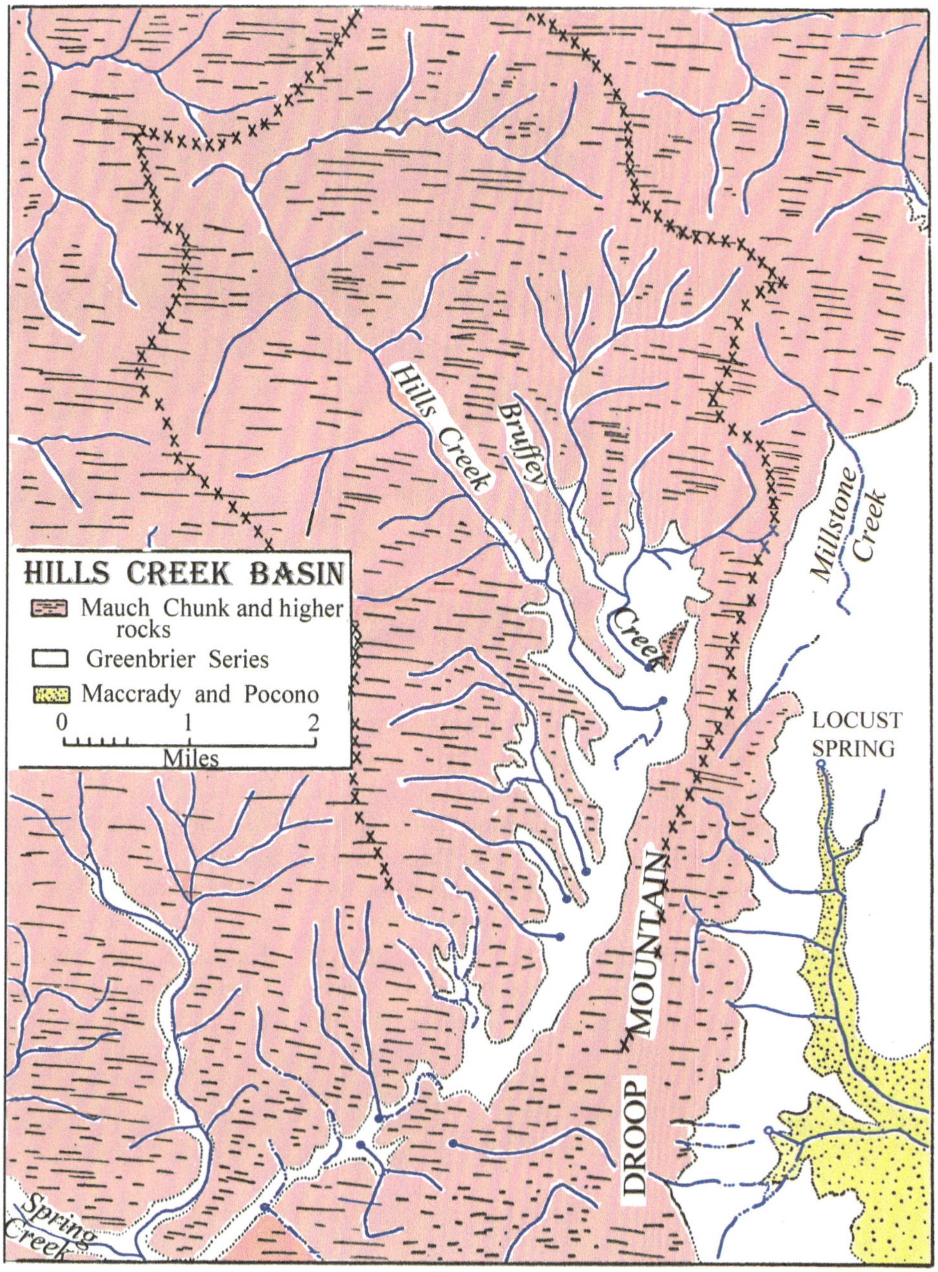

Fig. 7.13 Map of the Hills and Bruffey Creek basins

Bruffey Creek Cave (Fig. 7.14). There is a serious error on the Droop 7.5 min quadrangle. It shows Hills and Bruffey Creeks joining on the surface and ending in a spring. In reality, the two cave entrances are about 1200 ft apart.

7.7.1 Hills and Bruffey Creek Caves

The entrance to Bruffey Creek Cave opens into a 10–20 foot high and 15 ft wide passage that trends southeast for 500 ft to the junction with Hills Creek Cave. The entrance to Hills Creek Cave is often clogged with logs and other debris (electronic map M-7.8). Inside, the 8–20-foot high, 15-foot wide passage trends north for 200 ft to the junction with the Bruffey Creek passage. From the junction, the cave continues for several hundred feet as a 30-foot wide, 4–6-foot high passage filled with water to within inches of the ceiling. Finally the ceiling rises and the stream flows between gravel banks. The passage continues for about 2000 ft in a loop to the northeast and then to the south–southwest through two more pools. The stream passage bends sharply to the east at a large breakdown-filled section which appears to be following a fault. The steam sinks in its bed at the bend. No way through the breakdown has been found.

The details of the underground flow patterns in the Hills Creek Valley and the western edge of the Little Levels are

Fig. 7.14 Entrance to Bruffey Creek Cave. Photo by W.B. White

very complex, depend strongly on the magnitude of the discharge, and are not well understood. Under low-flow conditions, Bruffey Creek sinks in its bed upstream from the cave entrance and crosses under Droop Mountain to appear as the stream in Upper Hughes Creek Cave. This diversion path is likely perched on the Taggard Shale. Under low-flow conditions, Hills Creek sinks in its bed upstream from the entrance, flows to Cutlip Cave and from there southwest through the Friars Hole cave system (Chap. 8). Under moderate flow conditions both streams continue into the cave entrances so that a portion of the water goes directly to Locust Spring. Under high flow conditions, water backs up forming a lake in the blind valley which drains into Cutlip Cave and from there southwest down the Friars Hole System.

The gradient of the cave system is very low. The point where the stream is lost at the edge of the breakdown chamber is only about 30 ft below the cave entrance. A profile along Hills Creek and over Droop Mountain (Fig. 7.15) show the stream channel flattening out at about

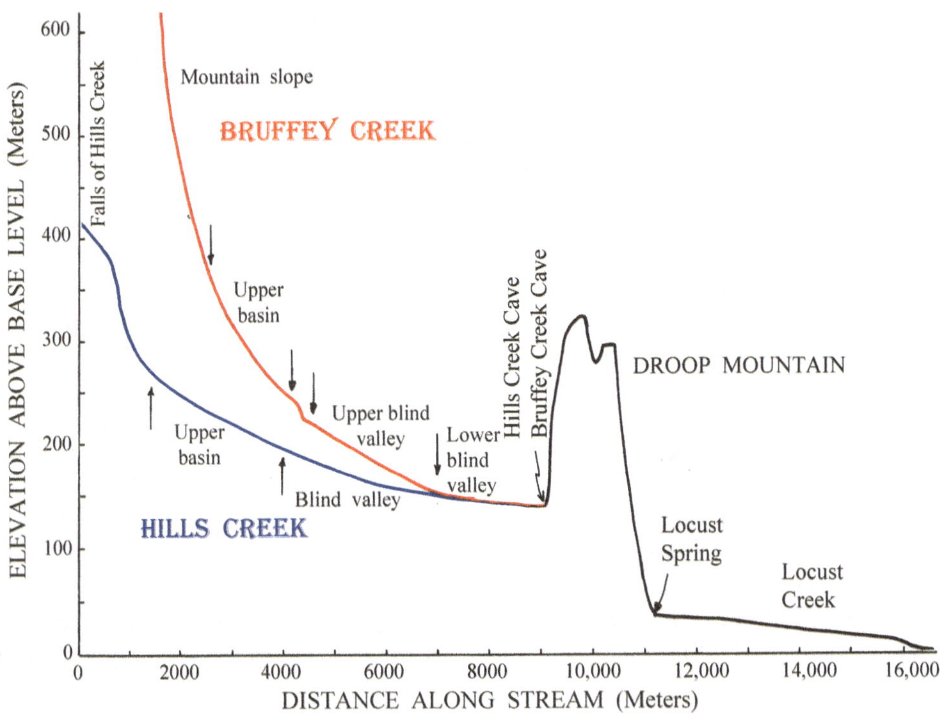

Fig. 7.15 Profile of Hills and Bruffey Creeks

2420 ft elevation at the final point of observation. Locust Spring is at an elevation of 2090 ft leaving a vertical offset of more than 300 ft to be explained. Between these points is the Taggard Shale acting as a confining layer and the unknown importance of the fault (or faults) that may be responsible for the extensive breakdown at the end of the Hills-Bruffey Creek system and at the upstream end of Upper Hughes Creek Cave (Medville and Medville 1991). Clearly there is something quite peculiar about the hydrology of the streams beneath Droop Mountain, unfortunately with access blocked by breakdown both up and down stream.

7.7.2 Martens Cave

Martens Cave is a fragment of master trunk conduit on the west side of Droop Mountain at an elevation of 2600 ft. The south entrance is a large hole in a 50-foot cliff in a large sinkhole. The north entrance is a wide opening in a low cliff about 300 ft linear distance from the south entrance. Between the entrances are about 500 ft of passage with widths up to 60 ft and ceiling heights of 10–30 ft (Fig. 7.16). A stream, Cave Run, flows through the cave from south to north. At the north entrance, the stream flows down a steep bank and joins Bruffey Creek. A breeze blows through the cave from south to north and was used by Davies (1960) for meteorological observations.

The reason for including this short cave in the discussion of the Little Levels is its unusual geographic and geologic setting. Martens Cave is in the Alderson Limestone with the stream perched on the Greenville Shale. The walls are smoothly but elaborately sculptured with many large ceiling pendants (Fig. 7.17). Passage shape and solution features indicate that Martens Cave was formed in the phreatic zone, by slowly percolating water along a flow path dictated by bedding plane partings between the Alderson Limestone and the Greenville Shale. This would require that the cave represents a time when base levels on the west side of Droop Mountain were well above 2600 ft, 500–600 ft above present day base levels.

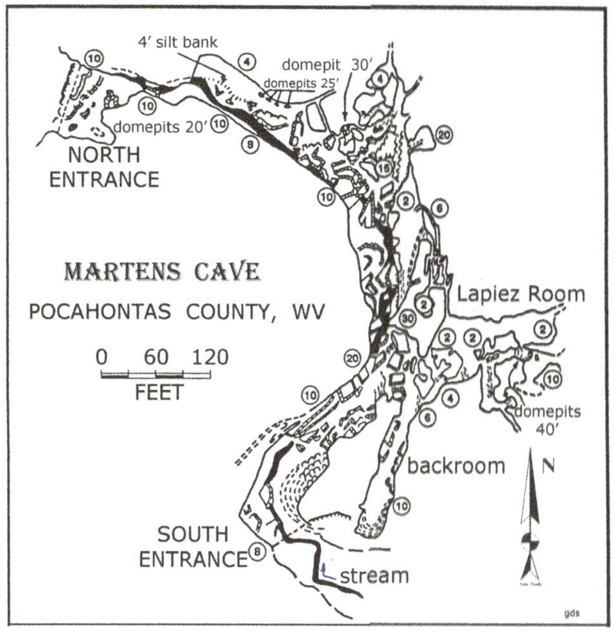

Fig. 7.16 Map of Martens Cave. From Storrick (1992)

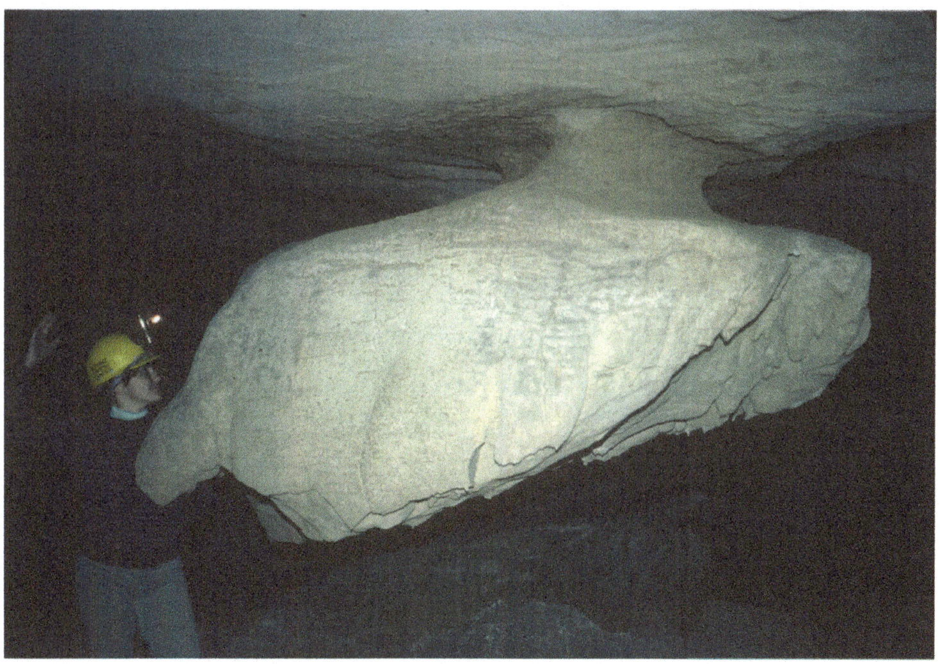

Fig. 7.17 Giant pendant in Martens Cave. Photo by W.B. White

References

Davies, W.E. 1949. *Caverns of West Virginia*, vol. 19, 353pp. Morgantown, WV: West Virginia Geological Survey.

Davies, W.E. 1960. Meteorological observations in Martens Cave, West Virginia. *National Speleological Society Bulletin* 22 (2): 92–100.

Medville, D. 1971. Cook pot, Pocahontas County, West Virginia. *Philadelphia Grotto Digest* 19 (2): 9–12.

Medville, D.M., and H.E. Medville. 1991. Structural controls on drainage beneath Droop Mountain, Pocahontas County, West Virginia. In *Appalachian Karst*, ed. E.H. Kastning, and K.M. Kastning, 11–18. Huntsville, AL: National Speleological Society.

Palmer, A.N. 1975. The origin of maze caves. *National Speleological Society Bulletin* 37 (3): 57–76.

Storrick, G.D. 1992. *The Caves and Karst Hydrology of Southern Pocahontas County and the Upper Spring Creek Valley*, vol. 10, 215pp. Barrickville, WV: West Virginia Speleological Survey Bulletin.

White, W.B., and E.L. White. 1991. Karst erosion surfaces in the Appalachian Highlands. In *Appalachian Karst*, ed. E.H. Kastning, and K.M. Kastning, 1–10. Huntsville, AL: National Speleological Society.

Wolfe, T.E. 1973. Sedimentation in Karst drainage basins along the Allegheny Escarpment in southeastern West Virginia, USA, 455pp. PhD thesis, McMaster University.

Zotter, H. 1965. Stream tracing techniques and results: Pocahontas and Greenbrier Counties, West Virginia—Part II. *National Speleological Society News* 23: 169–177.

8 The Friars Hole System

Douglas M. Medville and Stephen R.H. Worthington

Abstract

The Friars Hole Cave System in Greenbrier and Pocahontas Counties, West Virginia, at 73.4 km, is the longest cave in the Appalachian Highlands of eastern USA. The cave System, composed of three internal drainage complexes extending over a linear distance of almost 7 km, has had a long and complex evolution with most of it being over 730,000 years old and one dated speleothem having an age of over 1.67 million years. Although nearly the entire cave System is developed in the upper Greenbrier Group limestones (the Union and Pickaway limestones), two of the cave's three active drainages breach the Taggard formation, a major aquitard below the Pickaway and extending the cave's stratigraphic extent. The cave represents an evolutionary sequence of drainage Systems discharging to two springs 20-km distant from each other with high flows discharge to one and low flows to the other. Drainage patterns have shifted over time as the surface drainage is captured. An estimated age for the entire System is 4.1 million years.

8.1 Introduction

The Friars Hole Cave System is a 73.4-km-long complex cave developed in gently dipping Mississippian limestones. It is the largest cave in terms of surveyed passages in West Virginia. The cave is found in the Appalachian Plateaus Province in the eastern USA and is developed in moderately faulted northwest dipping limestones of the 217-m-thick Mississippian Greenbrier Group. The cave extends over a vertical range of 187 m and contains several currently active internal drainages that were formerly integrated but which now are separated (in terms of human exploration). The cave extends over a linear distance of nearly 7 km, underdraining several valleys above it. The floor plan of the cave shows currently active and abandoned flow paths and internal drainage divides.

Electronic supplementary material
The online version of this chapter (doi:10.1007/978-3-319-65801-8_8) contains supplementary material, which is available to authorized users.

D.M. Medville (✉) · S.R.H. Worthington
10701 Pinewalk Way, Highlands Ranch, CO 80130, USA
e-mail: medville@centurylink.net

8.2 Regional Setting

The Friars Hole Cave System is located in northern Greenbrier and southern Pocahontas Counties in eastern West Virginia (Fig. 8.1). The cave is in the central Appalachian Mountains, between the folded and faulted Valley and Ridge Province to the east and gently dipping sedimentary (primarily clastic) rocks of the Allegheny Plateaus Province to the west. The cave is on the west flank of the Browns Mountain Anticline, the largest fold of the central Appalachian Plateau. Although the primary orientation of the cave is along strike, extending for 7-km end to end, it also has a substantial development following the generally 2° dip to the northwest, with passages traversable for up to 1800 m in this direction. The cave underdrains both the Friars Hole Valley and adjacent valleys to the northeast and northwest of it, following the general regional strike of about N30°E. The Friars Hole drainage encompasses 95 km^2 and includes the surface drainages of Hills Creek, the Friars Hole Valley and infeeders to it, and Robbins Run to the west of the Friars Hole Valley.

The cave lies between linear ridges to the east and west (Droop Mt., Brushy Mt., Parker Mt.) with summit elevations

W.B. White (ed.), *Caves and Karst of the Greenbrier Valley in West Virginia*,
Cave and Karst Systems of the World, https://doi.org/10.1007/978-3-319-65801-8_8

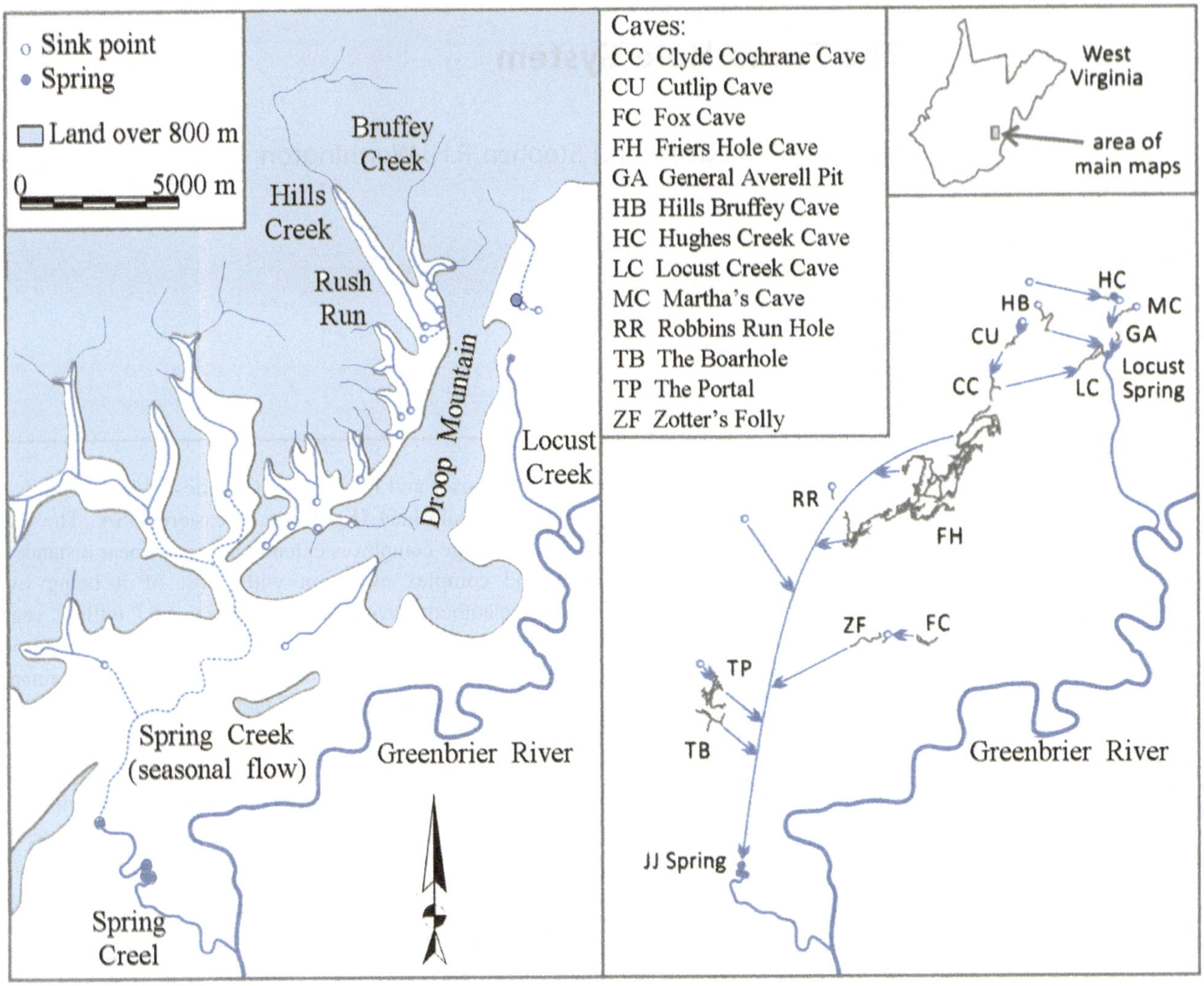

Fig. 8.1 Location, topography, and hydrology of the Friars Hole area. *Arrows* represent simplified flow paths of unexplored cave streams, confirmed by tracer tests

of 850–950 m, about 150 m higher than the valleys beneath which the cave is found. These ridges consist primarily of sandstones and shales of the Mississippian Mauch Chunk Series. Only the upper 20 m of the upper Greenbrier limestones and shales are exposed in a series of strike-oriented inlier valleys (Fig. 8.2). Seven of the nine entrances to the cave are found where streams sink in these limestone inliers. These entrances either consist of or lead to one of a series of vadose shafts dropping through the Union Limestone and into generally horizontal passages below at the contact between the Union and the underlying Pickaway Limestone. A plot of the cave showing its relationship to the local topography is shown in Fig. 8.3.

Streams flowing off of the higher Mauch Chunk clastics either sink into the highest member of the Greenbrier, the roughly 20-m-thick Alderson Formation, or into the upper part of the 48-m-thick Union Limestone below. The Alderson consists of interbedded calcarenites and calcilutites, primarily calacreous shales and siltstones and argillaceous and silty limestones. The basal member of the Alderson is called the Greenville Shale, a 4- to 9-m-thick aquiclude. Streams sinking on the Alderson rise at the top of the Greenville member, flow over it, and then sink again upon reaching the upper Union Limestone. All of the entrances to the cave are in the upper 10 m of the Union. Although the Union is the purest member of the Greenbrier Group, consisting of over 90% oolitic and muddy calcarenites, the majority of passages in the cave are found in the relatively impure 27-m-thick Pickaway Limestone below it, composed of alternating calcilutites with about 25% insolubles and calcarenites with 5–10% insolubles (Jameson 1985).

Both east- and west-dipping thrust faults are visible in many of the passages in the Friars Hole System. These faults

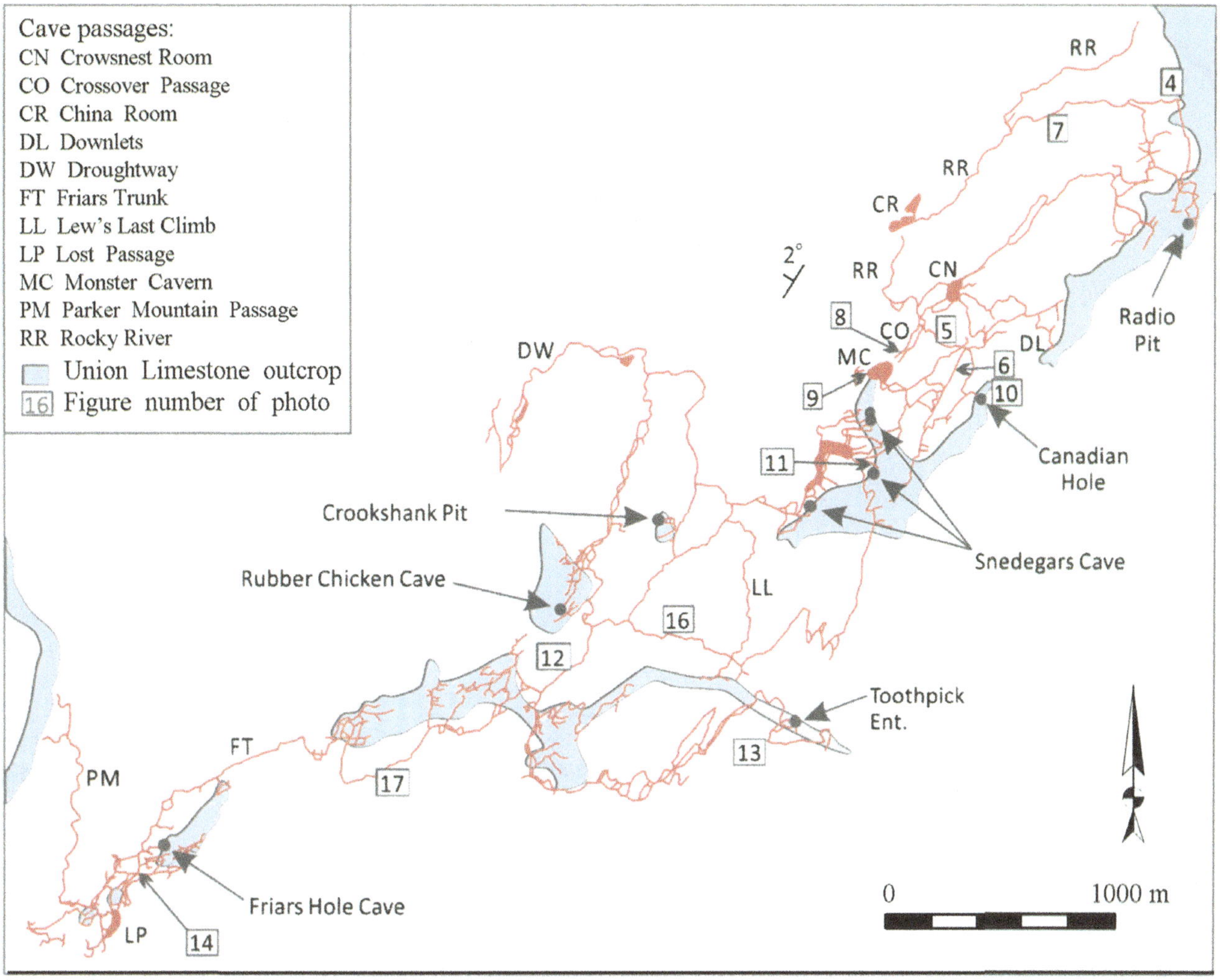

Fig. 8.2 Friars Hole Cave System passages showing surface outcrops and major features

generally strike at about N25°E and have low displacements. In three places in the cave, stream passages cross through the Pickaway Limestone and follow low-displacement faults, breaching the 7-m-thick Taggard Formation below it. The Taggard is a limestone-shale-limestone sequence that serves as an aquitard. It should be noted that all of the cave's drainage rises at a spring (JJ Spring, described below) that is perched on top of the Taggard Formation, and thus, the Taggard must be breached from below by the cave's drainage before rising at the JJ Spring.

8.3 Regional Hydrology

The regional hydrology of the area containing the Friars Hole Cave System as well as other caves is shown on the right side of Fig. 8.1. The solid lines show the traced flow paths of the cave streams in this area. Hills Creek is the largest of the sinking streams along the Friars Hole Valley. Under low-flow conditions, all of Hills Creek sinks in its bed and flows to Cutlip Cave, the upper of two aligned strike-oriented caves (Cutlip and Clyde Cochrane Sinks) before being seen again at the upper (NE) end of the Friars Hole System where it emerges from rockfall as Rocky River (RR in Fig. 8.2). Under higher flow conditions, Hills Creek is a partly losing stream with some of its flow sinking in its bed and entering Cutlip Cave, while the rest flows 500 m east and, with the flow of a parallel stream, Bruffey Creek, enters the Hills–Bruffey Cave, and passes beneath the north end of Droop Mountain in largely inaccessible passages. The stream is seen again in a sequence of caves on the east side of Droop Mountain: Upper and Lower Hughes Caves, Martha's Cave, and General Averall Pit, before flowing through the 3.26-km-long Locust Creek Cave and rising at Locust Spring at the base of the Greenbrier limestones on the east side of Droop Mountain, 2.25 km to the east of the Hills–Bruffey Cave entrance. Finally, under very high-flow conditions, the combined Hills and Bruffey Creek streams

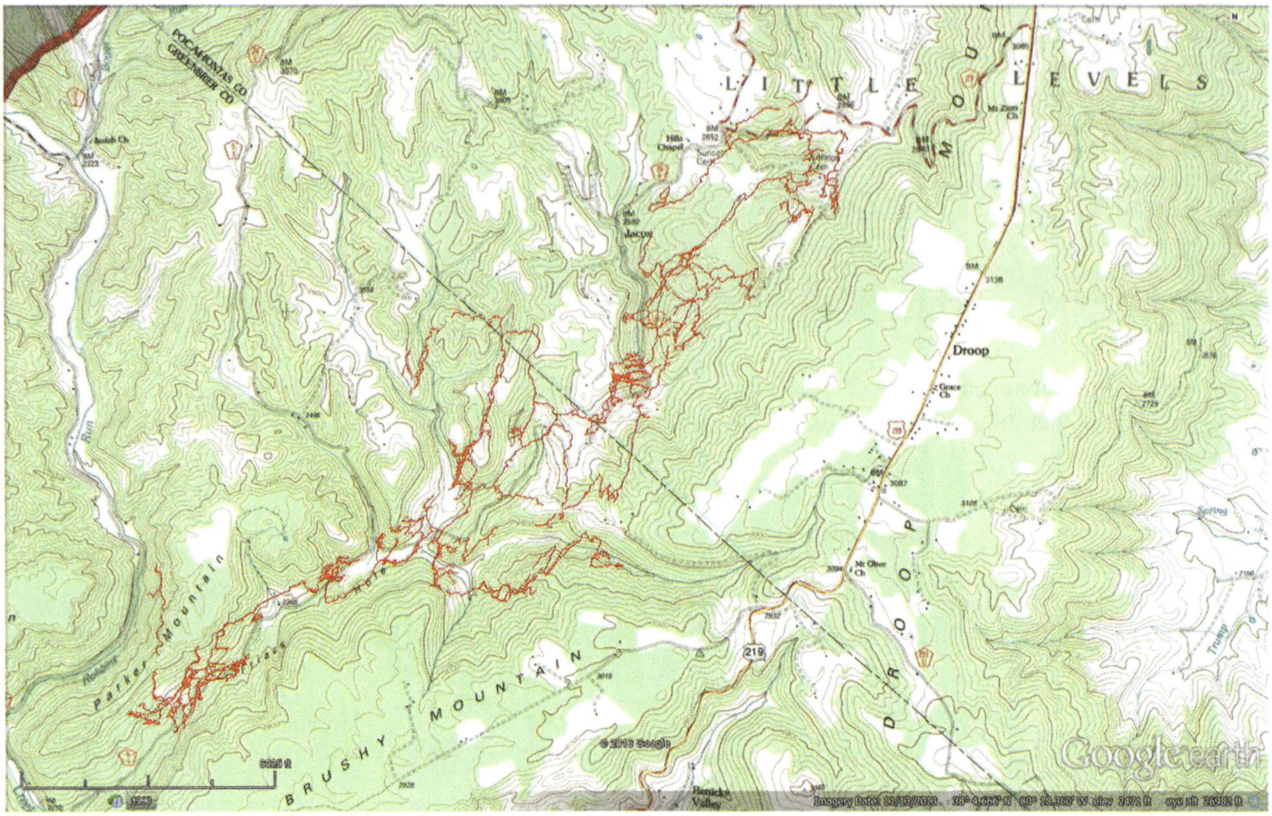

Fig. 8.3 Friars Hole Cave System line plot on topographic map. Base map: USGS Droop 7.5 min quadrangle

fill up the Hills–Bruffey Cave entrance and passages and back floods across the valley to the Cutlip Cave sink point.

The Rocky River (i.e., underground Hills Creek) stream as well as all of the other streams in the Friars Hole System rises 18 km to the southwest of the Cutlip Cave sink point at a spring (JJ Spring) on the east side of Spring Creek, a tributary to the Greenbrier River. JJ Spring is a perennial spring with a measured low-flow discharge of 12.6 ft^3/s (0.36 m^3/s) (Jones 1997). In addition to the Friars Hole Cave drainage, it is the resurgence for Spring Creek and several other streams sinking in a 265 km^2 area, including the Friars Hole drainage. Three other springs on Spring Creek, the Cannon Hole, the circulating Cenote, and Dale's Spring, are a few hundred meters upstream of JJ Spring and, under high-flow conditions, serve as overflow springs for the Friars Hole drainage, i.e., some of the Friars Hole water rises at these higher springs and flows on the surface while the remainder rises at JJ Spring (Jones 1997).

8.4 History of Exploration

A detailed description of the history of exploration of the Friars Hole System through the early 1980s can be found in another publication (Baker 1982) (see also Chap. 5). The text below is a very abbreviated summary with emphasis placed on those efforts that resulted in either the opening of entrances or the linking of major cave segments.

8.4.1 Snedegars/Crookshank

The centrally located dry and horizontal Snedegars Saltpetre entrance was used as access to the cave during the Civil War, where it led to passages used for Saltpetre mining. The earliest documented exploration during the twentieth century was in 1951 when explorers H. Ingalls and G. Moore descended the 30-m-deep Crookshank Pit 1 km to the southwest of the Saltpetre Entrance and explored passages that approached but did not connect to that entrance. The connection between these caves was subsequently made via a semi-sumped crawl in June 1964.

8.4.2 Friars Hole Cave

The southernmost entrance to the cave is a small opening at the bottom of a 15-m-deep doline located in the lower Friars Hole Valley, 3.5 km southwest of the Saltpetre entrance. This entrance was dug open by G. Titcomb and C. Schwab in August 1964. The opening led to a series of climbdowns and two 8-m pits followed by a 12- to 15-m-wide and 6- to 8-m-high master drain beneath the Friars Hole Valley. This passage contained a stream flowing southwest, along strike. Subsequent exploration and surveys by L. Bicking and others in the mid-1960s led to the discovery of 6 km of the cave. Although Bicking's surveying ended with his untimely death in a motorcycle accident, it was continued by C. and B. Williams. Subsequent exploration and surveys were carried out by numerous others with substantial discoveries made in the early 2000s by B. and S. Preaux and others in passages called Mauckland and Water World and by K. Owens and others in a passage called Big Water, as described under "Friars Hole System Description" below.

8.4.3 Rubber Chicken Cave

In March 1976, a sinking stream 2 km to the northeast of the Friars Hole entrance and 500 m SW of the Crookshank Pit entrance was excavated by D. Medville, B. Baumgartner, and G. Mothes. This led to a 3 m climbdown followed by a 22-m pit. A narrow passage at the base of the pit led to a 6- and 5-m-high passage containing a stream that flowed to the north, away from the Friars Hole stream. Surveys in 1976 resulted in 7 km of passage in a separate cave called "Rubber Chicken," named after a child's toy of that name found during the entrance excavation. In May 1976, the Rubber Chicken Cave was connected to the upper end of the Friars Hole Cave by cavers from McMaster University and then surveyed by US cavers C. and B. Williams, C. Hempel, and E. Strausser.

Comparison of the maps of Crookshank Pit and Rubber Chicken Cave indicated that the ends of two passages in these caves approached each other and were only 10 m apart. In the spring of 1976, parties entering each cave went to the respective ends of these passages and established voice contact. After removing rocks that separated the two passages, another connection was made. By the end of 1976, the combined Snedegars/Crookshank/Friars Hole/Rubber Chicken Cave System had over 24 km of surveyed passages and for simplicity, the entire cave was named The Friars Hole System.

8.4.4 Canadian Hole

A short but sporting wet cave 500 m to the north of the Snedegars Saltpetre entrance was visited in the summer of 1976 by cavers from Alberta, Canada. After descending a five-pitch entrance series, one of them (E. Neilsen) found a way up through breakdown at the cave's end. This led to a substantial canyon passage going downstream to the northwest (First Canyon). Within a year, 8 km of passages were surveyed and the previously short cave was re-named Canadian Hole. In subsequent years, a large majority of the exploration, study, and survey of this cave was carried out by cavers from McMaster University in Hamilton Ontario and by members of the Société Québecoisé de Spéléologie (D. Caron, A. Goupil et al.). The section of the cave discovered in 1976 was named the Alberta Extension. A 500-m-long wet boulder filled crawl (Almost Hell) led north from the end of the Alberta Extension and then opened into another substantial section of Canadian Hole called the Ontario Extension. Because trips to this part of the cave were long and tiring, a cave radio was brought into a passage near the north end of the Ontario Extension by S. Worthington and C. Pugsley, and communication with the surface was established. A 3-m dead bottom hillside pit a few meters from where the cave location was made then became the site of a multi-month dig effort to connect it to Canadian Hole, 39 m below. In May 1987, this effort succeeded, resulting in Radio Pit becoming the northernmost entrance to the Friars Hole System, 1.25 km northeast of the Canadian Hole entrance.

In September 1977, four McMaster cavers (J. Mort, A. Recklies, A. Thurston, O. Slupecki) entered Canadian Hole and, after surveying south through a series of low wet passages, found footprints in the mud floor and a survey station on the wall. The footprints and survey station had been left by US cavers who had previously surveyed north in a 2.3-km-long passage (The Highway) in Rubber Chicken Cave. With this connection, Canadian Hole was added to the Friars Hole System and, as of late 1977, the combined length of the cave had increased to 33 km.

8.4.5 Toothpick Cave

The 30–m-wide, 15-m-high, and 20-m-deep Amphitheater-like entrance to Toothpick Cave is at the bottom of a local drainage called Ravens Nest Hollow, 1.2 km south of the Snedegars Saltpetre entrance and separated from it by a 100-m-high western spur of Droop Mountain. This impressive entrance, named from the logs inside that had the appearance of toothpicks compared to the entrance size, leads to 150 m of vadose passage in the Upper Union Limestone, followed by a 20-m pit. Initial exploration of the cave was carried out in the early 1960s by V. Schmidt, R. Cope, W. White, L Bicking, and others. The stream at the bottom of the 20-m pit flowed west and into a mud-choked sump, limiting further exploration. During very dry weather in 1978, attempts were made to follow the Toothpick stream beyond the sump to see if it could connect to the main Friars Hole System, 1,100 m to the west. These efforts were successful, and in October 1978, three cavers (L. Baker, R.

Anderson, K. MacGregor) squeezed through low airspace passage and into much larger passage going to the southwest. A month later, a survey/exploration party (R. Anderson, L. Baker, P. Mothes) followed the Toothpick cave stream downstream for 1,400 m to a survey station at the upstream end of Rubber Chicken Cave, confirming a connection between Toothpick and the Friars Hole System and adding another 5.8 km of passage to it.

8.5 Cave Hydrology

Numerous streams sink in a series of inlets and blind valleys that are generally aligned along the N20°E strike of the Greenbrier outcrop on the west side of Droop Mountain. These sink points extend over a distance of 8 km and are in the upper few meters of the Union Limestone where the limestone is exposed in inliers between sediment filled surface drainages. Of the over 100 inlets to the cave, seven lead to entrances to the 73.4-km-long Friars Hole Cave System; three other entrances are dry.

After reaching these entrances, the streams flow short distances (up to 150 m) in vadose canyons in the upper Union before descending through the majority of the Union Limestone via one or more vadose shafts leading to the Union/Pickaway contact (Fig. 8.4). Inside the cave, numerous infeeder streams can be followed upstream to shafts carrying water from the surface, generally beneath the limestone inliers above. Passages at the tops of these shafts, where entered, are also vadose canyons in the Union, ending in surface fill or becoming too narrow to follow.

Fig. 8.4 Vadose shaft in Friars Hole Cave System (Trombone Aven). Photograph by Ron Simmons

Below these pits, the streams generally flow down dip to the northwest or to the southwest in high, joint-controlled vadose zone passages (Fig. 8.5) or gently dipping bedding planes serving as linking passages before turning to the northeast or southwest in lower gradient main drains. These drains are developed in the lower Union or just below the contact between the Union Limestone and the underlying Pickaway Limestone and can be followed for considerable distances along strike, following bedding planes and/or thrust faults that intersect these bedding planes. Examples of drains in the lower Union include the Friars Hole Cave trunk downstream of the Dung Ho Way, much of the Toothpick Cave trunk, the upper part of the Rubber Chicken Cave highway and further north, the Canadian Highway (Fig. 8.6). These are some of the larger diameter and most linearly extensive passages in the System.

As the System developed, the main drains migrated both down dip to the northwest and down stratigraphically, into the upper and middle Pickaway Limestone below. Examples include the lower part of the Highway in the Rubber Chicken Segment of the System, the Droughtway in the lower Crookshank Segment, and Skid Row/Columbus Avenue in the Canadian Hole Segment (Fig. 8.7). In a couple of places, these drains dropped further, passing through the lower Pickaway and Taggard Shales below it and into the Patton Limestone member of the Denmar Formation. Examples include lower Rocky River in the Canadian Hole Segment and the Water World passages at the southwest end of the Friars Hole Segment of the System.

Although most of the cave's passages are in the Pickaway Limestone, two long passages are found in the Union. One is the Lew's Last Climb passage, draining the Toothpick Cave Valley 600 m below the Toothpick Cave entrance. Dolines on the surface above the end of this passage may indicate that the Ravens Nest Hollow surface water sank in this area before headward capture by the current Toothpick entrance took place upstream. The Lews' Last Climb Passage was initiated along a low-angle thrust fault and consists of a high canyon in the lower Union Limestone that extends north for 700 m before dropping into the Pickaway Limestone at an 8-m pitch and then continuing to the Snedegars Cave stream.

The second long passage in the Union is the Parker Mountain Passage, a paleo-infeeder originating in domes

Fig. 8.5 First Canyon in Canadian Hole Segment. Photograph by David Bunnell. Used with permission

beneath the eastern hillside of Robbins Run, the next valley to the west of the lower Friars Hole Valley. Although now dry and almost sand filled, this passage can be traversed for over 1.9 km from the Robbins Run Valley before intersecting the top of the west wall of the lower Friars Hole Cave trunk passage, a few meters downstream of its junction with the entrance passage to this part of the System. A shallow phreatic half tube can be seen in the ceiling of the Friars Hole trunk extending beyond the Parker Mountain passage termination, 9 m above the floor of the Friars Hole trunk, indicating the former continuation of this passage.

The Friars Hole System contains three integrated internal drainages as described below.

8.5.1 Northeastern Drainage

This drainage, in the Canadian Hole part of the System, contains a large stream named Rocky River. This stream represents the part of Hills Creek that sinks in its bed above Cutlip Cave and flows through both Cutlip and another cave to its south (Clyde Cochrane Sinks) before being seen in the Canadian Hole part of the System, 120 m south of the downstream sump in Clyde Cochrane Sinks. Rocky River is a strike-oriented stream draining much of the flow from a 45 km^2 catchment to the north and is the perhaps the largest of the streams in the System. It flows for 1.4 km within the cave, is lost at a sump, and is seen again about 100 m downstream at another sump (Watergate). It continues to flow along strike for another 0.6 km before being lost in a boulder choke on a fault. Although Rocky River flows on the Pickaway Limestone for nearly its entire length, near its downstream end in the Rocky Horror passage (Fig. 8.8), it passes through the underlying Taggard Formation shales and into the upper Patton Limestone member of the Denmar Formation. This downstream end approaches, but is 30 m lower in elevation than the low point in the floor of the System's largest chamber. The appropriately named Monster Cavern is a 110-m by 55-m room containing a 32-m-high waterfall called Monster Falls (Fig. 8.9).

Monster Cavern is directly beneath the valley that contains the stream flowing into the Snedegars Cave North Entrance and is gradually pirating this stream. The vertical separation between the streambed and the highest point in the cave above Monster Falls is about 30 m. Under low-flow conditions, the surface stream sinks in its bed 200 m upstream from the North Entrance and is seen in the cave as Monster Falls. Under higher flow conditions, that part of the surface stream that does not sink in its bed flows into the North Entrance and then into the Snedegars Cave part of the System with more being lost in the first few meters of passage and flowing to Monster Falls.

Tributaries to Rocky River include water entering near the cave's northeast end at passages called Skid Row, Temptation Streamway, and Camp Inlets. Further south, the Canadian Hole entrance stream descends 35 m over three pits, passing through the Union Limestone (Fig. 8.10). This entrance stream then flows down dip on the Pickaway Limestone in a passage called First Canyon before joining Rocky River just above the Rocky Horror passage. The stream probably continues beneath Monster Cavern but cannot be followed and is not seen again in the cave.

Fig. 8.6 Canadian highway in Canadian Hole Segment. Photograph by Ron Simmons

Fig. 8.7 Skid Row in Canadian Hole Segment. Photograph by Ron Simmons

8.5.2 Central Drainage

The central drainage in the System is arguably the most extensive and contains numerous infeeder streams, five of which enter the cave at various entrances and all of which ultimately combine before ending at a sump 146 m lower than the cave's datum at the Saltpetre Entrance. Two of the streams flow into the Snedegars part of the cave. The intermittent stream flowing into the Snedegars North Entrance flows generally down dip for 200 m before entering the large chamber called the Amphitheater in the historic Saltpetre part of the cave (Fig. 8.11). After a few hundred meters, this stream is joined by another, flowing into the Snedegars Staircase Entrance. This stream then descends 35 m over four drops, about the same distance stratigraphically and vertically through the Union Limestone as does the Canadian Hole entrance stream.

Fig. 8.8 Lower Rocky River in Canadian Hole Segment. Photograph by Ron Simmons

Fig. 8.9 Monster Falls in Monster Cavern, Canadian Hole Segment. Photograph by David Bunnell. Used with permission

The third infeeder stream is the water flowing into the 33-m-deep Crookshank Pit. This stream flows north along strike for 250 m before merging with the two Snedegars streams. The combined streams flow north along strike on the Pickaway Limestone through a series of low airspace near sumps for another 800 m before merging with the remaining two infeeder streams entering the cave from the Rubber Chicken and Toothpick entrances, respectively.

The intermittent stream entering the narrow Rubber Chicken entrance near the top of the Union Limestone flows for a short distance and as noted above, drops over several pits passing through the Union, the deepest of which is 22 m. The stream then merges with a larger one flowing from the Toothpick Cave entrance, and the combined streams follow a series of large, strike-oriented passages for 1.9 km to the north before merging with the Snedegars/Crookshank streams (Fig. 8.12).

Finally, the Toothpick entrance stream follows the same patterns as the other streams entering the cave: flowing for a short distance in the upper Union Limestone and then dropping 20 m at a pit that takes it to the Union/Pickaway contact (Fig. 8.13). This stream then flows southwest along strike and then west, sub parallel to the down dip direction for 1.6 km before turning to the north and following a complex series of passages for another 1 km before reaching the junction with the entering Rubber Chicken stream. Along the way, several other infeeders join this stream, entering via domes at their upstream ends.

The junction of the Rubber Chicken/Toothpick stream and the Snedegars/Crookshank stream takes place at a low passage at the upstream end of a 30-m-long sump that has between 0 and 10 cm of airspace, depending on flow conditions. At the downstream end of this sump (Toms Sump), the stream flows north for 120 m and then turns to the west for 400 m in the larger passage. This passage continues to the southwest for another 700 m, passing beneath the last major chamber in this part of the cave, the 100-m-long and 15-m-wide Mint Room. The stream again turns to the north and then to the south before ending at a sump, 181 m lower than the highest entering stream at the Toothpick Entrance.

Fig. 8.10 Canadian Hole entrance. Photograph by David Bunnell. Used with permission

The Toothpick stream can be followed in the cave continuously for over 5.5 km to the final sump, following passages at the basal Union and upper Pickaway Limestones with much of the passage developed along the intersections of thrust faults and bedding planes. Controls on the development of this passage are discussed by Sasowsky et al. (1989).

8.5.3 Southwestern Drainage

The lower Friars Hole Valley is perched 75 m above local base level in Spring Creek, 1 km south of the southwest end of the cave. This hanging valley contains numerous dolines up to 10 m in depth that capture streams flowing in smaller side valleys. The deepest of these dolines is the one containing the Friars Hole Cave entrance. In addition, the Friars Hole Valley 1 km northeast of the Friars Hole entrance is a blind valley with sinking streams entering the trunk passage below in a series of domes and infeeder passages in the Union Limestone. The Friars Hole trunk passage, underdraining the lower Friars Hole Valley, is a 15-m-wide and an 8-m-high strike-oriented conduit containing an underfit stream.

Flow is from northeast to southwest with the stream dropping into and through the Pickaway Limestone (Fig. 8.14) at its southern end before being lost in boulders near a room bounded on its west side by a thrust fault (Avalanche Room). A 22-m-diameter room 100 m to the southwest, called Barbara's Room, is also developed along a

Fig. 8.11 Snedegars Saltpetre passage. Photograph by David Bunnell. Used with permission

Fig. 8.12 Rubber Chicken stream passage. Photograph by Ron Simmons

Fig. 8.13 Pit in Toothpick Segment. Photograph by David Bunnell. Used with permission

strike-oriented thrust with slickensides visible on the rooms' northwest wall. A boulder choked passage along this wall descends for 25 m, following the fault plane. At the bottom of this passage, a stream that is substantially larger than the Friars Hole trunk stream is encountered. Sumped at its SW end, the stream can be followed for 300 m to the northeast along a low, boulder filled passage aligned along the strike of the fault. This passage is very similar to the half-km-long Almost Hell passage in Canadian Hole, i.e., developed along a strike-oriented fault. A waterfall entering this passage about halfway along its length and containing surface debris is probably the Friars Hole trunk water, last seen sinking into boulders in a higher passage only 25 m to the south.

This large stream is also seen in another passage (Water World), reached via another route 200 m to the southwest. The two passages are aligned, are at the same elevation, and are separated by a sump. These stream passages are 100–110 m lower in elevation than the Friars Hole Cave entrance.

Fig. 8.14 Potholes above Pool Room in Friars Hole Segment. Photograph by David Bunnell. Used with permission

Taking into account the 2°NW dip, the location of the Friars Hole entrance about 5 m below the top of the Union Limestone, the down dip component of the distance from the entrance to the streams and the measured thicknesses of the Union and Pickaway Limestones and the Taggard Shale, the lowest streams in the cave are about 18 m below the top of the Patton Limestone member of the Denmar formation. The passages are also 32 m lower in elevation than the bed of Spring Creek, 1 km to the south. The vertical separation between the southwestern end of these streams, i.e., the lowest point in the cave, and the Friars Hole resurgence at JJ Spring is 21 m; the straight line distance between the two is 7.3 km and the gradient is 0.0029. The 21-m elevation difference implies that there may be substantial lengths of cave passage with airspace, although in high-flow conditions it may be completely flooded.

The large stream that is seen in passages at the southwestern end of the Friars Hole System contains water that has been traced from the Robbins Run valley, 2.3 km to the northwest and draining a 26 km^2 area (Jones, W.K., personal communication, 2016). Although it is possible that the water seen in the central and northern parts of the System also drains into this stream, this has not yet been confirmed by tracer tests.

8.6 Cave Paleohydrology

The paleohydrology of the cave is best illustrated by the passages to the west of the Canadian Hole entrance, where the development of the cave over time is the easiest to understand (Fig. 8.15). Many sink points have developed close to the top of the Union Limestone. Canadian Hole (Fig. 8.10) is one, and this and five other sinks that provide inlets to the cave are shown in Fig. 8.15. Passages from these inlets typically descend one or more pits (Fig. 8.13) to reach the lower part of the Union Limestone and the upper part of the Pickway Limestone, where most of the horizontal development has taken place.

Since the cave was first developed, Hills Creek has provided the largest surface catchment for the cave and has resulted in the formation of some of the cave's largest passages. Its initial sink point was at the Downlets, and from there the Highway, a large phreatic passage, drained the water along the strike (flow path 1 in Fig. 8.15; Figs. 8.6 and 8.16). As the creek eroded down into the limestone, new sink points developed further upstream, capturing the flow which then formed the lower strike-oriented conduits Crossover Passage and Rocky River (flow paths 2 and 3, respectively, in Fig. 8.15). Smaller sinking streams such as at Canadian Hole (Fig. 8.10) formed down dip passages that developed into vadose canyons (Fig. 8.5) and formed tributaries to the major strike-oriented passages. Progressive diversion to more efficient flow paths resulted in the complexity that is now seen in the cave.

The sinking streams that provide the catchments for the substantial flow through the cave also provide a substantial load of sandstone and clay. Active stream passages usually are floored by sandstone gravel and cobbles (Figs. 8.4, 8.5, 8.7, 8.12, and 8.16). Major floods in the cave raise water levels and overflow into normally inactive passages, leaving clay sediments (Figs. 8.6 and 8.17). In addition, some active stream passages also have banks of clay (Fig. 8.16). Where passages are wide, roof collapse can result in breakdown (Fig. 8.11). The largest void in the cave is Monster Cavern (Fig. 8.9), which has a volume of 350,000 m^3.

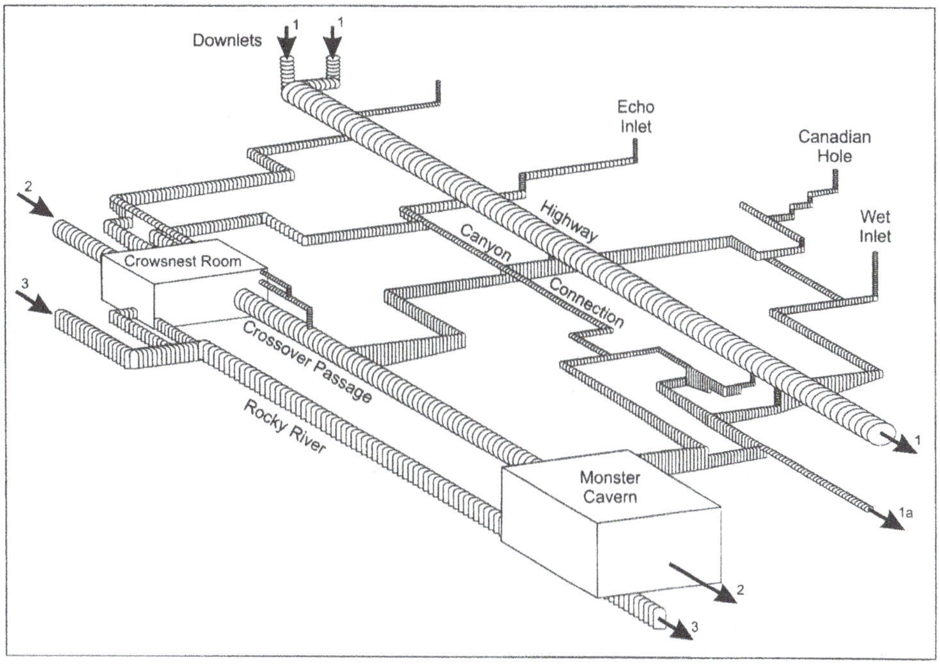

Fig. 8.15 Flow path sequence in Canadian Hole Segment

Fig. 8.16 The highway in Rubber Chicken Segment. Photograph by David Bunnell. Used with permission

A series of tracer tests have demonstrated the links between sinking streams, cave passages, and the resurgences at Locust Spring and JJ Spring (Fig. 8.1). These links represent unexplored cave streams, and it is clear from Fig. 8.1 that such unexplored passages are extensive. At the time when the earliest passages in Friars Hole System such as the Highway were formed, the higher base level and westward dip of the limestone would have meant that no limestone was exposed along the lower part of Spring Creek. Consequently, it seems probable that the Highway would have drained to the south under Droop Mountain to a resurgence along Greenbrier River (Fig. 8.1), and there must be extensive inactive passages that have yet to be discovered.

A detailed analysis of the development of the North Canyon of Snedegars Cave was made by Jameson (1985), and more generalized accounts of the whole cave appear in Worthington (1984) and Worthington and Medville (2012).

Fig. 8.17 Dung Ho Way in Friars Hole Segment. Photograph by Ron Simmons

8.7 Friars Hole System Description

A detailed description of a 73.4-km-long complex cave such as Friars Hole, containing passages at multiple levels and several internal drainages, is challenging. A passage by passage description would be both lengthy and lose the reader in detail. To avoid this, a more general, section by segment by segment description will be given, keyed to the five sheet map that accompanies this chapter (electronic maps M-8.1 to M-8.5). The description begins with the historic Snedegars/Crookshank part of the System, then continues with the large centrally located Rubber Chicken and Toothpick Segments to the south and west, then moves to Canadian Hole at the Systems' northeast end, and ends with the traditional Friars Hole Cave in the southwestern part of the System. Emphasis will be placed on major conduits, infeeders, and linking passages between these segments, taking note of the cave's geological setting and former (paleo) drainage routes that resulted in the cave's current size. The description is not intended to be a guide to navigating the cave but rather a way for the reader to visualize the cave's geography using the map as an aide. An earlier version of this description was provided in (Medville 1981).

8.7.1 Snedegars/Crookshank Segment

With the only three horizontal entrances in the cave, the centrally located Snedegars Cave, shown on map sheets B and C (electronic maps M-8.2 and M-8.3), is the most visited part of the System. The 3-m-high and 3-m-wide North (or Stream) Entrance takes an intermittent stream and leads to a high, joint-controlled canyon passage (North Canyon) that trends west in the Union Limestone (Jameson 1985). A second entrance just above and to the left of the Stream Entrance leads to a series of crawls and connects to the cave's historic Saltpetre section (described below). The North Canyon can be followed for 400 m to the top of a 4-m overhung pit at the Waterfall Room. A 1-m-high phreatic tube at the base of this pit leads to a 30-m-wide chamber called the Amphitheater, in the Saltpetre Trunk.

The Saltpetre Entrance to the System is about 300 m south of the Stream Entrance, on a hillside and on the west side of a field. This 5-m-high and 3-m-wide entrance can be followed west for 100 m and then turns to the northwest, descending over large breakdown blocks for 170 m to the Amphitheater Room passage. Near the top of the large breakdown slope, low passages lead north into the Saltpetre Maze, an extensive series of dry passages on three levels, ultimately connecting to the entrance above the North Canyon Stream, as noted above. These passages trend east, ending beneath the hillside between the North and the Saltpetre Entrances. At the base of the breakdown slope and on the Saltpetre Trunks's south side, a low inlet passage (The Druid Passage) can be followed upstream and to the east for 300 m before becoming too low to follow.

Beyond the Amphitheater, the cave continues south for over 300 m as a 5-m-high and 20-m-wide passage with now-dry inlet passages and an upper level paralleling it along the southeast side, e.g., Snedegars Dome and Formation Loop. Beyond this, the ceiling then lowers 100 m in an almost-choked, 0.3- to 0.6-m-high passage called The Cobble Crawl. At the western end of this crawl/squeeze, a

25-m-long pool with a half meter of airspace in low-flow conditions must be crossed. This pool, the Snedegars Sump, is the backwater from a stream entering on the north side of the passage at the lower (western) end of the pool. The source of the water is another entrance to the System called Snedegars Staircase. The Snedegars Staircase Entrance is at the bottom of a sink 350 m SW of the Saltpetre entrance. Streams rising at the base of the Alderson Limestone above flow over the Greenville Shale and into this entrance. Over the first 100 m, a series of four drops are reached (6, 4, 7, and 18 m, respectively) followed by another 270 m of the passage leading to the downstream end of the Snedegars Sump.

Downstream from the Snedegars Sump, the passage can be followed for 300 m to a junction with a crawl on the left side of the passage (Sloppy Crawl on the map) just before reaching a series of waterfalls. Continuing downstream for another 250 m, a T junction is reached. The passage to the right (downstream) can be followed north for 100 m to a very low airspace near- to total 40-m-long sump named Rick's Syphon. The passage downstream from this pool continues north and west for 400 m to a shorter pool called Doug's Sump. After another 150 m of low passage, a junction is reached with a large stream entering from the left, the Rubber Chicken/Toothpick Cave stream. The combined streams then flow north through another low airspace near-to total sump called Tom's Sump. The passage beyond this point is called The Droughtway and is described under the Rubber Chicken Segment below.

Returning to the T junction noted above, the passage to the left is dry and after 350 m, leads to the base of the 12-m-wide and 33-m-deep Crookshank Pit with canyon passages and dome complexes a few meters beyond the pit's base. The stream flowing into Crookshank Pit sinks into a narrow floor joint and is not seen again in the cave.

The passage called Sloppy Crawl can be followed south for 100 m to another junction. The passage on the left is the Lew's Last Climb passage and continues south for 700 m to an end at domes. The passage on the right is low and can be followed for 300 m to the southwest to "Breakdown Room" on the map and a connecting passage (Terrible Crawl) that also leads to the Crookshank Pit area. The somewhat larger passage beyond Breakdown Room is called the Promised Land Passage and, after 200 m, reaches a breakdown area with the Rubber Chicken Cave Segment of the System beyond. This is labeled "The Connection" on Map sheet E (M-8.5).

8.7.2 Rubber Chicken Segment

Because the original entrance to the Rubber Chicken Segment of the System is sealed, the description of this segment of the Friars Hole System begins at "The Connection" breakdown area at the south end of the Snedegars/Crookshank Segment of the cave System, shown on map sheet E (M-8.5). After following a moderately low passage south for 150 m, a T junction with a much larger passage is reached (Shark Room on map sheet E). This passage, called "The Highway," is a 2.3-km-long paleo-trunk passage that no longer contains an active stream (Fig. 8.16). This passage, averaging 6 m in height and width can be followed west for 150 m to an end at a junction with the north flowing stream passage in Rubber Chicken Cave.

The majority of The Highway is to the left (east) of the Shark Room. Following the Highway in this direction, it continues for over 1 km with minor infeeders along the way. A series of six bends, the Zig-Zags, are then reached where the passage crosses a series of the strike-oriented west- and east-dipping conjugate thrust faults. At the last (easternmost) of the bends, a paleo-downstream passage trends south for several meters but becomes too narrow to follow. At this junction, The Highway turns to the north in large dry passage for another 600 m to a canyon complex entering from the right (east) and ending at a series of vadose shafts (Domes Canyon on map Sheet C (M-8.3)). Although these are beneath the hillside across from and 170 m south of the Saltpetre Entrance, they do not connect to the surface. Beyond this canyon complex, the passage lowers with infeeders entering from the right. The passage continues north, joining the Canadian Hole Segment of the System as described below.

As described above, the western end of The Highway ends at a T junction with the north flowing combined Rubber Chicken/Toothpick stream. Going downstream, it sinks into a lower level. The passage continues north and after 200 m reaches an infeeder from the west that can be followed for 100 m upstream to the base of the Rubber Chicken entrance pit series. Continuing north from the entrance infeeder, a low passage is followed for 200 m to a complex area containing numerous streams entering from the southwest and northwest, beneath the valley containing the Rubber Chicken entrance. To the north of this infeeder complex, a 5-m-high and 5-m-wide dry passage continues north for over 700 m with the stream seen in a separate 1-m-high passage below. These passages eventually converge, continue north for another 250 m, and then join the Snedegars stream passage at the upper end of Tom's Sump as noted above.

The combined Snedegars/Rubber Chicken/Toothpick streams flow through the 60-m-long low airspace Tom's Sump and then turn west and down dip, entering a section of the System called the Droughtway (map sheet B). A 1- to 2-m-high passage can be followed for 750 m downstream through several low airspace pools before opening to the 20-m-high and 20-m-wide, 100-m-long Mint Room, one of the most remote large chambers in the Friars Hole System. The cave stream flows in a separate, lower passage and is rejoined at the western end of this chamber in 15- to 20-m-high passage that continues west and then abruptly turns north. After 350 m, the cave stream turns to the south, lowers, and within 70 m reaches a sump, 145 m below the cave's Saltpetre (datum) entrance. Based on the stratigraphic horizon of the ceiling of the Mint Room, the lower Union Limestone, the cave stream below is entrenched in the Pickaway Limestone and the final sump appears to be in the lower Pickaway, above the Taggard Shale.

From the lower (western) end of the Rubber Chicken Highway and turning left, the 3- to 6-m-high stream passage can be followed upstream to the south for 500 m to a passage junction where a dry passage enters from the west. This passage leads to a complex of high, joint-controlled infeeder passages, including the connection to the Friars Hole Cave Segment. Continuing upstream from this junction, the cave becomes considerably more complex with numerous loops and passages developed on several levels. Some of the side passages carry infeeder streams; these passages generally ending in domes up to 20 m in height and located beneath overlying sinks and sinking streams. The Rubber Chicken cave stream flows out of a low passage that leads to the Toothpick Segment of the System.

8.7.3 Toothpick Segment

The 30-m-wide and 15-m-high entrance to Toothpick Cave in Ravens Nest Hollow is near the top of the Union Limestone and as is the case with several other entrances, leads to a shaft dropping through much of the Union. This part of the cave System is shown on map sheet E. (M-8.5). From the base of this 20-m pit, passage can be followed upstream to a series of 10-m climbs, above which are infeeder passages leading toward the valley above. Downstream from the pit, the stream enters a 40-m-long near sump beyond which the passage opens to a series of wet crawls and domes for 400 m. The passage then opens to a 6-m-wide and 12-m-high canyon in the lower Union Limestone trending southwest. This passage, developed along joints and faults, can be followed for 1.4 km before lowering and reaching the far upstream end of the Rubber Chicken Cave Segment. In addition to the stream passage, this section of cave contains several km of dry upper-level passages and phreatic tubes that have not been completely explored, and a dense network of interconnected upper-level passages in the lower Union and Upper Pickaway that are aligned along strike, e.g., Elephant's Graveyard.

8.7.4 Canadian Hole Segment

Five hundred meters north of the Snedegars Saltpetre entrance, a small stream, rising at the base of the Alderson Limestone, flows over the Greenville Shale and sinks into the 12-m-deep pit that is the entrance to the Canadian Hole Segment of the Friars Hole System. Map sheets A and C (M-8.1 and M-8.3) show this part of the cave System. This pit is followed by several others descending through the Union Limestone. A short climb at the base of this entrance series leads to a 500-m-long passage at the base of the Union called First Canyon. This canyon and the passages that it leads to are collectively called "The Alberta Extension" as described below.

At the lower (western) end of first Canyon, a 2-m drop leads to the south trending Second Canyon ending at a dry chamber (Canyon Terminus) and a north trending dry passage called Cross Over Passage that can be followed for 250 m to the 100-m-long and 60-m-wide chamber called Crow's Nest Room. Passages on the west side of this chamber (Rubber Boot Route, Irv's Passage) lead west to the System's largest stream, Rocky River, i.e., the underground Hills Creek. This passage can be followed downstream along a fault for 200 m before being lost in rockfall in the Rocky Horror Streamway, 30 m below the lowest point in the breakdown-filled floor of Monster Cavern. Stratigraphically, the downstream end of this passage is below the Taggard Shales, only one of two places in the System where this is seen. Rocky River can also be followed upstream for almost 500 m to a sump called Watergate.

Two passages lead east from the Crow's Nest Room: A Neasy Stroll and McKeever's Passage, both of which contain inlet streams (Echo Inlet and the Downlets). McKeever's Passage also intersects a major strike aligned paleo-passage called the Canadian Highway, a northern continuation of The Highway in the Rubber Chicken Segment. Following this passage south for 350 m, it reaches an intersection with the upper end of Mud Canyon, a passage generally trending west for 450 m and ending at the east side of the 110-m-diameter and 75-m-high Monster Cavern, the System's largest chamber. About halfway along the length of Mud Canyon, a low passage leads south for 500 m through wet crawls before connecting to the far northern end of the Rubber Chicken Highway. The Canadian Highway continues southwest for another 500 m beyond the Mud Canyon junction before ending in breakdown only 50 m to the east of passages at the upper end of the Rubber Chicken Highway. Where it crosses over First Canyon, 10 m below, a hole in the floor of the Canadian Highway connects the two.

This part of the cave is fairly complex, containing linking passages, e.g., the French Connection between First Canyon and the Canadian Highway and the Canyon Connection between First Canyon and the lower end of Mud Canyon as well as a passage above the Canadian Highway called the Skyway.

On the north side of the Crow's Nest Room, a 120-m-long passage called 15th November Avenue leads to a 600-m-long, wet boulder choked and fault controlled crawl, appropriately named "Almost Hell." At the north end of this crawl, the passage becomes much larger and leads to the cave's northernmost passage complex, The Ontario Extension, shown on map sheet A (M-8.1).

Ontario Extension begins with Yonge Street, a large 200-m-long passage that leads to several km of passages, e.g., Almost Heaven and Temptation Streamway that end in several inlet complexes. Following a cave radio location effort to find a shorter way into this part of the cave, crawls at the bottom of a hillside pit were dug open. This led to the top of an 8-m-deep pit and below it, to the upper end of one of the inlet complexes in the Ontario Extension: the Dangling Dan Passages. This is the northernmost entrance to the Friars Hole System.

A passage at the north end of this complex, Tin Can Alley, intersects a 650-m-long, west trending stream passage called Skid Row. At its downstream end, it intersects a large strike-oriented stream, Rocky River II. The Rocky River II passage is also that part of Hills Creek that flows through Cutlip Cave and Clyde Cochrane Sinks to the northeast. This stream is an upstream continuation of Rocky River, seen in the Alberta Extension to the south. The stream can be followed for 500 m upstream before ending at a boulder choke and for 700 m downstream, ending at a deep pool beneath breakdown at the south end of a 30-m-wide chamber called Amphitheater II; only 120 m north of the upstream Watergate sump in Rocky River.

8.7.5 Friars Hole Segment

The entrance to the historic Friars Hole Cave, on map sheet D (electronic map M-8.4), is a small opening at the bottom of a 15-m-deep doline on the west side of the Friars Hole Valley, about 1 km north of Spring Creek. This leads to climb downs followed by two 8-m-deep pitches and an 80-m-long passage. This then opens to the cave's 15- to 20-m-wide and 5- to 10-m-high main conduit, a strike-oriented, NE-SW trending passage containing an underfit stream. The passage is developed along bedding partings in the basal Union Limestone.

Proceeding upstream, the passage can be followed for 1.2 km to an extensive complex of infeeder passages, bringing in water sinking in a blind valley in the middle Friars Hole drainage above. These passages climb through the Union Limestone and generally end in 10- to 15-m-high domes. Going upstream from this area, the ceiling lowers to 1–1.5 m, and the floor is covered in deep mud in a passage called "The Dung-Ho Way" (Fig. 8.17). After 300 m, the ceiling rises to 3–5 m and the passage continues for another 500 m to an infeeder entering from the passage's left (north) side, ending at a chamber called The Broken Room. Another infeeder enters after another 100 m and beyond this point, the passage becomes more complex, with an inter-braided upper-level passage. After another 400 m, a short climb is reached; this was the end of exploration of this cave in the 1960s and the place where it was connected to Rubber Chicken Cave in 1976. The north flowing Rubber Chicken stream lies another 400 m to the north of this point.

Proceeding downstream from the Friars Hole entrance passage, the cave becomes more complex, with passages extending from the middle of the Union Limestone down to the upper Patton Limestone member of the Denmar formation. Less than 50 m downstream of the entrance passage, a 10-m climb on the passage's west wall leads to the 1.9-km-long Parker Mountain Passage, an infeeder that in the past, carried drainage from the Robbins Run valley, 0.75 km to the west of the Friars Hole Valley, beneath Parker Mountain between, and into the Friars Hole trunk passage. The Parker Mountain Passage follows bedding planes to the southwest and then west before turning north in a low, sand floored passage before ending in domes beneath the western hillside of Robbins Run.

The downstream Friars Hole trunk below the entrance passage drops into the middle of the Pickaway Limestone in a series of wet climbdowns leading to the Pool Room, a major junction about 100 m south of the entrance passage (Fig. 8.17). An extensive multi-level network of infeeders containing over two km of passage enters from the east. Straight ahead, a large breakdown-floored passage leads south for 300 m to The Lost Passage, a 30-m-wide and 10-m-high passage floored with low sand and mud rises. This passage becomes smaller, ending at a sediment choke. At a slightly lower level, the passage can be followed for another 100 m to the cave's southern end, the appropriately named Hammer Passage, a breakdown-choked chamber climbing 25 m and apparently bounded at its southern end by a thrust fault striking NW-SE.

About half way between the Pool Room and The Lost Passage, a climbdown called Two Time Pit is reached. A low and wet passage below this pit leads west and then north to a substantial passage, Mauck's Discovery, a 650-m-long generally strike-oriented paleo-drain in the lower Pickaway Limestone that is now dry in all but its upper end. This passage, 8- to 10-m high and wide, abruptly ends at an intersection with a WNW-ESE trending passage, ending in a fault on its northwest end. To the southeast, the passage leads to a 30 vertical meter climbdown through the

Taggard Shales and then to a 75-m-long near sump beyond which is a substantial strike aligned stream flowing to the south in a passage called Water World. One hundred fifty meters downstream the stream passage becomes low and rock choked. The passage can be followed upstream for 500 m to a near sump. The downstream end of this stream is 142 m lower than the cave's datum entrance and at an elevation of 596 m is the lowest point in the System. The entire passage is in the upper Denmar Formation.

An upstream continuation of this stream called Big Water is reached via another route further north in the cave by following the dip of a fault on the west side of a chamber called Barbara's Room. After descending for 25 m following the fault plane downward, a low rock-filled passage similar to Almost Hell in Canadian Hole can be followed along the strike of the fault for 250 m to another segment of the large stream called Big Water. This is probably the same stream seen in Water World; the two streams are aligned and at the same elevation. This stream can be followed for another 150 m upstream to a mud-choked sump.

As noted under Cave Hydrology, the volume of water in these passages is considerably larger than that seen in the Friars Hole Cave stream above. At least some of this water is derived from the Robbins Run stream sinking in its bed 2.3 km northwest as per tracer tests conducted with fluorescein dye in 2015–16.

8.8 Summary

Passages summing 73.4 km in total length have been surveyed in the Friars Hole Cave System since the mid-1960s. The cave System, composed of three internal drainage complexes extending over a linear distance of almost 7 km, has had a long and complex evolution with most of it being over 730,000 years old and one dated speleothem having an age of over 1.67 million years (Worthington 1984). Although nearly the entire cave System is developed in the upper Greenbrier Group limestones (the Union and Pickaway limestones), two of the cave's three active drainages breach the Taggard formation, a major aquitard below the Pickaway and extending the cave's stratigraphic extent. Due to the presence of several long but low flooded passage segments that are entirely water filled in all but the driest of conditions, significant parts of the cave System remain to be explored, surveyed, and studied. These include the extensive Droughtway section of the cave System at the downstream end of the cave's central drainage and the Water World stream complex at the south end of the Friars Hole Cave part of the System; these are also the two deepest parts of the cave System. The potential thus exists for substantial additions to be made to the cave's known extent, subject to the accessibility of these areas.

Acknowledgements We thank the landowners, Aida and Robert Mothes, for their hospitality in allowing access to the cave System to the authors and to other researchers.

References

Baker, L. 1982. The Friars hole system: Highlights of the history of exploration. In *Captial Area Cavers Bulletin No. 1*, ed. R. Anderson, and L. Baker, 10–55. Washington, D.C.

Jameson, R. 1985. Structural Segments and the analysis of Flow Paths in the North Canyon of Snedegar Cave, Friars Hole Cave System, West Virginia, 437pp. MS Thesis, Morgantown, WV: West Virginia University.

Jones, W.K. 1997. *Karst Hydrology Atlas of West Virginia*, vol. 4. Charles Town, WV: Karst Waters Institute Special Publication.

Medville, D. 1981. Geography of the Friars Hole Cave system, U.S.A. In *Proceedings of the Eighth International Congress of Speleology. Bowling Green, Kentucky*, ed. B.F. Beck, 412–413. Huntsville AL: National Speleological Society.

Sasowsky, I.D., W.B. White, and D.M. Medville, 1989. The remarkably constant longitudinal profile of toothpick stream, Friars Hole Cave System, West Virginia, U.S.A. In *Proceedings of the tenth international congress of speleology*, 284–286. Budapest: Hungarian Speleological Society.

Worthington, S. 1984. The Paleodrainage of an Appalachian Fluviokarst: Friars Hole, West Virginia, 218pp. MS Thesis, McMaster University. Hamilton, Ontario, Canada.

Worthington, S.R.H., and D.M. Medville. 2012. Friars hole system, West Virginia. In *Encyclopedia of caves*, ed. W.B. White, and D.C. Culver, 334–340. Chennai: Academic Press.

Caves, Karst, and Science in the Buckeye Creek Cave Watershed

9

Gregory S. Springer

Abstract

The Buckeye Creek watershed is a 14 km^2 enclosed basin of which 12 km^2 drains through Buckeye Creek Cave to lower Spring Creek. The 1.6-km long stream passage is generally 6+ meters wide and 3+ meters high with the primary restrictions being the Gray Canyon near the entrance and partially flooded sewer passages near cave's downstream terminus. The passages below Turner Avenue are large trunks that are connected to the present stream passage by collapse features and solutional passages that may be remnant phreatic loops. Buckeye Creek grades to Spring Creek and the modern cave stream generally follows strike. The highest passages in Buckeye Creek Cave are at least 788,000 years old based on magnetic reversals found in cave sediments. Buckeye Creek Cave is being enlarged by corrosion, but abrasion and quarrying also play important roles. The abrasion is accomplished by sediment transported during floods. Three stalagmites were used to study local climates over the past 7,000 years. The most detailed time series record multiple dry periods lasting centuries. The "droughts" coincided with Bond Events, which were episodic periods of enhanced ice-rafting in the North Atlantic Ocean believed to have been triggered by protracted cooling.

9.1 Introduction

The Buckeye Creek watershed is a topographically enclosed basin draining Butler Mountain adjacent to lower Spring Creek. Buckeye Creek rises on Butler Mountain and flows into the watershed's central depression from which it is drained by Buckeye Creek Cave (BCC) (Figs. 9.1 and 9.2). Members of the West Virginia Association for Cave Studies have surveyed ~11 km of passages in the caves of the Buckeye basin, including 7.19 km (4.47 miles) in BCC. The watershed is the subject of *The caves and karst of the Buckeye Creek Basin*, a 1994 publication edited by George Dasher and Bill Balfour. Subsequent work by Greg Springer and others has produced multiple papers concerning the fluvial geomorphology of Buckeye Creek and paleoclimate records derived from stalagmites collected in BCC (Hardt et al. 2010; Springer 2004; Springer et al. 2003, 2008, 2010, 2014, 2015; Springer and Wohl 2002).

9.2 Exploration

The exploration of Buckeye Creek Cave is discussed in Dasher and Balfour (1994) and can be broken down into three major waves. National Speleological Society explorers pushed the cave's mainstream passage and adjacent upper levels during the 1950s and early 1960s, but made no attempt at systematically mapping the cave. Among these explorers were Jim Berry and Gordon Rutzen, after whom the largest upper level passage was named. However, the first systematic survey of BCC and Rapps was done by the

Electronic supplementary material
The online version of this chapter (doi:10.1007/978-3-319-65801-8_9) contains supplementary material, which is available to authorized users.

G.S. Springer (✉)
Department of Geological Sciences, 219 Clippinger Laboratories, Ohio University, Athens, OH 45701, USA
e-mail: springeg@ohio.edu

W.B. White (ed.), *Caves and Karst of the Greenbrier Valley in West Virginia*,
Cave and Karst Systems of the World, https://doi.org/10.1007/978-3-319-65801-8_9

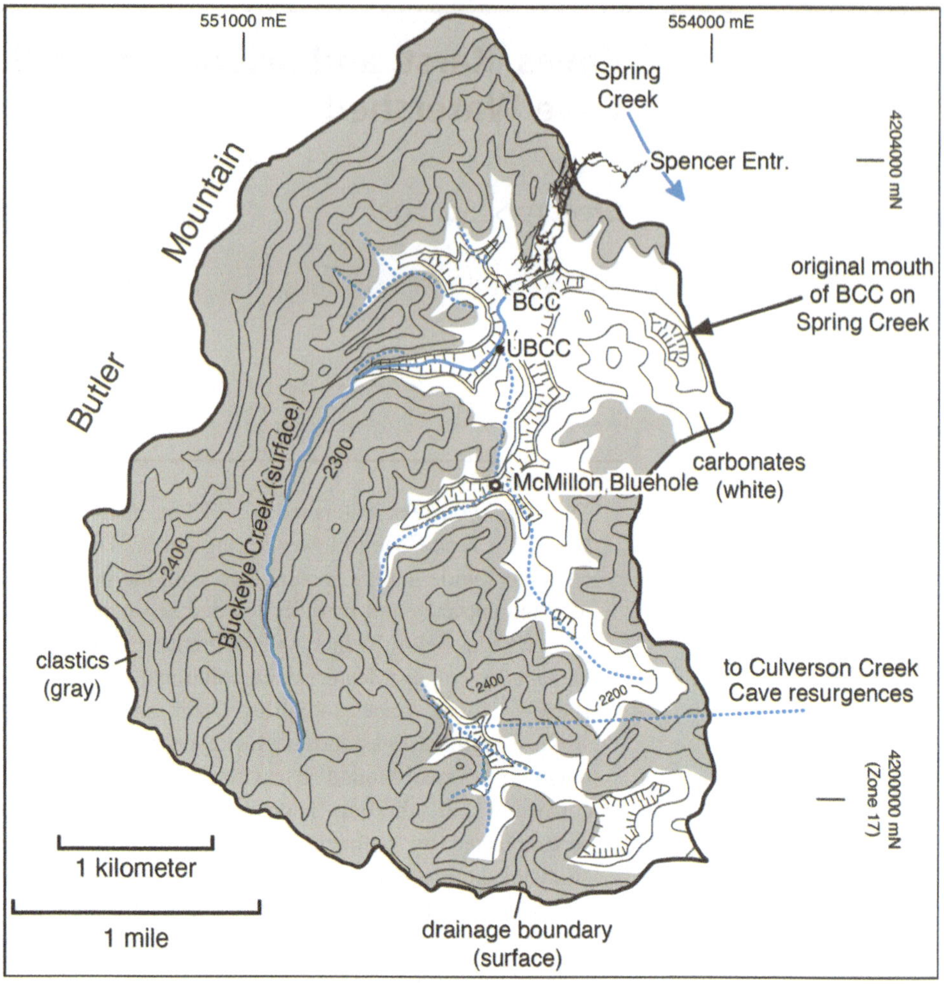

Fig. 9.1 Basic topography and hydrology of the Buckeye Creek watershed. The surface divide encompasses 14 km^2, but the southern-most tributaries have been captured and diverted to the Culverson Creek Cave springs. Buckeye Creek normally sinks where it reaches the carbonates and only follows its surface course during floods. *Dashed lines* indicate subsurface flow routes. Substantial flow passes through McMillon Bluehole and outs Upper Buckeye Creek Cave before flowing on the surface and into the main Buckeye Creek Cave entrance. The Buckeye waters reappear as talus-choked springs below the Spencer entrance, although during large floods water also flows out the entrance. Modified from Springer et al. (2003)

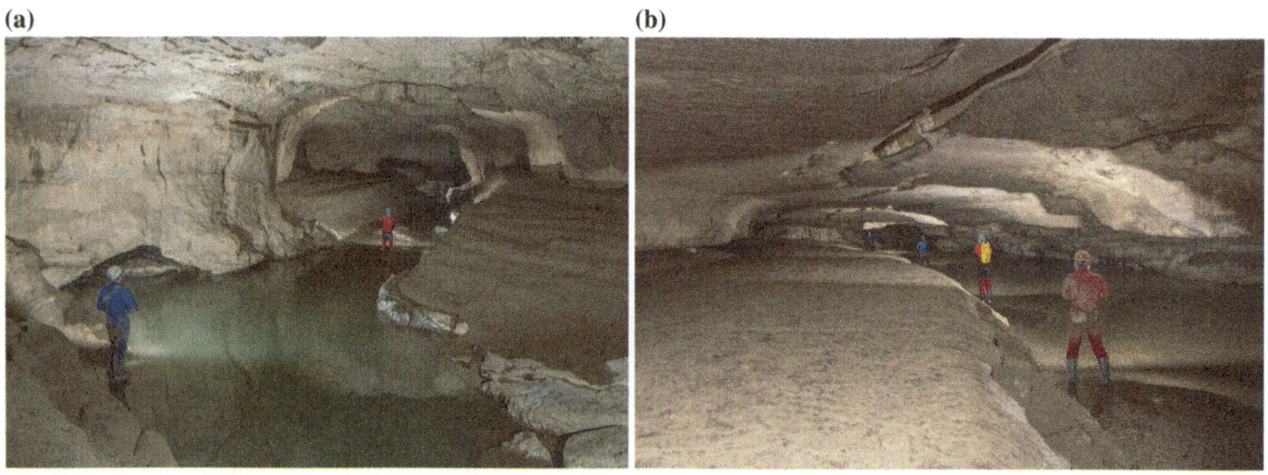

Fig. 9.2 **a** Stream passage just downstream of Buckeye Creek Cave entrance. The large silt banks were deposited after logjam formed 100 m downstream, but they began eroding away when the logjam was broken. **b** Typical stream passage in Buckeye Creek between the Gray Canyon and Watergate sump. Photos courtesy of Ryan Maurer

Baltimore Grotto in the early 1960s and led by Harvey Duchene. The Baltimore Grotto's map was not published and has since been lost (Dasher and Balfour 1994).

The second wave of surveying was by members of the West Virginia Association for Cave Studies (WVACS), who extended the work of the Baltimore Grotto and were the first to publish a map of BCC. These explorers included Phil Lucas and he was able to push low air (nearly sumped) passages at the downstream end of BCC and connect them to Spencer Cave. Subsequently, the low air passages became sumped because of a logjam near the Spencer entrance and Phil's connection was not surveyed for several decades (Dasher and Balfour 1994).

Exploration continued in BCC after the WVACS map was published, and one the cave's most ardent explorers, Bob Handley, discovered an upper level extension of Turner Avenue by climbing the 25-m high Prism Canyon. Bob led others to his discoveries and in the process they found Too Good To Last, which was the last major cave passage found in the system. However, few people were willing or able to do Bob's climbs and he later helped establish an easier route by digging mud out of undercuts at the top of canyon to create a series of crawlways leading from Turner Avenue to the new discoveries (Dasher and Balfour 1994). This is still the preferred route, although with its significant exposure extra care must be taken.

The third wave of surveying took place from 1985 to 1987 and was led by George Dasher in collaboration with WVACS. The entire cave was resurveyed and George produced an award-winning map that is much more detailed and complete than the previous maps (electronic map M-9.1). Although this effort was very thorough, the only major discovery was small, partially flooded stream ways near the Spencer entrance that were made accessible after Bob Handley and others removed a logjam that acted as a sump-generating dam.

A detailed map of Rapps Cave was produced in the early 1980s by Randy Rumer. His teams surveyed ~1.6 km of passages and his map is the basis for the Rapps passages shown in Fig. 9.3. When plotted together, BCC and Rapps were seen to be very close to one another and in 1992 Bob Handley and others established a voice connection between the two caves (Dasher and Balfour 1994). This connection has not been dug open or surveyed, so Rapps is still considered its own cave separate from BCC.

9.3 Basin Overview

The rim of the Buckeye Creek basin encloses 14 km^2, of which ~12 km^2 contribute to BCC. Dye traces have shown that ~2 km^2 of the basin's headwaters have been captured by subsurface conduits draining Culverson Creek Cave (Fig. 9.1). The watershed is typical fluviokarst with well-developed valleys whose waters have been captured by caves flowing under the adjacent hillsides. Of these piracies, the most notable (other than BCC) is diversion of southern tributaries into caves that feed McMillon Bluehole (a cenote-like, flooded sinkhole) via Hanna Caverns, Spout Cave, and inferred, but unentered caves to the southeast. Water flowing through the Bluehole next appears at an upstream sump in Upper Buckeye Creek Cave. The latter feeds two large springs beside the surface channel of Buckeye Creek. Under normal flow conditions, these spring waters are the only surface flow observed in the central depression. The waters cross the depression and flow directly into BCC.

The entrance to Buckeye Creek Cave is within a 7-m high headwall and is 3.5-m high and 2-m wide. The entrance constricts Buckeye Creek during floods, causing backflooding in the central depression. However, the 1.6-km long stream passage in BCC is generally 6+ meters wide and 3+ meters high (Fig. 9.2) with the primary restrictions being the Gray Canyon near the BCC entrance and partially flooded sewer passages near cave's downstream terminus (Fig. 9.3). The stream passage ends at a sump above which intermittently flooded passages lead to the Spencer Cave entrance of the BCC system. Under normal flow conditions, the BCC waters enter Spring Creek after rising through talus on the creek banks below the Spencer entrance, but during very high runoff water will flow out the Spencer entrance to form a 6-m wide, 1-m deep river dropping steeply to Spring Creek. The comparatively small passages near the Spencer Entrance cause extensive backflooding in BCC and sumping of long stretches of the cave stream.

The slopes of Butler Mountain are scored by very steep tributaries draining narrow sub-basins that end on large alluvial fans on the margins of the central depression. Except during very high runoff, the tributaries sink in alluvium- and colluvium-choked sinkholes and do not contribute surface flow to the central depression. Their collected waters are briefly seen in Nellies Cave and its surface spring, but the waters quickly return underground and are next seen in a side passage in BCC (see flow line in Fig. 9.1).

The central depression briefly hosted lakes during remarkable floods in 1940s and 2016. The lakes formed when runoff exceeded the receiving capacity of the BCC entrance and backflooding reached 1 km upstream along Buckeye Creek and a similar distance on the Race Track (Fig. 9.1). Such floods fill portions of BCC to its ceilings and the lakes may take several days to drain. However, the cave stream is not very flashy, although the author was trapped for 18 h in the cave's upper levels by a small flash flood in 2009. While care must be practiced, BCC is usually safe to visit during minor rainfall events.

Although unrelated to the ephemeral lakes described above, a terrace preserved just upstream of the BCC entrance includes nearly 6 m of laminated lake sediments showing soft-sediment deformation and graded beds. An attempt was made to radiocarbon date a twig obtained from the lake sediments, but the twig proved to be "radiometrically dead" (too old to date). Hence, the lake

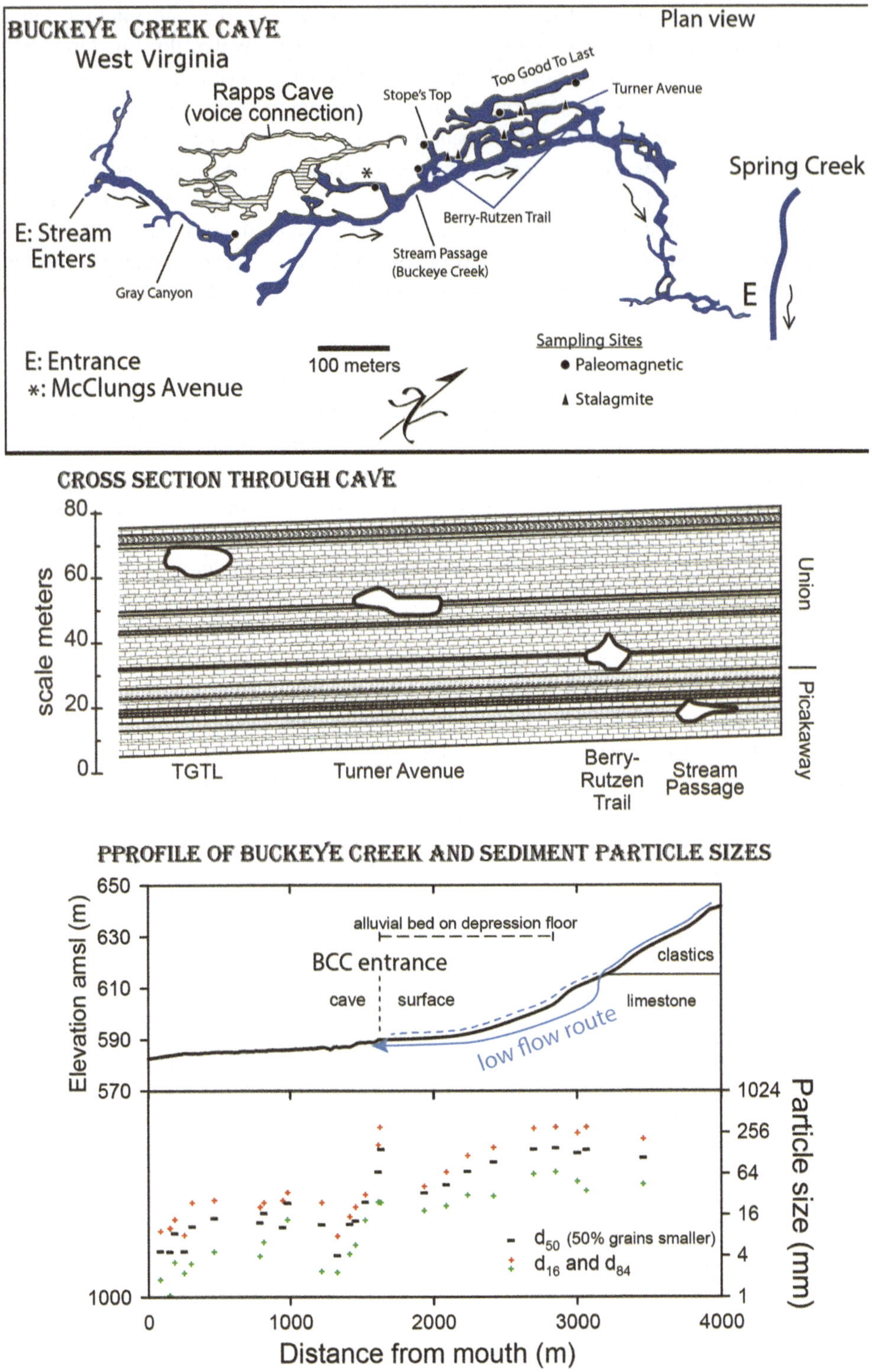

Fig. 9.3 *Top* Plan view of Buckeye Creek Cave and a cross section midway between the two entrances. The highest passage, Too Good To Last (TGTL), was the first to form. The cave subsequently abandoned TGTL, two succeeding levels, and several minor intermediate levels (*middle*). Stratigraphy from Worthington (1984). *Bottom* Longitudinal profile of Buckeye Creek along its surface and subsurface paths as surveyed using a total station. The stream has an extensive floodplain along the alluvial reach as it flows across the depression floor. *Bottom* Distribution of particle sizes along Buckeye Creek as determined from Wolman counts and sieving. Mean size peaks where gradient abruptly increases as stream enters depression and flows atop limestones. Peak at the cave entrance reflects spalling of large blocks from entrance cliff. Decline in particle sizes near downstream collapse is caused by slower floodwater velocities in backflooded passages. Modified from Dasher and Balfour (1994) and Springer et al. (2003)

sediments are almost certainly in excess of 47,000 years old. Unlike the recent flood-related lakes, this ancient lake must have been perennial. The flat Race Track, once used to race horses, is a remnant of the ancient lake bottom that ends in sandy deltas deposited on the margins of the lake. The largest delta is up valley from McMillon Bluehole and consists of at least 3.5-m of sandy sediments that prograded into the lake.

9.4 Development of Buckeye Creek Cave

A low gap marks the former mouth of Buckeye Creek (Fig. 9.1). Prior to its subsurface capture, Buckeye Creek flowed through this col and into Spring Creek, but after its incision into Greenbrier Group limestones, Buckeye Creek was diverted along strike to an upstream location on Spring Creek. Too Good To Last (TGTL) is the highest passage in BCC (Fig. 9.3), and its proximity to the top of the Union Limestone suggests it was the first major conduit diverting Buckeye Creek to the north. The passage's oval shape suggests a phreatic or epiphreatic origin, but it is unclear whether the passage contributed to a spring on the banks of Spring Creek (as the cave does at present) or if it flowed to an underground Spring Creek (as other local caves do at present).

Buckeye Creek abandoned TGTL for Turner Avenue and its continuances in Rapps Cave (Fig. 9.3). The latter cave consists of abandoned upper level passages correlative with those in BCC, but separated from BCC by collapses; a voice connection exists between BCC and Rapps, but the caves do not have a traversable connection. Access to Rapps Cave is extremely limited because of Native American remains in the cave, so the only significant research performed in Rapps was an archaeological survey led by Kim McBride and Sarah Sherwood (2006).

The passages below Turner Avenue are large trunks that are connected to the present stream passage by collapse features and solutional passages that may be remnant phreatic loops. The Berry-Rutzen Trail (BRT) passage is large (up to 15-m high and 18-m wide) and may reflect a period of stable base level. Whether the passage was adjusted to a surface or subsurface Spring Creek is uncertain. The modern cave stream is near the base of the Pickaway Limestone, which is underlain by insoluble shales that act as an aquiclude. The Taggard Shales outcrop on the surface between Spencer Cave and the former mouth of Buckeye Creek, which forces underground Buckeye and Spring creeks to rise and flow above ground.

Buckeye Creek grades to Spring Creek (Fig. 9.3), and the modern cave stream generally follows strike. However, the stream passage crosses several low amplitude geological structures whose dips are not reflected in the streambed. The structures were mapped by tracing a thin siltstone outcropping in the stream passage downstream of a thrust fault (Figs. 9.4 and 9.5). Nonetheless, passages appear to have initiated on bedding planes.

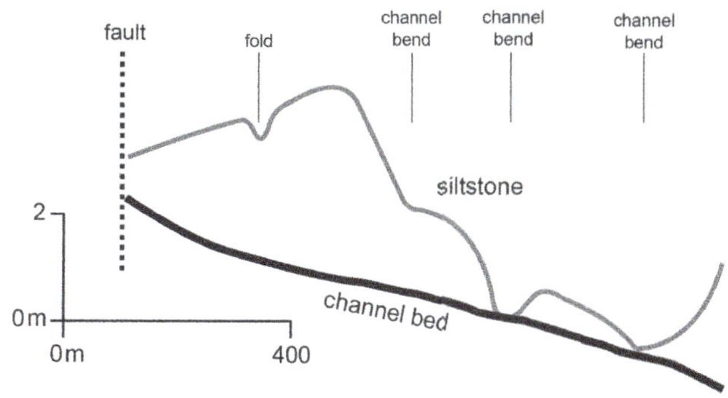

Fig. 9.4 Longitudinal profile of Buckeye Creek streambed and a dense siltstone observed in the walls and ceilings of the stream passage. The profile begins at the downstream terminus of a canyon ~300 m from the upstream entrance and ends at Spencer Cave. The stream has cut across and through prominent changes in dip, which may reflect phreatic development of the oval-shaped stream trunk along bedding planes. Modified from Springer and Wohl (2002)

9.5 Buckeye Creek and Base Level Lowering

The developmental history of Buckeye Creek Cave is recorded in its passage morphologies, sediments, and speleothems. Developmental histories are used, when sediment or speleothem ages are available, to estimate the rates at which surface streams have incised downward and to understand how major climate changes impacted Appalachian rivers, such as Spring Creek. These studies assume surface rivers act as base level to which caves contribute or grade. Minimum or maximum ages are known for several passages in BCC and these ages were used to approximate lowering rates along Buckeye and Spring creeks (Springer et al. 2015). However, calculating and interpreting lowering rates must be approached with caution when working in karst.

In the case of BCC, the "incision" rates calculated for Spring Creek can be interpreted in two different ways depending on whether we assume the upper levels of BCC contributed to a surface Spring Creek (as a spring) or to a subsurface conduit of Spring Creek (as a cave passage). Most studies assume that water table elevations were the same as channel bed elevations, but that need not be true in karst, and water table lowering rates in karst are not always equivalent to stream incision rates. As shown in Fig. 9.6, a passage in Buckeye Creek could lay substantially below the Spring Creek surface channel if the latter creek was also flowing in a vadose or epiphreatic cave. Notably, passages in The Portal—Boar Hole Cave System (PBHC), which is

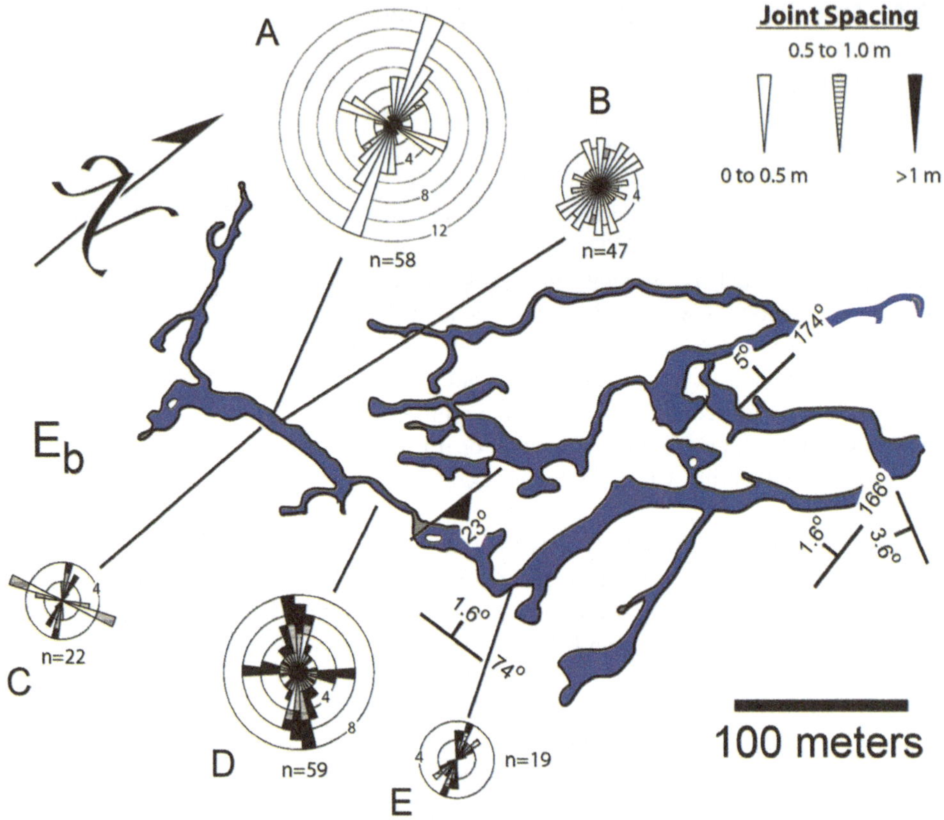

Fig. 9.5 Basic geology and fracture orientations in the southern portion of Buckeye Creek Cave. E_b denotes Buckeye Creek Cave entrance. Bimodal joint patterns reflect tectonic jointing for sites *A* and *C–E*. However, active wall collapse and stoping at *B* cause fracture orientations to be more evenly distributed

located 5 miles upstream, are developed below the elevation of the Spring Creek surface channel in the same rock units as the upper levels in BCC. The surface channel at PBHC only carries water during high runoff events and caves are adjusted to underground Spring Creek instead of the usually dry surface channel. A similar situation may have existed when TGTL, Turner Avenue, and BRT were forming.

If TGTL and other passages contributed to an underground Spring Creek, subsequent piracies to lower passages might lower the water table with or without significant surface channel incision. At present, Buckeye Creek reaches Spring Creek via a spring (bottom of Fig. 9.6), but an incision rate calculated based on the height of passage above the spring might underestimate surface channel incision because the abandoned cave passage was adjusted to a base level lower in elevation than the surface channel. Hence, it is best to speak of water table lowering rates when it is unknown whether the surface stream was once largely underground.

The highest passages in Buckeye Creek Cave are at least 788,000 years old. The clastic sediments in TGTL and Stope's Top, which is a continuation of TGTL, have reversed magnetic polarity. The reversed paleomagnetism indicates TGTL is at least 788,000 years old (Springer et al. 2015). Using the average elevation of samples in TGTL and Stope's Top, the water table lowered at a maximum of 36 m (120 ft) per million years or 36 mm ka^{-1}. Stalagmite ages calculated from lower passages are as great as 628,000 years and they yield maximum water table lowering rates of between 34 and 168 mm ka^{-1}. Hence, it appears quite likely that the water table lowering rate around Buckeye and Spring Creeks is less than 40 mm ka^{-1}. If water table lowering was driven by surface channel incision, then the local incision rate is similar to rates reported from elsewhere in the Appalachians.

9.6 Channel Process in Buckeye Creek Cave

Buckeye Creek Cave is being enlarged by corrosion, but abrasion and quarrying also play important roles in surface and subsurface Buckeye Creek. The abrasion is accomplished by sediment transported during floods. In addition to sand, silt, and clay, Buckeye Creek transports pebble to boulder size grains as bedload. The coarsest grains are found in the surface channel just upstream of the depression floor (Fig. 9.3), but the bedload is quickly reduced to gravel, which forms most of the streambed in BCC. The gravel can be carried in suspension and is frequently deposited several meters above the streambed.

Bed mobility in BCC changed significantly beginning in the late 1990s after a logjam was broken at the upstream end

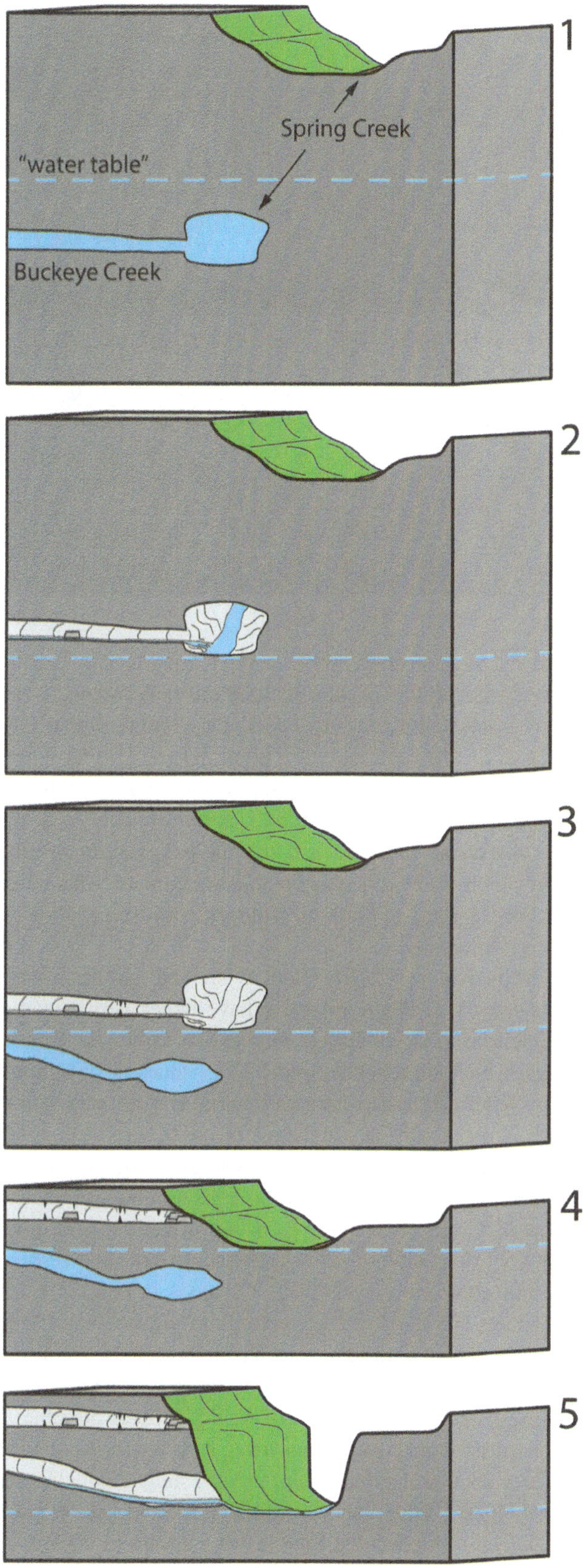

◄ **Fig. 9.6** Schematic illustration of one possible relationship between Buckeye Creek Cave and Spring Creek early in the cave's development. The highest passage, Too Good To Last, may have contributed to underground Spring Creek, as other nearby caves do. If so, the age of TGTL does not allow calculation of an incision rate for Spring Creek, only a water table lowering rate

of Gray Canyon (Fig. 9.3). The logjam had caused up to 3 m of upstream sediment accumulation and in the absence the logjam significant headward erosion occurred in the streambed, which in turn destabilized large banks of sediment extending almost to the cave entrance (top of Fig. 9.2). This is sending a wave of gravelly sediment through the cave and as of 2016 the bed is still very mobile and stream banks continue to erode. Although gravel dominates streambeds throughout BCC, several reaches are at least intermittently bedrock dominated. The most prominent of these is the Gray Canyon, which regularly accumulates up to 1 m of sediment that is subsequently scoured away to render the floors bare rock.

The Gray Canyon is generally less than 2 m wide and contains many sculpted forms, including scallops and lateral potholes (Fig. 9.7). The scallops record floodwater velocities of $\sim 2\ \text{ms}^{-1}$ ($\sim 6.5\ \text{fts}^{-1}$) within several meters of the bed. The lateral potholes are hemispherical depressions in passage walls that—like the scallops—are eroded by floodwaters (Springer and Wohl 2002). The lateral potholes range in size from 0.1 to 3 m (Fig. 9.8) and often contain scallops indicating upstream (recirculating) flow. The recirculation occurs as rotating vortices that redirect floodwaters into pothole interiors, which focuses erosion and accelerates growth of the depressions relative to adjacent walls.

Lateral potholes are not unique to BCC or other caves, but the BCC examples offer important insights into the mechanisms by which such sculpted forms grow. Scallops and sometimes-subtle etchings around fossils often record the direction of formative flows in the lateral potholes. Springer and Wohl (2002) report that recirculation appears to begin when the lateral potholes are one-quarter as deep (into the wall) as they are wide (along the wall) (Fig. 9.8). Numerical modeling shows that this is coincidently the approximate width/depth ratio at which lateral potholes are primed to erode quickly relative to the wall because of the geometries involved.

The largest lateral potholes have approximately teardrop shaped openings that are shallowest at their tops and deepest near their bases. Scallops record recirculating, descending flow in the largest lateral potholes, which may reflect whirlpool-like flow when the Gray Canyon is not entirely filled with water. The canyon is generally 6 m tall and scallops indicate rapid flow at near pipe-full, so the sculpted

Fig. 9.7 Picture of scallops and a lateral pothole (*above marker*) in Buckeye Creek Cave

forms presumably grow under a variety of flow conditions; they are not uniquely associated with open channel or closed conduit flow.

Scallops cover the ceilings of the Watergate near-sump, which is believed to become pipe-full several times a year. The scallops are large, indicating slow flow, but hydraulic modeling suggests discharge through the constriction is similar to that of the Gray Canyon (Springer et al. 2003; Springer 2004). Nonetheless, there is extensive backflooding upstream of the Watergate, as judged by high water marks observed after large floods. Pipe-full flow is not uncommon in the last kilometer of the cave stream passage because of small passage sizes near the Spencer entrance. The only portions of the stream passage that do not occasionally go pipe-full are those where the stream passage is intersected by large upper level passages that can raise ceilings in excess of 15-m.

9.7 Paleoclimate Records and Geoarchaeology

Climate histories have been constructed from stable isotopes in BCC stalagmites (Springer et al. 2008, 2009, 2014; Hardt et al. 2010). Globally, stalagmites are now one of the primary sources of terrestrial paleoclimate data because their geochemistry can reflect climate conditions outside of a cave and because they can be dated with high precision using uranium and thorium trapped in their calcite. The published time series from BCC extend to 125,000 years ago and are based on oxygen and carbon isotopes measured in stalagmitic calcite. These isotope records are believed to record above ground climate states because precipitation that falls can have distinctive ratios of ^{18}O to ^{16}O ($\delta^{18}O$) that reflect moisture sources and temperatures. Precipitation falls on soil, infiltrates, reacts with bedrock, and eventually feeds the drips that create stalagmites. Thus, the oxygen composition of precipitation is reflected in the stalagmites. This same principle applies to carbon isotopes, but the carbon is sourced differently.

Carbon isotopes reflect plant types and soil respiration above the cave. Plants typically have either very negative (−27 per mil) $\delta^{13}C$ values or comparatively low (−9 per mil) values, which are reflected in the $\delta^{13}C$ values in stalagmites fed by water that passes through the soil. But soil respiration, which is essentially the exchange of air between the atmosphere and soil pores plays a very important role by helping determine the amount of atmospheric CO_2 in the soil relative to low $\delta^{13}C$ in CO_2 respired from soil organisms and roots. Relatively dry conditions are associated with slow respiration and comparatively high $\delta^{13}C$ values in soil CO_2 and stalagmites and moist climate conditions are associated with low $\delta^{13}C$. Dry conditions also result in increased concentrations of the trace metal strontium (Sr), which comes at the expense of calcium (Ca) in drip waters, so it is possible to reconstruct wet and dry periods using $\delta^{13}C$ and Sr/Ca values.

Three BCC stalagmites were used to study local climates over the past 7,000 years (Springer et al. 2008; Hardt et al. 2010). The most detailed time series came from stalagmite BCC-002 wherein $\delta^{13}C$ and Sr/Ca values record multiple

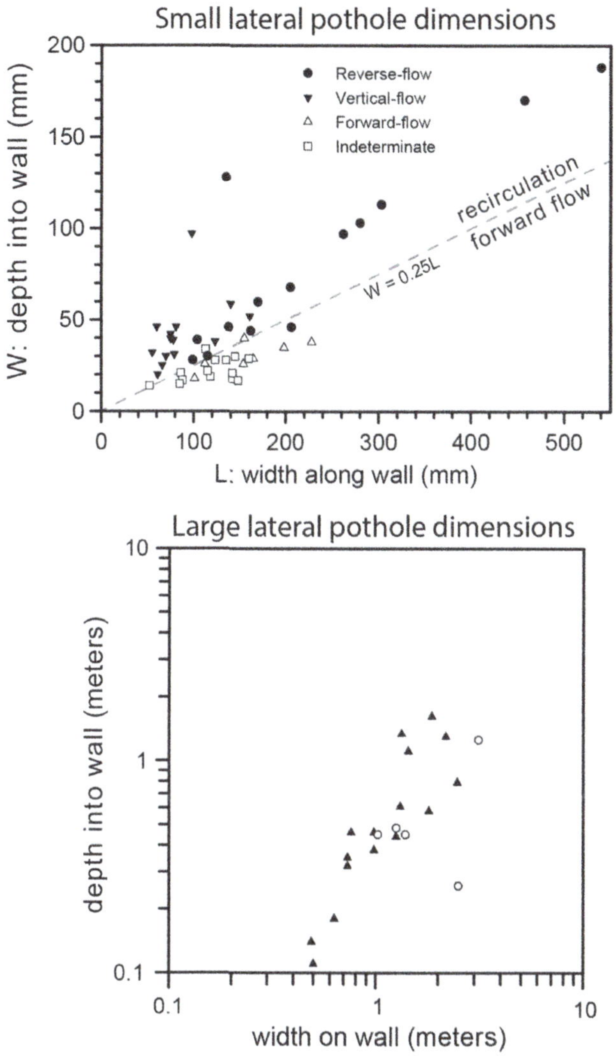

Fig. 9.8 Sizes of lateral potholes in Gray Canyon in Buckeye Creek Cave. Recirculation (upstream) flow begins in lateral potholes when depth is $\geq 25\%$ of width (*dashed line*). Modified from Springer and Wohl (2002). The lateral potholes show systematic scaling with depth increasing with width. Numerical modeling indicates the floodwaters in the potholes which are more efficient at erosion than waters flowing against the adjacent wall because water is directed against the pothole walls as water circulates in the upstream direction

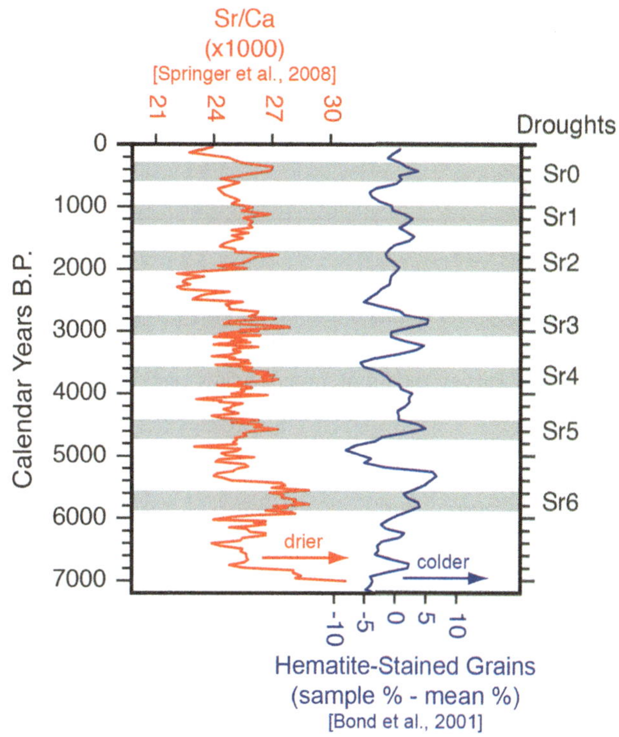

Fig. 9.9 Holocene climate variability record as Sr/Ca values in a Buckeye Creek Cave stalagmite. Springer et al. (2008) recognized seven droughts in the Sr/Ca and $\delta^{13}C$ time series (not shown). The droughts correlate with changes in the North Atlantic where enhanced ice-rafting is recorded as great abundances of hematite-stained grains. These climate events are believed to be related to solar output (Bond et al. 2001)

dry periods lasting centuries (Fig. 9.9). The "droughts" coincided with Bond Events, which were episodic periods of enhanced ice-rafting in the North Atlantic Ocean believed to have been triggered by protracted cooling. Thus, it appears that West Virginia droughts are associated with comparatively cool hemispheric temperatures and vice versa.

The BCC-002 climate record includes an anomalous increase in $\delta^{13}C$ since ~2000 years ago (Fig. 9.9). The increase is also seen in $\delta^{13}C$ values in stream sediments deposited in BCC, and charcoal concentrations in those sediments are highest ~800 years ago when Native Americans were using Rapps Cave. Springer et al. (2010) interpret the increased $\delta^{13}C$ values and charcoal as evidence that Native Americans were modifying the land above and presumably around BCC. This is compatible with an archaeological record generated from excavations in Rapps Cave, which uncovered tools, pottery sherds, and Native American bones. The archaeologists reported that peak land use by Native Americans in the watershed occurred ~800 years ago, which would explain the high charcoal concentrations because Native Americans made extensive use of fire to manage woodlands and clearings (McBride and Sherwood 2006; Springer et al. 2010).

The stalagmite BCC-010 yielded a climate record extending from 127,700 to 41,600 years ago, a time period stretching from the last interglacial (similar to the warm period of the last 10,000 years) to the middle of the last ice age. The record shows a striking correlation between global sea level and stalagmitic $\delta^{13}C$ values. High sea levels (similar to today) were generally associated with moist or wet conditions and low sea levels with comparatively dry conditions. However, millennial-scale dry periods are observed throughout the record and these events coincide with global events recorded elsewhere. Curiously, $\delta^{18}O$

values in BCC-010 closely mimic spring insolation over BCC with high insolation being correlated with low $\delta^{18}O$ values. The correlation is probably a reflection of changing precipitation seasonality in West Virginia in response to changes in the earth's tilt and precession.

9.8 Biology

Both cursory and detailed examinations have been made of the biology of the caves of the Buckeye Creek watershed. A general survey exists of invertebrates and vertebrates in caves of the Buckeye watershed (Fong and Culver 1994). The most thorough study examined subterranean species richness in the watershed and surrounding area (Schneider and Culver 2004), and a general study was conducted in the watershed. The study focused on invertebrates and experiments included placing baited pitfall traps in caves to attract cave dwellers. Eighteen species of cave-limited invertebrates were identified, including both terrestrial and aquatic species. Buckeye Creek Cave was proved to be particularly diverse and was assigned a high conservation priority. Perhaps uniquely, BCC also contains a cave-adapted sculpin that is the only known troglodyte in the Greenbrier Valley.

References

Bond, G., B. Kromer, J. Beer, R. Muscheler, M.N. Evans, W. Showers, S. Hoffmann, R. Lotti-Bond, I. Hajdas, and G. Bonani. 2001. Persistent solar influence on North Atlantic climate during the Holocene. *Science* 294: 2130–2136. doi:10.1126/science.1065680.

Dasher, G., and W. Balfour, Eds. 1994. *The Caves and Karst of the Buckeye Creek Basin*, vol. 12. Barrickville, WV: West Virginia Speleological Survey Bulletin.

Fong, D.W., and D. Culver. 1994. Invertebrate fauna. In *The Caves and Karst of the Buckeye Creek Basin Greenbrier County West Virginia*, ed. G.R. Dasher and W.M. Balfour, vol. 12, 41–42. Maxwelton, WV: West Virginia Speleological Survey Bulletin.

Hardt, B., H.D. Rowe, G.S. Springer, H. Cheng, and R.L. Edwards. 2010. The seasonality of east central North American precipitation based on three coeval Holocene speleothems from southern West Virginia. *Earth and Planetary Science Letters* 295 (3–4): 342–348. doi:10.1016/j.epsl.2010.04.002.

McBride, K.A., and S.C. Sherwood (eds.). 2006. *Report of archaeological investigations at [Unnamed Cave #14], West Virginia*. Lexington, KY: Kentucky Archaeological Survey.

Schneider, K., and D. Culver. 2004. Estimating subterranean species richness using intensive sampling and rarefaction curves in a high density cave region in West Virginia. *Journal of Cave and Karst Studies* 66 (2): 39–45.

Springer, G.S. 2004. A pipe-based, first approach to modeling closed conduit flow in caves. *Journal of Hydrology* 289 (1–4): 178–189. doi:10.1016/j.jhydrol.2003.11.020.

Springer, G.S., and E.E. Wohl. 2002. Empirical and theoretical investigations of sculpted forms in Buckeye Creek Cave, West Virginia. *The Journal of Geology* 110 (4): 469–481. doi:10.1086/340442.

Springer, G.S., E.E. Wohl, J.A. Foster, and D.G. Boyer. 2003. Testing for reach-scale adjustments of hydraulic variables to soluble and insoluble strata: Buckeye Creek and Greenbrier River, West Virginia. *Geomorphology* 56 (1–2): 201–217. doi:10.1016/S0169-555X(03)00079-5.

Springer, G.S., H.D. Rowe, B. Hardt, R.L. Edwards, and H. Cheng. 2008. Solar forcing of Holocene droughts in a stalagmite record from West Virginia in east-central North America. *Geophysical Research Letters* 35: L17703. doi:10.1029/2008GL034971.

Springer, G.S., H.D. Rowe, B. Hardt, F.G. Cocina, R.L. Edwards, and H. Cheng. 2009. Climate driven changes in river channel morphology and base level during the Holocene and Late Pleistocene of Southeastern West Virginia. *Journal of Cave and Karst Studies* 71 (2): 121–129.

Springer, G.S., D.M. White, H.D. Rowe, B. Hardt, L.N. Mihimdukulasooriya, H. Cheng, and R.L. Edwards. 2010. Multiproxy evidence from caves of Native Americans altering the overlying landscape during the late Holocene of east-central North America. *The Holocene* 20 (2): 275–283. doi:10.1177/0959683609350395.

Springer, G.S., H.D. Rowe, B. Hardt, H. Cheng, and R.L. Edwards. 2014. East central North America climates during marine isotope stages 3–5. *Geophysical Research Letters*, p. 2014GL059884. doi:10.1002/2014GL059884.

Springer, G., H. Poston, B. Hardt, and H. Rowe. 2015. Groundwater lowering and stream incision rates in the Central Appalachian Mountains of West Virginia, USA. *International Journal of Speleology* 44 (1). http://dx.doi.org/10.5038/1827-806X.44.1.9.

Worthington, S. 1984. The Paleodrainage of an Appalachian Fluviokarst: Friars Hole, West Virginia, 218pp. MS Thesis, McMaster University. Hamilton, Ontario, Canada.

The Culverson Creek Cave System

10

Philip C. Lucas

Abstract

Culverson Creek Cave is an active stream cave with a large surface catchment of 42 mi^2 (109 km^2). The cave is in the Union Limestone and discharges to springs along Spring Creek. Because of the large surface catchment and some partial blockages along the passages, the cave is subject to flooding. There are ten entrances, large passages, and numerous streams, with a total surveyed length of 20.1 miles (32.4 km).

10.1 Introduction and Historical Background

10.1.1 Overview

A surface stream called Culverson Creek drains the northwest corner of the limestone belt in Greenbrier County. Just west of the tiny community of Unus, the stream flows underground at the base of a steep escarpment (Fig. 10.1) and becomes the premier river cave in West Virginia—the Culverson Creek Cave System. The resurgence of this large underground stream is 4 1/4 miles to the east, issuing from a series of large springs along Spring Creek. The cave stream can only be followed for less than half that distance (2 miles) toward the resurgence. At this point, a large sump, called Dream Lake, is encountered which has not been successfully dived. Beyond the sump, a conjectured huge cave passage continues for another 2 1/4 miles, passing below the Buckeye Creek drainage basin. Even though the cave stream can only be followed for 2 miles, the accompanying side passages and tributary streams account for 20.1 miles of cave passages.

The Culverson Creek basin is a 42-square-mile closed karst basin (one huge blind valley). During exceptional flood events, the cave system becomes a metering point unable to accept all the flood pulse. When this happens, the Culverson Creek Cave entrance becomes submerged and a temporary lake begins to fill the valley (Fig. 10.2). This lake can become more than a mile in length (Fig. 10.3). Hypothetically, should the cave system be completely blocked, the extent of the lake would be considerably larger (Fig. 10.4).

The Culverson Creek Cave System is developed in the Union Limestone. There are ten entrances, large passages, numerous streams, and plenty of challenges for exploration and study particularly in the field of hydrology. There are many features and curiosities in this cave system. For instance, from the main entrance, the stream trunk passage extends for more than 1½ miles before being blocked by a huge log jam, where hundreds of saw logs and trees, some 75 ft in length, block the passage. Along the way, logs have jammed across the passage 60 ft above the stream. Sediment deposits more than 100 ft high have inspired names like Mudderhorn and Mud Everest. Other names such as Death Canyon, Psycho Siphon, and Dread Pool are indicative of the challenges this wet cave system offers. The depth reached by flood waters in some sections of the cave system is stunning, and the weather conditions certainly played a factor in the exploration and survey of this cave.

Electronic supplementary material
The online version of this chapter (doi:10.1007/978-3-319-65801-8_10) contains supplementary material, which is available to authorized users.

P.C. Lucas (✉)
587 Limestone Lane, Williamsville, VA 24487, USA
e-mail: Lucas@Virginiacaves.org

W.B. White (ed.), *Caves and Karst of the Greenbrier Valley in West Virginia*,
Cave and Karst Systems of the World, https://doi.org/10.1007/978-3-319-65801-8_10

Fig. 10.1 Main entrance to the Culverson Creek Cave system is where the surface stream, also called Culverson Creek, flows into a large cave entrance at the base of a large limestone escarpment which marks the end of a blind valley. The two figures seen in the entrance are Robert Handley and the Earl Thierry. They, along with Rod Scheer, were the first cavers to enter the Culverson Creek Cave system in 1953. Their figures were added to this modern entrance photograph using digital photographic techniques. Photograph adaption by Philip Lucas

Fig. 10.2 Cave system cannot accept the peak flow during extreme flooding conditions from the 42-mi^2 drainage basin. When this happens, part of the blind valley becomes a temporary lake, more than a mile in length. Photograph by William Jones. Used with permission

In the early years of exploration, there were three separate caves: Culverson Creek, McLaughlin, and Fullers. Each of these caves contained a section of the main Culverson Creek stream, but connections between these stream segments were blocked by impassable sumps and constrictions. Eventually, all the three separate caves were knitted together while carefully exploring all the many leads many of which were flood overflow routes (Fig. 10.5).

10.1.2 First Survey Trips

The first direct written record of a Culverson Creek Cave trip, when a survey was conducted, is briefly described in an Oct 1963 issue of the *Cavalier Caver*, Vol. 4. There were two surveying parties on this November 3, 1963, trip. There were eight members, presumably all UVA Cave Club members participating. One party surveyed a side passage near the

Fig. 10.3 Temporary lake is shown in *red* on this topographic map. The lake as shown here, is 1 mile in length. There have been flood events where the lake has extended much further up the blind valley

entrance that leads to the Fang Room. The other party surveyed from the Hairy Place "onward". On July 4, 1965, Charles Maus and John (Bud) Rutherford surveyed from the Culverson Entrance down to the Mudderhorn and into the Right Hand Passage. Two days earlier Charles Maus, Edward Bauer, Shelly Gordon and, Robert Malis surveyed from the Log Jam upstream to the Mudderhorn. The original survey notes of these trips and many others still exist.

During the summer of 1965, C. Michael Hamilton became very interested in Culverson Creek Cave, McLaughlin Cave, and the Fuller Cave Systems which were not yet physically connected, and began to gather material for his senior thesis. *The origin and geological relationship of the CCCS, Greenbrier Co., West Va., April 1969* was presented to the Colorado School of Mines in 1968. By that time McLaughlin Cave had been connected to Culverson Creek Cave, but the connection to Fuller Cave had remained elusive. Nonetheless, the thesis contained a map of all the surveyed passages in all the caves. This was the first map of the entire cave system and displayed a total of about 10 miles of passages including Fuller Cave.

The connection between Culverson Creek Cave and Fuller Cave took place on May 10, 1970, by Roger Baroody and Philip Lucas during a surveying trip beyond the Hurricane Lake Passage. This was a surprising connection because the Hurricane Lake Passage started as a small tributary infeeder, an unlikely route to find one's way down 180 vertical feet to the main stream passage a mile distant. With this connection, and other discoveries being made, the cave system was rapidly gaining a respectable length. It was becoming clear that a new map was needed. Philip Lucas accepted the role as coordinator for this decade long project. Soon after, William Royster became a co-coordinator for much of the project. William Balfour was a key member of the Culverson team for most of the project. Sandy Van Luik was a willing participant on many long trips and Bert Ashbrook joined the team near the end of the 10-year project. There were many others who braved the cold water and voluminous mud to survey this cave system. One epic trip was a coordinated effort by two parties: one entering the system through the Wildcat entrance and the other from the Fuller entrance with the objectives to rig the various drops,

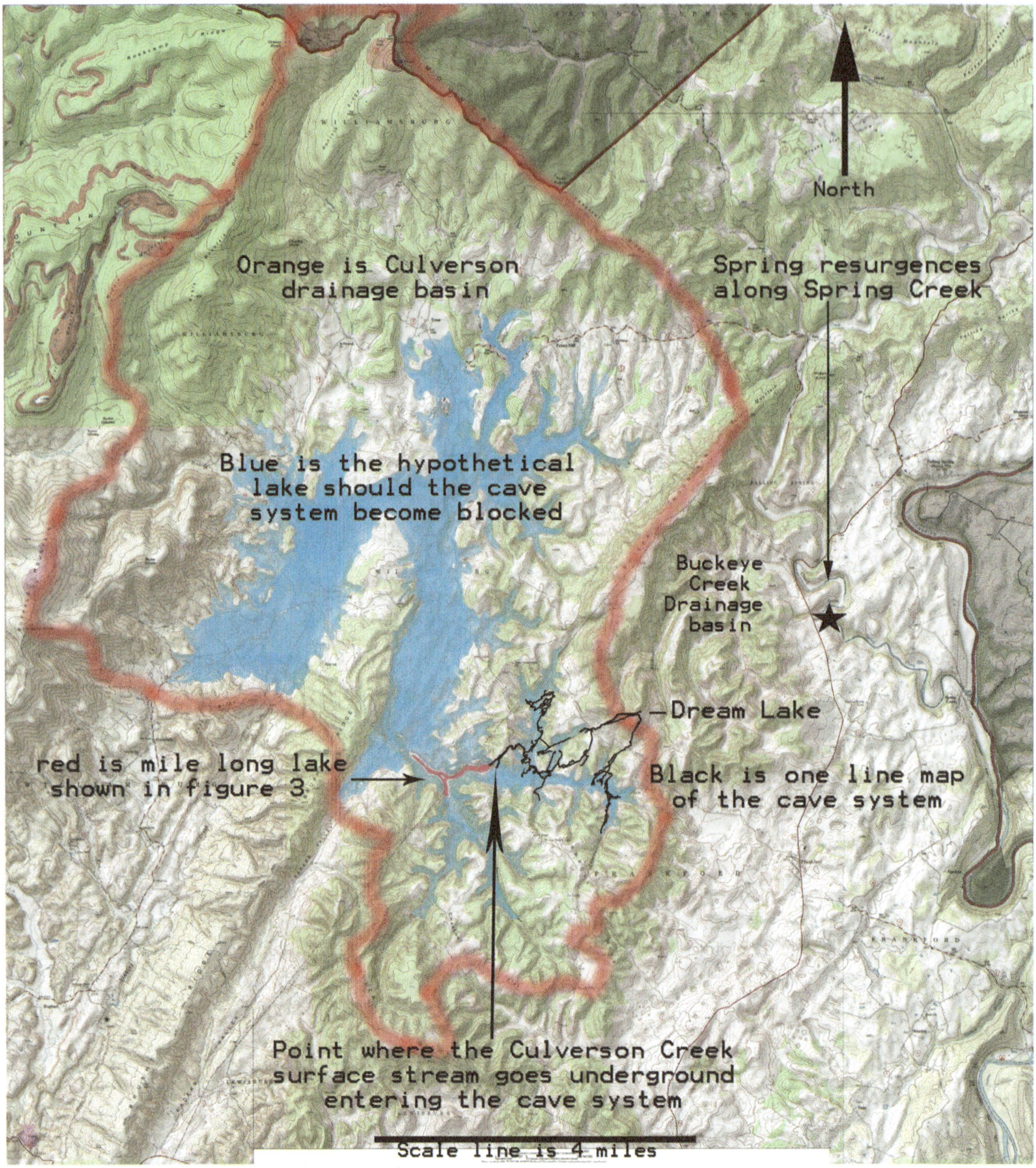

Fig. 10.4 Topographic map showing the entire 42 mi^2 drainage basin. The blue represents the size of the lake should the cave system become completely blocked. The resurgence for the cave system along Spring Creek is indicated by the star symbol which is over four miles to the northeast of the Culverson Creek Cave entrance

perform a dye trace from the Log Jam to the Alien Way, and survey the mile long Lower Culverson Trunk. Each party would exit the cave through the entrance the other party had entered. The two parties crossed paths on the top of Mud Everest on this difficult 5-mile journey with all objectives accomplished (Fig. 10.6).

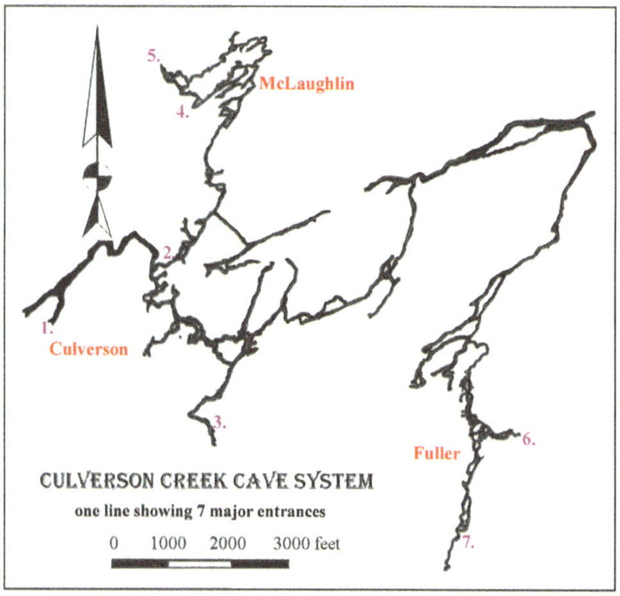

Fig. 10.5 Line map of the Culverson Creek Cave System showing the three separate caves: Culverson Creek Cave, McLaughlin Cave and Fuller Cave. They were eventually connected to form the cave system. Seven of the main entrances are shown. There are a total of nine entrances

10.2 Description of Major Passages and Connections

The sections that follow provide detailed descriptions of the cave. Maps of the area being described provide a guide to the layout of the cave passages and most of the place names that are used in the text. The complete map of the cave is provided in the electronic map file (map M-10.1).

10.2.1 Main Culverson Entrance to the Balcony (Map 10.1)

When free of flood debris, the main Culverson Creek Cave Entrance is inspiring. The opening is at the base of a 50-foot-high limestone escarpment. It is 20 ft high and almost 150 ft wide at the dripline, although it soon narrows to 80 ft wide. Culverson Creek plunges into the cave over a series of waterfalls and rapids, bottoming out in a long, deep lake. Beyond the lake, passage continues into big blackness (Fig. 10.7).

Rarely in the past 25 years has Mother Nature provided such an unobstructed view of the Culverson Creek Entrance. Usually, much of the entrance is obscured by a giant snarl of logs, limbs, and other debris. Each flood rearranges the logjam in its own way, so that the overall appearance of the entrance varies greatly from visit to visit (Fig. 10.8).

When the logjam is present, a careful traverse down through the logjam is needed to reach the other side which is a deep lake. The entrance lake requires swimming or the use of a flotation device (Fig. 10.9). After 30 ft, the stream bed becomes shallower. Wading is possible for the next several thousand feet.

About 500 ft downstream from the entrance, the passage enlarges to approximately 200 ft wide. Ceiling heights range up to 60 ft. A 30–40-foot-high mud bank sits along the left wall. At the top of the bank, large breakdown blocks project out of the mud (Fig. 10.10).

A side passages (Fang Passage) takes off to the northwest at this point. Averaging 50–70 ft wide and 10–30 ft high, this passage trends northwest for nearly 700 ft. A small stream has incised itself into the mud floor. The passage ends

Fig. 10.6 In November, 1974, two teams traveled the 5-mile journey between the Wildcat entrance and the Fuller entrance. The team members were Raymond Povirk, Rockwell Ward, William Douty, Charles Williams, William Jones, William Royster, and Philip Lucas. Photograph by Philip Lucas

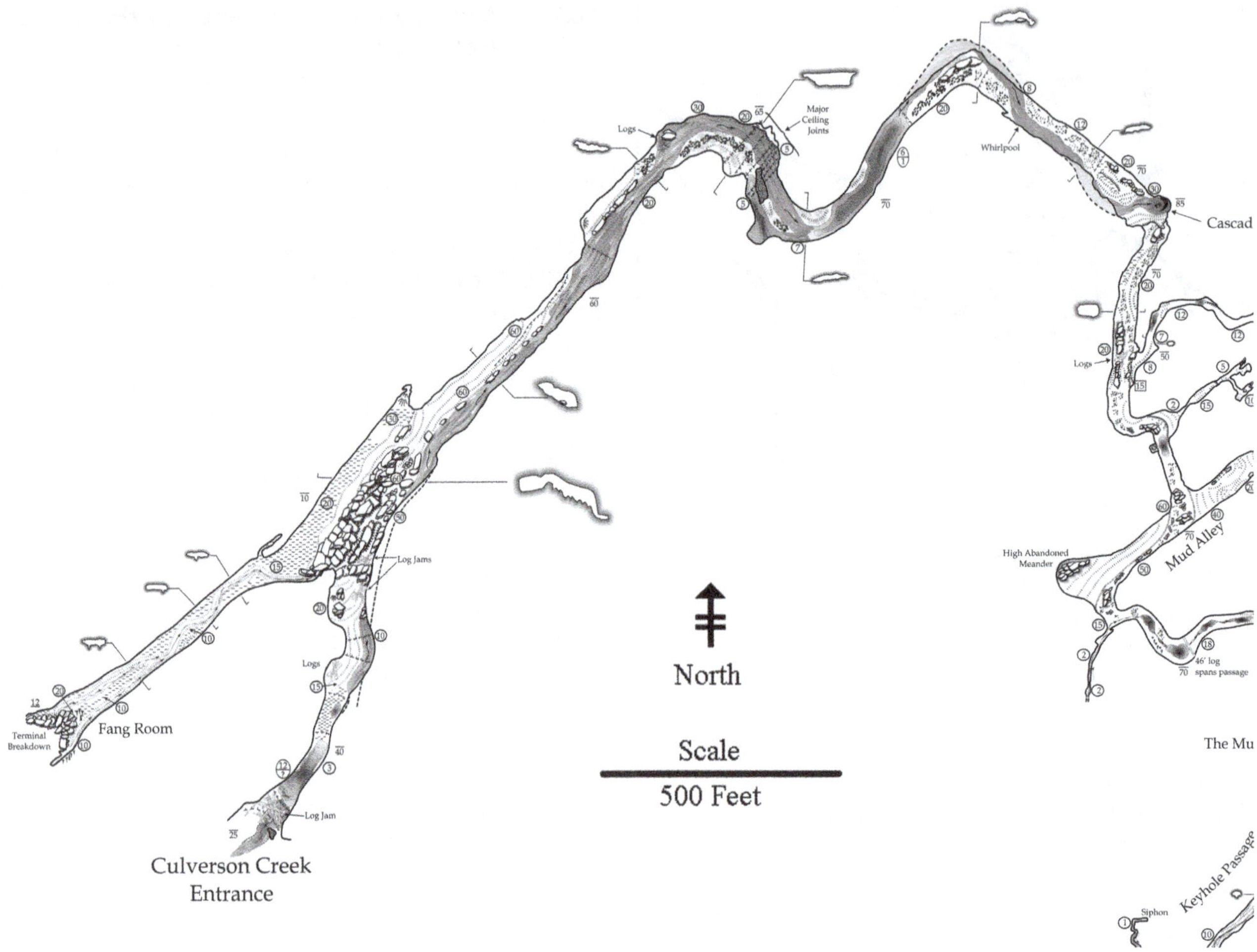

Map 10.1 Main entrance to balcony

at a huge breakdown choke. Leads can be followed for a short distance around both sides of the breakdown before choking off. It is likely that this collapse blocked a former entrance that once opened onto the hillside about 30–40 ft above the present entrance (Fig. 10.11).

Back at the main passage, the trunk continues for over 300 ft before the width narrows to 80 ft. It then continues, 80 ft wide and 60 ft high, for another 300 ft to a deep pool. It is possible to skirt the pool and then the passage narrows somewhat over the next 100 ft and the ceiling height reduces to 20 ft (Fig. 10.12). After another 200 ft, the passage takes a right hand turn. A small log jam is usually present here along the left wall. The passage turns another 90° after another 100 ft. In this bend, strong joint patterns can be seen in the walls. At one point, the passage height reduces to 5 ft. This "low" spot can be avoided by following the right wall to a parallel passage divided by a rock partition. However, this parallel passage requires wading through a pool where methane boils to the surface when rotting debris on the bottom of the pool is disturbed. After 80 ft the rock partition disappears and the passage becomes 50 ft wide and turns back to the northeast. For the next several hundred feet, the creek widens from wall to wall. Passage height averages about 6 ft. Water depths can be up to 4 ft. This is the smallest cross-sectional dimension along the main trunk passage until the log jam is reached (Fig. 10.13).

Finally, the ceiling height increases to 20 ft, and the passage makes an abrupt turn to the right and southeast. It continues in this direction for nearly 400 ft. Up until this point, the gradient of the stream has been very gradual, dropping only a few feet every hundred feet. Here, the gradient increases. The stream tumbles over a cascade, incises itself deeply in mud banks, and enters a deep pool (Fig. 10.14). This pool is known as the First Siphon. In dry

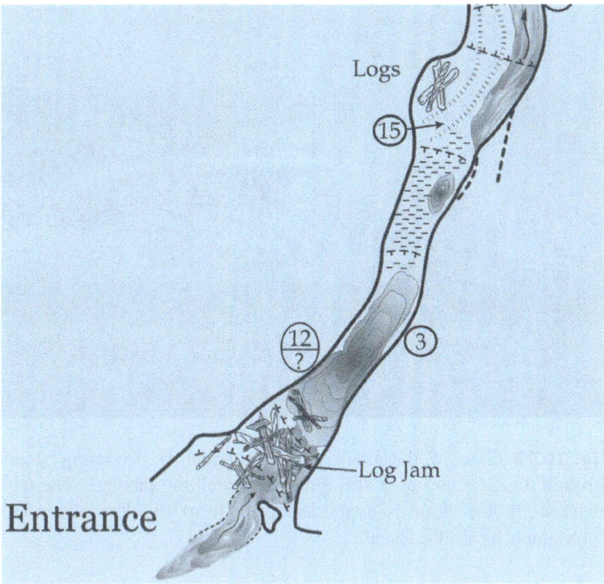

Fig. 10.7 Plan view of the Culverson Creek entrance shows the existence of the log jam during the survey when there was adequate room between the ceiling and the log jam for a relatively easy entry. There are times where the log jam completely fills the entrance requiring the excavation and rearrangement of logs to get through. There are other times when the entrance is nearly free of logs. It was reported in early June, 2016, to be free of logs. However, on June 23, 2016, 9 in. of rain fell and the resulting flood was estimated to have reached 40 ft above the entrance. Once the flood waters recede, there may be a new debris jam at the entrance once again

Fig. 10.8 Cave passage beyond the entrance log jam has a lot of flood debris scattered around. Some are huge logs jammed into cracks and crevices. Brian Williams is looking at one of the logs jammed in the ceiling. The photograph is titled; "Having a conversation with the Lord of Logs." Photograph by Philip Lucas

weather, the stream flows out under the left wall, through a tight fissure, heading in the direction of the McLaughlin section of the cave system (Fig. 10.15).

The trunk passage continues beyond the pool at the top of a steep 15-foot mud bank. During wet weather, water quickly overfills the siphon pool, and the overflow stream continues down the trunk passage. The passage narrows to 20–30 ft wide and high and continues with a silt-covered cobbled stream floor for 300 ft. Along this section are several logs "beached" on the mud banks or lying across the streambed. After times of flooding, there is also a chest-deep pool. Above the pool—and on the left wall at the ceiling—is the Balcony Passage. This leads to the Wild Cat entrance (Fig. 10.16).

10.2.2 Culverson Trunk from the Balcony to the Mudderhorn (Map 10.2)

At the balcony intersection, the Culverson Creek trunk is approximately 50 ft wide and 30–40 ft high. A cobblestone floor is interspersed with breakdown, mud banks, and sandy

Fig. 10.9 When there is a log jam at the entrance, the in-cave side of the log jam ends abruptly at the edge of a lake. This end of the lake is deep and must be negotiated by either swimming or using a flotation device to reach shallower waters. Photograph by Philip Lucas

Fig. 10.10 Further downstream the passage becomes quite large with a huge bank of breakdown on the *left*. Photograph by Philip Lucas

Fig. 10.11 This is a side passage that goes to the Fang Room which is a large breakdown termination. The passage would be about 50 ft larger if all the sediment was removed. Photograph by Philip Lucas

beaches. Approximately 250 ft downstream from The Balcony, a side lead takes off at the top of a steep mud bank. This lead pinches down to about 2 ft high, but after 10 ft opens up again into a walking passage, 20 ft wide and 15 ft high. There are copious mud mounds. About 150 ft down this passage, mud pits in the floor indicate some lower

Fig. 10.12 Size of the main stream channel is becoming smaller through this section but is still a nice size walking passage. The white material on the floor is foam left over from the last heavy rain. Photograph by Philip Lucas

Fig. 10.13 Through this section, the stream channel becomes lower and the stream spreads from wall to wall. Photograph by Philip Lucas

Fig. 10.14 Gradient of the stream has been gentle until it reaches this point where it cascades down through a jumble of breakdown. Photograph by Philip Lucas

Fig. 10.15 During low flow, the entire stream disappears into a small fissure in the north wall. This is a new piracy where the stream has been diverted through a new route into the McLaughlin side of the cave. The small passage it flows into is too small for human penetration. Photograph by Philip Lucas

Fig. 10.16 Balcony Passage connects the Wildcat Entrance to the Culverson trunk passage. Its intersection is about 15 ft higher than the floor of the trunk passage where there is usually a pool of water. Climbing down can be a bit tricky but it is generally free-climbed. Photograph by Philip Lucas

drainage. After 250 ft, the passage encounters mud-covered breakdown. Two small leads head to the right and southeast. Both are choked with mud and breakdown. One small room in the southernmost lead contains a highway road marker jammed into the mud bank. This road marker must have been washed into the cave during flooding.

Further down the main Culverson Creek trunk is a huge side lead extending only a short distance before becoming filled with sediment. Just beyond this point, the passage is filled with a deep pool which can be skirted on the right side without getting more than waist deep—if one is careful to maintain firm footing on a slippery limestone slope. To the right of the pool is a dripping dome pit that becomes a waterfall during heavy rains. About 100 ft beyond the previous side lead, the passage intersects a much larger segment of the trunk, now up to 60 ft high and wide. Another pool at this intersection is easily crossed by stepping on the tops of large submerged breakdown blocks. To the northeast from the intersection, the trunk continues for 150 ft over steep mud banks. This segment terminates in a huge mud fill.

Downstream and southwest from the intersection, the trunk continues for 500 ft with passage widths of up to 100 ft and ceiling heights of 50–60 ft (Fig. 10.17). A series of pools lined with large breakdown blocks sit along the left wall. A steep mud bank, reaching from the stream nearly up to the ceiling, lines the right side of the passage. The passage then turns south and narrows to 30 ft wide for the next 100 ft. A small tributary stream enters from a small side passage on the right and south. This tributary can be followed upstream for 250 ft, and the stream flows even during dry weather.

Beyond this side passage, the main passage turns east for 50 ft and then turns southeast. Scattered pools are encountered (Fig. 10.18). About 200 ft further is a large pool. A pair of logs spans the passage at the far end of this pool. Each log measures 46 ft in length. For the next 300 ft, the passage averages 40 ft wide and 20 ft high and has a clean-washed bedrock floor. Joint sets in the floor create striking designs (Fig. 10.19).

Eventually, a deep wall-to-wall pool is encountered (Fig. 10.20). Here the passage turns to the right and south. A side passage trends north–northeast and contains a small infeeder stream. This passage quickly narrows and within 100 ft leads to a series of tight mud tubes and crevices. These can be followed for several hundred feet to a series of small pockets, rooms, and side passages that eventually end in mud fill and collapse.

Forty feet beyond the intersection with this side passage, the main passage encounters a 6-foot-high waterfall. This is the Hairy Place. The undercut, unclimbable waterfall drops into a deep plunge pool (Fig. 10.21).

Map 10.2 Balcony to Mudderhorn

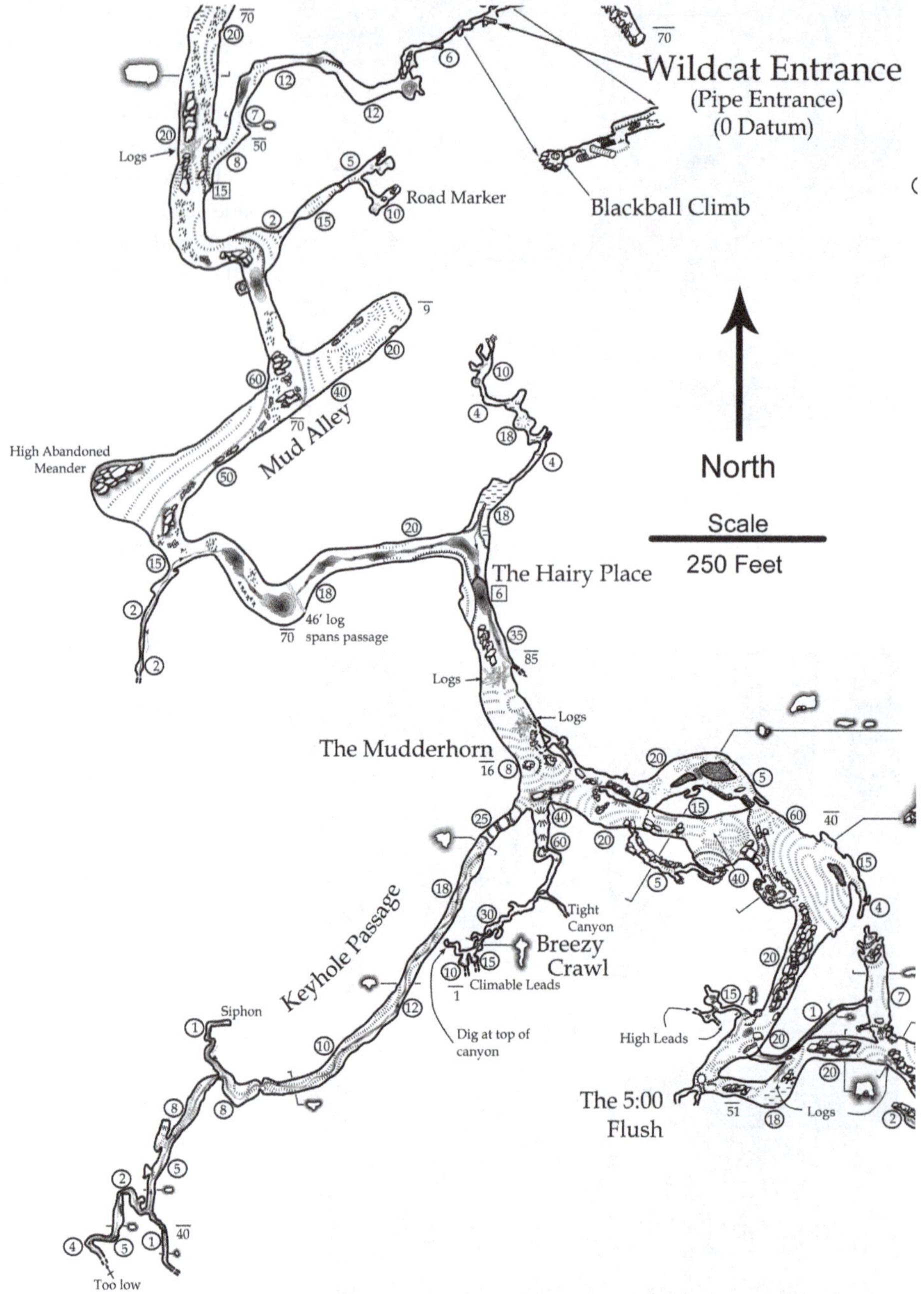

By climbing up a steep mud bank on the right-hand side of the passage, it is possible to traverse out above the waterfall and then down a steep mud bank beyond it. From the bottom of the mud bank, it is then necessary to climb down a rough wall for 15 ft to the downstream end of the pool (Fig. 10.22).

From the base of the waterfall, the passage continues for 100 ft to the base of a very large mud slope. The stream disappears here into a jumble of soupy mud and logs. As one climbs the slope, the mud firms up after 50 ft or so. The top of the mud slope is nearly 100 ft above stream level and slopes away on three sides. This is the top of the Mudderhorn. It was theorized for some time that the top of the Mudderhorn would be a place of safe refuge during a flooding event. Unfortunately, this theory was later disproved when leaves were found plastered to the ceiling at the very top of the Mudderhorn (Fig. 10.23).

Fig. 10.17 Culverson trunk passage gathers water from numerous small tributaries until it once again has a flowing stream. The passage becomes rather large in some sections and, although some mud banks have to be negotiated, most of the passage has a bedrock floor. Photograph by Philip Lucas

Fig. 10.18 This is one of the smaller cross sections in the trunk passage with William Balfour standing near a small ripple in the streambed. Photograph by Philip Lucas

Fig. 10.19 Somehow these 46 foot logs became wedged in the enlarged joints of the bedrock floor of the stream passage. The caver is William Balfour. Photograph by Philip Lucas

Fig. 10.20 Where the trunk passage makes this turn a deep pool stretches across the passage. Albert Grimm and William Balfour are carefully making their way around the inside corner of this meander. A steeply dipping and slippery limestone ledge demands a careful traverse; otherwise, a slip provides a quick descent into deep water. Photograph by Philip Lucas

Fig. 10.21 Roar of this waterfall can be heard from some distance away. Although the waterfall is only 6 ft, the plunge pool below is wide and deep. The water falls through a slot a few feet wide and the walls are undercut on each side. During moderately high flow, caution must be taken if approaching the lip of the waterfall, not to be washed over. Figuring out how to get past this obstacle was not obvious, and it was named the Hairy Place. Photograph by Philip Lucas

Fig. 10.22 Getting around the Hairy Place requires ascending a steep mud bank to the wall and then following the wall down to a point where a descent down the mud bank brings the caver to a series of ledges to down-climb to stream level. Photograph by William Balfour

Fig. 10.23 Not far downstream from the Hairy Place is the Mudderhorn. It is a large mound of muddy sediment about 70 ft high. Once, it was thought that the top of the Mudderhorn might be a safe haven from flood waters—until leaves were found plastered to the ceiling. Notice the pile of logs against the wall on the left side. They are about 40 ft above the channel where the stream flows. Photograph by Philip Lucas

10.2.3 Mudderhorn to the Log Jam (Map 10.3)

A caver with a strong light, looking south from the top of the Mudderhorn, can barely see an intersection of passages off in the distance. The main route slopes down a breakdown and mud bank before eventually leveling out into an area of undulating mud banks and mud-covered boulders. Immediately at the bottom of the slope on the left-north wall, the stream can be seen following a lower channel. This lower stream passage reconnects with the main passage 300 ft to the southeast.

Negotiating the next 1000 ft of passage will test a caver's ability to negotiate steep muddy banks and mud-covered breakdown ridges. Passage heights range from 15 ft to over 60 ft. Passage widths vary from 50 to well over 100 ft. Finding the easiest route through this large trunk passage complex can save lots of travel time and energy. Since this section of the cave floods, all signs of a trail are often erased. To further confuse matters, there are many side leads and lower passage segments that loop back into the main trunk.

About 800 ft from the Mudderhorn, at a sharp meander bend, a small trickle of water enters from a passage on the right-west wall. This side lead is known as the Five O'clock Flush. The name stems from an observation by William Royster and Philip Lucas on several trips through this section of the cave.

On the first trip, at approximately 5 pm, William and Philip had surveyed this short passage, which ended in a mud plug, and returned to the main trunk, to pack up their survey gear. Suddenly the small trickle quickly grew into a sizable stream making a considerable noise as it gushed from the side passage tumbling down into a small plunge pool. Being ever alert for floodwaters, this certainly got William and Philip's

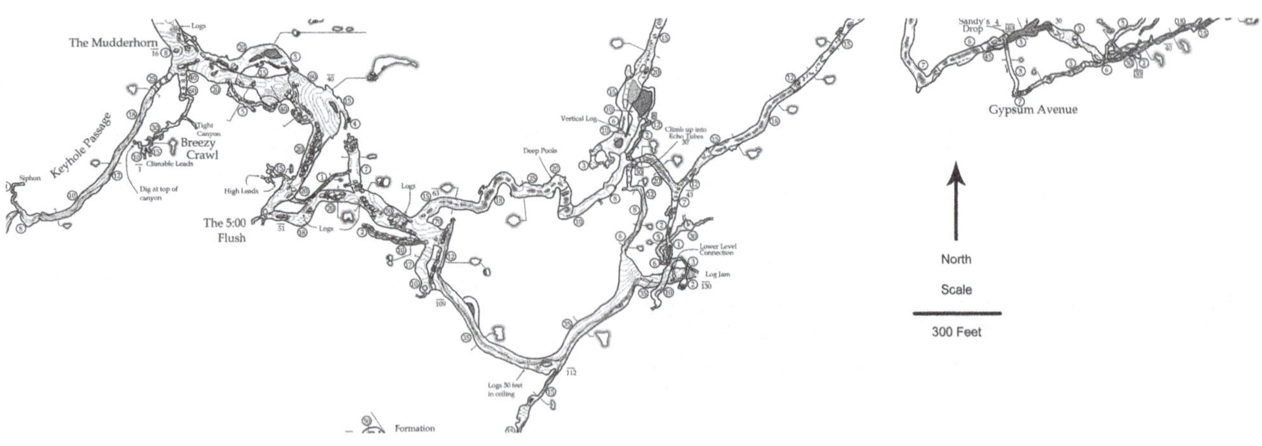

Map 10.3 Mudderhorn to logjam

attention. It even had a funky earthy smell that sometimes accompanies the flood pulse of a thunderstorm. But the weather forecast had called for no rain that day. As they nervously discussed the situation, the flow of water quickly decreased into a mere trickle again. Still wondering about the short-lived flood pulse, Philip and William continued their survey. On the very next trip, they passed the same small side passage at the same time in the afternoon. Once again the little trickle began to gush and then diminish. On several subsequent trips this phenomenon was again observed.

Reduction of the survey data presented some possible answers to the mystery. Overlaying surface features onto the cave map revealed that the little stream passage came very close to the trailer where the-then owners, Mr. and Mrs. Earl Hinkle, lived. The possibility of sewer leakage into the cave seemed logical. The increase in the amount of water flow and the duration might represent the draining of a bathtub or washer.

Downstream from the Mudderhorn, the stream can be seen in short segments of lower levels that are free of breakdown. Approximately 1400 ft from the Mudderhorn, a large passage takes off from the left wall at the top of a mud bank. This is the First Echo Tube. Ahead, another mud bank descends steeply for 70 ft to rejoin the stream. Travel from this point on is made at stream level. Occasional pools dot the passage, but seldom is the water greater than knee deep. After approximately 300 ft, a first-time visitor might be startled to see logs spanning the passage 40–50 ft above the floor. This is flood debris (Fig. 10.24)! Passage widths here are 30–40 ft. Ceiling heights range up to 60 ft.

Just beyond a climb over breakdown, a side passage carrying a small stream enters from the right. This is the Hinkle Unus Stream (Fig. 10.25).

After another 300 ft downstream in the main passage, the left wall becomes a steeply sloping mud bank. Where the passage makes a bend to the right, a large log nearly 40 ft long spans the passageway. At the top of the mud bank above this log, a small passage leads up to the Second Echo Tube.

Approximately another 150 ft downstream is the Log Jam. The Log Jam is a giant jumble of logs. Some are fencepost-sized. Others are large sections of sawed logs 3 ft in diameter. The Log Jam itself is approximately 25 ft wide and high. It completely fills the passage and seems to have plugged it up by the sheer number and volume of logs jammed together (Fig. 10.26).

Fig. 10.24 Logs jammed in the passage 40 ft above is a reminder to be aware of weather conditions when entering this cave. Photograph by Philip Lucas

Fig. 10.25 William Balfour stands on a breakdown block which marks the intersection with the Hinkle Unus passage and stream intersecting the Culverson trunk passage. Photograph by Philip Lucas

Fig. 10.26 Log Jam which completely blocks the master trunk passage. Photograph by Philip Lucas

Closer examination reveals that the ceiling descends immediately behind the Log Jam. The size and shape of the jam has changed from time to time over the years, and it has occasionally been possible to penetrate the jumble a short way. By following the ceiling downward, deep pools of water (filled with logs) were found. Apparently the Log Jam collected at a constriction in the trunk. And although during times of flooding the entire area is submerged, it does not appear that the Log Jam itself is the metering point. Metering is probably at some point downstream.

At the top of a steep mud bank on the north wall, just upstream from the Log Jam, a little waterfall enters from a small passage. This passage offers very muddy and torturous conditions. Going up a series of steep slopes and short drops, it eventually ties into the first Echo Tube near the top of the Third Ladder Drop.

10.2.4 The First Echo Tube (Map 10.4)

From the Log Jam Trunk, a short steep mud bank at the top of a larger mud bank leads into the First Echo Tube. The First Echo Tube begins as a phreatic tube-like passage. Mud banks on either side require the traveler to trudge down the center of the passage, through pools of water and sloppy mud. The mud is especially gooey and deep at the onset, and cavers must be careful to maintain some forward momentum lest they disappear up to the knees or even the waist. After several hundred feet, the pools contain more water than mud and travel is a bit easier. However, some of these pools are waist to chin deep.

The First Echo Tube follows a meandering course to the southwest for 700–800 ft. At this point, the muddy floor firms up and side passages enter from several levels.

On the right wall is a steep drop of nearly 40 ft. This is the Third Ladder Drop. At the top of the drop on the opposite wall the passage can be seen to continue. In the past, a traverse was made from the First Echo Tube to the Second Echo Tube across the top of this pit. The traverse required holding onto sections of rebar driven into the mud wall and quickly stepping across on mud footholds. Three or four traverses were made in this fashion in the early 1970s before the entire mud wall fell off after the last caver came across.

On the left and northwest wall are two passages that ascend up mud slopes and eventually reach sediment fills. Continuing down the main passage, the First Echo Tube takes on an oval shape. The floors are full of potholes. At several areas along the passage, the fine bedding of the limestone and an occasional small shale bed are visible. Pools are seldom over knee deep and there is little mud. The gentle meanders cause sounds to reverberate spectacularly. A loud shout will echo down the passageway for 10–15 s (Fig. 10.27).

Nearly 1500 ft beyond the Third Ladder Drop, the nature of the First Echo Tube changes abruptly. A short climbdown at the end of the echo tube leads to a seemingly endless area of massive breakdown that would appear to represent the collapse of a major trunk passage. It is possible to negotiate in several directions through the blocks, although the breakdown has not been pushed. Air movement in the First Echo Tube passes through this collapse. Flood waters also flow into the breakdown with no ponding.

The end of the First Echo Tube is approximately 325 ft southwest of the end of the Second Echo Tube. The Second Echo Tube also terminates when it hits a massive collapsed passage. Approximately 400 ft southwest of the end of the First Echo Tube is the surveyed end of the Williams Passage, which at this point is also encountering massive breakdown. The trend of the Williams Passage at this point is directly toward the Echo Tubes. A traverse between the Williams Passage and the Echo Tubes through the breakdown may well be possible. It is also likely that flood waters and airflow from both echo tubes exit via the Williams Passage.

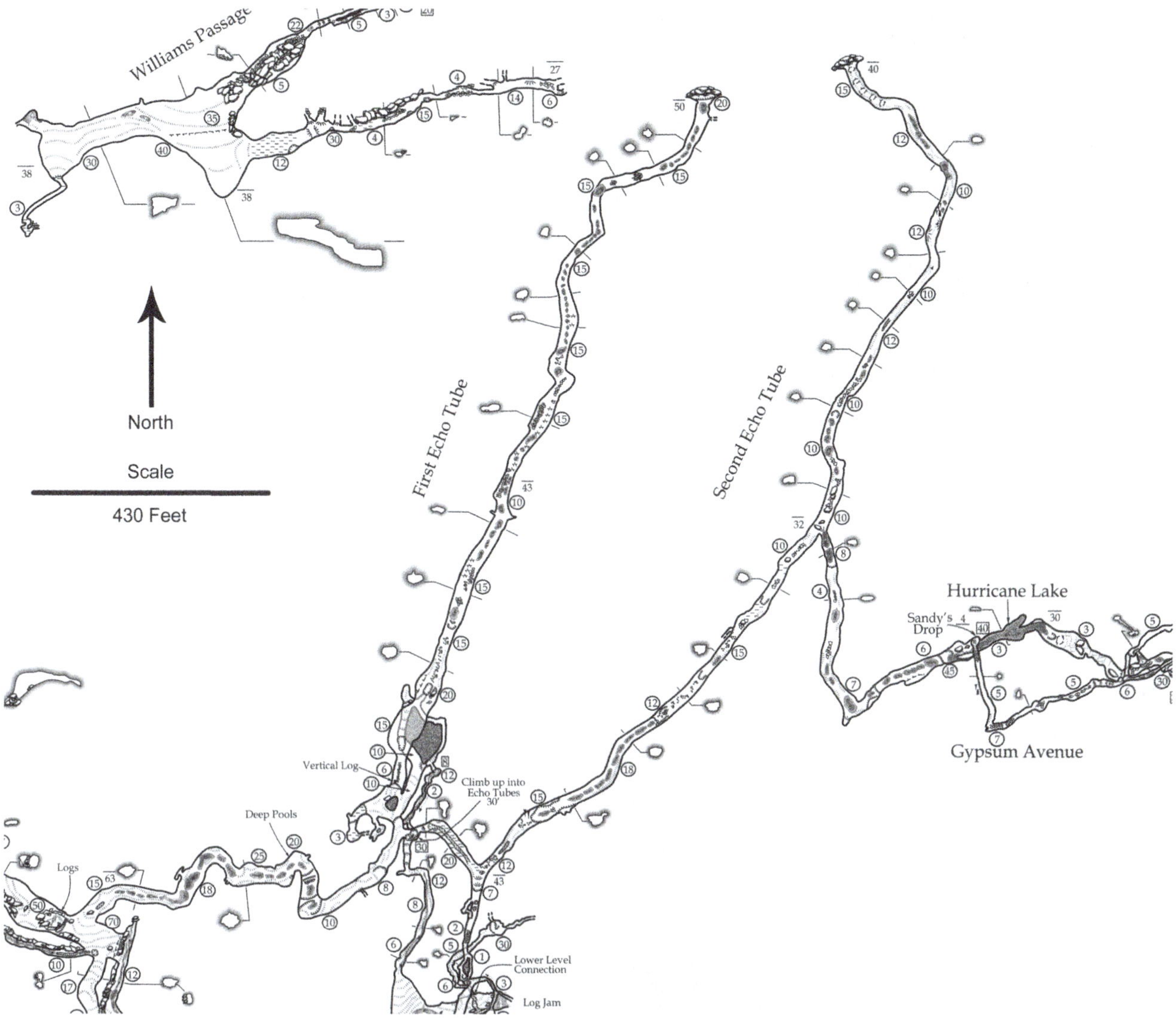

Map 10.4 First and second echo tubes to Hurricane Lake

The exact source and destination of the collapsed trunk is unknown. The potential for new passages is high. On the surface, a series of huge sinks that extend up through the sandstone caprock hint at the size of the passage that once existed below.

10.2.5 The Second Echo Tube

Just upstream from the Log Jam, a small passage at the top of a steep mud bank leads to the Second Echo Tube. Beginning as a sloppy muddy hands-and-knees crawl, this passage soon opens to a stoop walk and finally into walking canyon. Two hundred feet of walking passage brings one to the bottom of the Third Ladder Drop.

A polypropylene rope and aluminum rung ladder have been left permanently rigged at this drop. The ladder climb is vertical for 30 ft and then up a nearly vertical slope for another 20 ft. Prior to the installation of the ladder, this pit was free-climbed by the "lead" caver who would then belay the remaining party. Free-climbing this pit was difficult because of the slippery mud that covered the walls. When it became clear that many trips would travel beyond this point, the ladder was made and bolted into place by Philip Lucas (Fig. 10.28).

With no small relief, future parties could forego the necessity of free-climbing the drop. However, on the very next trip, when the team reached the base of the ladder drop, Philip Lucas was surprised and very disappointed. There was no ladder! The pit had to be free-climbed once again! When

Fig. 10.27 Second Echo Tube shown here has dimensions similar to the First Echo Tube. Both are long straight tubes, about 1500 ft long each, both ending in a wall of massive breakdown. Their name comes from the long reverberation of a loud yell which last many seconds. Photograph by William Jones

Fig. 10.28 Aluminum rung ladder was bolted to the wall at the top of the pit to eliminate the scary 40-foot free-climb up the pit. Above the pit is a steep muddy slope up to the Second Echo Tube. The ladder was later found, still attached to the wall, but instead of hanging down the pit was stretched out its full length up the steep mud slope. There had been a flooding event and the rising flood waters was swift enough to reverse the 40-foot ladder. Photograph by William Jones

Philip reached the top of the ladder drop, again much to his surprise, he found the ladder still bolted to the wall. But instead of hanging down to the passage below, the ladder ascended up from the bolt into the steep narrow mud-floored canyon above. All the way up to the top of the slope, it had gained another 50 ft in elevation. It was clear what had happened. The cave had flooded after the ladder had been installed. Flood waters backing up from the Log Jam had become deep enough and had enough current to wash the rope ladder up the pit and on up the slope until it was fully stretched out in the reverse direction.

The reversed ladder gives indication to the stages and depths of the flooding sequences in the Culverson Creek Cave System. The restriction at the Log Jam can accept only a certain amount of flow. Flood waters will then pond to a depth of nearly 150 ft before flowing out both Echo Tubes. The velocity is great enough to create potholes in the floors of the Echo Tubes and to reverse any dangling ladders. The bottom of the rope ladder is now securely tied down. It is cautioned to all visitors to this section of the cave not to untie the bottom of the ladder, and to watch the weather.

As a matter of interest concerning the corrosion of aluminum in a cave's atmosphere, it is noted that although this aluminum rung ladder has been in place for 35 years, and cavers are still using it today albeit with a belay. The aluminum rungs were made from 1/16 in. aluminum sheets folded into a U shape with a hole drilled and top of the U large enough to accommodate three polypropylene ropes ¼ in. in diameter. It was not expected that this thin-wall aluminum rung ladder would endure the cave's environment for very long. Perhaps the rate of aluminum corrosion varies from cave to cave.

From the top of the slope above the ladder, a narrow canyon soon joins a larger passage. This passage's oval shape characterizes the shape of the Echo Tube. In about 75 ft, a Y-intersection is reached. The right-hand lead can be followed down several small drops and tortuous mud passages. These eventually lead to an intersection with the Log Jam passage at ceiling level.

The left-hand fork is the main Echo Tube. Here loud shouts produce impressive echo reverberations that seem to indicate a passage that will "go forever." The nature of the Second Echo Tube is similar to the first, with shallow pools and potholes in the clean-washed phreatic tube.

Approximately 1000 ft down the Second Echo Tube, a side passage enters from the right. This is the Bath Tub Passage that leads to Hurricane Lake. Beyond, the Echo Tube continues with clean-washed walls, shallow pools, and only occasional small breakdown. The limestone is thin bedded with shale layers.

After another 1000 ft the Second Echo Tube encounters the same collapsed massive trunk passage that terminates the First Echo Tube. Just prior to this intersection, a series of enlarged joints can be seen in the ceiling and walls of the passage. Again it is obvious that flood water flow rapidly through this area in route to a lower level.

10.2.6 Bath Tub Alley and the Lake Passage

A side lead takes off from the Second Echo Tube. It looks nice, but a deep pool of wall to wall water discourages casual visitors. This pool is the first of a series of water-filled pothole-like depressions. Although none are more than 4 ft deep, their smooth slick, curved sides make it easy to slide down into these natural tubs and become completely soaked. Only with extreme caution can this 500-foot stretch of passage be negotiated without at least one member of the party slipping down and out of sight. Near the end of the pools, the passage reduces to a stoop walk, then gradually enlarges again into nice-walking passage. At this point a flowstone slope partially blocks the passage. The last pool continues beneath this flowstone. Above the flowstone is a room with a dome extending into the ceiling. This is Sandy's Drop, which leads up into Gypsum Avenue.

In the floor of this room is a fissure that can be down-climbed until it intersects the ceiling of a low wide passage just above the surface of the water. Air currents blow strongly over the water. This is Hurricane Lake. The passage is approximately 30 ft wide and 3–4 ft high and is nearly full of water. Air space ranges from almost non-existent to 12 in. (Fig. 10.29). The air at times blows across this lake with enough velocity to make little waves. By following a prominent joint in the ceiling you can traverse the 130-foot length of Hurricane Lake without removing your helmet.

At the upstream end of the lake, the passage turns right and southeast, and ascends gradually up through a shale layer. This is difficult to traverse because of the slippery surface and the 1-to-2-foot ceiling height. After several hundred feet, the passage bears left and east, and enlarges into a walking passage floored with deep pools and breakdown. One pool in particular is quite large and deep, and is filled with breakdown blocks below the water surface. In another 600 ft, the passage ascends through a series of ledges into a high-ceilinged room with a large breakdown slope. This is the Junction Room, where Gypsum Avenue enters from above (see the Gypsum Avenue description).

Continuing in an east–northeast direction, the passage follows a strong joint as a high narrow canyon. After 400 ft, the passage turns to the northeast and enlarges to 25 ft wide and 10 ft high with nice speleothems in places. This passage continues with the same dimensions for 400 ft to an intersection.

To the left at this intersection, a wide passage climbs a mud slope before leveling off. It then continues northeast for 200 ft before turning to the east. A slope down through breakdown intersects a lower passage that can also be reached from the previous intersection. This passage continues to the northeast, 25 ft wide and 10 ft high. A sticky mud floor indicates flooding from time to time. Near the end of this passage is a curious set of joints that create perpendicular partitions in the ceiling.

The right-hand passage, from the previous intersection, leads down through areas of massive breakdown and then divides. It is possible to continue down slope and find a route through more massive breakdown to a steeply sloping passage. At the bottom of this slope, another constriction is encountered. Twenty feet beyond this constriction, the passage suddenly intersects a 40-foot-wide and 30-foot-high passage. On the far side of this intersection is the Chocolate Drop, which drops down 30 ft to a stream passage, (Muddy Madness).

Steep mud slopes must be descended to reach the top of the Chocolate Drop. The passage, however, continues to the south beyond the Chocolate Drop as a 40-foot wide, 20-foot high passage with sloping muddy banks. Holes in the floor drop down to the stream passage below. About 150 ft from the Chocolate Drop, massive breakdown blocks the passage. A climbup into a left-hand passage just prior to the breakdown leads up into a large overhead room that is 40 ft wide,

Fig. 10.29 Hurricane Lake is a shallow pool 3–5 ft deep and 110 ft long. It was named Hurricane Lake for the strong breeze that created ripples on the surface of the lake. Air space above the water varies from 12 to 5 in. along a groove in the ceiling; otherwise there is only a couple inches. Photograph by William Jones

90 ft long, and 30 ft high. A high lead in the northeast corner of this room was not pushed.

10.2.7 Gypsum Avenue (Maps 10.4 and 10.5)

Gypsum Avenue is an unusual passage for the Culverson Creek Cave System. About 1400 ft long, it extends between Sandy's Drop and the Junction Room and overlies the Hurricane Lake Passage. Unlike just about every other passage in the system, it is doubtful whether Gypsum Avenue ever floods.

Gypsum Avenue was discovered from the Junction Room, where a steep exposed climbup gives access to this abandoned vadose passage. Beginning as a stooping and walking passage, the floor is covered with gypsum sand that is several inches deep in places. The passage continues as a meandering route and becomes strictly walking after several hundred feet. After 500 ft, the gypsum sands are replaced with scattered breakdown and smooth floors.

Two hundred feet further, a small crawlway on the left can be followed for over 50 ft before becoming too tight. After another 200 ft, the main passage reaches a series of short climbups and climbdowns. At this point a deeper channel can be followed for about 70 ft where a possible connection may exist with the Lake Passage below.

Beyond these climbs, Gypsum Avenue is a nice flat-floored walking passage. In another 350 ft, a 30-foot pit intersects with Hurricane Lake. This drop is usually rigged with a cable ladder. Gypsum Avenue is now the preferred choice of routes when traveling through the Hurricane Lake area.

10.2.8 Muddy Madness (Map 10.5)

At the end of the Lake Passage, the Chocolate Drop leads down into Muddy Madness. At this point, the Madness is a high stream passage that averages about 15 ft wide and 30–50 ft high. It is narrower at stream level.

The passage continues downstream for 500 ft with a gravel and mud floor. Occasional breakdown blocks, pools, and mud banks must be negotiated. In several places, steep mud banks slope down from upper levels. At certain points, it is possible to climb up these mud banks into a larger oval phreatic passage. Approximately 800 ft downstream, it is necessary to climb up into the upper level when the stream level becomes impassable. This upper passage requires careful side stepping on steep mud slopes. Some wiggling through breakdown is also necessary. Eventually, the stream level is rejoined by climbing down a nearly vertical mud slope. Beyond the slope the passage enlarges slightly but

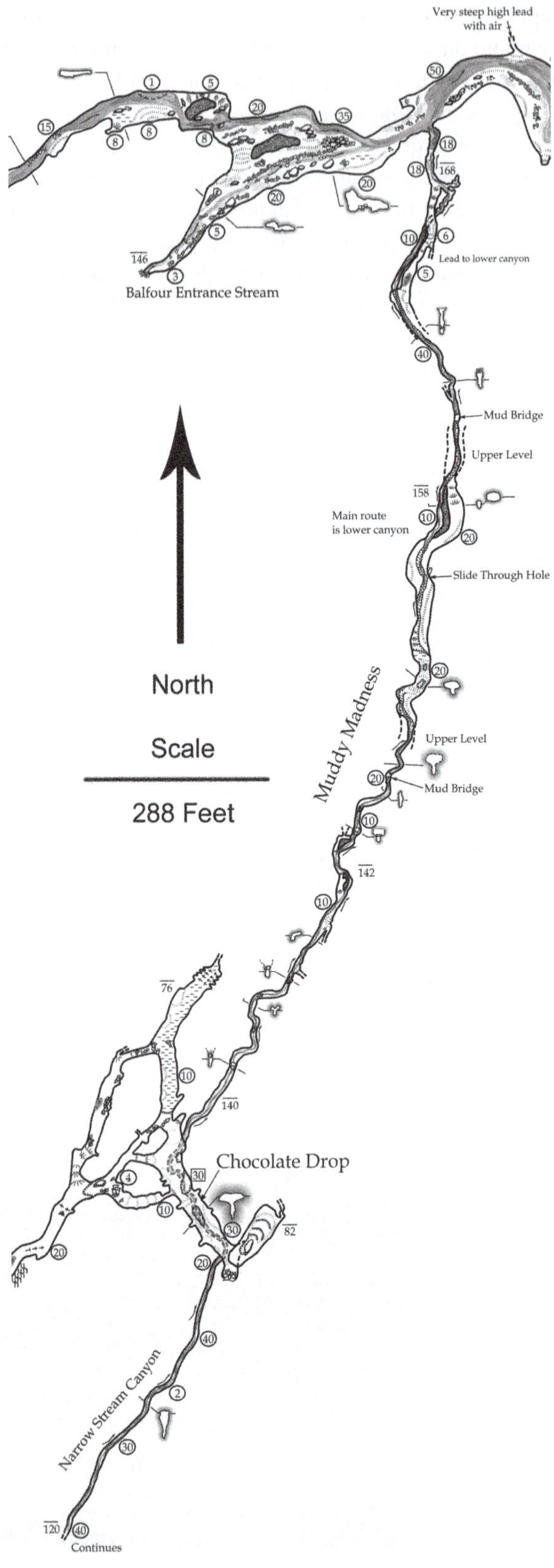

Map 10.5 Muddy madness

remains a muddy, narrow, tall canyon with a small stream. After approximately 500 ft, the passage meanders to the left becoming wider with a chest-deep pool which must be waded to continue. Around the corner is the blackness of a huge trunk passage and the roar of a large stream—the main stream of Culverson Creek.

10.2.9 Hinkle Unus (Map 10.6)

The Balfour (Hinkle Unus) entrance room itself is a rather spacious segment of stream trunk, with ceilings averaging 20–30 ft high and passage widths approaching 50 ft. In places high domes pierce the ceiling. The floor is covered with mud and small breakdown blocks. In the past, cattle would use the cool shadows of the entrance chamber (Fig. 10.30) to escape the heat of the midsummer sun. Fences inside the entrance room kept the more adventurous animals from investigating any further. Several years ago the fencing had been relocated restricting the farm animals from entering the cave. The current owner has converted much of this formerly muddy area into a stoned paved patio.

In wet weather, a stream enters from a passage from the left and south (Fig. 10.31). This passage begins as a walk-way, but within 100 ft reaches a constriction. Mud banks require some tight maneuvering. After wiggling over and through the stream, a slightly larger passage up to 2–3 ft high leads on for several hundred feet. Finally, the stream passage regains walking dimensions. After another 200 ft, the LL entrance is encountered. Exiting this entrance requires negotiating through a small collapse.

Back at the Balfour (Hinkle Unus) entrance room, big passage heads to the north, with the stream flowing over a cobblestone bed. After 150 ft, however, the cave abruptly narrows into a 20-foot-high and 2- to 4-foot wide canyon. This canyon meanders for 200 ft and then degenerates into a 2-foot high, wet, silt crawl. Because of manure produced by the cows that tend to hang out at the entrance, a heavy summer thunderstorm makes this crawl particularly disgusting and slimy. For this reason, this entrance, although it is the shortest route to the Echo Tubes and the Log Jam, is not frequently used.

The low silt crawl extends for about 300 ft until a small room finally provides some relief. The stream itself continues beyond into a lower channel, but a 15-foot climbup gives access to an upper passage. This passage extends for nearly 100 ft to an intersection. To the left the passage quickly ends in a mud plug. To the right the passage descends a steep mud slope until the stream is rejoined. Walking passage lasts for 150 ft. A low pinch may force the excavation of some cobbles if there has been a flooding event.

Immediately beyond this pinch, the ceiling height abruptly changes. An upper level is visible. A 10-foot climb leads up into several hundred feet of generally walking-sized passage that winds around in a higher layer of limestone. One room contains speleothems and a 50-foot high dome. Another 50-foot high dome is developed in a passage nearby.

The stream continues for nearly 300 ft beneath these upper levels as a 2-foot-high by 4-foot-wide crawl. At one point, a small alcove to the left leads to the bottom of a 40-foot-high dome. Gradually, the stream passage gets larger and higher and eventually reaches walking size. Two hundred feet later it intersects the Culverson Trunk. Here the Hinkle Unus stream joins another stream that flows from the Mudderhorn to the Log Jam. This confluence is approximately 450 ft upstream of the Log Jam.

10.2.10 Wildcat Entrance to the Balcony Overlook (Map 10.7)

The Wildcat entrance is in a large collapse sink with an imposing limestone headwall on the northern side (Fig. 10.32). The entrance itself is at the base of this headwall and is an 8- to 10-foot climbdown into a passage trending northeast and southwest. To the northeast lies the McLaughlin section of the system. To the southwest is the connection to the main Culverson Trunk.

Due its proximity to the edge of the entrance sink, the southwest passage is unstable. Michael Hamilton and others originally opened up this passage from the in-cave side in the late 1960s. It then collapsed in the mid-1970s. William Royster, Philip Lucas, and others opened it again in the early 1980s by digging down from the Wildcat Entrance. Over the years the entrance again became unstable and in 2006 the entrance was stabilized using a 30-in.-diameter plastic culvert. The culvert now provides access to the balcony passage and on to the main Culverson Creek stream passage (Fig. 10.33).

Approximately 30 ft into the passage is an abrupt down turn into a steep tight broken-rock crawlway. After 30 ft, this crawlway drops vertically into a small room with breakdown walls and floor. Against one wall in this room is a small hole that drops down through breakdown. Fifteen feet down is a small stream passage with pools on the floor. Rapidly moving floodwater keeps this passage clean-washed for the next 150 ft. At one point, it is necessary to climb over a short but very large log (Fig. 10.34).

The passage soon opens into a tube 6–12 ft high and 15 ft wide. It is almost perfectly oval with a small stream groove cut in the floor and continues as a pleasant walking passage for another 500 ft. There are mud slopes on each wall. Occasional pools will sooner or later force you to get your feet wet (Fig. 10.35).

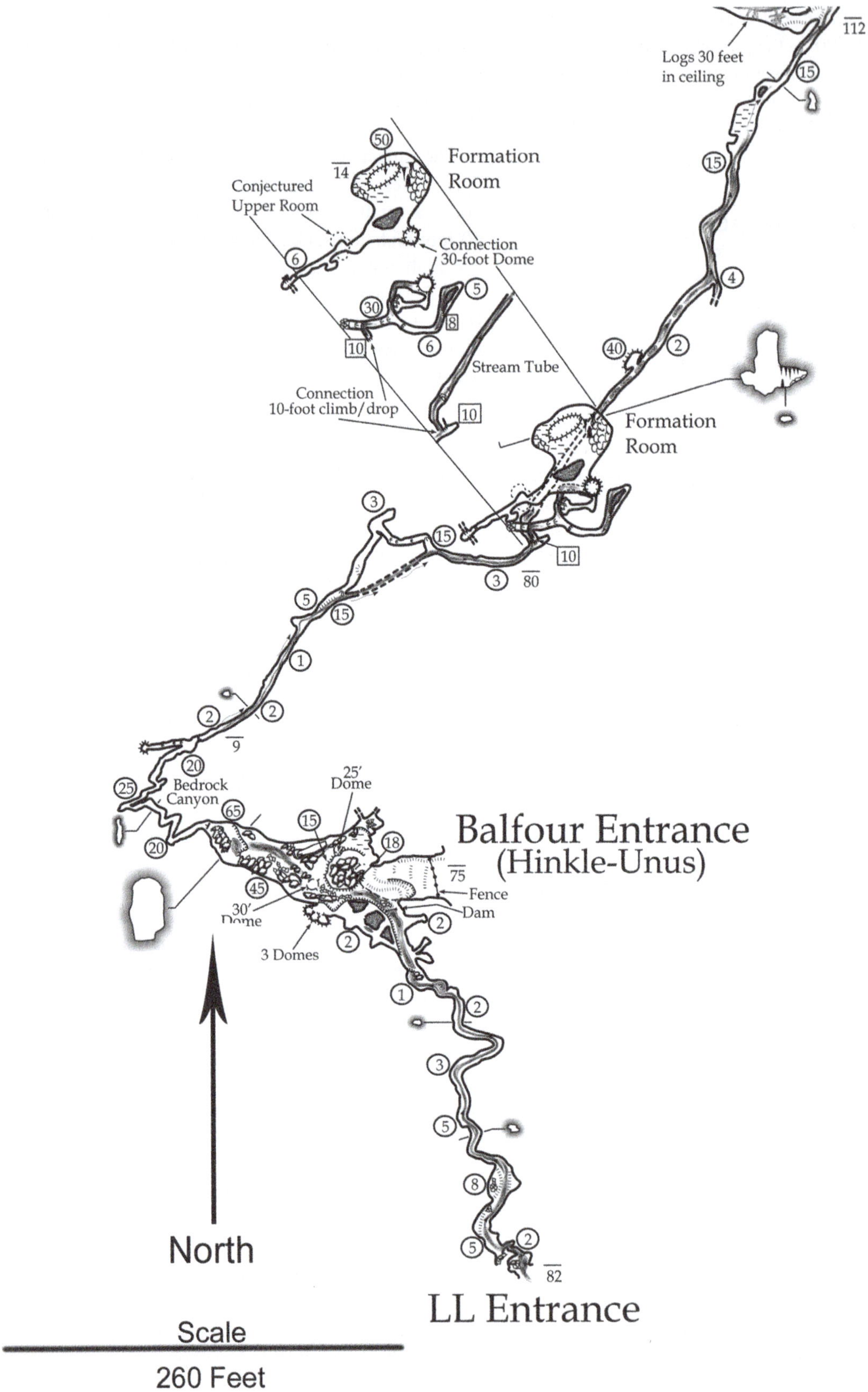

Map 10.6 Hinkle Unus

Fig. 10.30 Balfour-Hinkle Unus Entrance, located in a cow pasture, is where cattle would seek shade and cooler temperatures. Photograph by Philip Lucas

Fig. 10.31 View from inside the Balfour–Hinkle Unus Entrance shows the owner, William Balfour, eliminating a small stream which enters along one wall. Following this passage upstream 500 ft is the LL Entrance to the cave system. Photograph by Philip Lucas

Eventually, the passage intersects the Culverson Trunk at the Balcony. Here two windows look down about 30 ft into the trunk passage (Fig. 10.36). At first glance, the down-climb from the Balcony looks impossible without at least a rappel or handline. But a fairly straightforward route through the biggest window can be negotiated without much difficulty. The last 10 ft is a chimney in a small fissure which drops into a pool in the main trunk. This pool can be dry or up to waist deep.

During flooding, the view from the Balcony can be quite awe inspiring. At these times, the main Culverson stream begins to back up from the Log Jam, approximately one mile downstream. Given enough rain, the creek will eventually pond nearly to the ceiling at the Balcony, then overflow, and flow toward the Wildcat entrance. It is not clear whether this water somehow reaches the McLaughlin section of the cave under the bottom of the entrance sink, or if it flows into another passage on the other side of the sink toward parts unknown. It is this author's opinion that the latter is true and that there is a "missing" piece of trunk passage that carries the floodwater to Lower Culverson trunk. The lower end of this "missing" trunk would be the collapsed end of Alien Alley where breakdown now prevents further penetration.

10.2.11 Wildcat Entrance to the McLaughlin Entrance (Map 10.7)

After climbing down into the Wildcat entrance, a sloping fissure to the right (Fig. 10.37) leads down about 30 ft to a 50-foot wide breakdown-floored passage. Within 80 ft, sloping banks of sand and mud are encountered. Large slab breakdown is scattered about.

A short climbup leads into a hands-and-knees crawl. This leads to the top of a large trunk passage. The bottom of the passage is reached by descending a steep mud bank for nearly 100 ft (Fig. 10.38). This large trunk passage has a small stream flowing to the north.

The passage can be followed in the upstream direction for about 250 ft. At this point, further penetration is blocked by large breakdown. Downstream, the roar of big water can be heard ahead. Approximately 100 ft downstream, Culverson Creek gushes from three passages on the left wall. These passages can be followed back through watery channels and cascades for several hundred feet (Fig. 10.39).

At the point where Culverson Creek enters, the McLaughlin Trunk is between 50 and 60 ft high and 35–40 ft wide. It continues with similar dimensions for the next 600 ft. Throughout this section, the streambed can be followed, but in places climbing up into higher routes allows a more reasonable traverse (Fig. 10.40).

Six hundred feet downstream from where Culverson Creek first enters, it flows into what is known as J Station

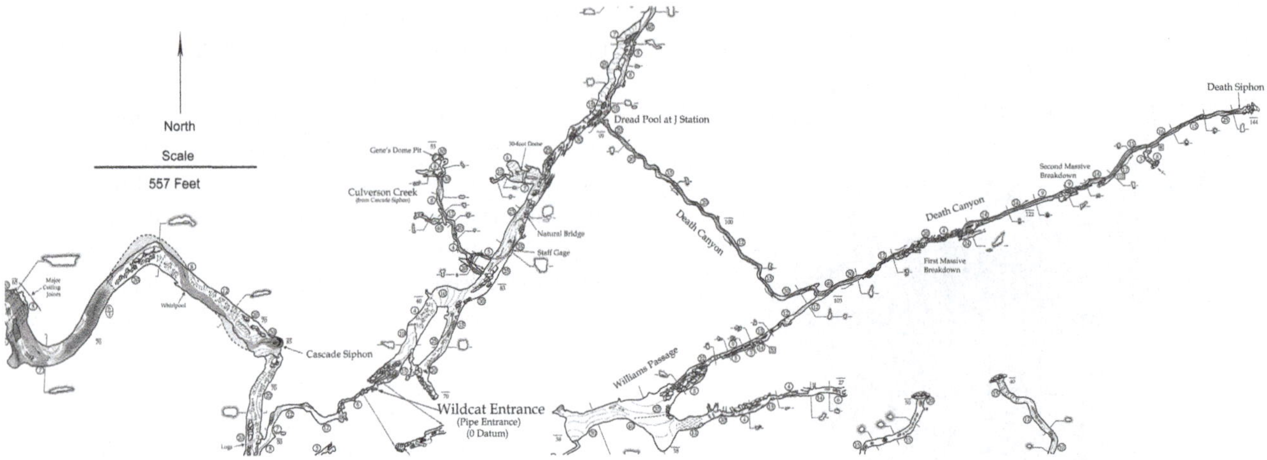

Map 10.7 Wildcat to Death Canyon

Lake or Dread Pool. Here the creek abandons the trunk, flowing off through a low area on the right-hand wall into Death Canyon. The McLaughlin stream, coming in from the north, joins with Culverson Creek at this point (Fig. 10.41).

Fig. 10.32 Wildcat Entrance is located at the base of a limestone escarpment near the bottom of a large sinkhole. Charlotte Lucas is above the Wildcat entrance. Photograph by Philip Lucas

The McLaughlin Trunk continues northeast beyond Dread Pool as a higher level. Intersecting passages lead down steep mud slopes to the lower level and the McLaughlin stream. Two hundred feet beyond Dread Pool, the levels join. One hundred and fifty feet later, they divide again. For the next several hundred feet, a series of upper and lower passages interconnect. Additionally, several side passages enter this somewhat complex area of the cave (Fig. 10.42).

The trunk does manage to continue, although somewhat smaller, averaging 10–30 ft high and wide. It trends generally to the north. After about 1300 ft, a series of plunge pools are encountered. The upstream plunge pool has a 5-foot waterfall. This waterfall can be very difficult to negotiate during high water.

The passage immediately above the waterfall is 12–15 ft wide. After 100 ft, it widens to 30–40 ft wide with 20–60-foot ceilings. These dimensions continue for the next 1200 ft. Several side passages branch off along this segment, both from the stream level and from upper levels. Twelve hundred feet north–northeast of the plunge pools the stream passage narrows to 8–10 ft wide and 20–30 ft high. The next 300 ft of passage follows a series of tight meanders, and the stream continues past a number of intersections. In the bend of one of the meanders, an unobvious side passage to the right leads to a narrow canyon passage that continues for over 500 ft. Another side passage enters from the right, carrying a small stream. This lead can be followed for several hundred feet and ends in a waterfall dome pit.

The main stream passage continues over a wet cobblestone floor for another 200 ft before finally reaching a long pool. Water is knee to chest deep with ceiling heights of 5–6 ft. The strange popping sounds made by the water as it slapped against low ledges led to the name Psycho Siphon. Continuing on for 200 ft, Psycho Siphon eventually gives way to dry

Fig. 10.33 Wildcat Entrance is critical as an easy access to both the Culverson and McLaughlin parts of the cave system. Unfortunately, the entrance was not stable and frost shatter caused the scree slope from the escarpment above to fill and collapse the entrance. To remedy the situation, the unstable area was enlarged, shored, and a 20-foot plastic pipe was inserted as the final stage of the new secure entrance. A recent update—flooding from the 9 in. rain of June 23, 2016, filled the cave to such an extent that water back-flooded out of the Wildcat Entrance blew the 20 foot length of culvert out of the entrance. Photograph by Charlotte Lucas

Fig. 10.34 Albert Grimm and William Balfour pass over a large log, and other debris that has been washed during a large flood event from the Culverson master trunk up through the Balcony Passage and into the McLaughlin part of the cave system. Photograph by Philip Lucas

Fig. 10.35 William Balfour stands in the round tube of the Balcony passage. Photograph by Philip Lucas

Fig. 10.36 Albert Grimm and William Balfour look through a portal between the balcony passage and the Culverson master trunk passage. Photograph by Philip Lucas

Fig. 10.37 Just inside the Wildcat Entrance, a fissure leads down a slope of breakdown into the McLaughlin side of the cave. William Jones and Albert Grimm are the cavers. Photograph by Philip Lucas

Fig. 10.38 Several hundred feet from the Wildcat Entrance and after scrambling over some breakdown, the roar of water can be heard ahead. A stoop walk over a slab of breakdown comes out at the top of a large passage with a steeply down-sloping mud bank. This is the McLaughlin trunk. Photograph by Philip Lucas

Fig. 10.39 Edward Bauer peers into one of the crevices in the wall of the McLaughlin trunk where the pirated Culverson Creek spills into the McLaughlin Cave. It has been witnessed during flooding, that Culverson Creek comes out of the wall with such velocity that it nearly reaches the opposite wall. On the right wall is an old staff gauge. William Balfour and William Jones are the cavers standing next to the staff gauge. Photograph by Philip Lucas

Fig. 10.40 McLaughlin trunk downstream from the old staff gauge is washed nearly free of mud by the Culverson Creek stream during high flow conditions. The cavers lighting the way are Albert Grimm, Edward Bauer, and William Jones. Photograph by Philip Lucas

Fig. 10.42 In many places, the McLaughlin trunk passage is clean-washed limestone bedrock with fast flowing water. Here, Edward Bauer climbs a natural bridge while William Balfour watches. Photograph by Philip Lucas

Fig. 10.41 Dread Pool is where the Culverson Creek leaves the McLaughlin trunk channel and turns to the northeast into Death Canyon. Water levels in the Dread Pool can vary, and it is usually necessary to stoop beneath an undercut wall to get into Death Canyon. A mat of leaves and flood debris rest on the bottom of the Dread Pool. When stepped on bubbles of methane rise to the surface. Photograph by Philip Lucas

Fig. 10.43 McLaughlin Entrance is an opening where a surface stream flows underground in the bottom of a sinkhole. The entrance is usually covered with flood debris. Photograph by William Balfour

Fig. 10.44 Caver passing through the Dread Pool. Years ago carbide lamps were used. These lamps produced a bright flame. This was somewhat of a hazard if the carbide lamp flame ignited the methane bubbling up from the rotting flood debris on the bottom of the pool. Photograph by Bill Jones

passage, and the breakdown room inside the McLaughlin Entrance is quickly encountered. The McLaughlin Entrance itself is a very dynamic hodgepodge of logs and flood debris that changes greatly from year to year (Fig. 10.43).

10.2.12 Death Canyon (Map 10.7)

The Dread Pool is where the main Culverson stream takes a right turn into Death Canyon. Here is a rather large pool reaching a depth of perhaps 6 ft. During flooding, 2 or 3 ft of leaves and debris would accumulate in the bottom of this pool. In summer when the flow Culverson Creek is still warm, the decay of the debris is active. Bubbles of methane gather on the bottom of the leaves, trapped in a tangle of stems and twigs. Back in the "good old days" when cavers use carbide head lamps, first-timers passing through the Dread Pool could get quite a surprise as they stepped on this decaying debris releasing a bubbling and boiling discharge of methane. If they held their heads low to go under the low ceiling, the carbide flame would ignite the bubbles of methane. The sudden flash of flame would singe eyebrows, eyelashes, and produce squeals of surprise and fright. Even veterans of this area would amuse themselves by holding the carbide lamps above the water while stomping the methane laden debris generating great bursts of yellow flame that illuminating the dark surface of the Dread Pool (Fig. 10.44).

On the other side of the Dread Pool near duck-under is the beginning of Death Canyon. This is probably one of the more famous sections of the Culverson cave system. It is a somewhat improbable major cave passage that, geologically, is a youthful piracy of the main Culverson Creek stream. Beyond the Dread Pool duck-under, the canyon immediately becomes a 40-foot-tall Canyon with polished scalloped walls with a cobblestone floor over which flows the swift flowing Culverson Creek (Fig. 10.45). Its width averages between 6 and 8 ft and continues in a southwest direction and a nearly straight line for over 800 ft. Then it abruptly turns to the northeast for the next 1400 ft where it reaches Death Siphon. After turning northeast, a side passage about 100 ft downstream intersects Death Canyon as a steep but climbable ascent to a narrow fissure that soon opens into a broad and muddy passage called Williams Passage. This passage turns to the northeast roughly paralleling Death Canyon for several hundred feet before encountering massive breakdown. This breakdown can be negotiated for several hundred feet, and it is thought by some that a persistent push in this direction might lead to a connection at the northwest end of the First Echo Tube.

From the Williams Passage intersection, Death Canyon continues for approximately 200 ft as a clean-washed, mud-free, canyon passage. For the next 200 ft, breakdown is encountered scattered along the Canyon passage. This breakdown, however, can be negotiated fairly easily with only a few spots where climbing up and over and down through cracks is necessary. This breakdown lasts for about 200 ft until open canyon passage is reached once again. The next 300 ft is easy strolling until once again massive breakdown is encountered. This breakdown is not so easily negotiated and persists for about 100 ft. Then it is open canyon passage again with only an occasional breakdown

Fig. 10.45 Albert Grimm in a typical section of Death Canyon passage. The strong current of the stream and scalloped walls are typical of a canyon rapidly downcutting. Photograph by Philip Lucas

block for 500 ft. At that point, massive breakdown once again becomes prevalent. At the base of the breakdown is the terminal siphon of the Death Canyon Passage. This siphon may well be a complete sump, but because of strong currents and a narrow passage, no one has been able to verify this. To this author's knowledge, no diver has attempted to penetrate this point. It has been reported that, just prior to the siphon, the massive breakdown can be ascended to the northwest for 100 ft or further. The distance between the Death Canyon Sump and the Double Siphon in Lower Culverson is 800 ft.

10.2.13 The Woodson Entrance to McLaughlin Section (Map 10.8)

The Woodson entrance is comprised of several portal openings into a passage, although only one is "reasonable" for cavers (Fig. 10.46). This entrance leads to a 25-foot-wide passage floored with breakdown that continues for over 400 ft. Several side leads are encountered over the length of this passage. One of these leads off to the left and east, rising up over a series of rimstone dams. The surface of the "final" pool comes to within an inch of the ceiling. On the initial survey trip, echoes heard beyond the pool indicated more passage. The pool was drained on a subsequent trip with a 1-in. plastic hose, allowing easy access into the passage ahead. As it turned out, this passage rejoined the main stream passage at a point just beyond where the last survey had ended.

The main Woodson stream passage continues on downstream through a 3-foot high crawl and then into a pooled area. The next 500 ft is generally 10 ft high and 6 ft wide. Eventually a 15-foot waterfall is reached. This requires a rope or short ladder. At the base of the waterfall, there is a small corkscrewing passage. This corkscrew spills out into the top of a 20-foot wide by 50-foot high flowstone slope. A hand line is helpful at this second climbdown.

From the bottom of the flowstone, the stream passage averages three to 12 ft wide and 6–40 ft high. The passage eventually intersects with the McLaughlin stream passage after nearly 2000 ft of walking. There are a number of side passages between the 15-foot waterfall and the McLaughlin intersection. Most come from higher levels and some are quite extensive. Not all have been fully explored.

10.2.14 Fullers Entrance to Bypass Area (Map 10.9)

The Fuller Stream first disappears into a series of segmented cave passages and karst windows, approximately 700 ft up Thorny Hollow from the main Fuller's entrance (Fig. 10.47). The stream is small in normal weather. Most of the passage segments are walking-sized with large wide entrances. The last downstream segment does require some stoop-walking and minor breakdown crawls before emerging into the bottom of the Fuller entrance sink.

There are actually two entrances in the main Fuller entrance sink (Fig. 10.48). One is at the bottom of the sink and it takes the stream. The other is about 30 ft higher and is a window into the top of the canyon leading away from the lower entrance. This upper entrance is the original phreatic tube that forms the top of the canyon. It resembles a giant keyhole at this point, with the roundish upper tube continuing above the canyon as far as one can see. To add to the complexity of the Fuller sinkhole entrance, the small stream in the bottom of the sinkhole entrance is beyond the drip line of the western side of the sink which is a vertical limestone escarpment. The stream can be followed in the upstream direction for several hundred feet where it emerges in a wide

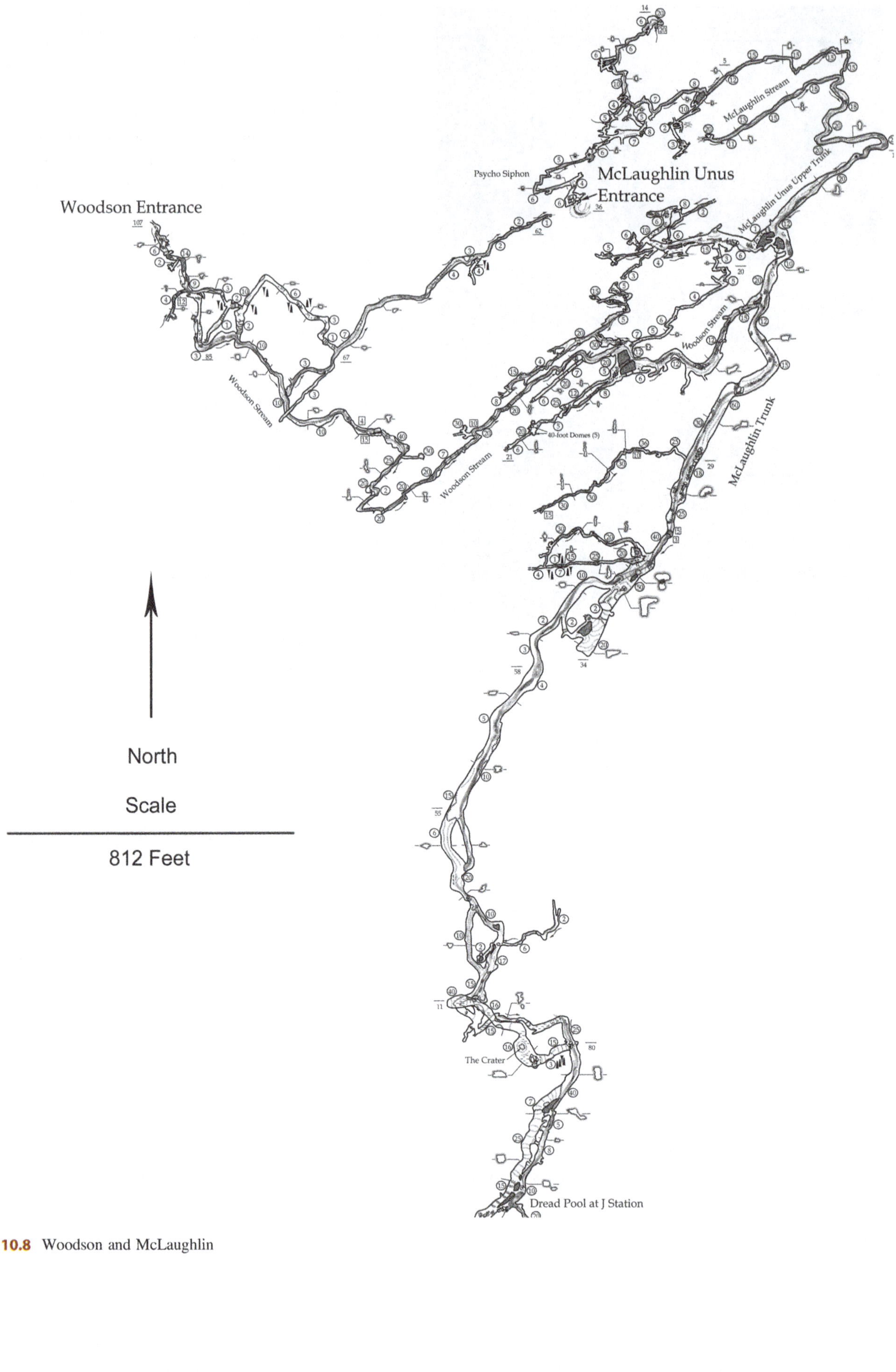

Map 10.8 Woodson and McLaughlin

Fig. 10.46 Small surface stream flowing down a hillside ravine until it reaches the limestone, is quickly diverted underground at the Woodson Entrance. Photograph by William Balfour

entrance. This entrance is actually a part of a elongated karst window as described in the above paragraph.

Continuing downstream from the main Fuller sinkhole entrance, the bottom of the canyon passage averages 6–10 ft wide, although there are numerous places where it is necessary to turn sideways to continue downstream. The ceiling varies from 30 ft to over 60 ft high. In many places, the top of the passage cannot be seen due to the three-dimensional meandering of the lower canyon.

The phreatic tube in the top of the canyon passage is generally 15–20 ft high and wide. It can only be reached in certain locations and then only followed for short distances. Approximately 100 ft downstream from Fullers entrance, a side passage enters from the right-east wall. This passage begins as a tiny meandering infeeder, but quickly widens to 20 ft. Massive breakdown accompanies 50-foot ceiling heights. Above the breakdown at one point, it is possible to see daylight filtering in. Using this entrance would not be easy.

Approximately 300 ft from the main entrance, a high canyon passage enters from the right and east wall. Climbing up 30–40 ft accesses the Oh No, No Truck Passage. A phreatic tube approximately 800 ft long, the Oh No, No Truck Passage has several intersections with the top of the main Fuller Canyon. Following the Oh No, No Truck passage to the south for 200 ft leads to the Oh No, No Truck Entrance. Downstream in the main canyon from Oh No, No Truck passage, the stream plays "peek-a-boo," sometimes flowing in the main passage while at other times pirated into a lower route. Soon the main canyon itself begins to braid on various levels. Interconnections between levels are common, and the area can be confusing if the wrong route is taken.

One prominent upper passage, the Finger Canyon, leads to a series of domes and climbs. A fairly large dome, 30–40 ft in diameter and 60 ft high, is visible through a small window. The sound of falling water echoes from beyond, but no entrance into it seems possible without enlarging the window.

Four hundred feet downstream of the Finger Canyon is a confusing network of upper levels and lower level canyons with the base-level stream passage. For 250 ft, the stream flows from pool to pool through a 2-foot high cobble crawl. This area can flood quickly. An upper route, the Bypass, avoids the crawl and can be reached by climbing straight up from the stream canyon at Station 79 (marked on the wall). Beyond an area of breakdown, a canyon passage provides a walking and climbing route back down to the main stream passage. Explorers of Fuller Cave would do well to know the exact route of the upper level bypass. It provides a safe route around the stream crawl in case of a sudden afternoon thunderstorm. On the other side of the low stream crawl, all routes seem to converge into a single passage 10–15 ft wide and 6–15 ft high (Fig. 10.49).

10.2.15 Fullers Stream Passage—Bypass to Waterfalls (Map 10.10)

On the downstream end of the Bypass, a side passage to the right enters carrying a small stream. This is the stream from the SSS entrance. Approximately 100 ft downstream from the bypass is an intersection of upper canyons. To the left is Mason's Lost Passage, which can be followed for several hundred feet before tying into passages entering from the Peterbilt Trunk.

Back at stream level, the main stream passage continues as a marvelous meandering canyon full of plunge pools, pot holes, and occasional rapids (Fig. 10.50). Flowstone enters at many points along this route, enhancing the beauty of the canyon (Fig. 10.51). A few pools make getting wet to your navel (or higher) a possibility, but nonetheless this 2000-foot section is some of the most spectacular stream canyon in West Virginia (Figs. 10.52 and 10.53).

Approximately 500 ft downstream from the bypass is an area of passage intersections. Some lead up into the Peterbilt Passage, while others lead to small parallel canyons. Approximately 1000 ft downstream from the bypass is a series of side passages along the left (west) wall. These lead into a confusing upper network of canyon passages. Known

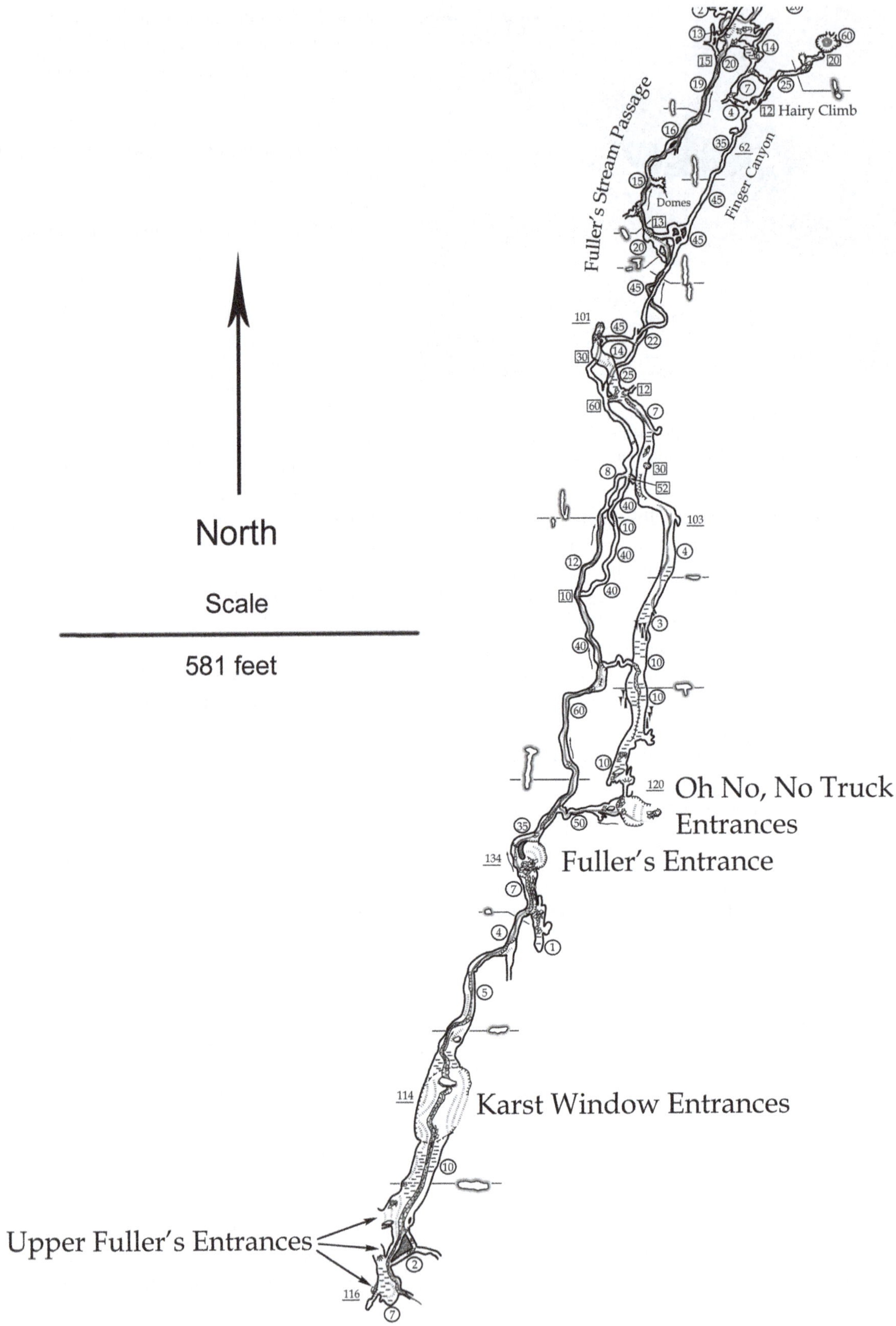

Map 10.9 Fuller to bypass

Fig. 10.47 Upstream from the Fuller entrance is a series of karst windows and cave entrances. This entrance, the Fuller Karst entrance, receives the small stream flowing from Thorny Hollow. The cave passage and stream flow through the Fullers entrance which is actually yet another karst window. Photograph by William Balfour

Fig. 10.48 William Balfour is using a hand line to descend into the Fuller Cave entrance. Unfortunately, in years past, the cave was used as a dump to dispose of trash which included the old truck body seen in this photograph. Photograph by Clifford Lindsay

Fig. 10.49 Fullers Stream canyon is mostly clean-washed limestone. If you do not mind wet feet, it is a joy to explore. Juliette Balfour is the caver traversing the canyon. Photograph by Philip Lucas

as the N Survey, this section lies lower than, but very close to, the upstream end of the Cataract Avenue. Several attempts have been made to connect this area to Cataract Avenue without success.

Continuing downstream in the main canyon, areas of downcutting through the chert can be seen. Excellent examples of prismatic jointing can also be seen in the bedrock ledges. About 1500 ft from the bypass, the canyon begins to meander severely. Eventually, as the explorer continues downstream, the roar of a waterfall can be heard. Twenty feet upstream of the first falls is a small gravely beach where vertical gear can be unpacked and assembled. Ahead, the stream plunges over the top of the First Drop as a nearly circular column of water to the plunge pool below (Fig. 10.54).

10.2.16 SSS Entrance and the Peterbilt Passage (Map 10.10)

Although there are several SSS "entrances," only the southernmost gives passable access to the Culverson Creek System. Lying near a fence line that separates a patch of woods from the nearby cow pasture, this entrance leads to a low, wide stream crawl (Fig. 10.55). After crawling for about 70 ft, 60 ft of walking passage leads to a stream entering in from the right and east. This stream has gone below ground via the northern SSS entrance. This stream leads down a narrow meandering canyon. A series of three to 6-foot-high waterfalls and many cascades makes the next several hundred feet "sporting." The passage quickly loses

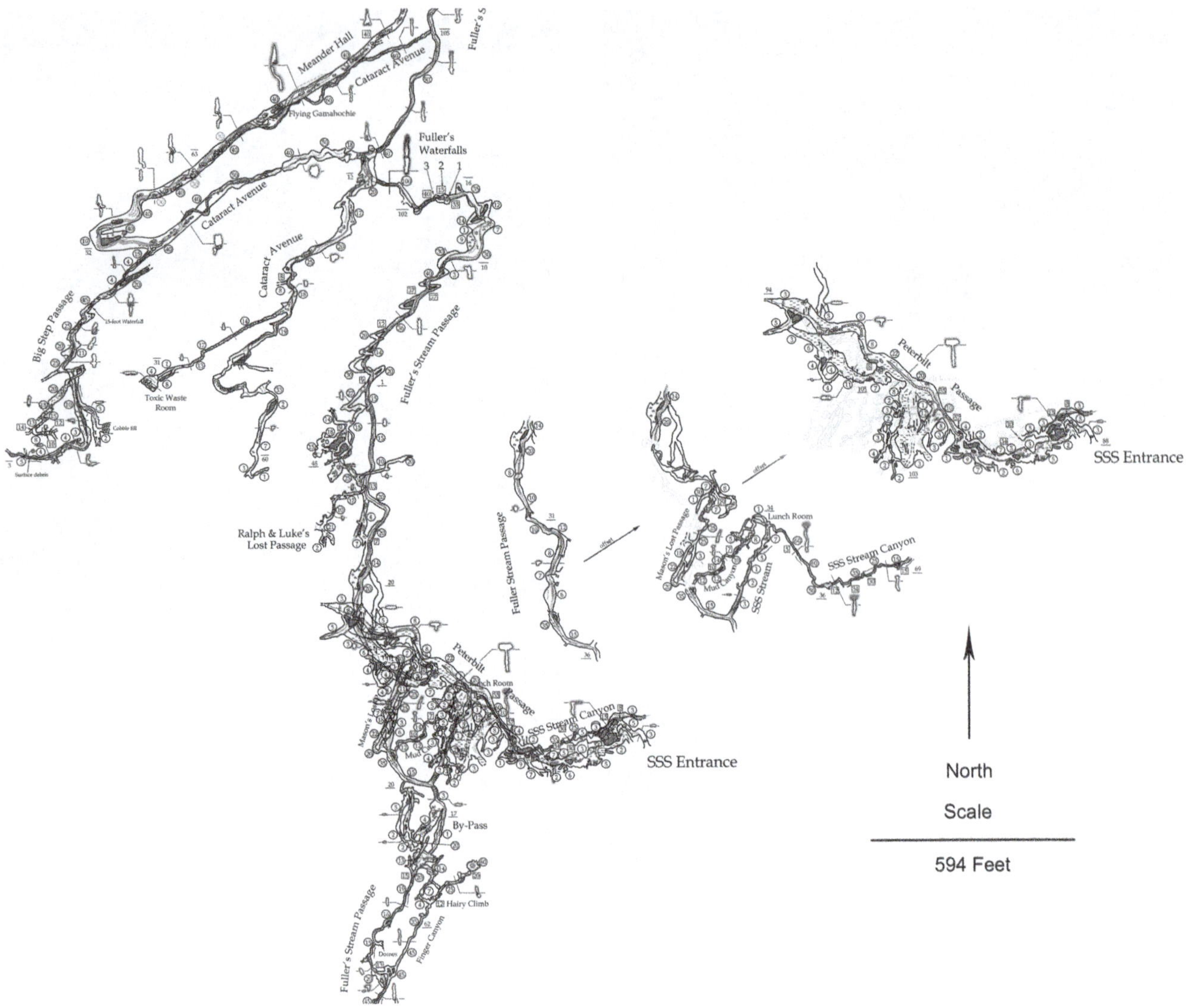

Map 10.10 SSS fuller to waterfall

about 70 ft of elevation before it eases off into a more gentle gradient.

The stream disappears under the left wall just before reaching the Lunch Room and then reappears inside the Lunch Room. The Lunch Room can be followed downstream for approximately 120 ft until it becomes choked with cobblestones. A small passage leads west out of the Lunch Room to a series of upper levels crawls, climbs, and narrow canyons. These passages can be negotiated for several hundred feet. Ultimately, the main Fuller Stream Passage is reached, just downstream of the bypass area.

Back near the entrance, where the stream begins to downcut, a crawl leads off to the left. It rejoins the canyon in about 50 ft. This same point can be reached by staying high in the canyon as the floor begins to fall away. Also at this point, a short climbup leads into another crawlway. Fifty feet down this crawl is another intersection. Straight ahead the passage rejoins the top of the stream canyon.

Turning left at the intersection leads to a short crawl that soon opens into a low, wide mud-floored passage (Fig. 10.56). This passage unwinds into a confusing network of phreatic channels, some of which are highly decorated. Continuing to the west, after about 350 ft, the "main" passage widens and then intersects a large passage. This is the Peterbilt Passage, with ceiling heights up to 30 ft and a width averaging 50 ft. It also has a 4-foot-wide canyon that is 70-ft-deep meandering along the floor for the first 400 ft of passage (Fig. 10.57).

A large left-hand passage 6–10 ft high and 50 ft wide leaves the Peterbilt Passage at the intersection and finally loops back 100 ft "downstream." Peterbilt continues for another 200 ft before it again divides into three branches.

Fig. 10.50 William Balfour is carefully stepping from the lip of one pothole to the next in this stretch of the Fuller Canyon. The *light gray color* of the Union Limestone makes this passage especially beautiful. Unfortunately for the photographer, muddy water has caught up with the caving party before the photograph could be taken. Photograph by Philip Lucas

The right branch contains a series of meanders. After approximately 400 ft, this branch ends in a too tight crawl. The left branch runs parallel to the right branch and connects with it at two points. The left-hand passage has a number of very beautiful speleothems and is quite photogenic. The Peterbilt Section as a whole is probably the most highly decorated area in the Culverson Creek Cave System (Fig. 10.58). The third branch is a lower canyon. An easy down-climb through a fractured zone, this passage leads approximately 100 ft to an intersection. A left-hand side passage leads in the direction of Mason's Lost Passage and eventually connects with the Fullers Stream Passage. Continuing straight ahead from the intersection, a canyon can be followed at both upper and lower levels. The upper level eventually ties into the two main parallel Peterbilt branches. The lower level continues for several hundred feet until reaching an area of collapse. No penetration has been made beyond this point.

Fig. 10.51 There are places along the stream canyon that have beautiful speleothems flowing down from the narrow canyon above. William Jones looks up at one such location. Photograph by Philip Lucas

The canyon containing the SSS Stream meanders below much of the Peterbilt Passage. Although the canyon is up to 70 ft deep, it is narrow enough to easily step over in most places. A rock dropped at one of these "step-over" will clatter for quite a while as it bounces from wall to wall before splashing into the stream (Fig. 10.59).

10.2.17 Lower Fullers Canyon—Waterfalls to the Breakdown (Map 10.11)

The first waterfall in the Fullers Canyon drops 30 ft into a plunge pool that is 25 ft in diameter and over 6 ft deep. Stainless steels bolts installed on the left wall at the top of the drop make it possible to descend without hanging in the waterfall. Incidentally, all bolts installed in the waterfall series are stainless steel. At the bottom of the falls, a ledge along the left wall avoids the depths of the plunge pool. Immediately on the other side of the pool is a 10-foot cascading waterfall. This climb is no problem during low water,

Fig. 10.52 Scott Olson admires another section of flowstone that adds to the canyon's beauty. Photograph by Philip Lucas

Fig. 10.53 As the stream canyon progresses downstream, it gradually becomes larger. William Jones looks over his shoulder at the various layering in the Union Limestone and at the large chock stone wedged between the canyons walls. Photograph by Philip Lucas

but a bolt on the left wall allows a short rappel in case of higher water.

Just beyond the bottom of the cascade, the stream plunges down a 40-foot waterfall. An exposed, 15-foot traverse on ledges along the left wall leads to a wide "belay" ledge. Another bolt allows a clean 25-foot drop to a landing ledge. A route then leads along the tops of large blocks of breakdown to a final 15-foot drop and the floor of the canyon.

The next 1500–2000 ft comprises some of the most outstanding vadose stream passage in West Virginia. In many places, the stream flows over limestone bedrock and the entire passage is washed clean. Only an occasional piece of breakdown mars the open conduit run of the Fuller Stream. In several places, water cascades down into potholes, and at one point the stream drops 15 ft in a series of cascades. The ceiling of this passage probably reaches 100 ft in height, but is difficult to measure due to the three-dimensional twisting of the canyon. This passage probably allows the fastest travel time in the Culverson Creek Cave System.

Fig. 10.54 There is a series of three waterfalls that occur in the Fuller Canyon about a mile from the Fuller Entrance. At the first waterfall, the stream flows through a V-shaped notch, forming a round column of water that plunges 30 ft into a deep pothole. This photograph was taken during low flow conditions. When the stream is flowing a bit deeper, negotiating the waterfalls becomes a bit more "sporting". Photograph by Philip Lucas

Fig. 10.55 Jessica Lindsay is entering the SSS Entrance. Photograph by Philip Lucas

Fig. 10.56 Kneeling below a ceiling of anastomoses, Jessica Lindsay patiently poses, providing scale for the setting. Photograph by Philip Lucas

Continuing downstream, breakdown becomes more numerous. Eventually, it becomes necessary to climb up and over breakdown chokes. Then layers of mud begin to coat the tops of the breakdown. Finally, an area of massive collapse seemingly blocks the way. But by following fissures between the breakdown near stream level, a way eventually leads 450 ft to the other side of the collapse. These passages are confusing, torturous, and tiring, but better than a newly discovered overhead route that requires pre-rigging from the downstream direction.

In 1994, an upper passage was found from the downstream end of the Breakdown Section. A series of mud slopes eventually lead up to a large walking passage 30 ft wide and up to 60 ft high. Called High Hopes, it was felt that a bypass to the Breakdown Section had been found. Indeed,

Fig. 10.57 At this point the Peterbilt Passage is wide and spacious. William Royster peers down a narrow slot canyon cut in the floor of the Peterbilt Passage. The slot canyon is 70 ft deep. Photograph by Philip Lucas

after 700 ft the floor falls away into a steep slope and then a vertical drop to the stream below.

The upstream end of High Hopes is nearly 80 ft above the Fuller stream. A bolt was set and a rope left to provide easy access into the High Hopes for future trips. Unfortunately, it was soon determined that the extreme muddiness of the drop made ascending the rope far more of an effort than it would be to negotiate the Breakdown Section in the first place.

10.2.18 Lower Fuller Canyon from Breakdown to Culverson Intersection (Map 10.11)

The passage characteristics change abruptly downstream of the First Breakdown. Passage dimensions reach 40 ft in width and nearly 90 ft in height. It is now often necessary to climb above the stream over large, mud-covered breakdown or to traverse large mud slopes.

At a point nearly 300 ft downstream of the First Breakdown, the passage suddenly closes down again, reducing to

Fig. 10.58 Clifford Lindsay looks up at a dripline of flowstone curtains and stalactites that decorate this section of the Peterbilt passage. Photograph by Philip Lucas

6 ft high and 10–12 ft wide. An upper-level meander with flat muddy floors bypasses this section. Beyond the "restriction," the passage soon regains its former dimensions. It is possible to walk at stream level for the next several hundred feet.

Eventually, an area of collapse known as the "Slippy Poo" is reached. Here a section of very slippery mud-slickened boulders demands full concentration. Any slip would result in a slide down through gaps in the boulders and into some deep pools below. Luckily, the Slippy Poo extends only for 50 ft, when a section of leisurely walking stream passage saves the day. This passage continues unobstructed for the next several hundred feet.

The next obstacle is the Water Maze. It first appears that an overhead traverse may be necessary. But by carefully following the stream through fissures in the breakdown, a way can be found through the Water Maze into more wide spacious passage beyond. It is an easy stroll downstream for the next 800 ft. At this point, a mud bridge requires a quick climbup and climbdown. Beyond the bridge it is more easy strolling for the next 800–1000 ft to the intersection with the Lower Culverson Trunk at the Dragon's Breath Room (Fig. 10.60). What had seemed like large passage up until this point is now put into a different perspective, dwarfed by the blackness of Lower Culverson.

10.2.19 Cataract Avenue (Map 10.11)

Cataract Avenue intersects the main Fullers Canyon about 800 ft downstream of the Fuller Canyon Waterfalls. A small stream cascades out of Cataract Avenue and joins the Fuller stream. Beginning as a narrow canyon, the first 100 ft or so of Cataract Avenue is fairly straightforward but tight. The top of the canyon continues upward until it meanders out of sight. At one point, a large breakdown block must be climbed up and over. For the next several hundred delightful feet, one cascade after another is encountered. In places the

Fig. 10.59 Taking a big step across 70-ft-deep slot canyon, Clifford Lindsay does not look down. Photograph by Philip Lucas

canyon stays very narrow. Black voids loom overhead. After about 1000 ft, it is possible to climb up into the overhead passage. Near this point, a passage with a small stream enters from the right and south. This goes to the Big Step Lead.

Continuing upstream in the bottom of the canyon, the distance separating the canyon and the large overhead passage becomes less and less until the two passages become one. Here the passage averages 25 ft high and approximately 40 ft wide. It continues with these dimensions for the next 1000 ft, at which point the passage begins to divide and become smaller. After another 400 ft, the passage narrows even further in an area where many dry pools of crystals decorate the floor. Soon after, the passage crosses up through the "crappy layer" and pinches in collapse.

The large paleo upper passage can be followed back downstream beyond the point where it merged with the lower canyon. Traversing this passage requires stepping across the narrow stream canyon at several places. The stream can be heard cascading far below. The upper passage soon becomes 40 ft high and 40 ft wide and continues over breakdown slopes for the next 1300 ft. Sharp meandering can be seen in this stretch of passage, especially in the ceiling channels. At one point the passage takes an abrupt left-hand turn and then nearly doubles back on itself with a sharp right-hand turn. In this sharp meander, the inside bend of rock has broken and the end of what otherwise would be a huge rock pendant nearly blocks the passageway. At this point, the passage has become nearly 50 ft high but has narrowed to approximately 20–30 ft wide. It continues straight for nearly 800 ft to a point where another intersection is reached. This intersection is the top of the Fuller's stream canyon. Although the stream is 80 ft below, mud in Cataract Avenue indicates back flooding from the Fullers Stream. Cataract Avenue appears to merge into the top of the Fuller Canyon. It can be seen to continue beyond, but without a floor.

Back near the merge point between the large upper passage and the lower canyon sits the Big Step Lead. This lead marks a clear division between the upper "original passage" and the younger canyon that has downcut approximately 70 ft to the present level of the Fuller Stream. Some climbing is involved to gain access to the Big Step Lead. A little side stream comes out of a too small crack, but a climbup into an upper canyon leads to a scramble across the top of a large breakdown block. This passage continues, but gradually loses its floor until finally it is necessary to drop into the stream. The water is pooled a couple feet deep in a narrow "hands-and-knees' passage for 40–50 ft. At this point, it is necessary to chimney up approximately 15 ft over a restriction into a larger overhead passage. Continuing upstream in this passage, a 15-foot waterfall guards the opening to a passage 30 ft high and 15 ft wide. This is the Big Step. In order to negotiate the waterfall, it is necessary to traverse out on a small ledge and then take a "big step" across to the top of the waterfall. Several hundred feet of passage has been explored beyond this point. No surveying has been done, and several leads remain. Several hundred feet beyond the Big Step, William Royster and Bert Ashbrook used a walkie-talkie to make voice contact with Clifford Lindsey on the surface. They were able to maintain contact for some distance while crawling down a passageway. This passage is apparently beneath the hard surface road in front of Clifford Lindsay's garage.

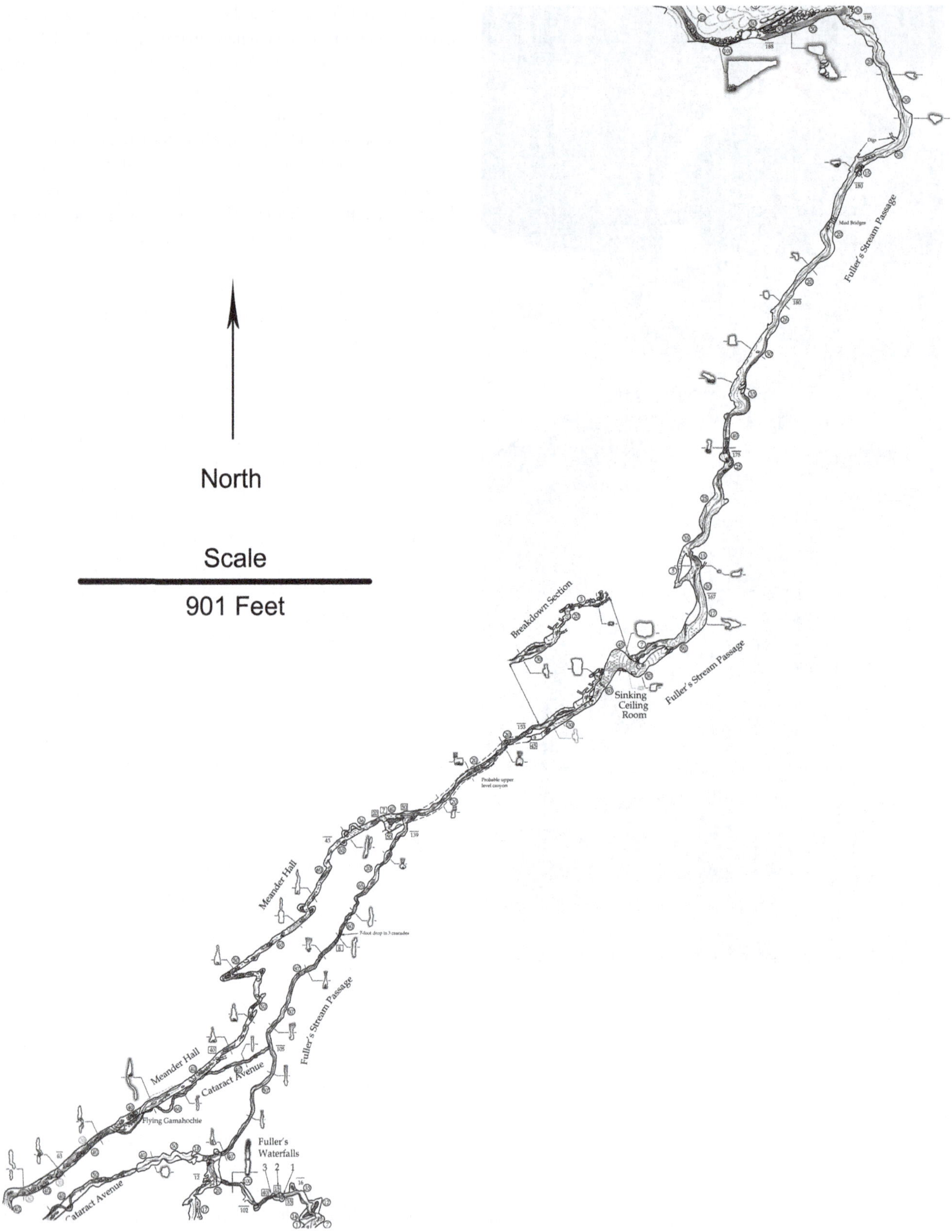

Map 10.11 Waterfall to Dream Lake

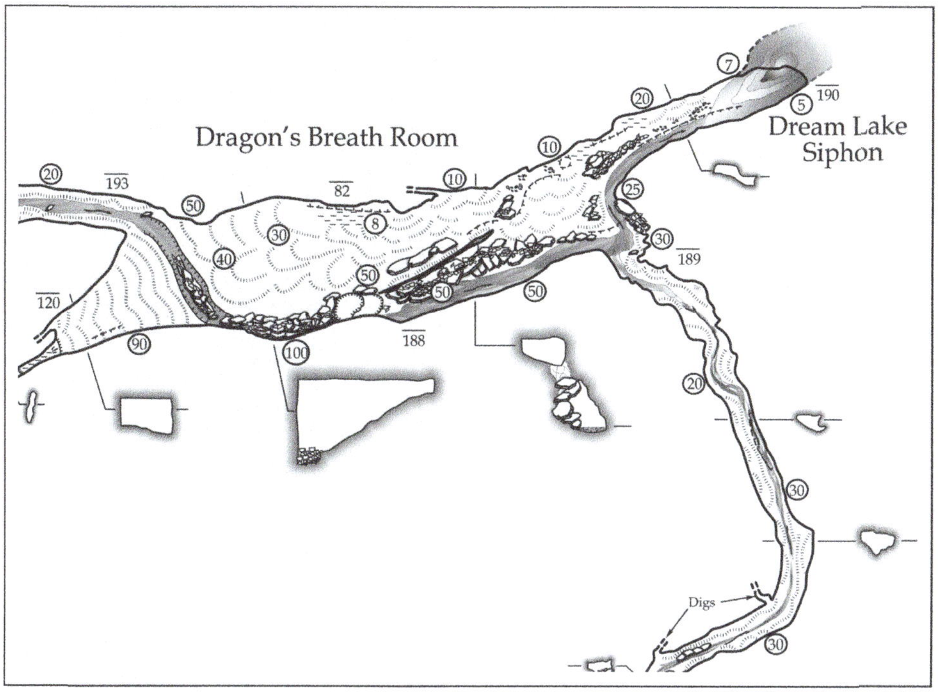

Fig. 10.60 This illustration of the Dragon's Breath Room and Dream Lake will have to take the place of photographs; there are none. During the summer and fall, the Culverson Creek stream retains enough heat to generate enough water vapor to create a foggy atmosphere. This area contains some of the largest volume in the cave system. Frustratingly, the hundred-foot ceiling descends into a mere 5–6 ft at the sandy shore of Dream Lake and then continues to plunge below the lake surface. The large passage continues under water beyond Dream Lake

10.2.20 Lower Culverson Trunk: Dream Lake to Double Siphon (Map 10.12)

The furthest point downstream on the underground course of Culverson Creek that exploration has penetrated is Dream Lake. It is, of course, the lowest point in the cave system, 165 ft below the main Culverson Creek entrance. It is at the north east end of the Dragon's breath room where ceiling heights of 100 ft descend to a mere 7 ft at the Sandy Beach of Dream Lake with a passage width of 40 ft. The sump "lake" extends about 50 ft to a point where the ceiling gradually descends to the water.

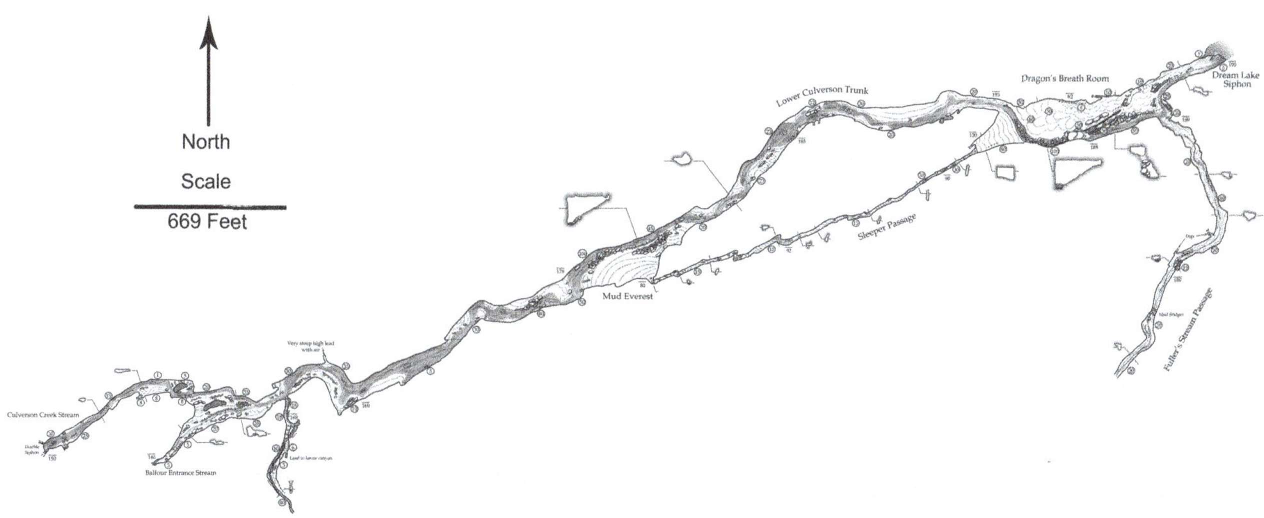

Map 10.12 Lower Culverson

Ronald Simmons attempted to dive the sump in circa 1992 without success. Although the creek was at near normal levels, the water was a bit murky due to recent rains. Ronald's first attempt was to go deep and follow the main channel. He reached a depth of about 30 ft but found himself in a huge passage with no discernible channel to follow. He also had no points to attach his diving tagline. His second dive attempt was to follow along the flat ceiling of the submerged passage which was only a few feet lower than the surface of Dream Lake. He attempted to swim in a straight direction, but being unable to see his compass clearly, he ended up swimming in a large circle, and it reappeared at the far end of Dream Lake. Although frustrated by the murky water, he did establish that the passage continued in very large dimensions.

The description of the lower Culverson trunk will start at the Dragons Breath Room and describes features in the upstream direction to the upstream sump of Double Siphon and the breakdown collapse at the end of Alien Way.

The Dragon's Breath Room is one of the larger spaces in the cave system. Its floor is mostly huge blocks of breakdown most of which are covered in large mud banks reaching some 60–80 ft above the stream level. The stream finds its way through the breakdown and, although difficult, it is possible to follow the stream the entire length beneath the Dragon's Breath breakdown mountain. During the summer the Dragon's Breath Room is filled with a hazy atmosphere. This is because Culverson Creek still retains a lot of heat, causing water vapor, even though it has been underground for 2 miles. Heading upstream from the western end of the Dragon's Breath Room, the Lower Culverson Creek Trunk varies from 40 to 100 ft wide and from 12 to 80 ft high. Wading upstream, the deeper pools can be avoided by following one side of the passage or other. Sloping mud banks predominate. Here and there are scattered breakdown blocks.

After 1200 ft, a roar announces that the stream is rushing through massive breakdown along the right wall. Following the stream through the breakdown looks to be difficult—in fact, it has not been attempted. However, a huge sloping mud "mountain" along the left wall provides an alternate route. This area of the trunk passage has become very spacious.

The mud mountain is called Mud Everest. Its sides are steep. The top of the mountain is nearly 100 ft above the stream, and the opposite wall at this point is almost 150 ft away. The ceiling is nearly flat. During the summer, the warm waters of Culverson Creek create a misty atmosphere that makes it impossible to see the opposite wall and floor. The volume of this passage seems immense (Fig. 10.61).

Near the top of Mud Everest is a small flat area where the Sleeper Passage begins. The entrance to this passage is not obvious from the stream below. Strong air currents blow from this lead.

Fig. 10.61 Taken by the photographer near the summit of Mud Everest, this image shows the large passage looking downstream toward Dream Lake. Ceiling heights here are approximately 100 ft. Photograph by William Jones

To continue upstream from Mud Everest, it is necessary to descend the upstream slope. Continuing to wade upstream, the passage averages 30–70 ft wide. There are large mud banks and scattered breakdown. At a point about 500 ft upstream of Mud Everest, the ceiling height abruptly changes from 30 to 8 ft and then 5 ft. Deep pools make it necessary to nearly swim in several places. Several trips have found thick patches of yellow foam on the surface of these pools. On some occasions, it has been necessary for the lead caver to sweep and blow the foam aside to make an open route through this unusual obstacle.

About 1200 ft upstream from Mud Everest, the passage makes an abrupt right turn and the ceiling heights again raise to 30 and 40 ft. There are possible high leads at this bend and at the next bend in the passage, but no attempt has been made to climb them. After another 100 ft, the passage makes a 90° left-hand turn to the southwest. A canyon enters along the left-south wall about 100 ft further.

This canyon is known as the Muddy Madness Passage. It is guarded by a deep pool that appears to be over 6 ft deep. To enter the Muddy Madness, stay on the right side of the pool, where the water is only about chest deep. The Muddy Madness Passage leads to the connection with the Culverson Creek, Wildcat, and McLaughlin sections of the cave system.

Just upstream from this intersection, the main stream flows down a scree slope in a series of rapids. In another 200 ft, Alien Alley enters from the left and southwest wall in a Y-type intersection. Continuing to the right (west) at the Y, the stream passage becomes somewhat smaller, averaging 50–20 ft wide and has several deep pools that must be negotiated. Breakdown becomes more apparent about 500 ft upstream from the Y and eventually blocks further progress. A deep pool, known as the Double Siphon, guards this final massive collapse. A passage appears to go in two directions beyond the deep pool.

Alien Alley begins as a broad, 70–80-foot-wide passage with a breakdown-covered floor. Continuing in the passage, the breakdown becomes more massive. The ceiling also begins to lower, and further progress becomes very difficult after 250 ft. A small stream flows out of Alien Alley, although it is usually hidden beneath the breakdown floor. This stream has been dye traced from the Log Jam.

10.2.21 The Sleeper Passage (Map 10.12)

One end of the Sleeper Passage opens at the top of Mud Everest in the Lower Culverson Trunk. The other end enters the upstream end of the Dragon's Breath Room.

The Sleeper Passage was first entered from the Mud Everest side by early explorers who traversed about 300 ft to a deep pool of water. The survey of the Sleeper Passage was started in 1995 by climbing a steep mud slope at the west end of the Dragon's Breath Room. Over 100 ft of elevation is gained going up this slope. At the top of the slope is a 15-foot-wide passage trending west–southwest. This passage quickly narrows into a 4-foot-wide by 3-foot-high canyon that continues for several hundred feet. At several points, holes in the floor and mud bridges interrupt the passage.

After 400 ft, the canyon widens into a room 15 ft wide and 70 ft long. At the far end of the room is a pool which requires wading. Beyond, the passage continues for 200 ft and is 10 ft wide with sloping banks on the left wall. Another large deep pool, about 60 ft long, must be waded carefully in order to avoid complete immersion. Beyond this pool, the passage continues as a 15-by-15-foot walking passage which leads to yet another deep pool. Much of this pool can be avoided by chimneying along a ledge. The next 300 ft of walking passage continues west–southwest over mud and breakdown before making an abrupt right-hand turn to spill out onto the top of Mud Everest.

10.3 Floods in Culverson Creek Cave

10.3.1 Flood at the Balcony Passage

After a big flood event, a group of cavers entered the Wildcat Entrance and traveled down the balcony passage in order to witness the magnitude of flooding in the normally dry abandoned trunk passage that eventually leads to the log jam. Halfway down the balcony passage, low vibrations and booming sounds as from a not so distant thunderstorm could be heard and felt. At the lip of the balcony overlook, the frothing brown water of the Culverson stream could be seen racing by. The creek at this point was 20 ft deep and 50 ft wide. From the downstream direction, there were long deep tuba like notes that morphed into a series of weird rumblings. The vibrations from this were similar to those vehicles with an amped-up sound system that vibrates the windows of nearby vehicles. Suddenly, there would be loud booming sounds as if a huge gong were struck. Occasionally, there would be trumpet like sounds. As this was happening, the creek was steadily rising. There were a few explanations as to what might be causing the strange sounds and none seemed to be adequate.

10.3.2 Flooding of the Cave System

An unusual aspect when the cave system begins to fill during flooding is the number of overflow tubes that carry flood waters and the sequence in which these overflow tubes are utilized as the various metering points are exceeded. Figure 10.62 shows what is known about some of these overflow routes. Figure 10.63 show the extent of the flooding in the cave when the Culverson Creek Cave entrance becomes completely flooded. This figure does not show how much of the cave become flooded when the flood level is such that a temporary lake backs up from the cave entrance, extending more than a mile up the blind valley.

Some of the metering points within the case system are known. However, some metering points can only be speculated. One speculation is that there is a metering point beyond Dream Lake that causes the entire system to be eventually flooded during big flood events. Since it is more than 2 ¼ miles to the resurgences along Spring Creek, there is ample opportunity for such metering points to exist. Perhaps the metering point beyond dream Lake is where Culverson Creek cuts down through the Taggard Shale.

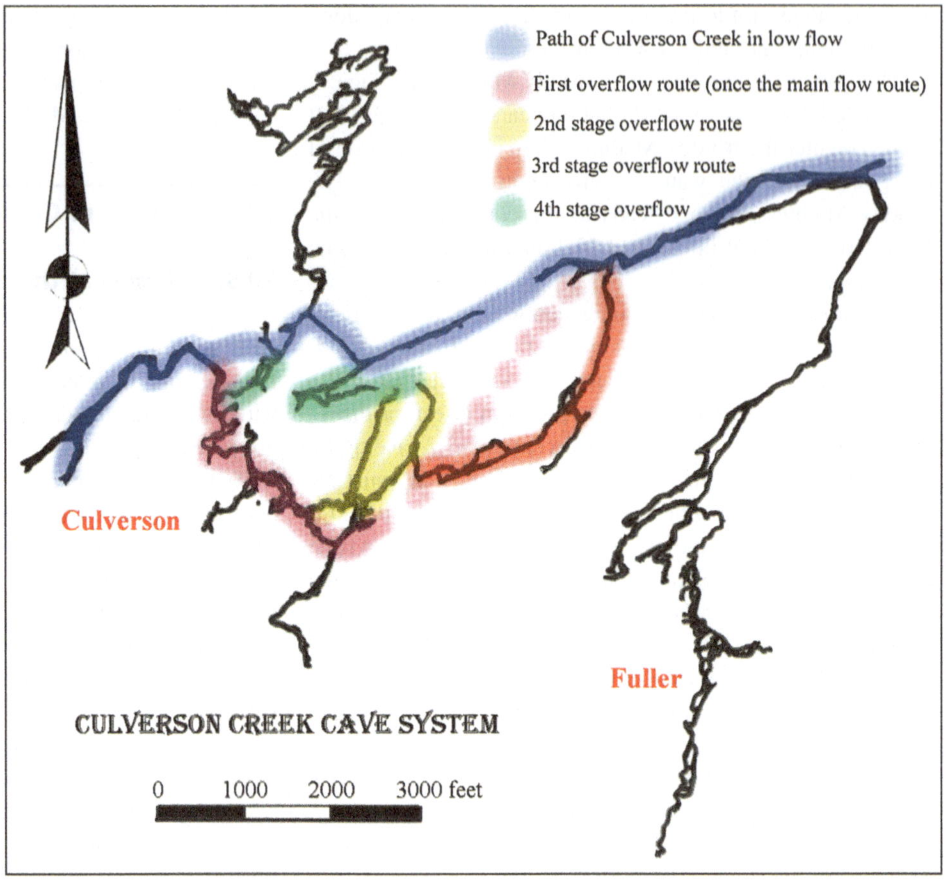

Fig. 10.62 During major flood events, there is a complex series of metering points and overflow routes that dictate how all the cave floods. This illustration shows what is presently known about some of this. More is known about the overflow routes than is known about where the metering points might be located. For instance, we know that the new piracy just upstream from the Balcony Passage, is a metering point because we have seen it "in action". However, it is unknown WHETHER the log jam is actually a metering point. It would be extremely risky, or perhaps deadly to get a visual confirmation of that

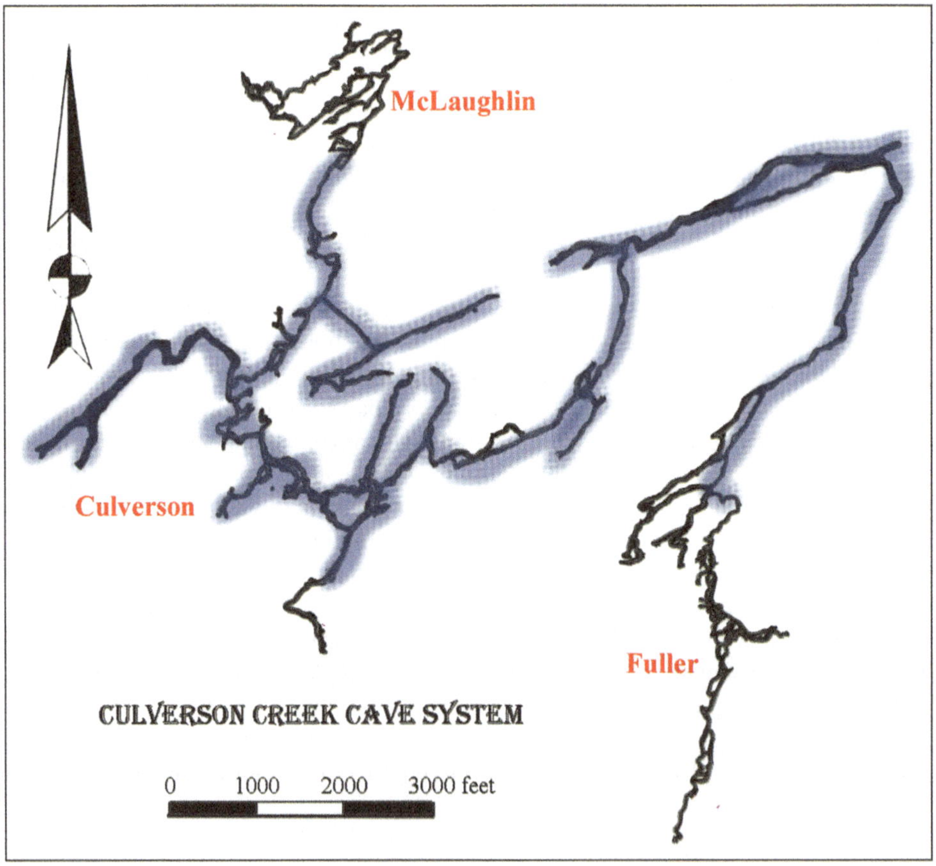

Fig. 10.63 This illustration shows how much of the cave's passages becomes flooded during a flooding event when the main Culverson Creek entrance goes underwater. Of course, the cave becomes even more flooded when the temporary lake becomes deeper extending further and further up the blind valley

Since the cave is so prone to flooding, caving parties should be keenly aware of the weather forecast before they enter the cave. Because the drainage basin is so large, there can be heavy rains in the upper basin creating a flood pulse while there is no rain at the cave. Should a party enter the cave system, they should be aware of rising water levels or a change in the sound of running water while in the cave.

10.4 Conclusion

In conclusion, the Culverson Creek System is a beautiful but difficult and sometimes dangerous cave to enter. It has both beauties not only with spectacular flowstone in the upper levels but those wonderful sculpted shapes of the vadose stream passage. It is truly a Greenbrier County marvel.

Acknowledgements Many people helped with the Culverson project, and without their help it would have been impossible—thanks to all. Those who helped on the survey, over one hundred, are listed on the Culverson map. Two friends I would like to especially thanks are William Royster for his development of the place names within the cave system along with the text of the Rockwell Ward story of Dream Lake and William Balfour who helped in many ways including his assistance drawing the map.

10.4 Conclusions

The Contact Caves of Central Greenbrier County

11

William M. Balfour

Abstract

Six major cave systems are developed along the eastern border of the basal Hillsdale Limestone (Greenbrier Group) and the underlying Maccrady Shale between the town of Lewisburg and Spring Creek in central Greenbrier County. The total combined surveyed passage is just over 128 km (80 miles). The Hole is the northernmost of the contact caves and drains north to Spring Creek. Ludington, McClung, Maxwelton Sink, Benedict, and Wades caves drain southwest to Davis Spring. These caves are developed in the lower section of the Hillsdale Limestone with many passages entrenched by vadose erosion by up to 12 m (40 ft) into the shale. Recharge is mostly from small surface streams that flow westward on the Maccrady Shale and sink at the contact with the limestone. The pattern of passages is typically dendritic with an overall trend sub-parallel to the regional strike. Folding and faulting influence passage orientation in some of the passages, but the perching effect of the underlying shale forces the conduits downward along the local dip.

11.1 Introduction

The Mississippian age Greenbrier Limestone forms a large karst plateau that stretches uninterrupted for nearly 16 km (10 miles) and is over 5 km (3 miles) wide in central Greenbrier County, West Virginia. The area described in this chapter is bounded on the south by the town of Lewisburg, on the west by a series of clastic hills including Weaver Knob and Anderson Knob, on the north by Spring Creek, and on the east by Maccrady shale (Fig. 11.1). The Greenbrier River and Spring Creek are entrenched several hundred meters below the surface of the plateau. This eastern contact is unique from a geologic aspect and has allowed large cave systems to form in generally the same manner. The caves line up along strike adjacent to the contact and stretch the entire length of the plateau.

On the east side of the plateau, surface drainage on the Maccrady Shale flows east down into the entrenched valley of the Greenbrier River or west to the contact zone with the overlying Greenbrier Limestone. The western contact zone has produced this unique karst area characterized by the large contact cave systems. The surface water on the Maccrady Shale roughly follows the dip of the strata along the western limb of the Sinks Grove Anticline until it encounters the limestone where it immediately sinks. This sinking water enters into the caves via small dolines, cave entrances, and in some cases, classic blind valleys with cliff faces up to 30 m high. Cave development continues along the contact with the main stream passages developed in the Hillsdale Limestone (the lowest formation in the Greenbrier Group). The contact passages are characterized by stream entrenchment into the underlying shale (Fig. 11.2), in some places up to 12 m (40 ft). The entrenchment can form large, wide, walking passages or narrow shale canyons depending on the gradient and the size of the stream.

The term "contact karst" is occasionally seen in the European literature (Gams 1965; Mihevc 1994) and is used

Electronic supplementary material
The online version of this chapter (doi:10.1007/978-3-319-65801-8_11) contains supplementary material, which is available to authorized users.

W.M. Balfour (✉)
3954 Carrol Hill Road, Frankford, WV 24938, USA
e-mail: bal4karst@hotmail.com

W.B. White (ed.), *Caves and Karst of the Greenbrier Valley in West Virginia*,
Cave and Karst Systems of the World, https://doi.org/10.1007/978-3-319-65801-8_11

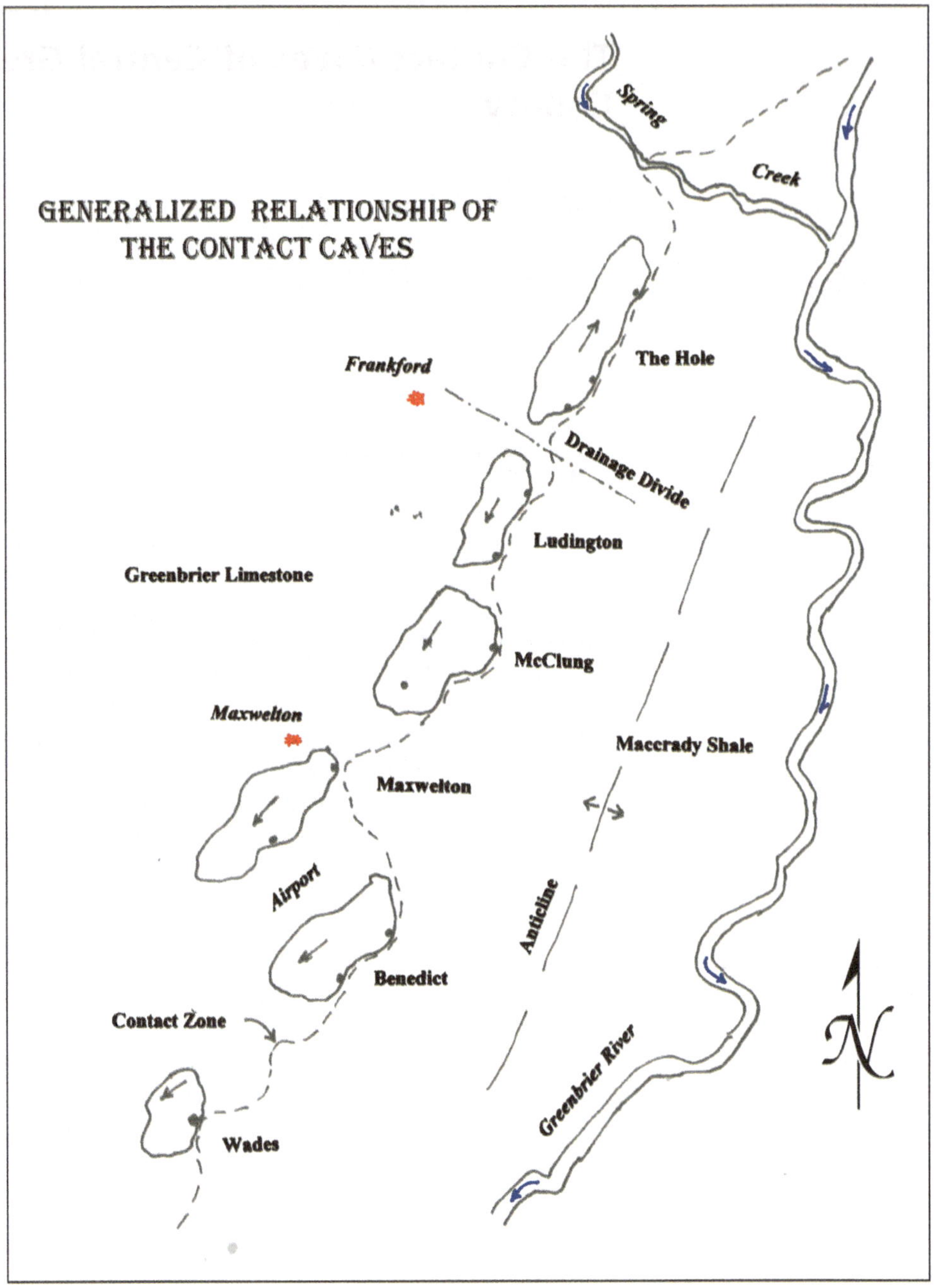

Fig. 11.1 Generalized relationship of the contact caves

to describe karst settings where most of the recharge is allogeneic flow coming from adjacent non-carbonate rocks. Typical sink points are at blind valleys where chemically aggressive water from surrounding clastics first encounters soluble units. However, the defining feature of the contact caves is passage development that may down cut into the Maccrady Shale by up to 12 m below the bottom of the limestone. The six largest contact caves are described in the sections that follow (electronic map M-11.1). Detailed maps of three of them, Benedicts, Maxwelton Sink, and McClungs Caves, are given in electronic maps M-11.2 to M-11.4.

The term "contact caves" was probably first coined by John M. Rutherford (Rutherford 1967) and was originally applied to six caves starting with Wades to the south and ending with The Hole to the north. The Hole was discovered in 1960 and was explored and mapped by members of the West Virginia Association for Cave Studies (WVACS). Organ Cave and several other caves south of the Greenbrier

Fig. 11.2 Typical contact canyon, with limestone ceiling and shale walls—The Hole. Photograph by W.K. Jones. Used with permission

River are also considered contact caves and are described in other chapters of this book.

11.2 General Geology

The structural setting for the contact caves is on the west limb of the Sinks Grove Anticline and the east side of a broad synclinorium and is characterized by minor undulations with an average 3–5° dip to the west (Heller 1980; Price and Heck 1939). Groundwater flow is generally downward along the stratigraphic dip but is sometimes deflected by permeable units along faults or other structural features (Palmer 1974). Locally, some faulting has been observed within the caves and at least one major thrust fault has been observed in two adjacent caves (Fig. 11.3). Surface evidence for faulting and folding is not readily seen due to thick soil cover. The limestone and shale appear to be in fault contact for at least part of the study area. Heller (1980) mapped the Lewisburg Fault and described it as a southeast dipping reverse fault. Wells et al. (1967) suggested it is a normal fault between Lewisburg and Caldwell. This fault does not appear to extend as far north as Spring Creek but may affect the development of caves along the contact just north of Lewisburg.

The central Greenbrier County contact caves are developed almost exclusively within the 30 m (100 ft)-thick Hillsdale Limestone with vadose entrenchment into the underlying Maccrady Shale. A detailed stratigraphic section from Ludington Cave was presented by Palmer (1974).

Fig. 11.3 Drag fold associated with thrust faulting. McClung Cave. Photograph by W.K. Jones. Used with permission

The Hillsdale Limestone is relatively pure but contains beds of nodular chert near its base. The unit is generally massive biomicrosparite until the bottom 3 ms (9 ft) and the section grades to a shaly limestone containing undular shale laminae with a local thick weathering rind of clay and silt. The underlying Maccrady Shale is generally red shale that contains lensatic siltstone units and some thin calcareous zones. The top of the Maccrady Shale is an incompetent limey green shale that grades downward to a flaggy red shale (Palmer 1974). The upper 12 m (40 ft) of the shale is sometimes brecciated and recemented with calcite which makes it easily erodable by the aggressive cave streams.

Palmer (1974) estimated that 60% of the passages in Ludington Cave began at the Hillsdale/Maccrady contact. The contact caves formed under a combination of both phreatic and vadose conditions. Passages that are orientated along components of the dip started under phreatic conditions but have been modified by aggressive vadose action of surface streams that sink along the Hillsdale/Maccrady contact. Depending on the size of the stream that is coming in from the contact, the sink point can be a shallow sinkhole or a very large well-developed blind valley. Once underground, these streams follow the limestone–shale contact down dip.

Most of the entrances to the contact caves involve a sinking stream that begins on the Maccrady Shale and flows on the surface down dip until it intersects the limestone where it immediately sinks into the cave. In the case of Maxwelton Sink Cave, a large spectacular blind valley is developed with an amphitheater cliff face over 30 m high with the entrance at the base (Fig. 11.4). This entrance is usually clogged with flood debris and talus material. During periods of flooding, a large lake tens of meters in depth will back up the valley for over a kilometer. The Green Sink at Ludington Cave is another blind valley; it is not as large as the one at Maxwelton but is equally impressive. No physical entrance has been excavated at this blind valley, but several passages do come very close to the surface at this point.

The six caves that make up the central Greenbrier County contact zone are all formed in a similar pattern. In addition, there are other large cave systems in the area that are formed in the same way; they include Organ Cave and Windy Mouth Cave also in Greenbrier County and Scott Hollow Cave in Monroe County. Another contact zone in central Monroe County has the same geology, but no caves have been opened or excavated as of yet. They will be there; it is just a matter of time before someone finds (or digs open) an entrance.

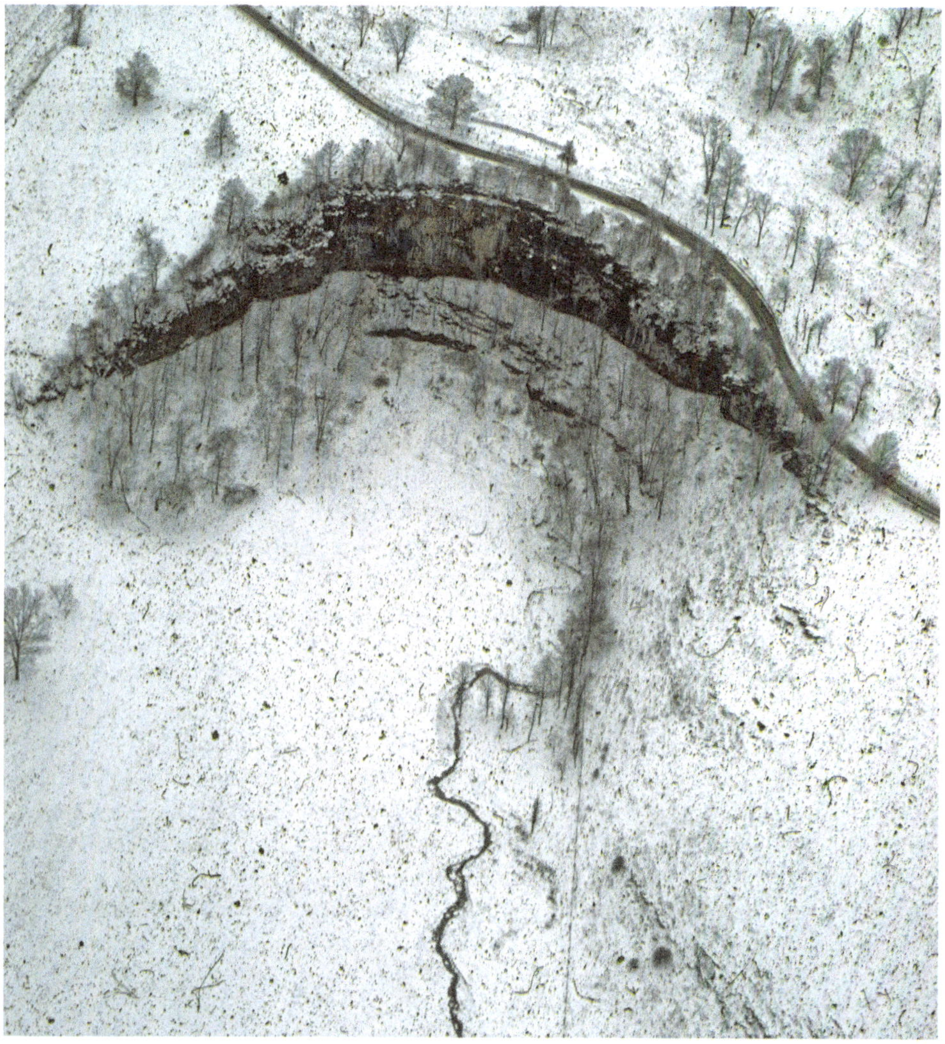

Fig. 11.4 The blind valley at Maxwelton containing the blocked Cove Creek entrance to Maxwelton Cave. Photograph by W.K. Jones. Used with permission

11.3 Contact Passage Morphology

There are three things that control the shape and size of contact passages: the size and velocity of the underground stream, the gradient of the stream, and the makeup of the immediate top of the shale (Balfour 1973). A large stream and a gentle gradient will usually produce a large rectangular cross section, scouring out the shale on the side of the passage and allowing areas of breakdown to occur. A smaller stream will produce a smaller cross section. A high gradient stream will tend to carve a canyon into the shale that can be several meters deep characterized by waterfalls and plunge pools. Again, based on the size of the stream the canyon can be fairly wide or narrow and sinuous. The third thing that influences passage formation is the immediate contact zone with the limestone. The Maccrady Formation consists of red shale, siltstone and mudrock that are very resistant. However, in some areas the top 6–8 m are cemented loosely with calcite that is easily dissolved and eroded by the vadose stream action. In addition, there are areas within this zone that are brecciated and appear to have been disturbed during deposition.

Contact passages usually follow components of the dip of the strata. Typically the dip is between 2 and 5°. However, faulting has been observed in McClung Cave and Ludington Cave with major implications within the latter (Palmer 1974). Where faulting has occurred, passage formation along major drag folds complicates the overall picture.

11.4 Description of Caves

11.4.1 Wades Cave

The entrance to Wades Cave is in a sink developed along the contact north of the golf course and to the east of the bowling alley. A stream flows from the clastic hills adjacent to the sink. This stream sinks into a swallet developed in the sediment-floored bottom of the sink. The entrance is in a rock outcrop about 6 m above the base of the sink and is a very small hole that is body sized. Squeezing down through this hole, you come to a 3 m downclimb that bottoms out in a rockpile. Just a few meters further, you encounter the stream. The next several hundred meters is mostly stoop walking in the stream through various sizes of potholes and plunge pools and is developed along the strike to the north (Fig. 11.5). After this entrance section, the stream passage turns west to follow the dip and rapidly enlarges to a spacious trunk channel averaging 7–8 m in height. Shale banks up to 3 m feet high are exposed in the walls. This passage continues for over a kilometer and is interrupted only by a few areas of massive breakdown where intersecting passages are encountered. Several small infeeder streams join the main passage with one apparently coming from the sink at the golf course about 1.5 km away. The discovery of golf balls in the stream seems to indicate this connection. Two other infeeders are coming in from the entrance sink and are probably high water overflow routes.

About 300 m from this junction the main stream intersects another major stream coming from the north. Just beyond this confluence, the combined stream sumps. Going north in the other stream, another sump is encountered after another 300 m. Wades Cave was surveyed by Robert Amundson and members of WVACS in the 1960s, and it contains about 5.6 km (3.5 miles) of mapped passage but is not completely explored. Visitation was always extremely sporadic, and the cave has been off limits for many years.

11.4.2 Benedict Cave

There are two entrances to Benedict Cave, the main Benedict Entrance and the Persinger Entrance, located about 600 m apart. Both entrances are at the end of shale valleys that terminate at the limestone contact, and both are small openings that give access to crawling passage. The Persinger Entrance is owned by the West Virginia Cave Conservancy. The cave was surveyed by Roger Baroody and members of WVACS in the 1960s and resurveyed by Bill Douty in the late 1970s. Benedict Cave contains 23.7 km (14.7 miles) of mapped passage.

The Benedict Entrance is a low crawl in the stream that is approximately 180 m in length. After this crawl, Benedict Creek intersects two other infeeding creeks and immediately enlarges into a breakdown room called No Parking. The passage enlarges into a shale canyon and turns north for 200 m to another junction with another infeeding creek at the Mercury Room. Benedict Creek turns west for 400 m as an impressive trunk passage averaging 10 m high and 5–10 m wide. This trunk passage then turns north for another 200 m with numerous side passages feeding into it. A major stream confluence is developed where the Sinking Creek trunk passage intersects Benedict Creek. Sinking Creek is a large trunk passage that can be followed upstream for over 800 m before an upstream sump is reached. Sinking Creek is the continuation of the Persinger trunk passage that contains Persinger Creek coming in from the other entrance. There is a sump between Persinger Creek and Sinking Creek.

Fig. 11.5 The entrance passage in Wades Cave. Photograph by W.K. Jones. Used with permission

The Sinking Creek trunk passage is a large breakdown-floored passage that is up to 10 m high and 20 m wide.

About 150 m beyond the junction with Sinking Creek, the Benedict trunk passage intersects the master drain for the entire cave. This trunk passage is called Sizeable Creek, and it comes in from the northeast along the strike. From this point, all of the water in the cave is flowing to the southeast in a low-wide cobble-filled trunk passage. This passage is flood-prone and continues to lower until a terminal sump is reached at a point about 600 m from the confluence. The water has been dye-traced to Davis Spring along the Greenbrier River.

Going northeast in upstream Sizeable Creek, the passage can be followed for 2 km before its terminus in breakdown. Several infeeder trunk passages are developed along the master drain and are coming in down dip from the east/northeast. The three main creeks are Mediocre Creek, Turkey Creek, and Trash Creek. Trash Creek and Turkey Creek come together in an area called the Confluent Room about half way up Sizeable Creek. Trash Creek is a very large breakdown-floored trunk passage that extends to the northeast for approximately 1 km before ending in a terminal breakdown at the contact. In places, it is over 30 m wide. Turkey Creek is a much smaller trunk passage and is mostly an incised canyon until it reaches its terminus in a large breakdown room (Fig. 11.6). Numerous upper level routes and abandoned phreatic tubes are developed throughout this area. The North–South Highway is one such abandoned upper level that cuts across to Mediocre Creek.

Mediocre Creek is a more than 1 km shale canyon that intersects Sizeable Creek about 200 m upstream from the Benedicts confluence. This passage also trends east/northeast going up dip in places and along strike in others. Upper level development is not as extensive but does occur. The North–South Highway coming from Turkey Creek is a major shortcut. A continuation of this passage on the south side of Mediocre Creek gives access to the Persinger section of the cave.

The Persinger Entrance is a small hole below a headwall at the end of a small blind valley. This hole gives access to a body-sized tube that is somewhat awkward but quickly opens into a 1 m high stream way that intersects a large trunk passage after about 40 m. The Persinger trunk passage is a pleasant dirt-floored stream channel with occasional breakdown areas. There is an abundance of infeeder passages and upper level development along the Persinger stream (Fig. 11.7). Persinger Creek abruptly ends in a sump after a few hundred meters. The sump is approximately 50 m long and on the other side is the continuation of the trunk passage, Sinking Creek.

Just before the sump, a maze of upper level crawls and small canyons is developed to the west. First Creek is 600 m long and ends in breakdown. A north-trending crawl intersects Rib Bone Creek and eventually intersects Mediocre Creek. This intersection is the only connection from the Persinger side of the cave to the Benedicts side. An 8 m drop has to be negotiated to make this connection, and a rope or cable ladder is required.

11.4.3 Maxwelton Cave (Aka Maxwelton Sink Cave)

Maxwelton Cave is the classic example of a contact cave. Cove Creek drains several square kilometers adjacent to the Greenbrier Valley Airport and is underlain primarily by red shales of the Maccrady Formation. Flowing southwesterly the creek immediately sinks at the contact with the basal portion of the Hillsdale Limestone. The sink is in a rubble pile formed from sluffage off an adjacent cliff face. At 100 m beyond the sink point, the imposing cliff face rises as a semicircular blind valley up to 30 m high. At the base of this cliff is the original entrance to the cave. Cavers opened this entrance in the early 1970s, and a corrugated metal pipe was placed in the excavated crawl. Many kilometers of passage were surveyed during the short time the entrance was open. A hurricane flooded the entrance in 1972 putting it under 10 m of water and backing a lake up the valley for a kilometer. Once the lake drained, there was no sign of the entrance or the pipe and the area had been filled with more rock and flood debris. Over the next 6 months, a concerted effort to reopen the cave eventually proved successful. However, within two weeks the area again was hit by massive flooding and the entrance was sealed once again, even worse than the first time. At that point, efforts to continue exploration were dropped and the cave was declared finished with about 15 km surveyed and a map was drawn.

In the early 2000s, a local caver (Dr. James David Scott) became interested in seeing if another entrance could be excavated on his property. Using micro-gravity techniques on the surface and trying to locate 1970s vintage cave survey proved to be time-consuming, but ultimately an anomaly was identified as possible passage. This passage was next to a surface sink, and a dig was instigated. It became apparent quickly that heavy equipment would be needed so a bulldozer and track hoe were mobilized. After many stops and starts, a hole was opened that blasted air. However, that hole was deemed too unstable and an adjacent area was excavated and broke into actual known cave passage. A 12 m vertical pipe was inserted into the shaft and backfilled. The entrance was named the Scott Entrance after the owner of the property. This gave access to the cave, and within a year the West Virginia Cave Conservancy purchased the property from Dr. Scott for permanent access. A resurvey of the cave was initiated and is still ongoing.

During this time, another caver bought the Cove Creek entrance at auction and a dig was initiated at the original

Fig. 11.6 Typical contact canyon in Benedict Cave. Photograph by Ryan Maurer. Used with permission

entrance. Heavy equipment was also brought into work on this project. Within a year, the Cove Creek entrance was reopened and a few trips were made before another flood sealed the entrance again. To date, no attempt to reopen it has been made.

The Turnpike was a low boulder and breakdown crawl from the Cove Creek entrance. After 140 m, this passage dropped 2.3 m into a large well-decorated trunk passage carrying the main Cove Creek stream (Fig. 11.8). At this junction, the main Cove Creek trunk averages 8 m high and 10 m wide. Going upstream, there are numerous branches that all end in breakdown against the blind valley cliff face. Going downstream, the Cove Creek trunk splits into two and sometimes three distinct parallel passages. These parallel

Fig. 11.7 Large trunk passage near Persinger entrance of Benedict Cave. Photograph by Ryan Maurer. Used with permission

Fig. 11.8 Large flowstone near the original Cove Creek entrance —Maxwelton Cave. Photograph by Ryan Maurer. Used with permission

passages represent abandoned flow routes from the east with the main creek flowing in the most westerly passage. These passages are all developed along the strike of the beds.

Approximately 365 m downstream from the intersection of the Turnpike and Cove Creek, the first major infeeder is encountered. H Canyon is a contact canyon developed mostly in the Maccrady Shale and is coming in down dip from the surface. In places, this canyon is up to 15 m deep with a small stream flowing in the bottom. After 350 m, a large breakdown room is encountered. Several passages lead upstream from this room, and all are blocked by terminal breakdown near the surface outcrop. At one time, the farthest east passage was open to the surface. This entrance was named the Airport Entrance due to proximity of the airport runway. However, the entrance was short-lived as a factory was built and the opening was covered by a parking lot.

Going downstream from the confluence of H Canyon, the main Cove Creek passage enters a low section called Le Mudge where water and mud predominate in about a meter-high section. There are several higher parallel passages that bypass this section. After Le Mudge, the Cove Creek trunk makes an abrupt turn to the north for about 100 m before it turns southwest along the strike again and opens into the Hall of Kings which is a large breakdown-floored passage.

Just downstream from the Hall of Kings is the intersection of the stream infeeder coming from the Heaven passage. Going up this passage, there are a series of upclimbs up to 15 m that need to be traversed before the upper level passage termed Heaven is reached. Once in the Heaven passage, it is a relatively easy 300 m walk to the base of the pipe at the Scott Entrance. At this time, the Scott Entrance is the only access to the cave. So getting down to the Cove Creek trunk 700 m from the Scott Entrance involves several climb downs and some rope work. Heaven was named because of the abundance of crystal formations and helictites (Fig. 11.9). The major flood of 2016 destroyed many of these delicate formations in an area that had never seen flooding of this magnitude.

Continuing downstream in the Cove Creek trunk about 200 m, the stream flows over an impressive 12 m waterfall (Fig. 11.10). Just before the falls is an upper level bypass that opens into an extensive series of parallel passages that tie back into the main drainage several hundred meters downstream. Where the upper levels tie back, there is a major infeeder passage entering from the south. This infeeder leads to a complicated junction where a passage goes upstream to the southwest and another passage goes northeast to Thunder Hall. The southwest passage eventually gets to small to traverse after 300 m. Climbing up into Thunder Hall involves some canyon traversing and exposed climbs. A large 20 m dome named Thunder Dome has been bolted and more passage discovered that is presently being mapped.

Beyond the area of the upper level development above the main Cove Creek the cave becomes a single master trunk (Fig. 11.11). The passage is wide and 3–4 m high with areas of breakdown. After about 300 m, there is a 6 m waterfall that opens into the continuation of the stream passage which is named the Cascade Causeway. The Cascade Causeway is a low-wide stream passage that is alternately floored by pools of water and cobbles. It ranges from 1 to 3 m in height. After about 750 m, a sump is reached at Gasoline Alley.

In early 2017, a major discovery was made when a breakdown blowhole was pushed at the end of the GOD passage which is a parallel passage just to the west of Cascade Causeway. Several hundred meters of muddy low phreatic tube was encountered beyond the constriction, and a low near sump was encountered. Just beyond this near sump, the passage intersected a large river passage going both upstream and downstream (Fig. 11.12). This river passage is estimated to have at least three times the volume of Cove Creek and is 15 m wide and 8–10 m high. Over 1300 m have been surveyed in the downstream section which continues. Upstream has been surveyed for 1100 m and also continues. Dye tracing has confirmed that this river passage (now named Sweetwater River) is the master trunk carrying the water from McClung and Ludington Caves and perhaps the water from several other significant caves (Higginbothams and Savannah). With this discovery, Maxwelton has grown to 24 km of mapped passage with the potential to add many more kilometers.

11.4.4 McClung Cave

McClung Cave is the next major cave north of Maxwelton. The main entrance is located in an open field. A small wet weather stream drains a shallow swale that ends at the entrance of the cave. The entrance is 3 m wide and 2 m high and opens into a large flat-floored room. Beyond this room, a large passage meanders for 100 m to a breakdown area beyond which the cave takes on a classic contact appearance. The small entrance stream has down cut into the underlying shale and formed a tight sinuous canyon. This is best traversed somewhere along the middle of the canyon along crumbly ledges of shale. The depth of the canyon varies from a few meters up to 10 m. The top of the canyon is a classic phreatic tube developed in the limestone that is up to 10 m wide. The canyon section persists for approximately 500 m until a major collapse is reached. First Breakdown marks a major intersection of upper level passages that fan out in several directions. The entrance stream flows under the breakdown and emerges on the other side into a large trunk passage. It is possible to follow the stream under the breakdown, and it is also possible to climb up and through the breakdown to continue downstream. At the top of the breakdown on the right is a small passage that has an old iron gate. This passage is the gateway to the northern half of the cave.

Dropping back down to the stream level, the characteristic canyon enlarges into a more uniform passage that is occasionally floored with breakdown (Fig. 11.13). Several strike-orientated passages intersect the down dip canyon just beyond First Breakdown some of which carry small infeeder creeks coming from the north. There is one phreatic strike tube developed on the south side of the canyon, and a climb up into it gets into Batbone Crawl. Batbone Crawl extends 220 m to intersect with another down dip canyon complex called Freeman Avenue. Just before this intersection, there is a crawlway that branches to the right off of Batbone Crawl which eventually leads to a very tight constriction that opens into the WVACS Room, one of the most highly decorated rooms in the contact caves.

Continuing down the main canyon, the passage is mostly large walking trunk that averages 5–10 m high. One low area

Fig. 11.9 Helictites and aragonite crystals in the Heaven passage, Maxwelton Cave. Photograph by Ryan Maurer. Used with permission

can be negotiated by a short bypass crawl. After 500 m, another major breakdown area is encountered. Second Breakdown is negotiated by climbing up and over the constriction. Several side passages are developed at this point and go south to join with the paralleling canyon in Freeman Avenue. A hole in the breakdown leads back down into the continuation of the entrance stream canyon. This is a large passage that ends after 200 m at the intersection with Chocolate Avenue. Chocolate Avenue carries the main drainage of the cave, and the water has been traced from the Thunderbolt Passage of Ludington Cave to the north.

Going downstream in Lower Chocolate Avenue, the passage is wide and contains large mud banks associated with the stream. The height averages 5 m with evidence of

Fig. 11.10 Waterfall in the Cove Creek trunk, Maxwelton Cave. Photograph by Ryan Maurer. Used with permission

flooding. After 400 m, a low near sump has to be negotiated before the passage enlarges again. This is short-lived as a terminal sump is soon reached. This sump marks the low point in the cave and is 103 m below the main entrance. Going upstream in Chocolate Avenue, the main stream reaches a low area after 300 m that is an impassable gravel-filled sump. However, just before this sump there are a series of strike-orientated passages that intersect in an area called Third Breakdown. These passages eventually connect into a low-wide avenue termed the Meatgrinder. The Meatgrinder trends northeast for approximately 1500 m and averages 1–2 m high.

There are two major ways to get into the northern half of the cave. One is via the Meatgrinder, and the other is

Fig. 11.11 Cove Creek trunk near the Hall of Kings, Maxwelton Cave. Photograph by Ryan Maurer. Used with permission

Fig. 11.12 Chris Coates (foreground) and Carl Amundson stand in the newly discovered and surveyed Sweetwater River in Maxwelton Sink Cave. Photograph by Nikki Fox. Used with permission

through the gate that opens into the Gateway back at First Breakdown. The trip through the Gateway is the preferred way.

The initial area just past First Breakdown is a complex of intersecting phreatic tubes and vadose canyons that are mostly strike orientated but developed in different vertical horizons. These passages are developed well above the Maccrady Shale. Occasional domes funneling water from the overlying sinkhole plain are encountered in this section. There are several small infeeder streams coming in from the eastern contact zone including Beetle Run. These infeeders are small walking size canyons that terminate in breakdown or domes.

The Wind Tunnel is a phreatic tube that is developed along the last infeeder and is the connection passage with the northern section of the system. It is a dry gypsum-encrusted tube that is 1–2 m high for its entire length of 300 m. As the name implies, it carries a good breeze. The Wind Tunnel intersects a large trunk passage named Liberty Bell Avenue. To the right (east), the passage is a dirt and breakdown-floored trunk for about 150 m where it ends in a series of rooms and flowstone with one formation called the

Fig. 11.13 Typical contact canyon—McClung Cave. Photograph by Ed McCarthy. Used with permission

Liberty Bell. Going straight from the Wind Tunnel intersection, the passage is completely floored with large breakdown that is not easy to traverse (Fig. 11.14). After about 100 m, a 12 m drop ends the upper trunk and drops down into the Crystal Canyon section.

There are several hundred meters of cave below the drop. Of note, there is a well casing that intersects one of the passages coming straight down the middle of the 7 m high passage (Fig. 11.15). At one time, there was another entrance (Kidd Entrance) that was open for a couple of years when the Ludington surface stream collapsed into the cave. This collapse was about 15 m deep and captured the surface stream. Subsequent collapse, flood debris, and silt refilled the entrance allowing the surface stream to renew its previous course and flow unimpeded into the main entrance of Ludington Cave.

Back at the intersection of the Wind Tunnel and Liberty Bell Avenue, there is a climb down to the left that opens into what is called the G Trail. This is a pleasant walking canyon that carries a trickle stream for a portion of its length

Fig. 11.14 Liberty Bell Avenue, McClung Cave. Photograph by Ed McCarthy. Used with permission

Fig. 11.15 Well casing in Crystal Canyon, McClung Cave. Photograph by W.K. Jones. Used with permission

(Fig. 11.16). G Trail meanders west for several hundred meters until a pit is developed in the floor that drains the stream into a lower level. The pit is about 10 m deep and drops into several hundred meters of small unpleasant canyon passages that eventually are not traversable. Stepping around the pit in the floor, the main passage turns north as a nice silt-floored walkway named Megalithia. There are several junctions off of Megalithia that lead to many hundred meters of passage. After 300 m, Megalithia intersects the Christmas Passage. Going left into the Christmas Passage, the passage trends back toward the entrance of Ludington Cave and comes within 120 m horizontally of the entrance. Halfway out the Christmas Passage is a complex area that contains a 14 m drop to a large breakdown chamber. This room opens into two large breakdown passages that trend north: Big Muddy and Little Muddy. Both of these passages terminate in breakdown and mud mountains and show evidence of back-flooding.

Back at the intersection of Megalithia and the Christmas Passage, there are a series of small passages developed to the left (west). The Mindbender and Wafflefoot Alley intersect the Meatgrinder after 350 m. This forms a several kilometer loop going back to Third, Second, and First Breakdown.

Fig. 11.16 G-trail area in McClung Cave. Photograph by Ed McCarthy. Used with permission

There are two main ways to access the southern portion of the system. One is via Batbone Crawl that opens into Freeman Avenue, and the other is through the Lightner Entrance. The Lightner Entrance was discovered in 1985 and is now the only viable entrance since the main McClung Entrance is usually closed by the landowner. The West Virginia Cave Conservancy owns the Lightner Entrance and has an open cave access policy.

The Lightner Entrance is a small slot in a rock face on the side of a large sink. It is the only natural entrance to a contact cave that is not located at the contact. A 3 m climb down leads to the top of a 15 m drop that bells out at the bottom. A small drain hole at the base of the drop opens immediately into a 7 m drop along a flowstone wall into a well-decorated room. Beyond this room, about 30 m of mostly crawling passage and a squeeze that was excavated leads to another few meters of passage and then another 3 m drop that requires a handline into a lower level passage. The lower level passage ends in a flowstone-filled room to the right and massive breakdown to the left. Pushing through the breakdown via the Champaign Squeeze, the passage opens up and very quickly intersects the Tufa Trail. Going upstream in the Tufa Trail is a continuous climb up rimstone dams for 300 m until a large room is reached with waterfalls coming in from the ceiling.

Going downstream in the Tufa Trail, you climb down the rimstone. After about 300 m is the intersection with the Freeman Avenue trunk. Freeman Avenue is a large breakdown-floored passage that ends in a sump 400 m downstream from this intersection. Upstream, Freeman Avenue parallels the main entrance canyon for almost a kilometer enlarging as it goes upstream and up dip. It is developed along the Maccrady contact but does not have the classic canyon development because of the modification by breakdown. The upstream terminus is in massive breakdown.

Batbone Crawl is about 800 m upstream from the confluence of the Tufa Trail on the left wall of a large breakdown chamber. Batbone Crawl is the major access point to the majority of the cave. A couple hundred meters beyond Batbone is a large junction with a major trunk coming in from the right (south). This passage is called the JEB Stuart Passage and leads to over a kilometer of infeeding canyons developed along the Maccrady contact that drain several surface sinkholes. This area is called the Seven Fingers because the map portrays the surveyed canyons in a hand-shaped fashion (Fig. 11.17).

McClung Cave has a surveyed length of over 26 km and a depth of 103 m. It is a very popular cave and is continuing to be explored.

11.4.5 Ludington Cave

The main entrance to Ludington Cave is in a small blind valley. A creek flowing on the Maccrady Shale flows into this entrance that is approximately 4 m wide and 3 m high. The entrance passage retains these dimensions for a couple of hundred meters until a 13 m (the Old Drop) drop is encountered. Below this drop, the passage encounters the Maccrady Shale zone and decreases in size. This section of the cave is named Old Ludington because it was the first

Fig. 11.17 Seven fingers contact canyon, McClung Cave. Photograph by Ed McCarthy. Used with permission

section to be explored. One hundred fifty meters from the bottom of the drop the main passage reaches a low area that is sumped shut with gravel fill. About 50 m below the drop, a small infeeder canyon averaging 4 m high comes in from the right wall. Going up this canyon for about 200 m, a 10 m waterfall (Half Way Falls) is encountered. Above the waterfall is a complex of infeeder passages coming in from the surface, one of which actually opens in the side of the main entrance.

Approximately 40 m before the Old Drop, an overflow canyon is developed on the left wall. This canyon is named the Polar Passage because of the amount of air that moves through it during the winter. It averages 1–3 m in height for about 300 m before another drop is encountered. The New Drop is a 9 m deep fissure along a drag fold with chert-encrusted walls that drops down to the level of the Maccrady Shale again. A small canyon is developed at the contact and extends 200 m to a junction with a larger passage on the right. This larger passage is the continuation of the main entrance passage that is sumped just downstream of the Old Drop. Several hundred meters of complex canyon maze is developed off of this section going back up dip toward the surface. Continuing downstream the passage gets continually larger and is floored with breakdown in places. The walls are composed of Maccrady Shale. After 200 m, the passage lowers and widens out, turning abruptly west for 50 m before it intersects the Thunderbolt Passage, the master trunk for the cave.

Going downstream from this junction, the Thunderbolt Passage is a large gravel-floored passage that gradually lowers from walking height to stooping. After 450 m, the ceiling lowers to the point that traverse is no longer possible. There is a sump at this point called Odd Man Crawl that is composed of water and gravel that fills the passage. The stream in the Thunderbolt Passage is the main drain for the entire cave and has been dye-traced to Third Breakdown in McClung Cave where it emerges as Chocolate River.

Going upstream, the Thunderbolt passage enlarges into a spacious trunk channel carrying the stream. The floor of the passage is cut into the Maccrady Shale with mud banks and cobble deposits scattered in the meanders of the stream channel. In places, large breakdown blocks have fallen from the ceiling and sides of the passage (Fig. 11.18). The average dimensions are 10 m × 10 m for over 800 m, and the passage follows the strike. At 800 m, there is an infeeding canyon coming into the Thunderbolt Passage from the left (west). This is the only western infeeder at the same level as the trunk within the system. This passage is an incised stream canyon within the Maccrady Shale and has a small ceiling channel developed in the limestone. This stream canyon gradually climbs stratigraphically out of the contact zone and after about 100 m the passage is completely developed in the limestone. This canyon remains a comfortable walking height for another couple of hundred meters before it gradually lowers and becomes a very low wet crawl that has not been pushed. The canyon is approximately 365 m long.

The Thunderbolt Passage in this area is floored with breakdown for much of its length. On the right side (east) is a large chute of breakdown that extends up the flank of a major drag fold that is associated with thrust a fault that is developed 40 m above the Thunderbolt Passage (Palmer 1974). This thrust fault parallels the strike and is developed

Fig. 11.18 Thunderbolt Trunk, Ludington Cave. Photograph by Ed McCarthy. Used with permission

on the limb of the Sinks Grove anticline approximately 1.6 km to the east. Climbing up this chute, the Maccrady Shale turns vertical until it intersects the thrust fault and opens up into a large breakdown chamber called the Belfry (Fig. 11.19). Strata exposed in the Belfry are nearly horizontal and the room is up to 20 m high. At the east end of the Belfry is an infeeder stream passage that within 30 m reaches a surface insurgence. This insurgence is the Bell Entrance and is subject to flooding and siltation; at times it had been a tight muddy crawl, but it had occasionally been a walking canyon. In the early 1990s, a collapse occurred at the entrance that allowed the passage to silt completely shut. Efforts to reopen the entrance stalled when an unfriendly landowner bought the property, and the entrance remains sealed at this time.

Northeast of the Belfry a series of breakdown-floored passages open into the Boulder Room, the largest room in the cave system. The Boulder Room is a large trunk segment developed along the strike and is floored with large blocks of breakdown. It averages 20 m wide and 15 m high and is 300 m long. Numerous side passages are developed on the east side of the room, and they lead back close to the surface contact. Several of these are highly decorated with speleothems due to their proximity to the surface.

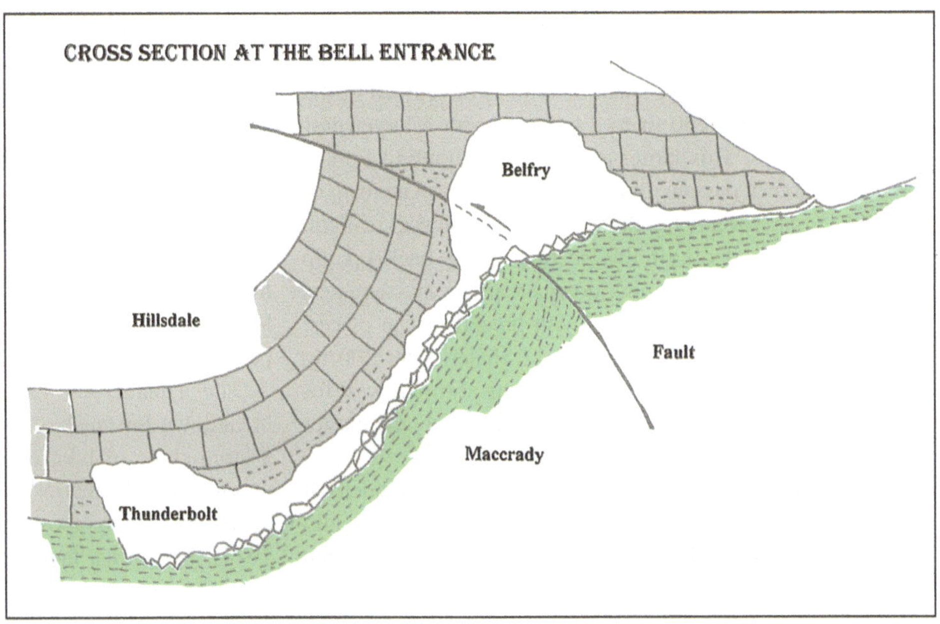

Fig. 11.19 Cross section at the Bell Entrance, Ludington Cave

Within the breakdown floor of the Boulder Room, there are several openings that lead down to a series of parallel passages. These passages contain several hundred meters of breakdown-floored channels. There are also a couple of breakdown chutes, again developed along the drag fold, that empty into the Thunderbolt Passage. The Thunderbolt Passage continues upstream for another 500 m averaging 5–8 m high before eventually branching into several smaller infeeder tubes coming in from surface sinks.

In 2010, a blowing hole on the side of a sink a few hundred meters north of the Bell Entrance was excavated. After much effort and 200 m of extremely tight and wet passage, a connection was made to a hole on the side of the Boulder Room. This new entrance is called the Yates Entrance and now allows entry into the northern portion of the system once again.

A blind valley is developed midway between the Bell Entrance and the main Ludington Entrance. This is the Green sink where a surface stream sinks at the base of a cliff face reminiscent of the Maxwelton cliff but somewhat smaller. Two parallel strike-orientated phreatic tubes terminate at the southern edge of this sink. Boo Boo Boulevard and the Long Hall are thought to be abandoned inflow channels (Palmer 1974). Boo Boo Boulevard is a low-wide cobble-floored tube, while Long Hall is a large breakdown-floored trunk segment. The Green sink is directly over top of the Thunderbolt Passage but is 35 m higher. Water entering from the sink flows rapidly down a series of minor faults and steep dip slopes associated with the main thrust fault (Fig. 11.20). A vertical maze of passages is developed within this section.

From the junction of Boo Boo Boulevard and Long Hall, the passage continues southwest along strike as the Long Hall Crawl. It is a wide cobble-floored tube averaging 1–2 m in height for 230 m to the intersection with X Canyon. Going upstream in X Canyon, there is approximately 500 m of mostly walking height stream canyons developed along the contact. All passages end in surface breakdown. Going downstream in X Canyon Shale Pit is reached after 150 m. Shale Pit is a 10 m drop into a small canyon that within 150 m empties into Lower Ludington just south of its confluence with the Polar Passage.

From the intersection of Long Hall Crawl and X Canyon, the main passage continues southwest and after 200 m opens into the Star Room. To the east of the Star Room are a series of interconnected passages on various levels that trend up dip toward the surface contact. Several of these passages are carrying small streams that all converge in the Star Room. A large canyon exits the Star Room on the west side, and this passage is called Lower Ludington. Lower Ludington intersects X Canyon after 150 m and then continues another 200 m to its confluence with the Thunderbolt Passage.

Ludington Cave has approximately 17 km of surveyed passage. It was originally surveyed by members of the West Virginia Association for Cave Studies in the late 1960s to a length of 8 km. A subsequent resurvey led by Joseph Caldwell and WVACS members in the early 2000s doubled the known extent of the cave.

Fig. 11.20 Waterfall in Thunderbolt passage of Ludington Cave, coming from the Green Sink. Photograph by Ed McCarthy. Used with permission

11.4.6 The Hole

The Hole is the northern most contact cave and the only one that does not drain south to Davis Spring. The Hole resurges at springs associated with Burns #2 Cave along the banks of Spring Creek. From a geomorphic standpoint, The Hole is the least complex contact cave and exhibits classic maze development associated with uniform geologic conditions.

The Hole has three main entrances and a recently discovered fourth entrance that was excavated. The Boggs Entrance is the southernmost and the most upstream entrance to the system. It is a large entrance at the end of a blind valley with a surface stream entering the cave. A little over a kilometer to the north is the Perkins Entrance, a small stream sink in a pasture just below a stock pond. The Gibbs Entrance is approximately 2.2 km north of the Perkins Entrance and is the northern most entrance. The Gibbs Entrance is another small hole on the edge of a sink with a small surface drainage associated with it. The newly excavated Perkins #2 Entrance is not far from the original Perkins Entrance but leads into a different section of the cave.

The cave system consists almost entirely of down dip vadose canyons carrying active streams from the surface and phreatic tubes following the strike. These phreatic conduits usually intersect the stream canyons above the shale layer near the ceiling and are developed entirely within the limestone. The down dip canyons are incised into the Maccrady Shale up to 10 m. The dip of the strata ranges from 3 to 5° and is fairly uniform throughout the entire system. This has allowed a fairly predictable speleogenesis to occur (Fig. 11.21). The cave system exhibits its majority development along the strike, and the parallel phreatic tubes offer the major connections between the various sections of the cave. Where canyons are numerous, maze sections are developed. Vertical development above the contact zone is not common, and most of the passages in The Hole are developed within the lower 20 m of the Hillsdale Limestone member and the underlying Maccrady Shale.

The Gibbs Entrance is a gated small opening in a shallow wooded sink. The entrance opens into a small stream channel less than 2 m high. Within 30 m, a small pool (the Bathtub) is encountered that requires immersion before the passage opens up into the Long Room. The Long Room is a remnant of a strike-orientated phreatic tube that has been modified and enlarged by breakdown near the surface contact. The entrance stream flows across this room and into Barefoot Creek, a down dip canyon developed at the shale contact. Numerous plunge pools are developed as the water cascades quickly down slope. Approximately 100 m down this canyon, there is an intersection in the ceiling with a wide phreatic tube. A 3 m climb gives access to the Crossover. The Crossover is about 12 m long, and then there is another 3 m drop into the Adjacent Stream Passage. This stream passage is completely separate from Barefoot Creek, and the passage has larger dimensions (Fig. 11.22). The Adjacent Stream Passage trends along strike and picks up incoming water from the surface contact via several down dip canyons such as Spike Street, Bullwinkle Boulevard, and the Upper Adjacent Stream Passage.

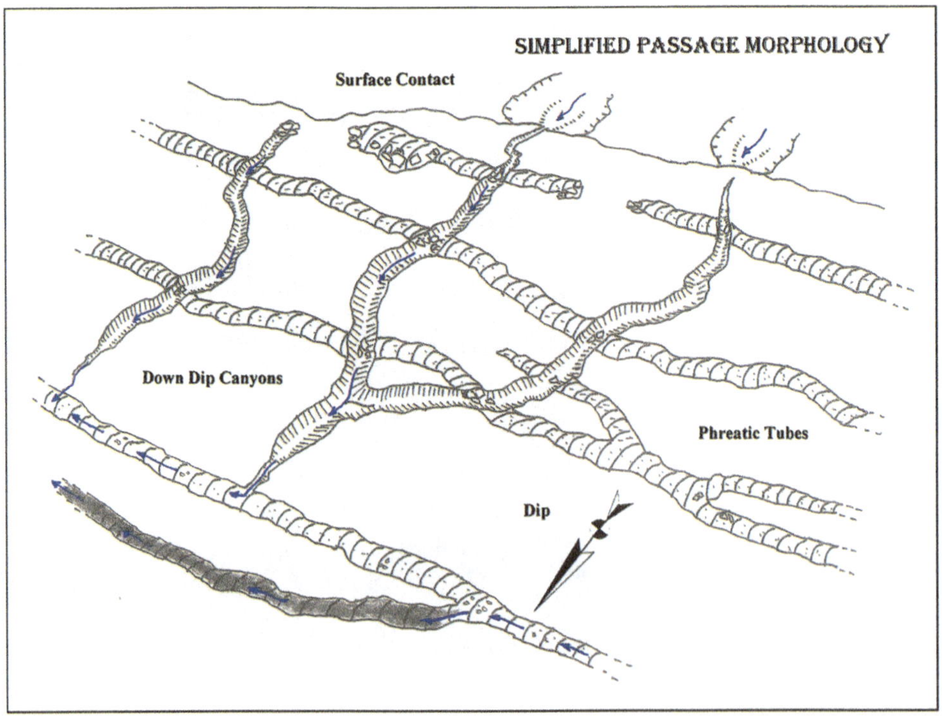

Fig. 11.21 Simplified passage morphology for the hole

Fig. 11.22 Adjacent stream passage in the hole. Photograph by Ed McCarthy. Used with permission

Downstream from the Crossover in Barefoot Creek, the canyon enlarges with the addition of the phreatic tube in the ceiling. Ceiling heights average 6–7 m. After 70 m, the stream exits via a small hole in the floor and an area of dirt-floored crawlways is encountered. Turning north leads into what is called the North Maze. The North Maze consists of hundreds of meters of low phreatic tubes that are parallel to each other and spaced about 30 m apart. They are cut by down dip canyons carrying small trickle streams coming in from the surface. In one section, there are eight parallel tubes along strike, each perched just above the limestone/shale contact. Each one represents a former base level and with subsequent lowering new tubes formed. A ninth tube (Shale River) exists in this area and is completely sumped indicating perched phreatic conditions. Shale River is the master drain for the entire cave and in this section is visible only for a few meters at the northern most point in the cave before it reaches a terminal sump.

Near the end of the Adjacent Stream Passage is a remnant tube named the Broken Crockery Passage. This passage is floored with loose slab breakdown for 150 m before dropping into the north side of the South Maze Stream Passage. Going downstream, there are numerous tubes intersecting at ceiling level on the left side of the canyon. These phreatic tubes are the beginning of the South Maze and are spaced every 10 m. They are wide, dirt or cobble-filled passages that average a meter in height and are interconnected on many levels. As with the North Maze, they represent former water table levels. The South Maze is more dense than the North Maze. The South Maze Stream Passage empties into a segment of Shale River (Fig. 11.23). Just downstream from this confluence, Shale River intersects the Maccrady Shale along the strike. This portion of Shale River follows the contact which abruptly rises 12–15 m above stream level. The stream has down cut a narrow canyon within the shale at this point that is up to 12 m deep. This canyon is about 170 m long and ends as abruptly as it began when the limestone/shale contact returns to level at the beginning of the canyon. Shale River is no longer entrenched in the shale and flows in a wide 2–3 m high tube.

Going southwest out of the South Maze, there are a couple of parallel phreatic tubes. The larger one is named the Long Northeast. The Long Northeast is well named and is the main passage that ties the entire system together. It averages 1–2 m in height 3–4 m in width and is floored with cobbles, sand, and bare bedrock. The Long Northeast extends from the South Maze to Echo Canyon near the Boggs Entrance and has a linear extent of just over 1 km. There are a number of parallel tubes (No Name Northeast, Nice Northeast, Nasty Northeast, Bear Wallow Way, Pittsburgh Section Northeast) that are developed adjacent to the Long Northeast, but none of them are continuous.

Fig. 11.23 Shale River in The Hole. Photograph by Ed McCarthy. Used with permission

Shale River is similar to the Long Northeast except it has water and sumps in several places, it can be followed for 1.3 km. The Long Northeast is the easiest passage to traverse to get from one section of the cave to the other.

The Long Northeast and the other northeast-trending tubes are all cut by down dip stream canyons formed at the contact. The Blue Canyon, Crumbling Canyon, Perkins Canyon, and the Third Stream Canyon are the more notable ones with each having tributaries that extend to the surface contact (Fig. 11.24). The Perkins Canyon breaches the surface just below a farm pond in the base of a sink. This entrance gives access to the middle portion of the system. These canyons cut the northeast-trending tubes at right angles and trend down dip to intersect Shale River. On approach to Shale River, the gradient lessens and the stream rises at the contact zone. The canyon aspect of the passage disappears and the passage is smaller and developed completely within the limestone. Tower Canyon and Echo Canyon mark the end of the Long Northeast. Echo Canyon leads upstream to Bear Wallow Way, which in turn intersects Second Stream Canyon. All of these canyons drop into Shale River. A series of small parallel tubes are interconnected on the south side of Second Stream Canyon and form the Boggs Maze. The Best Way Northeast and Frosted Avenue both cross First Canyon on their way to the Boggs Entrance. A 3 m ladder climb from Frosted Avenue drops into the large entrance room. The Boggs Entrance is the only natural entrance to the cave; the other entrances were excavated.

Members of the West Virginia Association for Cave Studies originally surveyed The Hole in the 1960s. It was the first major cave system found that was not known prior to that time. In the late 1970s, the Pittsburgh Grotto of the NSS discovered what was named the Pittsburgh Section. After mapping that portion, it was decided that the entire system needed to be remapped for better quality. Led by Charlie Williams and other WVACS members, over 70 trips were made and 37 km were mapped. Water quality measurements were taken by Pasquarell and Boyer (1995) in the 1990s in The Hole watershed to study the relationship between agricultural practices and bacteria and nitrate levels in cave waters. A seasonal (summer) increase in bacteria counts correlated with increased cattle grazing on the sinkhole plain above the cave.

Fig. 11.24 Phreatic tube crossing down dip contact canyon —The Hole

References

Balfour, W.M. 1973. The origin of contact passages developed at the contact between the Greenbrier Series and The Maccrady Series, Greenbrier County. *West Virginia. Region Record* 3 (1): 14–17.

Gams, I. 1965. Types of accelerated corrosion. In *Problems of the speleological research*, ed. O. Štelcl, 133–139. Prague: Czechoslovak Academy of Sciences.

Heller, S.A. 1980. A hydrogeologic study of the Greenbrier limestone karst of central Greenbrier County, West Virginia, 421pp. PhD Thesis, West Virginia University.

Mihevc, A. 1994. Contact karst of the Brkini hills. *Acta Carsologica* 23: 100–109.

Palmer, A.N. 1974. Geologic influence upon cave–passage orientation in Ludington Cave, Greenbrier County, West Virginia. In *Proceedings of the fourth conference on Karst geology and hydrology*, ed. H.W. Rauch, and E. Werner, 33–40. Morgantown, WV: West Virginia Geologic and Economic Survey, Morgantown.

Pasquarell, G.C., and D.G. Boyer. 1995. Agricultural impacts on bacterial water quality in Karst groundwater. *Journal Environmental Quality* 24: 959–969.

Price, P.H., and E.T. Heck. 1939. *Greenbrier County*, 846. Morgantown, WV: West Virginia Geologic and Economic Survey County Report.

Rutherford, J.M. 1967. Initial explorations of the contact caves of the Great Savannah (abs.). *National Speleological Society Bulletin* 29 (3): 111.

Wells, D., J.J. Renton, and R.L. Perkins. 1967. Normal faulting in the Caldwell-Lewisburg area of Greenbrier County, West Virginia. *West Virginia Academy of Science Proceedings* 1966 (38): 155–158.

Hydrology of the Sinking Creek/Muddy Creek Karst Basin

12

William K. Jones and William M. Balfour

Abstract

Muddy Creek represents the resurgence of Sinking and Hughart Creeks with a catchment of about 40 mi^2 (104 km^2) in central Greenbrier County, West Virginia. The combined flow of Sinking and Hughart Creeks appears in a large karst window just south of I-64. This karst window floods to depths of 80 ft (24 m) or more during high water and is locally known as "Interstate Lake." This water reappears at Piercys Mill Cave (the head of Muddy Creek) under low-flow conditions. Water level loggers were established at the two sinking streams, the stream in the karst window, and the two cave resurgences to monitor the response of the karst drainage system to storm events. With rising water levels, a threshold is reached in Piercys Mill Cave at a discharge of about 60 cfs (2.55 cms) and excess water is shunted to Piercys Cave about 2400 ft (750 m) downstream. The flow from Piercys Mill Cave does not exceed a maximum value of about 200 cfs (5.66 cms). The straight-line distance from Sinking Creek to Piercys Mill Cave is about 2.5 miles (4 km) and tracer tests showed a range of travel times between 15 h under average flow conditions to 85 h at very low flow.

12.1 Introduction

The purpose of this study was to examine the response of the Sinking Creek karst drainage basin to storm events and to document the flooding phenomenon in Interstate Lake at Alta. The recent availability of relatively inexpensive water level loggers has resulted in new hydrographs from many karst drainage systems around the world (Kresic 2013; Kovačič 2010; Field 2010). Conventional time-series analysis (Haan 1977; Davis 2002) and new statistical methods have also been applied to the analysis of spring hydrographs (Mangin 1984; Petrič 2002; Turk 2010).

Electronic supplementary material
The online version of this chapter (doi:10.1007/978-3-319-65801-8_12) contains supplementary material, which is available to authorized users.

W. K. Jones (✉)
P.O. Box 356Warm Springs, VA 24484, USA
e-mail: wkj30@hotmail.com

W. M. Balfour
3954 Carrol Hill Road, Frankford, WV 24938, USA

Sinking Creek flows south from Williamsburg, WV for about 9 miles (14.5 km) and sinks at the Boxley Quarry just north of Interstate-64 at Alta (Figs. 12.1 and 12.2). Hughart Creek drains a smaller area to the west of Sinking Creek and also sinks at the Boxley Quarry (Fig. 12.3). The combined flow from these Sinking Creeks is exposed in a large karst window just south of the interstate called "Interstate Lake" (Fig. 12.4). The water sinking at ponors at the south end of Interstate Lake flows south for 1.4 miles (2.25 km) and resurges at Piercys Mill.

Muddy Creek is a tributary to the Greenbrier River at Alderson, West Virginia. The creek rises from two caves at Piercys Mill. The head of Muddy Creek is a stream flowing from Piercys Mill Cave (Figs. 12.5 and 12.6). Additional water, especially at higher flow levels, comes into the creek from Piercys Cave (Fig. 12.7) about 0.45 miles (0.75 km) west of Piercys Mill.

Water level loggers were installed in December, 2013, at Sinking and Hughart Creeks at the sink points in Boxley Quarry and at the sink point in the bottom of the Interstate Lake karst window. Level loggers and conductivity loggers

W.B. White (ed.), *Caves and Karst of the Greenbrier Valley in West Virginia*,
Cave and Karst Systems of the World, https://doi.org/10.1007/978-3-319-65801-8_12

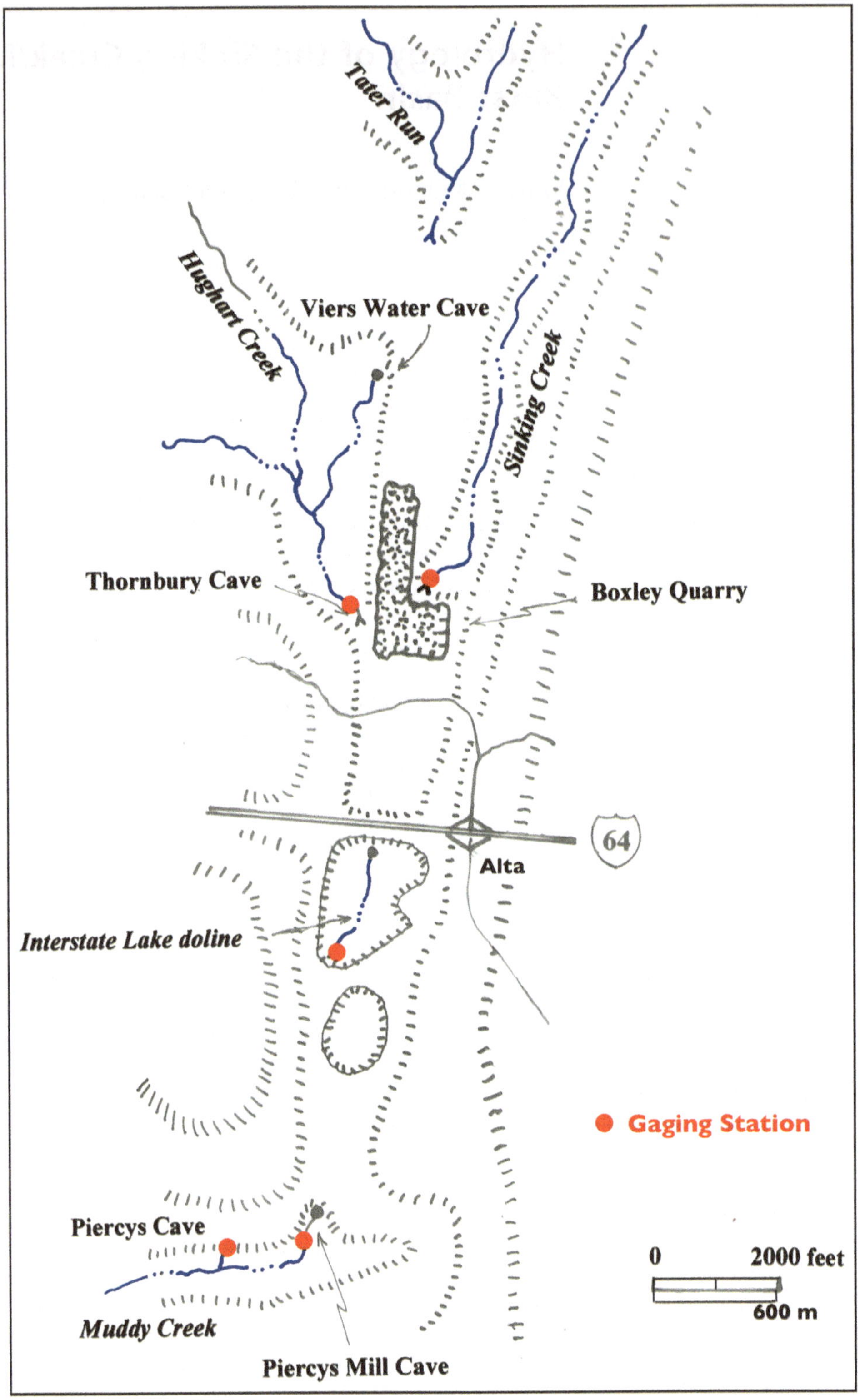

Fig. 12.1 Sketch map showing the sinks, karst window and resurgences

were placed in the outlet stream below Piercys Mill Cave and just inside the entrance to Piercys Cave. Data were collected for the winter and spring months of 2014 and 2015. Stage—discharge rating curves were prepared for Piercys Mill and Piercys Caves. Rating curves could not be defined for the two sinking streams or Interstate Lake due to backwater during

Fig. 12.2 Aerial photograph looking south showing overview of the stream sinks at Boxley Quarry at Alta. Interstate-64 is between the quarry and Interstate Lake. Photograph by W.K. Jones

Fig. 12.3 Aerial photograph showing the sink of Hughart Creek (*right center*) and Sinking Creek (*top left*). Photograph by W.K. Jones

higher flow levels at these sites (Figs. 12.8 and 12.9). Three quantitative water-tracing tests were also conducted to define the relationship of the sinking streams and springs.

12.2 Geology and Karst Features

12.2.1 Geology

The Sinking Creek drainage follows the western most outcrop of the Greenbrier Group limestones in Greenbrier County (Fig. 12.10). The upper carbonate units (Alderson and Union limestones) crop out in a band on the west side of the Williamsburg (Brushy Ridge) anticline (Price and Heck 1939). The width of the carbonate outcrop in the Sinking Creek valley averages about 0.5 miles (0.9 km). The Greenbrier Group rocks are bounded to the east by the older Maccrady Shale and Pocono Group sandstones. Younger Bluefield Group sandstones form the western boundary and may be in fault contact with the limestones, at least in the Alta area. A digital terrain model of the area is shown in Fig. 12.11.

Fig. 12.4 Aerial photograph of the Interstate Lake karst window under low-flow conditions (looking north). Interstate-64 is at the *top* of the photograph and the spring (resurgence) of Sinking and Hughart Creeks is just below the interstate. A least four ponors (sink points) can be seen at the *bottom* of the photograph. Photograph by W.K. Jones

12.2.2 Interstate Lake

Interstate Lake is a local term for the large closed depression just south of Interstate 64 at the Alta exit. At lower flow conditions, this is a karst window with the combined flow from Sinking and Hughart Creeks emerging from a spring at the north end just below the interstate and sinking at the south end at several ponors about 1700 ft (530 m) south of the spring (Fig. 12.4). The rim of the depression is at the 2200-ft contour, and the floor is at 1930 ft for a depth of about 270 ft (82 m). The depression slowly floods and drains over a period of several days following heavy storm events. The maximum flood depth recorded during this study was 79.7 ft (24.3 m) on March, 2015. The calculated water volume is about 155,000,000 ft^3 (4,400,000 m^3) at this level (Fig. 12.12).

The elevations of the sink points in the quarry are about 2000 ft (610 m), so at full flood the water level in Interstate Lake is above the elevation of the ponors in the quarry. The backwater in the quarry sink is about 40 ft deep or 30 ft above the water level in Interstate Lake under these conditions, so there is plenty of hydraulic head to drive the flow through the system. The two resurgences are at an elevation of 1700 ft (518 m) at Piercys Mill Cave and 1640 ft (500 m)

Fig. 12.5 Aerial photograph showing Piercys Mill Cave at the head of Muddy Creek (*top*) and the clear water from Piercys Annex Cave ponded by the weir at Piercys Mill. Photograph by W. K. Jones

at Piercys Cave. This gives about 300 ft (100 m) of hydraulic head from Interstate Lake to the resurgences.

12.2.3 Piercys Mill Cave

Piercys Mill Cave is in a pocket valley and is the head of Muddy Creek. During low-flow conditions, all of the water from Sinking and Hughart Creeks resurges at Piercy Mill Cave. The stream passage (Fig. 12.13) in the cave is about 900 ft (274 m) long, and the stream appears to flow from a low passage at the base of massive breakdown. The description by Davies (1965) based on a map from 1945 shows the water entering the passage from a 10 by 24 in. (25 by 61 cm) hole near the ceiling. This was described as a "meter hole" and was believed to regulate or limit the amount of water flowing through the cave. The breakdown at the end of the stream passage may have hidden this feature —it was not observable during this study. The cave has a side passage to the east that contains large rimstone dams (Fig. 12.27) that overflow in wet conditions and this water is a small tributary to the main stream. The stream channel is lined with fine sediments that are readily mobilized at increasing water levels (see electronic map M-12.1).

Discharge from Piercys Mill Cave ranged from less than 4 cfs (0.11 cms) to 230 cfs (6.5 cms) during this study. Some

Fig. 12.6 Entrance to Piercys Mill Cave. Photo graph by W. K. Jones

Fig. 12.7 Aerial photograph showing the entrance to Piercys Cave. Photograph by W.K. Jones

Fig. 12.8 Aerial photograph showing Sinking Creek ponded to a depth of 18 ft above the pressure transducer on March 8. 2015. Photograph by W.K. Jones

Fig. 12.9 Aerial photograph showing Interstate Lake ponded to a depth of 50 ft on March 8, 2015. Photograph by W.K. Jones

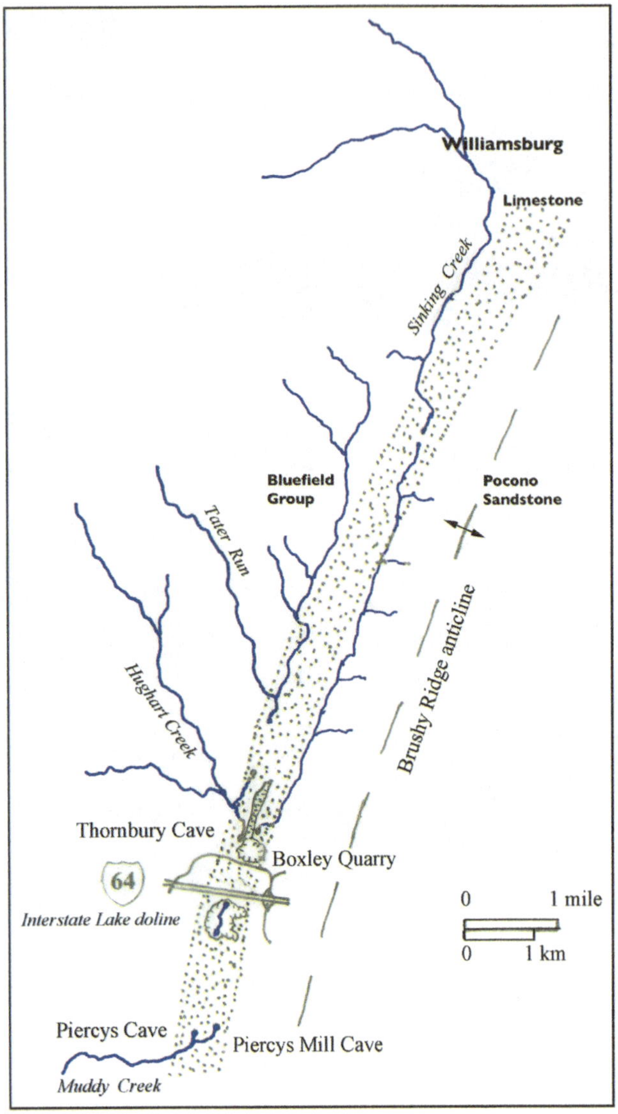

Fig. 12.10 Sketch map showing the geology of the Sinking and Hughart Creek valley

natural feature in the karst plumbing system that connects Interstate Lake to Piercys Mill Cave keeps the storm discharge at a relatively constant level with excess water shunted to Piercys Cave at a threshold of about 60 cfs (2.5 cms) at Piercys Mill Cave.

Piercys Annex Cave is located on the east side of Muddy Run just downstream from the entrance to Piercys Mill Cave (Fig. 12.5). A small amount of water discharges from this cave but this water is apparently locally derived and is not from the Interstate Lake drainage. Piercys Annex Cave is above the old dam that served as the control for the gaging station used in this study, so a small amount of the measured flow was from this cave. The flow from this cave showed minimal response to storm events and was not observed to run muddy.

12.2.4 Piercys Cave

Piercys Cave is on the north side of Muddy Creek about 2400 ft (750 m) west of Piercys Mill Cave. The first 700 ft from the entrance is a very straight stream passage trending N30E and averages 30 ft wide by 20 ft high (Fig. 12.14). The cave ends in breakdown with a couple of side leads 4,800 ft from the entrance. Piercys and Piercys Mill Caves have a complex hydrologic connection, but a physical connection of the two caves has not been found (Fig. 12.15). The two caves basically parallel each other following the stratigraphic strike and separated by about half a mile (700 m) (see electronic map M-12.2).

Low flow discharge is less than 0.30 cfs (8.5 L/s). This water is from autogenic recharge and does not appear to be connected to the sinking streams or the Interstate Lake karst window. During storm events, the water levels in Piercys Cave rise rapidly, even before the transfer threshold is reached in Piercys Mill Cave. The maximum discharge observed was 180 cfs (5.1 cms). The flow in Piercys Cave also reaches an approximate maximum discharge following storm events, but the recession starts well ahead of the recession at Interstate Lake and Piercys Mill Cave.

12.3 Hydrology

12.3.1 Tracer Tests

Several tracer tests have been conducted in this drainage basin over the years. The early tests were done under lower flow conditions and established that water from Hughart and Sinking Creek reappears in the stream exposed in the Interstate Lake karst window and then at Piercys Mill Cave. Travel time was less than 24 h under normal flow conditions. No dye was recovered at Piercys Cave during these tests (Jones 1997). Three tracer tests were done as part of this study.

Fluorescein sodium dye (400 g) was injected in Sinking Creek on April 3, 2014. The dye concentration peaked in about 16 h at Piercys Mill Cave and 28 h at Piercys Cave (Fig. 12.16). Discharge at Piercys Mill Cave was 76 cfs (2.2 cms) and 1 cfs (0.028 cms) at Piercys Cave during the dye recovery period. This was the first tracer test to demonstrate that Sinking Creek water reappears at both caves under higher flow conditions.

A test under extremely low-flow conditions was started on September 25, 2014 with fluorescein sodium (200 g) injected in Sinking Creek and Rhodamine WT (100 g) injected in Hughart Creek. Peak dye concentrations at Piercys Mill Cave occurred at 90 h for the Sinking Creek water and 140 h for the Hughart Creek water (Fig. 12.17). No dye was recovered at Piercys Cave. Discharge was estimated at about 2.0 cfs (0.028 cms) at Piercys Mill Cave.

Fig. 12.11 Digital terrain model of the study area. Imagery by Mercedes Dahler and Ira D. Sasowsky, University of Akron

Fig. 12.12 Photograph of Interstate Lake looking south from the interstate, March 21, 2015. Water level is 77 ft above the pressure transducer

The third tracer test was stared on June 11, 2015 with 200 g of fluorescein sodium in Sinking Creek. The tracer concentration peaked in 9 h in Interstate Lake (not ponded) and 29 h at Piercys Mill Cave. No dye was recovered at Piercys Cave. Discharge was about 10 cfs (0.28 cms) at Piercys Mill Cave and 0.4 cfs (0.01cms) at Piercys Cave.

Fig. 12.13 Photograph of the stream passage in Piercys Mill Cave. Photograph by Ryan R. Maurer. Used with permission

Fig. 12.14 Main stream trunk passage in Piercys Cave. Photograph by Philip Lucas. Used with permission

12.3.2 Storm Response

Stage was recorded using Onset (Hobo) water level loggers at five stations within the Sinking Creek—Muddy Creek basin from December 22, 2013 through May 5, 2014 and from October 1, 2014 through June 11, 2015. Note that all of the loggers were above the water at very low flow during the first monitoring interval, so the available record is only used to examine storm events. The stations were:

1. Sinking Creek at Boxley Quarry at Alta. Gage attached to tree below the dam and about 200 ft (60 m) upstream

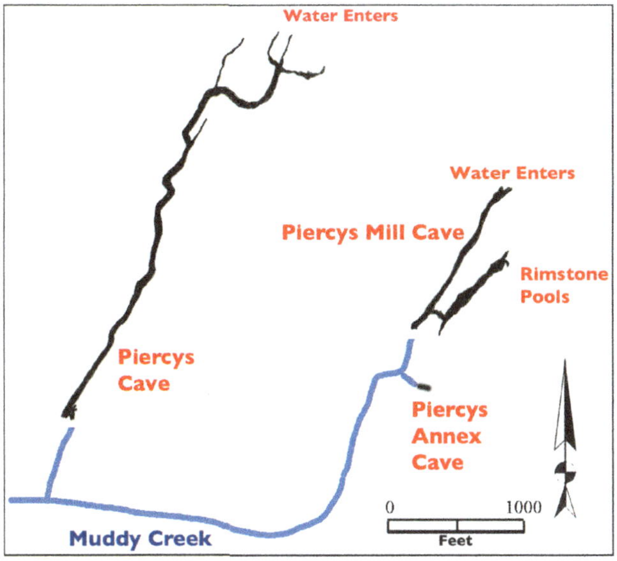

Fig. 12.15 Map of Piercys Mill and Piercys caves

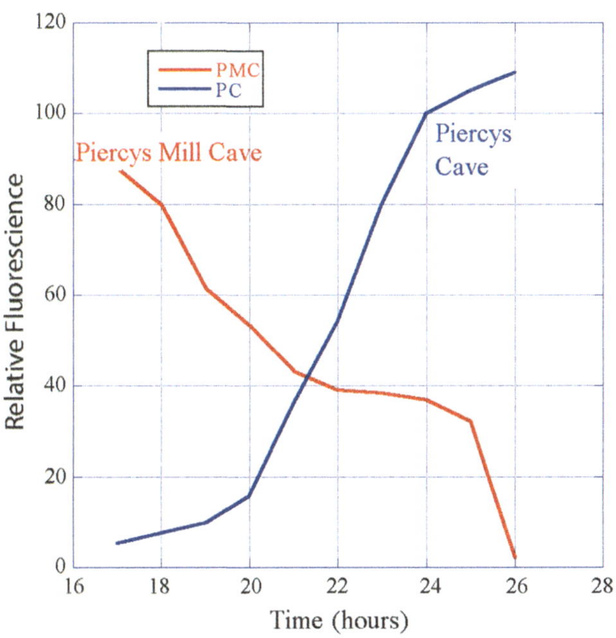

Fig. 12.16 Recovery of fluorescein sodium dye from Sinking Creek at Piercys Mill and Piercys caves. Time is in hours from dye injection. Times to peak concentrations were not captured during the sampling interval but are estimated at 16 h to Piercys Mill Cave and 28 h to Piercys Cave

from the sink point. Gage was out of the water at very low flow and in backwater from the ponor at high flow. No discharge data are available.

2. Hughart Creek at Boxley Quarry at Alta (started January 14, 2014). Gage attached to a tree about 100 ft (30 m) upstream from the ponor. The timing of response to storm events was identical to Spring Creek so only the

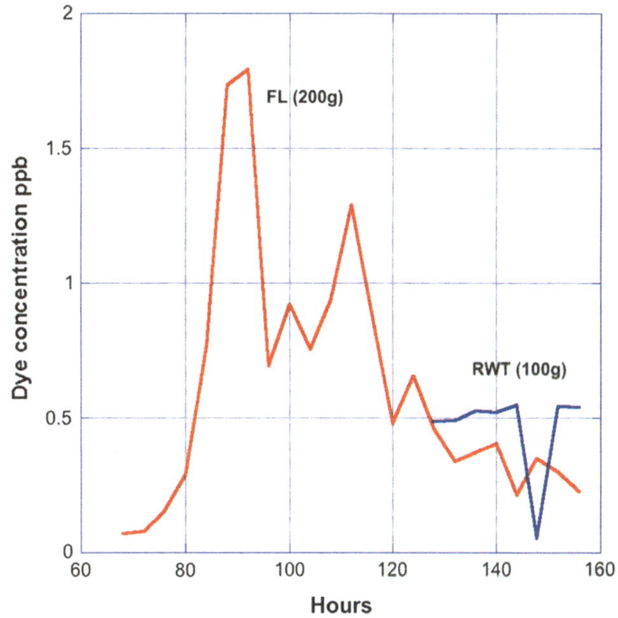

Fig. 12.17 Recovery at Piercys Mill Cave of fluorescein sodium (FL) from Sinking Creek and Rhodamine WT (RWT) from Hughart Creek. Dyes were injected September 25, 2014 under very low-flow conditions. Time is hours from dye injection

Spring Creek data were used in the analysis presented in this report.

3. Interstate Lake at Alta. Logger attached to an iron stake driven into the ground in the lower ponor at the south end of the karst window.
4. Piercys Mill Cave. Loggers for water level and specific conductance attached to upstream face of weir 500 ft (150 m) downstream from the cave entrance and downstream from Piercys Annex Cave. Four discharge measurements were used to prepare a stage/discharge rating for this site.
5. Piercys Cave. Loggers for water level and specific conductance attached just inside the cave entrance. Four discharge measurements were used to prepare a stage/discharge rating for this site.

An additional logger was placed in the old mill to correct water levels for fluctuations in atmospheric pressure and a tipping bucket rain gage was established at the mill. The recording interval at all of the stations was 15 min for the first monitoring period and 30 min for the second period.

Storm Event March 17–25, 2014: Precipitation of 0.58 in. (15 mm) over March 16 and 17 generated a storm pulse that resulted in some back flooding at the ponors on Sinking and Hughes Creeks but did not result in significant ponding in Interstate Lake where water reached a maximum depth of 1.1 ft (0.335 m). The water level peaked in Sinking Creek at 10:00 on the 20th at a gage height of 1.84 ft (0.561 m).

Water levels at the downstream stations peaked within an hour of each other about 24 h after the maximum was reached in Sinking Creek. The maximum discharge at Piercys Mill Cave was 127 cfs (3.6 cms) and 24.5 cfs (0.70 cms) at Piercys Cave.

The hydrographs for this storm show rather typical rising limbs starting a 16:30 on the 19th and peaking 17 h later at Sinking Creek and 41 h later at the other gaging stations (Fig. 12.18). The recession limb of hydrograph for Sinking Creek is fairly typical for a small storm event, but the drop in water levels is delayed at the other three stations. Interstate Lake and Piercys Mill Caves appear to hold water in storage, probably due to constrictions in the underground plumbing network partitioning the flow between the two caves. The water level in Piercys Cave starts to fall at the same time as at Interstate Lake and Piercys Mill Cave, but the last part of the recession curve is much steeper than for the other stations.

Storm Event February 3–12, 2014: Precipitation of 1.15 in. (29 mm) on February 2 through 4 generated a storm pulse with maximum stages of 23 ft (7 m) at Sinking Creek, 37.6 ft (11.5 m) at Interstate Lake, 2.32 ft (0.71 m) at Piercys Mill Cave, and 4.19 ft (1.28 m) at Piercys Cave. Maximum discharge was 185 cfs (5.24 cms) at Piercys Mill Cave and 82.5 cfs (2.34 cms) at Piercys Cave. The maximum stage at Interstate Lake was reached 80 h following the peak at Sinking Creek. Maximum water levels lagged Sinking Creek by about 8 h at Piercys Mill Cave and 16 h at Piercys Cave (Fig. 12.19).

The hydrograph for Interstate Lake shows a very uniform curve for filling and emptying the doline. The water levels at Piercys Mill Cave reached an approximate maximum range of 2.2–2.4 ft (0.67–0.73 m)) and remained in this range for 103 h followed by an abrupt and rapid recession.

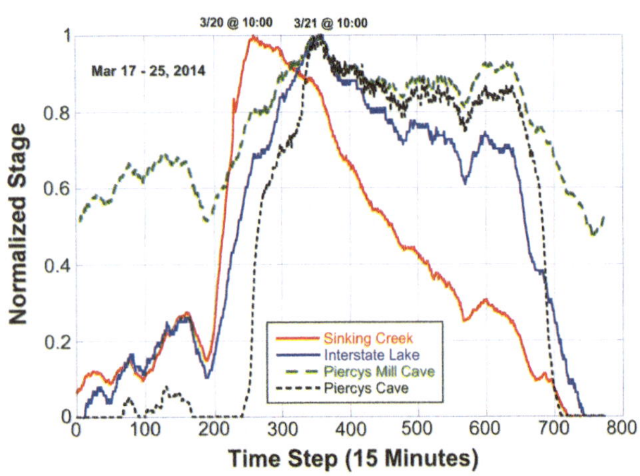

Fig. 12.18 Normalized hydrographs for a storm event that did not result in ponding in Interstate Lake. Even a relatively small storm results in a delayed recession at Interstate Lake and the two caves on Muddy Creek

The water levels in Piercys Cave quickly rose to a maximum range of 4.0–4.2 ft (1.22–2.28 m) and remained in this rage for 51 h. A gradual recession started 20 h before the peak stage was reached in Interstate Lake and two days before the recession appeared at Piercys Mill Cave.

Storm Event March 1–April 2, 2015: Multiple storms in March of 2015 generated the highest water levels observed during the study period. Three significant peaks were generated in March by 2.29 in. (58 mm) of rain on March 3–6, 1.22 in. (31 mm) on March 11–15, and 0.63 in. (16 mm) on March 27. Water levels at all the stations remained high for the month (Fig. 12.20). Multiple peaks are discernible on the hydrograph for Sinking Creek but are not obvious for the other stations. The maximum water levels for March, 2015 were:

1. Sinking Creek—41.83 ft (12.75 m) on 3/12 @ 10:30
2. Interstate Lake—79.66 ft (24.28 m) on 3/19 @ 8:00
3. Piercys Mill Cave—3.30 ft (1.00 m) on 3/11 @ 16:00; Q = 183 cfs (5.18 cms)
4. Piercys Cave—5.69 ft (1.73 m) on 3/14 @ 4:00: Q = 105 cfs (2.99 cms)

Water was ponded in Interstate Lake from March 5 through the first of April. Water levels at Piercys Mill Cave mimicked the pattern in Interstate Lake. Stage at Piercys Cave was very steady until recession began on March 25, several days ahead of falling water levels in Interstate Lake and Piercys Mill Caves.

Electrical conductivity and stage at Piercys Cave for March, 2015 (Fig. 12.21) show a typical inverse relationship with event water diluting the older more highly mineralized water during the storm. The increase in conductivity at the start of the storm is typical of the "flushing effect" where some older water is pushed out storage at the very start of the rising limb of the hydrograph.

12.4 Discussion and Conclusions

It is difficult to formulate a conceptual model for this karst basin that explains the storm response at the various gaging stations. The data sets are complicated by significant back flooding at the sinks of Sinking and Hughart Creeks. Using only the tracer test results the simple explanation is that water levels reach a certain maximum at Piercys Mill Cave due to a constriction in the conduits and excess water is shunted to Piercys Cave. If this is the full explanation then the discharge at Piercys Cave should correlate very well with the water levels in Interstate Lake. The amount of hydraulic head generated by high water levels in Interstate Lake should drive the discharge at the caves and the recession limbs of the hydrographs should

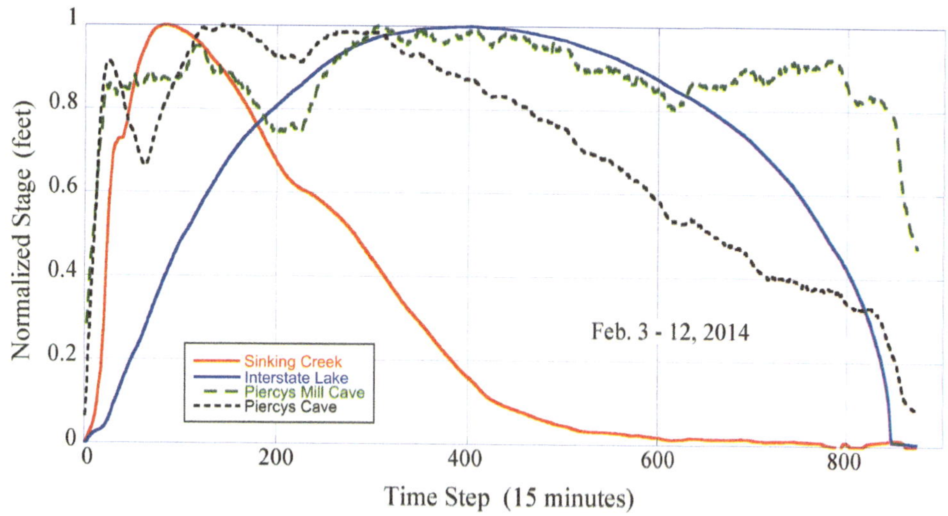

Fig. 12.19 Normalized hydrographs for a storm event that produced ponding to 37.6 ft (11.5 m) in Interstate Lake. Note the smooth filling and emptying of Interstate Lake and the recession at Piercys Cave commencing before the maxima are reached at Interstate Lake or Piercys Mill Cave

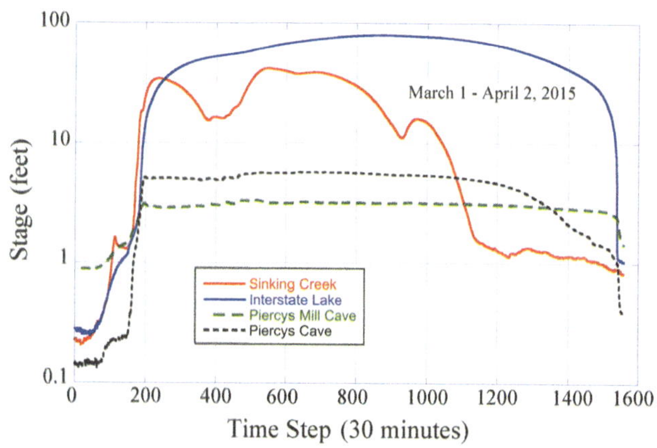

Fig. 12.20 Hydrographs (stage on log scale) for March, 2015. Multiple peaks in Sinking Creek are not reflected at the other gauging stations. The highest water level recorded at Interstate Lake during this study was 79.66 ft (24.28 m) on March 19

track the recession from Interstate Lake. This holds generally true for Piercys Mill Cave, but the storm water recession at Piercys Cave starts consistently ahead of the crests at Interstate Lake and Piercys Mill Caves (Fig. 12.19).

There is a good correlation between the maximum stage at Sinking Creek and Interstate Lake (Fig. 12.22). For a given amount of ponding in Interstate Lake, there is some correspondence to the maximum discharge at the two caves (Fig. 12.23) and to the water levels at Sinking Creek (Fig. 12.24). However, the timing of the maxima in Interstate Lake and the two springs do not match very well. There is also ponding at the two sinking streams in the quarry, so the water levels at the caves, especially at Piercys Cave, may be more closely correlated with Sinking and Hughart Creeks than with Interstate Lake (Fig. 12.25; Table 12.1).

Some if not all the dyes injected in Sinking and Hughart Creeks pass through Interstate Lake so the flow should be partitioned between the two caves. If the flow at the two caves is only related to the water level in Interstate Lake then the storm hydrographs should show the water levels at the four gaging stations rising at the same time with some lag for the downstream stations following the crest at Sinking Creek. The start of the rising limb of the Piercys Cave should be somewhat delayed from Piercys Mill Cave or show an acceleration when the apparent 90 cfs threshold is reached at Piercys Mill Cave. However, this is not the pattern for most of the storm events.

If there is no significant ponding in Interstate Lake or backwater at the stream sinks, the storm hydrographs (Fig. 12.18) appear typical with maxima reached almost simultaneously at the downstream stations although the recession limbs show a gradual start and then a sharp decline in water levels. The hydrographs for the two caves appear to follow the water levels in Interstate Lake.

The storm response for events that triggered significant ponding in Interstate Lake was much different from the pattern show in Fig. 12.18. The hydrograph for the storm event starting February 3, 2014 (Fig. 12.19) is representative:

1. Water levels in both caves rise rapidly in response to rainfall and show a first crest ahead of the crest in Sinking Creek. This may be due to back flooding at the stream sinks delaying the maximum water levels at these sites.
2. The rise and fall of the water levels in Interstate Lake are remarkably smooth and uniform on both the rising and falling limbs of the hydrograph.
3. The maximum water level at Piercys Cave is reached shortly after the crest of Sinking Creek and recession starts well before maxima are reached in Interstate Lake or Piercys Mill Cave.

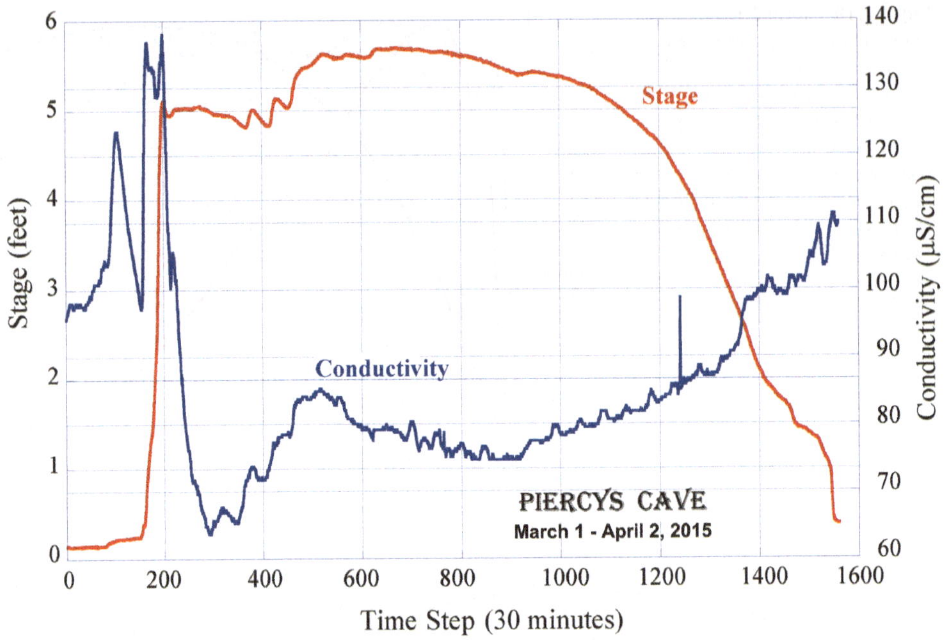

Fig. 12.21 Graph showing stage and electrical conductivity at Piercys Cave for March, 2015. Note the initial rise in conductivity at the start of the storm hydrograph as older more mineralized water is flushed from storage

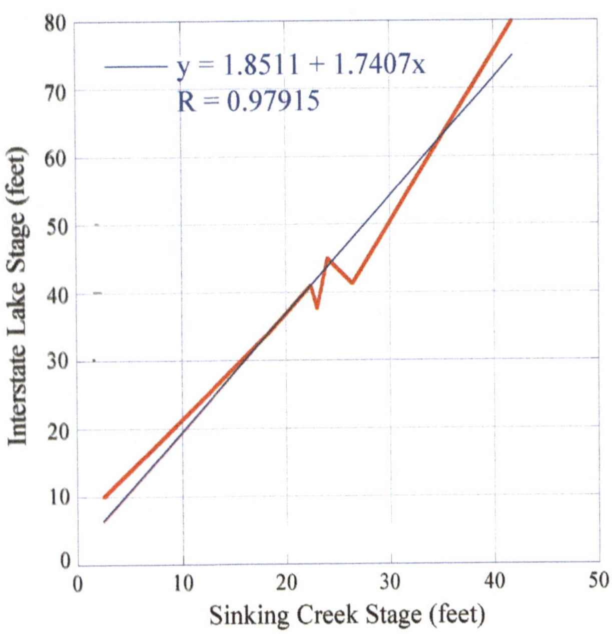

Fig. 12.22 Graph showing relationship between maximum stages at sinking creek and Interstate lake

4. When the water level in Interstate Lake dropped to about 18 ft (5.5 m) from a maximum of 38 ft (11.6 m) rapid recession was triggered at both caves.

The early storm response at the different stations is generally straight forward. An example of the first 80 h of a storm starting March 2, 2015 is presented in Fig. 12.26. The first response in water levels at the gaging stations was 11 h at Sinking Creek, 16 h at Interstate Lake, 16.5 h at Piercys Mill Cave, and 25 h at Piercys Cave. The threshold for the excess water arriving from (or at least diverted from) Piercys Mill Cave to Piercys Cave appears at the 63-h mark and represents a discharge of about 60 cfs at Piercys Mill Cave.

The storm response at Piercys Mill Cave was generally well paired with the water levels in Interstate Lake. However, the hydrographs for Piercys Cave show little correspondence with Interstate Lake. The storm hydrographs consistently show an early start to recession at Piercys Cave. This occurs before recession has started at Interstate Lake or Piercys Mill Cave and on at least one occasion before the crest was reached in Sinking Creek. Recession started at Piercys Mill Cave after Sinking Creek crested for most of the stronger storm events. The water levels in Sinking Creek had fallen to the 9–13-ft range (3–4 m) range before recession started at Piercys Cave for four of six significant storms.

The tracer tests indicate that water from Hughart and Sinking Creeks pass through Interstate Lake and resurge at Piercys Mill Cave under low-flow conditions and at both caves at higher levels. Based on the storm response of the system, it appears that Piercys Cave has a limited (if any) connection to Interstate Lake and much of the flow is directly from the sinking streams. Back flooding at the two sinking steams creates considerable storage and delayed recessions without even considering the volume of water stored in Interstate Lake during storms.

Both Piercys Mill and Piercys caves are significant caves even without considering the hydrology of the area. A side

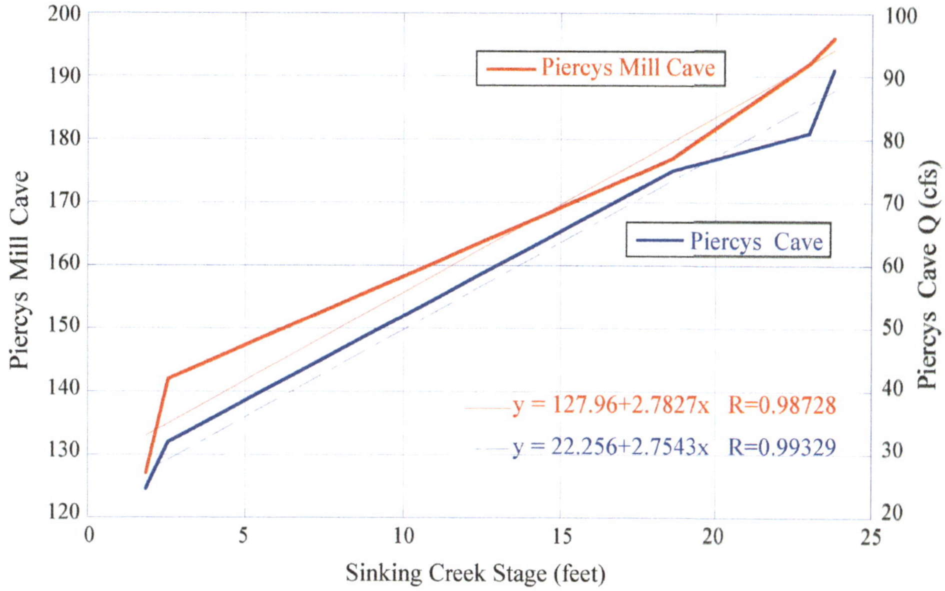

Fig. 12.23 Rating curves for discharge at Piercys Mill and Piercys caves based on stage in Interstate Lake

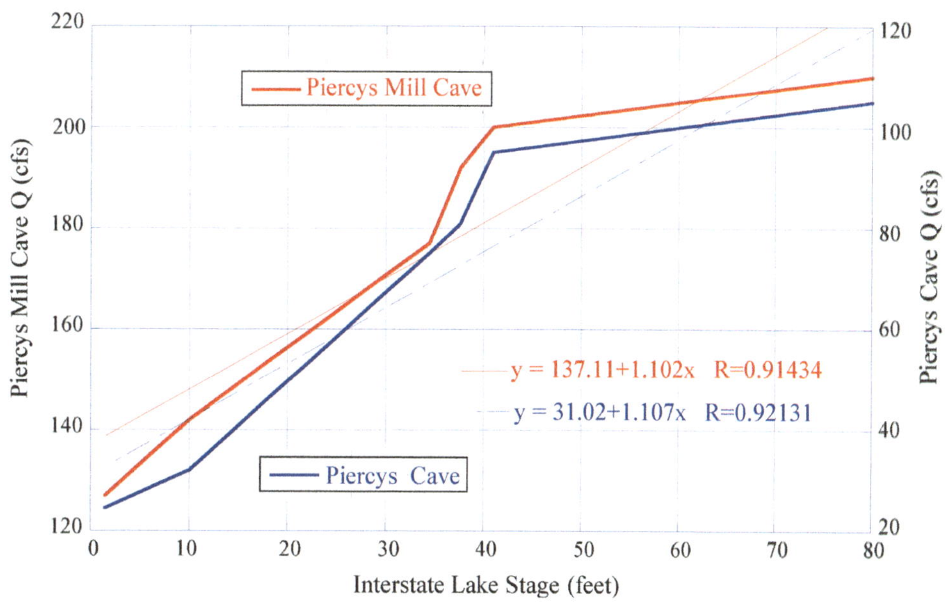

Fig. 12.24 Rating curves for discharge at Piercys Mill and Piercys caves based on stage at Sinking Creek. Data from 2015 were not used because the Sinking Creek gage had been replaced and the earlier stage datum was not available

passage to the east of the stream in Piercys Mill Cave contains some of the largest rimstone dams in West Virginia (Fig. 12.27). The main stream trunk in Piercys Cave is unusually straight for over 1000 ft and is a classic underground steam—especially at high water levels (Figs. 12.28 and 12.29).

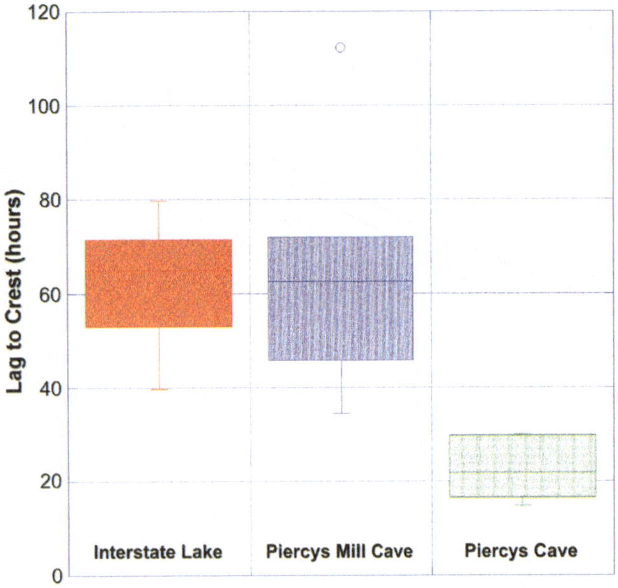

Fig. 12.25 Box plots showing lag times (hours) following storm crests at Sinking Creek

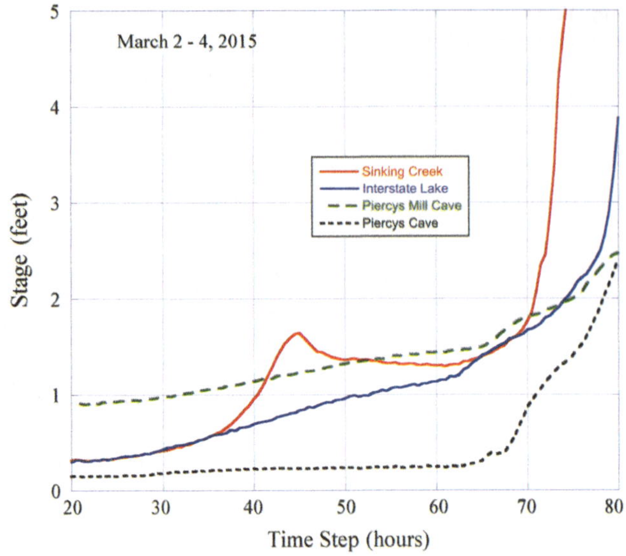

Fig. 12.26 Hydrographs showing early storm response at the gaging stations

Table 12.1 Summary statistics for lag times in hours to maxim A at downstream gaging stations following the maximum stage at Sinking Creek

	Interstate lake	Piercys mill cave	Piercys cave
Minimum	39.75	34.5	14.75
Maximum	79.75	112.25	30
Points	6	6	6
Mean	62.333333	64.958333	22.416667
Median	65	62.625	21.75
Std. deviation	14.305302	26.870717	6.4665808
Variance	204.64167	722.03542	41.816667
Std. error	5.8401151	10.969924	2.6399705
Skewness	−0.45227889	0.79760219	0.13438762
Kurtosis	−0.889778	−0.23698098	−1.5049324

Fig. 12.27 Photograph showing rimstone dams in Piercys Mill Cave. Photograph by David Bunnell. Used with permission

Fig. 12.28 Photograph showing the normal maximum water level, about head high, in Piercys Cave. Photograph by David Bunnell. Used with permission

Fig. 12.29 Photograph showing main stream trunk passage in Piercys Cave. Photograph by David Bunnell. Used with permission

Acknowledgements Support for this study was provided by a grant from the Cave Conservancy of the Virginias. We wish to thank the various landowners and Boxley Quarry for granting access to the study sites and Carroll Bassett for helping with the installation of the data loggers. Philip C. Lucas, David Bunnell, and Ryan Maurer provided some of the in-cave photographs.

References

Davies, W.E. 1965. *Caverns of West Virginia*, vol. XIX A, 119–122. Morgantown: WV Geological and Economic Survey.

Davis, J.C. 2002. *Statistics and data analysis in geology*, 638. New York: Wiley.

Field, M.S. 2010. Simulating drainage from a flooded sinkhole. *ActaCarsologica* 39 (2): 361–378.

Haan, C.T. 1977. *Statistical methods in hydrology*, 378. Ames, IA: The lowa State University Press.

Jones, W.K. 1997. *Karst hydrology Atlas of West Virginia*, vol. 4, 114pp. Charles Town, WV: Karst Waters institute, Special Publication.

Kovačič, G. 2010. Hydrogeological study of the Malenščica karst spring (SW Slovenia) by means of time series analysis. *Acta Carsologica* 39 (2): 201–221.

Kresic, N. 2013. *Water in Karst*, 708. New York: McGraw Hill.

Mangin, A. 1984. Pour une meilleu reconnaissance des systèms hydrologiques à partir des analyses corrélatoire et spectrales. *Journal of Hydrology* 67: 25–43.

Petrič, M. 2002. *Characteristics of recharge-discharge relations in karst aquifers*, 154. Postojna, Slovenia: Carsologica.

Price, P.H., and E.T. Heck. 1939. *Greenbrier County*, 846. Morgantown, WV: West Virginia Geologic and Economic Survey County Report.

Turk, J. 2010. *Dynamics of underground water in the karst catchment area of the ljubljanica springs*, vol. 11, 136pp. Postojna, Slovenia: Carsologica.

Organ Cave

13

Raymond Cole

Abstract

Organ Cave with a length greater than 38 miles (62 km) is the second longest cave in the Greenbrier Karst. It is located beneath a fragment of the Big Levels karst surface bounded by the Greenbrier River to the north, Second Creek to the south and west, and shales and sandstones to the east. Organ Cave is a contact cave with the active drainage cut into the underlying Maccrady shale. It displays a dendritic drainage system with multiple streams merging to a common resurgence on Second Creek. The geologic setting of the cave, both stratigraphic and structural, is presented in considerable detail. The main trend of the cave is along the Caldwell Syncline. Chert horizons in the Hillsdale Limestone and steep folding and minor faulting have all served to guide passage development. The cave has a long exploration history extending back to the mid-twentieth century.

13.1 Introduction

The Organ Cave Plateau is located in Greenbrier County under the community of Organ Cave (Fig. 13.1). To date, ten entrances to the system have been discovered, and 38.452 miles of passage have been mapped. The Organ Cave System is one of 46 caves, which have been found on the Organ Cave Plateau. The surface of the plateau is generally at an elevation of 2200 ft above sea level; however, some residual hills rise another 100–200 ft and several sinkholes are over 100 ft deep. The Greenbrier River and its tributary Howard Creek bound the plateau on the north. The Greenbrier River also forms the plateau's western edge. Second Creek and its tributary Carpenter Creek bound the plateau on its south side. These boundaries cut into the plateau causing escarpments of 400–600 ft. White Rock Mountain runs in the north–south direction at elevations of up to 3100 ft above sea level and forms the eastern boundary.

The Organ Cave Plateau is underlain by limestones of the Middle Mississippian Greenbrier Group. On the Organ Cave Plateau only its two lowest members, the Hillsdale and Sinks Grove Limestones, remain; the upper units have been completely eroded away. The limestones are about 200 ft thick in the area of the Organ Cave System. The cave-bearing limestones lie stratigraphically on the clastic red sandstone and shale sequences of the Maccrady and Pocono Formations which outcrop on the eastern surface of the plateau and along its north and west escarpments.

The Organ Cave Plateau is located near the eastern edge of the Allegheny Plateau physiographic province. Structurally, the plateau consists of gentle anticlines and synclines that are the result of the Appalachian orogeny. The more severely folded rocks of the Valley and Ridge Providence are only nine miles to the east. Locally, the main stream passage (Hedricks and Big Canyon) of the Organ Cave System is formed along the Caldwell Syncline, which plunges toward Second Creek at approximately 4° along a bearing of N30°E. Near the structural axis, the limestones are occasionally steeply dipping and overturned. There is also local faulting, which has influenced cave development.

Electronic supplementary material
The online version of this chapter (doi:10.1007/978-3-319-65801-8_13) contains supplementary material, which is available to authorized users.
Raymond Cole: Deceased.

R. Cole (✉)
Alexandria, VA, Egypt

W.B. White (ed.), *Caves and Karst of the Greenbrier Valley in West Virginia*,
Cave and Karst Systems of the World, https://doi.org/10.1007/978-3-319-65801-8_13

Fig. 13.1 Location map of the Organ Cave Plateau. Segment of US Geological Survey 1:100,000 Lewisburg sheet. Note this is a metric map with 50-m contour interval

The Organ Cave Plateau receives about 37 in. of precipitation annually which is evenly distributed throughout the year. Small surface streams drain from the shale into blind valleys and dolines and sink near the limestone contact. This drainage and direct percolation circulates through an underground drainage network to exit at a series of springs along Second Creek and the Greenbrier River. The underground streams have downcut 10–30 ft into the shale at a few points within the caves, but the shale generally acts as a relatively impermeable boundary.

Various fauna are found in the caves on the plateau, and the Organ Cave System is the type locality of several troglobitic species of crustaceans and snails. A variety of vertebrate fossils have also been discovered including those from armadillo, mastodon, and saber-toothed cat. Organ Cave was mined for saltpeter during the early 1800s and the American Civil War. While accounts of the cave date back to the late 1700s, modern exploration did not begin until 1948. Much of the early exploration of the Organ Cave System was conducted by members of the Charleston Grotto of the National Speleological Society (NSS) and the West Virginia Association for Cave Studies (WVACS). Members of the District of Columbia Grotto (DCG) of the NSS continued the exploration and survey of the system during the 1970–1982 time period. Since then members of both organizations have worked to further the project. As a result of these efforts, detailed maps of the caves were developed.

13.2 Historical Background

13.2.1 History of the Region

Organ Cave lies in the Irish Corner District of Greenbrier County named after its early settlers many of whome were Scotch Presbyterians from the Province of Ulster in the northern part of Ireland. They settled the area just prior to the start of the American Revolutionary War in 1776 (Humphreys 1926; Rice 1970). Prior to their settlement, the land was a popular hunting area for the Indians who camped here in large numbers. The French claimed the region before the French and Indian War (1756–1763), but did not establish settlements. There were repeated attempts by the English to settle Greenbrier County between 1749 and 1768, but the French and Indians killed or drove the settlers out (Stuart 1971). In the early 1800s saltpeter, an essential ingredient of gunpowder was mined from Organ Cave. The Old Virginia Saltpetre Route terminated at Uniontown (now Union, WV), and "it is very probable that Uniontown was a collecting and trans-shipping center" for Organ Cave and the other saltpeter-producing caves in the area (Faust 1964). Prolix (1837) describes a visit he made into the cave and mentions the existence of a saltpeter section. The cave's 1812 Room, just off the main entrance passage, is the site of saltpeter mining reputed to have occurred during the War of 1812. It still contains the decaying remains of several saltpeter vats.

The Saltpetre Room on the current commercial tour route was used for saltpeter mining by the Confederacy during the American Civil War (1861–1865). There are over 37 vats in the room, and some of them are in excellent condition. During the war, water was carried in wooden buckets from the Main Organ Stream and dumped into the vats. After dissolving the calcium nitrate in the saltpeter dirt, the water was collected in wooden troughs and carried out of the cave in buckets and placed in kettles along the entrance road. This mother liquor was mixed with the leachings from wood ashes, filtered, and then boiled to crystallize potassium nitrate, an essential ingredient of gunpowder (Sively, G., personal communication, 1985).

Cave visitors of the early 1800s not only mined saltpeter, they explored the cave. Several dates can be found written on the wall of the Upper Stream Passage more than a half mile from the Organ Entrance.

13.2.2 Organ Cave as a Tourist Attraction

Organ Cave was named after a calcite formation which resembles a pipe organ in a large auditorium (Fig. 13.2). The speleothem is about a half mile into the cave and is easily accessible by tourists. Early descriptions of it indicate that it was formed by "white stalactites" and gave "an exact reproduction of a large pipe organ-at least by striking on the different pipes notes of remarkable purity and strength are reproduced" (Cole 1917).

The first recorded owner of Organ Cave was John Gardner who was granted the land from the Commonwealth of Virginia in 1783. Ownership of the cave passed to Thomas Cox in 1819 and then via his heirs to John Rogers in 1822. By then, the resorts at White Sulphur Springs (Greenbrier County) and Salt Sulphur Springs (Monroe County) were well established with turnpikes between them and the tide-water counties of eastern Virginia. Stage coaches regularly ran between these resorts hauling both passengers and mail. Often the coach would stop at John Roger's Organ Cave. One visitor of that era wrote:

> The mouth of the Organ Cave is situated nearly under the road, at the bottom of a deep ravine, which seems as if it had formerly discharged a large stream of water into the cave… The approach is very romantic, descending the steep and wooded side of the ravine, by a zig-zag path which leads by a easy slope to the black and yawning chasm. The preparation for exploring one of these cyclopean caves consists of a supply of pitch-pine sticks, faith in your guides, and folly in yourself. The sticks are about two feet long and each one as thick as a thin finger; fifteen or twenty of which being held together in the hand, and fired at the upper end, make the best of torches, will burn bright for two

Fig. 13.2 The Rock Organ with Ray and Susan Cole. The Rock Organ is the best known speleothem in Organ Cave System, and the cave is named after it

> hours, and distinctly show the floor, sides and roof of the cave through the palpable obscure. Little magazines of sticks are judiciously left at intervals of a Quarter of a mile, as you penetrate deeper and deeper into the bounds of the land, to replenish from time to time the moribund luminaries. (Prolix 1837)

John Rogers sold the land containing the cave in 1836 to James Robinson who willed it to his daughter Elizabeth Boone, wife of George Boone. They in turn deeded the property to their son, James H. Boone, in 1878. While primarily a farmer, Boone opened the cave commercially in the early 1900s when every third visitor was given a candle to light the way for groups walking through the cave. In 1914 Boone completed the development of built-up pathways through the commercial section of the cave and installed Edison light bulbs powered by a Delco generator and 72 storage batteries (Sively, personal communication, 1983). Cole (1917) later wrote:

> Organ Cave has been made an object of interest to the tourist at considerable cost and labor by its owner, J. H. Boone. A perfect lighting system for lighting the cave has enhanced the scenes of the underground auditorium, some of which are extensive. Good walks lead to any and all points of interest along shore lines of miniature lakes, and through many long subterranean avenues and vast underground caverns. Hundreds of electric lights illuminate the darkened passageways and some thousands of dollars have been expended to make visits here of worth and long to be remembered. As a natural wonder it has to be seen to be appreciated.

James Boone sold the cave in 1926 to George Carter and Seth Sively of Sweet Chalybeate, VA, who moved onto the land in early 1927. While they were primarily farmers, they opened the cave each summer for tourism. The cave was also used for community dances and church services. The photograph of the Organ Cave Entrance circa 1927 (Fig. 13.3) shows the entrance building used by tourists to store their umbrellas (it was very fashionable then to carry one) and obtain candles and candle holders for their trip into the cave. By this time a dam had been installed in the commercial section of the cave to create a lake for trout as a tourist attraction. When George Carter died in 1942, his share of the ownership was willed to his daughter, who was then the wife of Seth. Upon the deaths of Seth in 1964 and Edna in 1978, ownership passed to their children, George Sively and Lorraine Sively Withrow. George Sively and his wife Lee continued to open the cave to tourists each summer, and Lee was principally responsible for the tourist operation. After the death of George Sively the property was put up for auction in 1994. Local residents Janie and Sam Morgan purchased the cave entrance and an adjacent parcel of land. The Morgans made improvements in the cave lighting, Civil War interpretation, and built a new gift shop and have kept the cave open for commercial tours. They also offer trips into the wild portion of the cave.

In the 1950s the US Government built a storage shed in the Organ Cave Entrance room and stocked it with survival

Fig. 13.3 Early entrance photograph circa 1927. Photograph reprinted from a postcard

supplies so it could serve as a shelter in case of nuclear war. However, the supplies soon degraded in the cave environment. More recently, the entrance room has been the scene of community dances and slide shows. The tourist entrance known as Organ Cave is now known to be one of the ten entrances to the Organ Cave System. None of the other entrances has been developed or used for any purpose other than caving, and most are rarely entered.

13.2.3 History of Cave Exploration

While areas of Organ Cave were well known in the 1700s, there is no record of systematic exploration prior to the late 1940s, when members of the newly formed Charleston Grotto of the National Speleological Society became interested in the Organ Cave Plateau. They began by exploring some of the other local caves in the hope that one of these would lead into an extensive system. On a photograph trip for *Life* magazine in 1946, they visited Hedricks Cave which was then considered only a small but pretty cave. During that trip Alice Williams (in a bathing suit for camera) negotiated a wet crawl and returned to report that the cave opened up and continued (Rutherford and Handley 1976). Bob Handley and Bob Flack returned to Hedricks Cave on December 12, 1948, for another photograph trip and to check the passage Alice Williams had found. During that trip Bob Handley found a bypass around the "water hazard" which had previously marked the end of the cave and the two cavers proceeded south along Hedricks Stream past its junction with the Masters Stream and "downstream to point where [the] ceiling is very low," the Hedricks Sump.

The following April, Bob Flack and Bob Handley returned with others to survey from the Hedricks Entrance to the Masters–Hedricks Junction. They then proceeded to explore up the Masters Stream until they found the North Entrance (Masters Entrance).

> [Then] on May 23, 1949 the three Charleston cavers Bob Flack, Bob Barnes and Bob Handley accomplished a long-hoped-for feat and virtually the 'dream of every caver' when they emerged from the commercial Organ Cave entrance amidst electric lights, tourists and the consternation of the cave owners after a three and one half hour underground journey from the Hedricks Cave entrance. [The] one way journey was made in record time through a fast trip down the known passage to its end, and a fortunate choice among the three possible routes. Added to the known sections of Organ Cave, the Hedricks mileage makes this network probably the most extensive known cave east of Kentucky. As Mac McGriff, Sarah McFarland, Charles Wray and Don Engel will testify after an 11-hour trip 'looking out' [for] additional possibilities, there are still a wealth of unexplored passages awaiting further investigation. (Flack 1949)

The discoveries continued at a fast pace. Over the July 15, 1949, weekend Bob Barnes, Don Engel, Bob Flack, Bob Handley, Mac McGriff, Sara McFarland, and Glen Musser found the Flack Room and Floyd Collins Avenue.

Bob Handley was one of the original explorers of the Organ Cave System. He explored the Hedricks Maze section of the cave in 1949 and is shown climbing Handley's Climb (Fig. 13.4), a 20-foot flowstone-covered vertical wall leading to The Great White Way, Handley Room, and the Waterfall Room in a series of trips. That October, Bob Flack, Bob Handley, and Siegel Workman discovered and explored the Cyclops Hall area, and in November Bob Flack, Bob Handley, Bert Ash, and Earl Walters explored Hedricks Maze including the Sarver Room and Bone Room. A fourth entrance, Sively #2, was rediscovered, having been known during the early 1800s (Humphreys 1926), at the end of a passage above the Rock Organ. The stream in Foxhole Cave (which lies above the Organ Cave System) was also dye-traced and found to feed into the waterfall of the Waterfall Room (Rutherford and Handley 1976). Thus ended the initial phase of exploration (Fig. 13.5).

Fig. 13.4 Bob Handley ascending Handley's Climb circa 1949. Photograph by Flack (1949)

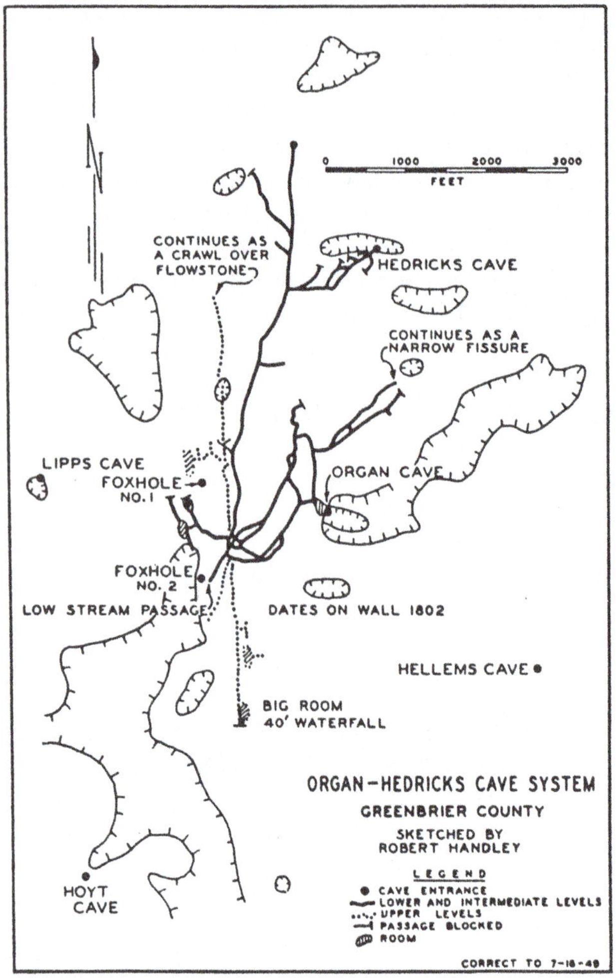

Fig. 13.5 The Organ–Hedricks Cave System as of July 1949. From Davies (1958)

The second phase began in 1951, when Handley, this time accompanied by John Rutherford, made more major discoveries. First in Humphreys Cave (for several years erroneously referred to as Erwin Cave) and then in Lipps Cave, lying to the west of Organ–Hedricks, they found drops which led to 40-ft-high stream passages. The Humphreys and Lipps trunk passages were quickly found to form a second major drainage net roughly parallel to, but a considerable distance west of, the Organ–Hedricks drainage. A second Lipps Entrance (Deems Entrance) was found nearby. Upstream exploration in Lipps–Humphreys was blocked by massive breakdown in each passage. About this same time, additional passage, continuing beyond the Waterfall Room, was found in Organ–Hedricks by David Bowen. This passage is a continuation of the Foxhole Cave drainage route, which formerly crossed the Waterfall Room at ceiling level. It quickly led to the Bowen Room, containing a 100-ft drop down the face of a huge flowstone cascade. At the bottom of the drop, the Big Canyon was found. This was a 40–60-ft-high and 6–10-ft-wide passage, which headed south for about 3500 ft until it was blocked by a sump. Ironically, David Bowen never descended the drop. Near the end of Big Canyon, a long rimstone crawl, named the Meatgrinder, led to the Rutherford Room and crawls beyond. Although the promising leads began to dwindle once again, the concept of the Greenbrier Caverns (Organ Cave System) had now been born needing only a connection of Lipps–Humphreys with Organ–Hedricks.

The third phase of work began with mapping in Organ–Hedricks by Earl Thierry. Because many of the early explorers had moved away from the area, the search for the connection proceeded slowly until Hugh Jones and Conrad Revak, two budding young cavers from Charleston, had to be rescued from the drop in Lipps after becoming stuck. Undaunted by their awkward introduction to the system, they eagerly joined Handley in the search for the connection by a systematic check of all leads in the northwest part of Organ–Hedricks. Working off the Handley Room, they ultimately pushed a series of rocky crawls westward into Jones Canyon, another stream passage which lay between and parallel to the two parts of the presumed system. Then, working through a maze of crawls, the three cleared a breakdown choke and finally made the connection with Lipps in September of 1958 (Rutherford and Handley 1976).

By now, the cave system was well known in the caving community and caving traffic was at an all time high. Unfortunately, some cavers failed to treat the landowners and their property with respect; as a result, caver–landowner relations gradually deteriorated until most of the entrances were closed.

In the summer of 1958, several cavers left the rails off the pit opening that is the Hedricks Entrance. A calf fell in causing the landowner inconvenience and loss of property. Subsequently, friends of the owner wired a bale of old fence wire securely in the entrance to prevent access—in lieu of it being sealed. In the summer of 1960 several determined thoughtless spelunkers cut their way through the reel of wire and entered the cave. Other cavers camped on the owner's land without permission. As a result of these incidents, the owner has sealed the cave permanently with fieldstone and concrete.

In 1959 cavers camped in and near the Lipps entrance without permission leaving a mess. On another occasion, inexperienced spelunkers stranded themselves down a pit and required rescuing. Because of incivilities at the commercial Organ Entrance, the owners installed a gate and lock. In the winter of 1960 cavers entered the Organ Entrance without permission and, upon emerging, built a bonfire adjacent to the souvenir stand (Stafford 1961).

In 1964 Hartline (1964) told of other incidents and reported that the local landowners had closed the entrances to Organ, Foxhole, Hedricks, Lipps, and Hellems due to the actions of discourteous cavers since 1958.

In 1961 the West Virginia Association For Cave Studies (WVACS) was formed by many of the cavers who had been active in exploring the Organ Cave System (which they called Greenbrier Caverns) to promote the exploration and study of all West Virginia caves. During WVACS's early years exploration in Organ Cave System was suspended until landowner relations improved. In 1964 WVACS members Charles Maus and Henry Stevens quietly resumed surveying in the less heavily traveled Lower Lipps portion of the cave. Their goal was to resurvey the system so as to bring the survey up to WVACS's new standards. The strategy they adopted was to first survey the trunk passages in order to create a multiple loop baseline from which to branch out. In the process they found new passage to the west of the Lower Lipps Stream and a connection from the lower end of Jones Canyon to the Rutherford Room and the Big Canyon. Surveying continued well into 1970. Over 16 miles of passage was resurveyed in comparison with the 23 miles mapped in the previous survey effort. During this period, John Fisher, Jack Gravenmier, and Bob Amundson maintained the survey data in a computer database.

The year of 1970 was a period of transition between the survey efforts of WVACS and DC Grotto members. Originally, DC Grotto members led by Pat Moretti were just going to survey the route between the commercial entrance and Lipps, but they became more interested in the cave. After much frustration and negotiation between the two groups, they met in September 1970 and agreed to a joint venture to study the cave system with full cooperation and exchange of working teams. The goal was to publish a joint map in a future West Virginia Speleological Survey Bulletin (Stevens 1982). As part of the agreement WVACS turned over its current survey data, computer program, working maps and briefed DC Grotto members on exploration potentials. However, because of a lack of a mutual spirit of cooperation, the two organizations did not actively join forces. Communication remained open on a limited basis during the early 1970s, and the joint venture was not consummated until the next decade (Fig. 13.6).

Trips to Organ Cave by DC area cavers were frequent throughout 1970 with large survey efforts occurring every other week. But as winter approached, progress was threatened by cold weather which made camping in the Organ Cave parking area increasingly unattractive. Then George Sively offered the use of an old tenant house in his back pasture. The DC area cavers moved into the house that December and started the refurbishment effort which was to last through most of 1971 (Fig. 13.7). As 1970 ended, Pat Moretti announced that DC Grotto members had mapped 33 miles of the Organ Cave System. Throughout 1970 and 1971 the principal focus of DC cavers was the resurvey of the cave's main passages. This fifth phase of exploration was led by Pat Moretti (then DCG chairman), Bruce Boss, Ray Cole, Dick McGill, and John Walker. In February 1971 the National Speleological Society recognized their efforts and chartered the Organ Cave Project with Pat Moretti and Bruce Boss as co-chairmen.

By July 1972 the main passages had been resurveyed and exploration of North Upper Level Organ (NULO) was well underway. Bruce Boss and Fred Stork's trips to Erwins Cave during this period were especially grueling as they surveyed to the base of the Third Waterfall. Gradually, many of the original DC cavers became less active and were replaced by newcomers Mike Dyas, Dave Engel, Jim Gildea, Larry Lilly, Dick Nigon, Tommy Shifflett, Cady Soukup, and Paul Stevens. Bruce Boss and Ray Cole became project co-chairmen in September, 1973, when Moretti resigned to take a job out of the area; Boss also resigned later that year when he moved to California. Late in 1973 Purgatory Way was surveyed and the Erwins and Sively #3 entrances were connected to the system. Throughout 1974 the project continued to be active with the Four Domes area of Lower Lipps one of its principal focuses. Jim Borden and Mike Dyas surveyed most of Lewis cave in the 1974–1975 period. As unprocessed survey data piled up, the project began to lose momentum. Early in the period Moretti quickly distributed copies of the map of recently surveyed passage. After he moved from the area, those exploring the cave no longer saw the fruits of their labor. The crisis was solved when Dick McGill and Paul Stevens replotted all the data in the project's computer database and, with the help of the others remaining on the project, pasted the individual line plots together to construct a new baseline plot of the cave. Then, over the next several years the map of the cave was redrawn based upon the integrated line plot. Missing survey data were duplicated by resurveying the passage. In 1975 Paul Stevens was named project co-chairman. 1976 was a very active year. George Dasher led the resurvey of Lipps Maze, the northern parts of Jones Canyon and Left Hand Passage area. Rich Hall, Liz Hall, Paul and Lee Stevens discovered the Great Escape entrance. Forrest Wilson and Sheck Exley dove the 270-foot-long Bowen Sump with scuba gear and surveyed several thousand feet of virgin passage. Rich Hall, Lee Stevens, Tom Vines and John Rue followed by connecting the Borehole with NULO on New Year's eve, opening up NULO to further exploration. The following year Sheck Exley and David Morrow repeated the Bowen Sump dive and surveyed to David's Dungeon, an impenetrable sump. Most of the remaining activity in the cave concentrated on field checking the map and resurveying areas where the map was poor. The resurvey of the Revak Room area was one of the largest efforts, and it was led by Tom Kaye. The project was also very active at home with evening and weekend work meetings to sketch cave passages using baseline *Calcomp* plots.

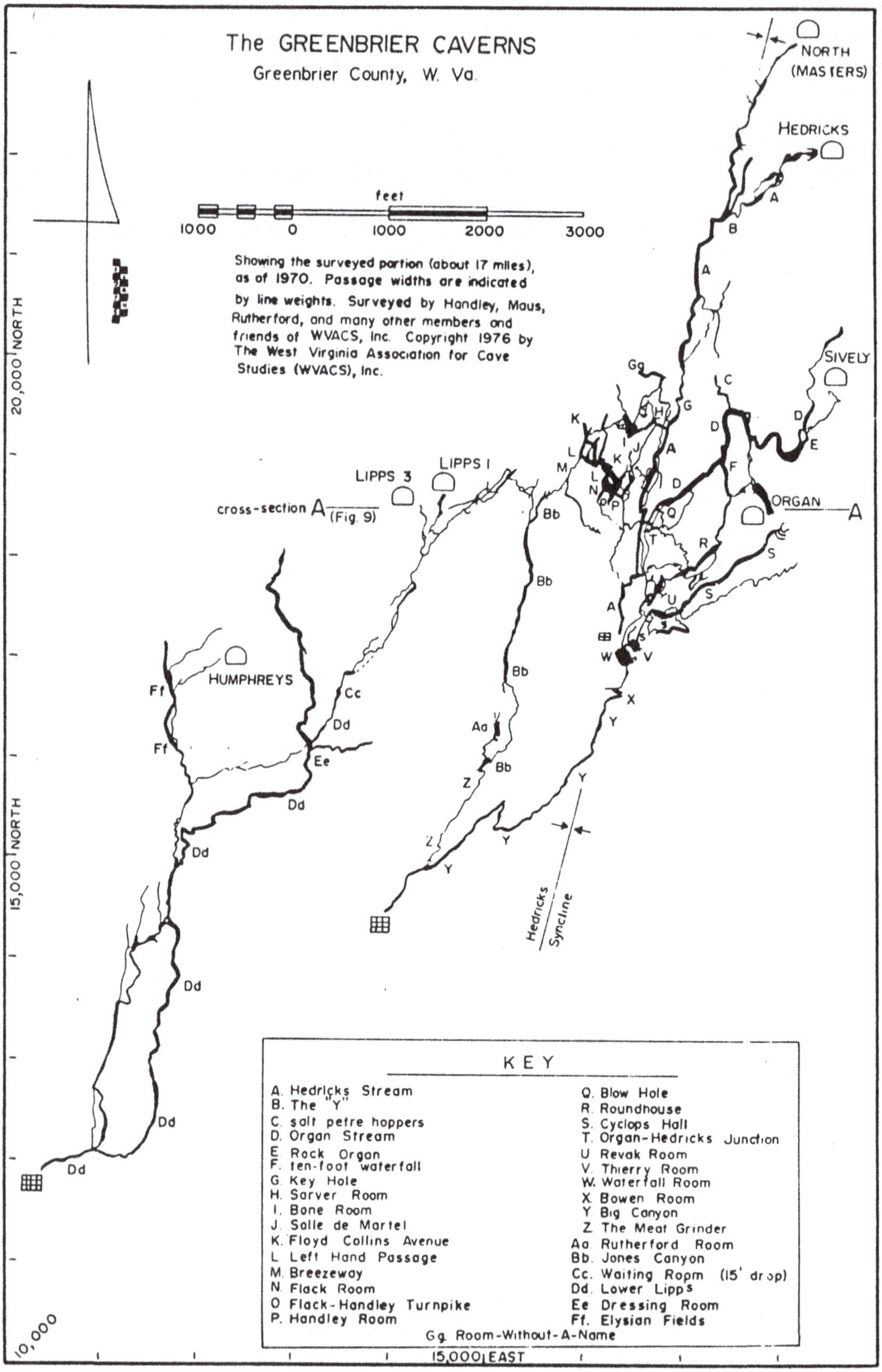

Fig. 13.6 Greenbrier Caverns (Organ Cave System) as of 1970. From Handley and Rutherford (1976)

Fig. 13.7 Organ Cave fieldhouse

Fig. 13.8 Tommy Shifflett using an electronic distance meter (EDM). A theodolite and EDM were used to create an accurate surface survey

In May 1977 Paul Sevens led a major effort to create a database with each survey station having a unique name. During the next 2 years he keypunched all of the cave survey data into a format acceptable for input into a computer processing program by Bob Thrun. The entire cave survey database was integrated together for the first time and could be run on a computer simultaneously. Bob Thrun was able to provide computer processing support to the project for a while, but his other commitments soon prevented continued support. During 1979 and 1980 the project was without computer processing support, but Don Major came to the rescue by converting Bob Thrun's program called *CMAP* to run on a mainframe computer. Don Major provided computer support for the project through 1982 when Bob Hoke assumed the responsibility.

During the late 1970s and early 1980s, work in the caves continued with George Dasher leading the resurvey of Cricket and Foxhole caves; Tom Kaye leading the dig into the Raccoon Room; and Forrest Wilson unsuccessfully attempting to dive the Organ resurgence at Second Creek. Rich Hall, Bob Hoke, Larry and Klaye Lilly and Paul and Lee Stevens continued to field check the map compiled from surveys done in the early 1970s. In the process many passages were resurveyed. Ray Cole and Tommy Shifflett led a theodolite survey on the surface to accurately tie together the cave entrances (Fig. 13.8).

Later, Ray Cole designed a cave radio and tied the theodolite surface survey to the Bowen and Lipps sumps, Lipps–Humphreys Junction, and entrance to Fat Man's Misery (Fig. 13.9). The more accurate surface surveys and cave radio locations formed constraints to "fix" the position of the major passages, so they would remain fixed as additional surveys were added. Dave West led another series of efforts to connect Organ with Foxhole, but without success. Dave West and George Dasher also attempted to dig a bypass around Bowen Sump, but that was abandoned after many trips. Chris Welsh and Bob Anderson were more successful in their attempt to get into a hanging lead above the Masters stream and in the process discovered Novice Odyssey in 1980.

By the end of 1982, the survey of the Organ Cave System and surrounding caves was essentially completed, though unattractive leads remained scattered throughout the system and nearby caves. The large project weekends at the fieldhouse were no more. Since 1982, study of the cave was done by small groups on an infrequent basis. During this period, George Deike and Bill Jones made frequent trips to the cave to develop their interpretation of the cave's geology and hydrology. George Dasher completed the survey of Cricket Cave; Bill Balfour led the survey of Whites Cave; George Dasher, Sonja Ostrander, and Paul Stevens surveyed Hellems Cave; Ray Cole, Paul and Lee Stevens surveyed Sheep Shelter Cave; and Fred Grady searched for vertebrate fossils. Ed Swepston and Giff Lindsey continued the pursuit of an Organ–Foxhole connection with a dig just north of the Revak Room. Bob Hoke and Paul Stevens continued to debug the cave survey database, and Paul Stevens inked the maps and

Fig. 13.9 Pat Moretti operating the Organ Cave radio at the Lipps–Humphreys Junction. Photograph by Paul Stevens. Used with permission

led the editing of the West Virginia Speleological Survey Bulletin. During 1985 and 1986 the maps and manuscript were extensively reviewed by members of both DC Grotto and WVACS. The final result was the publication of the West Virginia Speleological Survey Bulletin 9 titled *Caves of the Organ Cave Plateau* (Stevens 1988). This publication and maps represent the efforts of more than 400 cavers over half a century. The detailed map sheets were scanned and are included in the electronic map files (M-13.1).

13.3 Physical Description of Organ Cave

Perhaps the principal features that distinguish the Organ Cave System are its long, large stream passages littered with breakdown. One of the best examples is the Organ Main Stream Passage, in the tourist section of the cave. Another notable feature of the system is its large rooms, such as the Waterfall Room and Handley Room. However, the dominant impression left with most visitors to the wild portion of the cave is its length, 38.5 miles of passage. With ten entrances, a variety of "through trips" are possible. The most common through trip is between the Organ and Lipps Entrances. The system also has areas with a maze of passages to confuse the newcomer, for example Lipps Maze. To attempt a description of such a large cave, it is convenient to divide it into individual sections (Fig. 13.10).

With the exception of the Rock Organ area, the cave is not known for its speleothems; however, it does contain some good examples scattered throughout the cave, often in remote areas. The Organ Cave System has a significant collection of well-preserved saltpeter vats from the Civil War period. These are located in the tourist section of the cave (Fig. 13.11).

13.3.1 Masters–Hedricks (Lower Level) Section

The Masters–Hedricks (Lower Level) Section of the cave system consists of four miles of passage, most of which is large stream passage at base level. It is bounded on the north by the Masters, Hedricks, and Great Escape entrances and includes the Masters Stream Passage, Hedricks Stream Passage, and the Borehole which all join at the Masters–Hedricks Junction, a large gravel-floored room. The combined passage (Hedricks Stream Passage) proceeds downstream and south along the axis of the Caldwell Syncline for almost a mile to the Hedricks sump with occasional side passages formed by tributary streams. Much of the passage is floored by large breakdown blocks and in two locations by shale collapse piles. At its junction with the Organ Main Stream Passage, the passage is large with a gravel floor. Silt banks line the Hedricks Stream Passage as it approaches its sump.

The most northern entrance to the cave system, the Masters Entrance, is located at the low end of a shallow elongated sink just south of the surface drainage divide for the Organ Cave System basin. The entrance crawl is normally dry; however, a stream is encountered just beyond a climbable ten-foot drop. The cave proceeds downstream as a large breakdown passage for 250 ft before degrading into a short belly crawl in the shallow stream. After another 400 ft of walking passage the ceiling again drops forcing a second short belly crawl in the stream.

Now 750 ft from the entrance and wet, the caver is rewarded by a large display of rim stone pools which mark the intersection of a stream to the northwest and the difficult, overhung climb into the Novice Odyssey section. A short distance downstream from the rimstone dams is the Totem Pole, a tall column in the center of large passage, and 750 ft later, the Masters stream joins with the Borehole and Hedricks Stream Passages.

The Hedricks entrance was sealed with concrete in 1960 and remains blocked. The nearby Great Escape entrance is extremely tight and covered by debris. Just inside, the cave opens into a large well-decorated room. The 1,000 ft of passage between the entrance room and Masters–Hedricks Junction is mostly breakdown-floored with a stream occasionally visible. Along the way is the Big Room which, because of its breakdown and low ceiling, serves to confuse the route.

The Masters–Hedricks Junction also serves as the downstream end of the Borehole, a very large trunk passage going north for 400 ft before ending in a massive breakdown collapse. A low passage at the base of the wall on the east edge of the breakdown pile can be followed up through the breakdown into North Upper Level Organ.

The Hedricks Stream Passage proceeds downstream to the south of the Masters–Hedricks Junction as a large

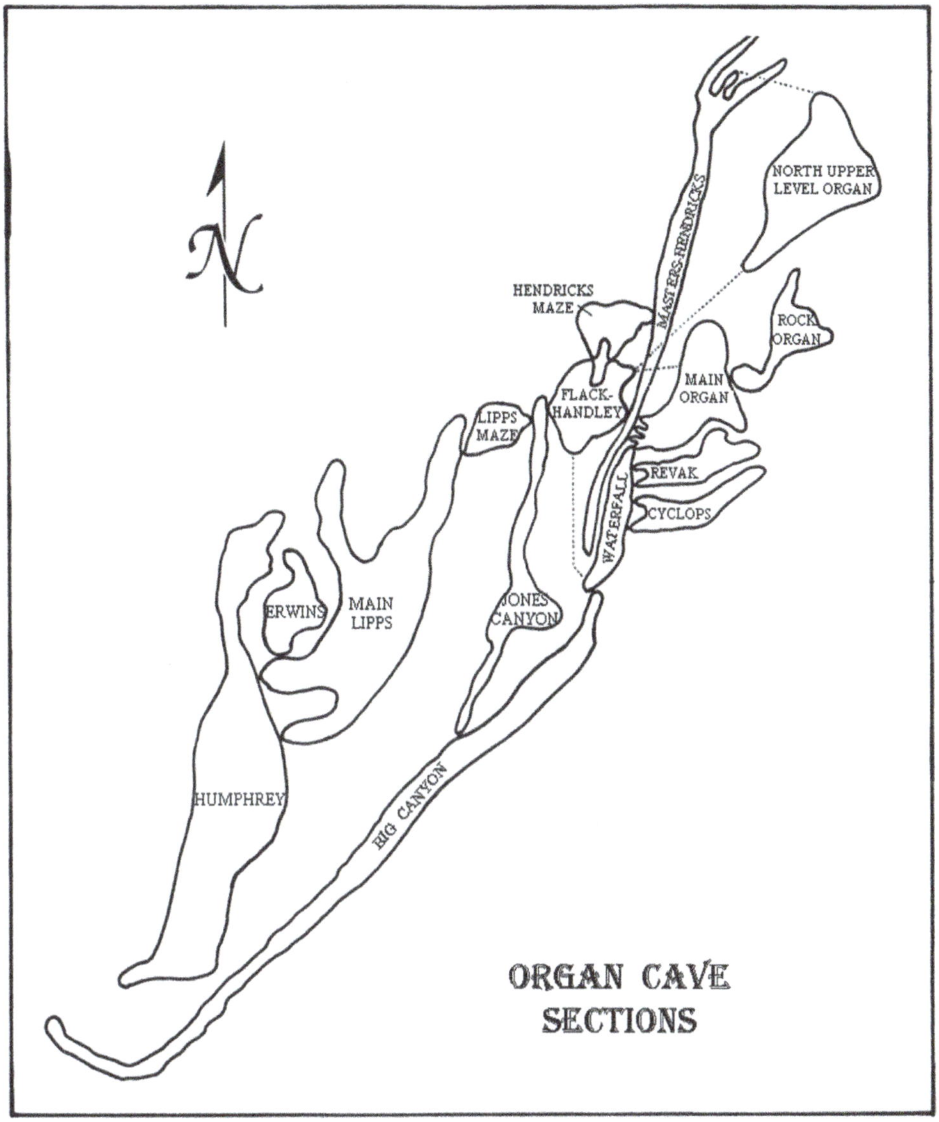

Fig. 13.10 The sections of the Organ Cave System

breakdown-floored stream passage until it is interrupted by Shale Room #2 with its massive shale pile. Along the way is a dry side lead to the southeast, the Paleo Saltpetre Passage, which runs 1,000 ft in a direct line toward the Saltpetre Vat Passage in the tourist section of Organ Cave, gradually becoming too low to continue just 250 ft short of the vats. Along the way, the Paleo Saltpetre Passage intersects the Rimstone Ceiling Passage, a 1,500-foot-long tributary stream passage tending northeast which joins Hedricks Stream Passage 500 ft south of Shale Room #2.

After traversing 1,000 ft of breakdown-filled stream passage south of Shale Room #2, the Hedricks Stream Passage is again interrupted by a shale collapse just before reaching its junction with the Hedricks Maze stream coming in from the northwest. The junction area is notable for its gravel floor and no breakdown. There are two routes to the west into the Hedricks Maze area of the cave, and the south route via Sarver Room is the easiest. Continuing past the Hedricks Maze stream junction, it is another 1,100 ft downstream through monotonously familiar breakdown passage before one first encounters a west lead toward the Fun and Flack Rooms and then the Organ–Hedricks Junction with the Organ Main Stream coming in from the east. Again, the junction area is characterized by its gravel floor and lack of breakdown.

Just 400 ft further down Hedricks Stream Passage, one intersects the Old Saltpetre Route, an overflow route for water entering the Organ Entrance during periods of high flow. Normally, this passage is dry. Hedricks–USP Connector, a climb up to the Upper Stream Passage leaves this junction just inside Old Saltpetre Route.

With only 700 ft to go before its sump, Hedricks Stream Passage continues with its typical breakdown floor, proceeds through the Slickenside Room (Slate Creek enters low from

Fig. 13.11 Saltpetre Vat in Organ Cave

the northwest), and then changes character dramatically. The last 300 ft is gravel-floored with silt banks along the side and free of breakdown. The ceiling gradually lowers until one is forced to belly crawl in water a short distance before reaching a large room where the stream sumps.

13.3.2 North Upper Level Organ (NULO) Section

Above the Masters–Hedricks (Lower Level) Section lies NULO with its 3.2 miles of passage. At its northern end NULO is a maze of small- and medium-sized passages, connected to the rest of the system by a vertical route through the breakdown at the north end of the Borehole. The principle passages of north NULO are the vadose Right and Left Hand Perpendicular Passages, which have small streams that join to form the Upper Stream Passage. Just downstream of the junction of the Right and Left Hand Perpendicular Passages, the Upper Stream Passage is a narrow crawl opening at the top of a waterfall into a small room. With skill, the drop can be negotiated without rope. The passage proceeds south, downstream past the tributary East Passage, to join the rest of the cave system at the Sand Room.

The northern end of NULO is a maze of small muddy crawls and canyons with incised streams that form a dendritic pattern feeding the Upper Stream Passage. Serpentine Way, a dry walking passage with an uncluttered floor, is unusual for the area and forms the route from the top of the Borehole breakdown climb into the maze. The Left Hand Perpendicular Passage (LHPP) is reasonably large for its northern most 900 ft, but breakdown slows one's pace. It then degrades into a stream crawl before joining the Right Hand Perpendicular Passage (RHPP). The RHPP trends upstream to the northeast from its junction with LHPP as a deep narrow canyon which can be traveled laboriously along its top for 1,000 ft before enlarging into trunk passage. It intersects a maze of passages at its northern end, one of which proceeds west to LHPP.

In the ceiling of the Waterfall Room below the LHPP–RHPP Junction, a passage heads south to East Passage forming what appears to be the ancient LHPP–RHPP water route. Downstream of the waterfall, the Upper Stream Passage degrades into a breakdown crawl shortly before intersecting a small tributary stream to the northwest coming from Peggy's Tear Room. After 400 ft of shallow stream walking passage, the Upper Stream Passage is joined by the stream in East Passage. Just north of this junction is a short narrow vertical squeeze followed by a short pool where the ceiling drops enough to require a caver to get wet. East Passage trends northeast for 1,500 ft as canyon passage before becoming too tight to continue. It breaks into several parallel passages along its course which rejoin near its end. The junction of East Passage and Upper Stream Passage has been located by radio fix to be 15 ft east of the Hedricks Stream Passage and 30 ft higher than the Hedricks Stream. From this junction, cavers could hear another party below in Hedricks Stream.

The Upper Stream Passage proceeds south from its junction with East Passage for 1,300 ft as a shallow stream walk initially with no breakdown, though the final section requires canyon straddling. At one point it is interrupted by a large block of breakdown and can be traversed only by climbing the block and belly crawling along its wet top.

13.3.3 Hedricks Maze Section

The 1.7 miles of passage in the Hedricks Maze section are dominated by the large breakdown-floored Bone and Sarver Rooms and is connected to the rest of the system by: (1) Rat Alley Crawl, a low, straight eastward lead leaving the Bone Room at its ceiling and entering Floyd Collins Avenue at its ceiling; (2) Octopus Alley, a straight southern lead leaving the Sarver Room at its ceiling and entering the Fun Room at its ceiling; and (3) two separate stream passages which join the Hedricks Stream Passage in a gravel-floored area just downstream from a shale breakdown pile. North of the Sarver Room is a maze of passages dominated by the very large breakdown-filled Room Without A Name (RWAN). Considerable surface debris has been found at the upstream terminus of the RWAN and Car Room stream passages. Their termini are near the head of a blind valley on the surface which contains a small wet weather sinking stream.

Cavers typically enter the Hedricks Maze section using the route from the Hedricks Stream Passage to Sarver Room. The route begins 10 ft off Hedricks Stream behind a breakdown pile along the west wall and quickly enters Sarver Room via a short down-climb. Just before descending the drop into the Sarver Room, a belly crawl can be followed up to Octopus Alley, the dry walking passage between Fun Room and Sarver Room. Octopus Alley enters at the ceiling of the Sarver Room with a 30-foot drop to the stream below. To enter the Bone Room to the west, one must climb breakdown and negotiate a crawl through steeply folded rock strata. While the Bone Room is quite large, its very large breakdown blocks and irregular floor and ceiling hide the fact well. A perched stream passage enters the room's ceiling and provides a small waterfall. This passage can only be entered with the aid of a scaling pole. Rat Alley Crawl takes off high along the west wall and proceeds as a hands and knees crawl over to Floyd Collins Avenue; unfortunately, a short stretch of the crawl is typically muddy. Just south of the entrance to Rat Alley Crawl, a protrusion of the Bone Room gradually becomes stream passage ending under the breakdown pile of Salle de Martel. Going upstream into the passage to the northwest, the route ends in the Car Room when the passage becomes a narrow slit. Old Pepsi bottles and other surface debris were found here, and the room was named when the original explorers heard periodic rumbles which they attributed to passing cars on US 219. North of the Bone and Sarver Rooms is a maze of passages leading to the breakdown-littered and steeply sloped trunk passage known as Room Without a Name (RWAN). Upstream the passage gradually becomes a crawl with much surface debris from the blind valley across US 219 from Mr. Sarver's house. Along the west wall of RWAN are two crawls which lead to the top of Handley's Climb (Fig. 13.4) and the Great White Way with its white gypsum-covered walls.

13.3.4 Rock Organ Section

The Rock Organ section consists of one mile of passage, mostly breakdown-floored stream passages at base level. The Sively #2 and #3 entrances form its northern boundary. The Sively #2 entrance is a short pit requiring rope; the low passage at the base of the pit is often plugged by silt and surface debris. The Sively #3 entrance requires a tight squeeze to enter a confusing breakdown crawl before finding a large stream passage a short distance from the entrance. The Sively #2 and #3 stream passages proceed south (downstream) to join at the Rock Organ at a pile of breakdown. The "Rock Organ" (Fig. 13.2), namesake for the cave system, is a large drapery formation near the north end of the tourist tour. A large, meandering breakdown-floored stream passage (Organ Main Stream Passage) results and leaves the Rock Organ area to join the rest of the cave system near the dam in the tourist section of Organ Cave. Electric lights and a built-up tourist path are present between the dam and the Rock Organ.

13.3.5 Main Organ Section

The Main Organ section of the cave contains 2.7 miles of passage including most of the tourist section of Organ Cave, including the Saltpetre Room. Its eastern boundary is the Organ Entrance. A surface stream flows into the entrance during wet weather and sinks at the west end of a large entrance room; during drier weather the stream sinks in the entrance breakdown. The stream reappears in the 1812 Route and joins the Organ Main Stream at a ten-foot waterfall just downstream from the Saltpetre Room. The tourist trail leaves the Organ Entrance Room via a short route which enters at the ceiling of Organ Main Stream Passage downstream from the Rock Organ. Organ Main Stream Passage proceeds as a large breakdown-floored passage until just before its junction with the Hedricks Stream, at which point its low ceiling makes an upper bypass route more convenient for cavers. The Main Organ section also joins the remainder of the system through several leads which leave the Organ Main Stream Passage at its ceiling and ascend until they ultimately join the continuation of the Upper Stream Passage at the Throne Room.

The tourist entrance to Organ Cave is directly under highway WV 63 at the base of a large east facing escarpment which forms the terminus of a broad valley. The 75 foot wide by 25 foot high entrance opens into a 250-foot-long room of similar dimensions. The tourist trail follows a built-up path along the north side of the room leaving its west end through a 450 long walkway that leads to Organ Main Stream about 800 ft downstream from the Rock Organ. The tourist trail splits into two paths; one goes to the Rock Organ and the other proceeds downstream along Organ Main Stream for 300 ft past a dam and enters the Saltpetre Room via Fat Man's Misery. The Saltpetre Room contains 37 vats, some of which are in excellent condition (Fig. 13.11).

While the tourist trail ends here, Organ Main Stream continues downstream from the entrance of the Saltpetre Room as a large stream passage with little breakdown. After 450 ft it intersects a high lead entering from the south which carries the Organ Entrance Stream forming a ten-foot-high waterfall at the junction. Proceeding down Organ Main Stream the route becomes littered by large breakdown. Discovery Passage leaves Organ Main Stream at its ceiling along its south wall and continues as mostly walking passage to the "T"-Room where it intersects Straddle Alley. Straddle Alley is a keyhole-shaped passage to Organ Main Stream intersecting it just before it joins the Hedricks Stream Passage. Gypsum Passage leads to the south of T-Room as a crawl to the 1812 Room and A-Trail leads to the west of T-Room ascending in a mixture of walking and crawling passage to the Upper Stream Passage at the Throne Room.

Along the south wall of the Organ Entrance Room is a 250-foot-long crawlway intersecting the Rotunda Room at its ceiling. The passage to the north of the Rotunda Room captures the Organ Entrance Stream and carries it in walking passage to the waterfall at Organ Main Stream just downstream from the Saltpetre Room. The passage to the south of the Rotunda Room forms the Old Saltpetre Route, a large breakdown-floored passage which proceeds through the 1812 Room and hence to Hedricks Stream Passage. While saltpeter was mined in the 1812 Room perhaps even earlier than 1812, only the badly decayed remnants of a few vats remain.

13.3.6 Waterfall Section

The 1.6 miles of passage in the Waterfall section joins the Main Organ and Big Canyon sections of the cave. Two isolated cave sections (Revak and Cyclops) branch off it to the east. The section includes the continuation of the Upper Stream Passage starting at the Throne Room and disappearing into breakdown at the base of the Waterfall Room. The Waterfall Room is the largest room in the cave. It is over 100 × 150 ft in size, and its floor is covered by large breakdown. The Foxhole Cave stream and another stream of unknown origin enter at the ceiling of the Waterfall Room to form the waterfalls. At the ceiling, at the opposite (south) end of the Waterfall Room, is The Highway, a dry passage to the top of Bowen Drop (into Big Canyon). The Thierry Room and its feeder stream passages lie above and to the north of the Waterfall Room.

The Upper Stream Passage is joined by the A-Trail at the Throne Room (named for a rock shaped in the form of a chair/throne) and proceeds south for 1200 ft as 10-by-10-foot breakdown-floored stream passage. Along the way it intersects the entrance to the Revak Room and the normal route to Cyclops Hall. Just before entering the base of the Waterfall Room, a short side lead to the east goes to the base of the Thierry Room (which can be entered by ascending through very unstable breakdown) and PD-60. A waterfall descends through the Thierry Room breakdown into this passage and into the hole in the floor at PD-60. The hole opens into a tight twisting passage down to the Cyclops stream. Upstream and through a short sump is Cyclops Hall.

Just before the Upper Stream Passage reaches the base of the Waterfall Room, the stream disappears into the east wall never to be seen again. However, a stream from the Waterfall Room appears 50 ft down the passage along the east wall, crosses the passage and continues to the west in a breakdown-choked lead along the base of the Waterfall Room. The Waterfall Room is a 150-foot-long, 100-foot-wide, and 50-foot-high breakdown-floored void. The waterfall along its north wall is fed by two streams, the east one of which has been traced to Foxhole #1 Cave. The east stream passage above the waterfall can be followed for 300 ft before becoming too low to continue. To the side of the east stream passage above the waterfall is the cone-shaped breakdown-floored Thierry Room with two tributary stream passages feeding it. The west stream passage above the waterfall can be followed upstream for 200 ft before being blocked by a flowstone bar.

The stream passages feeding the waterfall enter the north wall of the Waterfall Room at its ceiling. On the opposite side of the room, The Highway, an apparent continuation of these passages, leads from the south wall of the Water fall Room at its ceiling to the top of the Bowen Drop and a 130-foot vertical descent into Big Canyon.

13.3.7 Revak Section

The Revak section is an upper-level maze of 1.5 miles of passages connected to the remainder of the system along the Upper Stream Passage between the Throne Room and the Waterfall Room. The section is named for its only large room, the breakdown-floored Revak Room. Surface debris has been found in the Raccoon Room, another section room located near the surface, by the Organ Entrance.

The Revak Room is normally entered from the Upper Stream Passage via a large short side lead which ascends steeply up its breakdown-covered floor. A large low side passage leaves the Revak Room and proceeds north to a highly inclined route (Revak Silo) down to Upper Stream Passage. Another low side passage leaves the Revak Room and proceeds south to Crowbar Crawl, a low, dry, alluvium-filled passage, and a small room where the Foxhole #1 stream enters one side and exits on the other. Along the west wall of the Revak Room and at its ceiling is Moretti's Turnpike, a passage which climbs steeply before ending in breakdown from above. Moretti's Turnpike, Crowbar Crawl, and the top of Revak Silo are the closest section of Organ Cave to Foxhole #1 Cave.

The passage leaving Revak Room to the east goes about 400 ft to the 15-foot-wide, 40-foot-long mud-floored Cathedral Room. At the northeast end of the Cathedral Room is Walker's Climb which leads to a small crawlway which continues almost to the Organ tourist entrance before ending. A hole in the floor of the Cathedral Room leads to a 470-foot cobble crawl that enters the Racoon Room. At the far end of the Racoon Room is a chimney which may lead to the surface just south of the Organ Cave gift shop.

The final Revak Room lead is a phreatic tube that begins near the entrance to Crowbar Crawl and crosses over the Revak Room, which can be seen through cracks in the floor, then intersects the passage between Revak and Cathedral Rooms at its ceiling. About 25 ft into this tube, it is possible to slide under a false floor and enter a well-hidden crawlway that trends toward the Waterfall Room area. The passage becomes larger and soon contains a stream. The Hard Roll and Walker's Ramble are located in this area.

13.3.8 Cyclops Section

The 1.8 miles of passage in the Cyclops section are dominated by the breakdown-covered Cyclops Hall, with headwaters under the Organ Cave parking lot. The passage gradually becomes smaller and lined with silt as the stream flows west and approaches its sump located near the base of the Waterfall Room. It connects to the Waterfall section at PD-60 via a tight ascending passage just beyond its initial short sump. Entry to the Cyclops section is usually done via a lead off the Upper Stream Passage. The lead enters Cyclops Hall through a small "eye" in the ceiling.

The lead to Cyclops Eye leaves the Upper Stream Passage 150 ft south of the entrance to Revak Room. As it meanders 600 ft to the Eye, the caver is forced through a 50-foot-long belly crawl, the Drag Strip, and then a long canyon straddle. A rope or cable ladder is required to negotiate the 20-foot drop at the Eye to the floor of Cyclops Passage.

From the Eye, the large breakdown-floored Cyclops Hall proceeds northeast for 1200 ft before splitting into two large passages which both quickly end. Southwest of the Eye, the large breakdown-floored Cyclops Passage continues for 400 ft before changing into a muddy silt-covered stream passage which gradually forces cavers to crawl through a short sump and on to the connection with a chimney up to PD-60.

At the ceiling on the wall of Cyclops Hall opposite the Eye is a lead to the top of the Pool Room. However, cavers normally enter the Pool Rom from below by proceeding southwest from the Eye to the side lead just before Cyclops Hall becomes muddy. The Pool Room rimstone pools are filled by a 25-foot waterfall fed by the stream from Crystal Palace Gallery, 500 ft to the south. The route to Crystal Palace Gallery is a tall stream canyon which ends upstream of the Gallery in a flowstone wall that has yet to be climbed.

At the floor of the wall of Cyclops Hall opposite the Eye is Davis Folly, a stream passage, initially breakdown-choked, that goes 1300 ft to the northeast before breaking up into various small leads. The stream passage has one dry side lead, Gypsum Way, which goes 1,000 ft toward the Pool Room before ending.

13.3.9 The Flack–Handley Section

The Flack–Handley section contains 4.9 miles of passage. It is a complex maze of passages with over 200 ft of vertical relief. The section includes the large, overlapping Flack and Handley rooms and two large, overlapping trunk passages: Left Hand Passage and Floyd Collins Avenue. It is bounded on the east by the lower-level Hedricks Stream and Hedricks Maze sections and the North Upper Level, Organ, Main Organ and Waterfall sections. On the west it connects to the north end of Jones Canyon (Fig. 13.12).

Fig. 13.12 Flack–Handley Turnpike. Ray Cole crossing the Flack–Handley Turnpike from the Handley Room into the Flack Room. Photograph by Bruce Boss

The following description and the accompanying map consider the Flack–Handley area as divided into two levels: (1) Fun and Flack Rooms and Floyd Collins Avenue—Lower Level; (2) Handley Room and Left Hand Passage—Upper Level.

Cavers entering the area from Hedricks Stream Passage enter the lower level near the Fun Room or via Hedricks Maze and the Rat Alley Crawl. Cavers entering the area from Jones Canyon to the west enter the upper level at Left Hand Passage. Cavers entering the area from Upper Stream Passage to the east enter the upper level via the Sand Room. Cavers can traverse between levels only via the Flack–Handley Turnpike.

The Flack–Handley area can be entered from Hedricks Stream Passage using the first passage going west, just a short distance upstream from the junction of Organ and Hedricks Streams. The breakdown passage gradually ascends and after 200 ft turns north up Slate Creek to reach the base of the Fun Room 150 ft later.

Slate Creek continues as a large stream passage going north 500 ft before ending in a breakdown pile at Salle de Martel. About 200 ft into it are ceiling leads along the west wall that lead up to Octopus Alley and a short low passage to the Centrifugal Room. The Centrifugal Room can be climbed to its ceiling to join a belly crawl that opens on an overlook on the east wall of the Flack Room. Octopus Alley continues north to an overlook at Sarver Room and south to the upper level of the Fun Room. The Fun Room is a two-level room, the top level of which has a floor you must see to believe. The floor is contoured with an intersecting pattern of 3-foot-deep grooves forming a series of rounded "tables." A northwest route from the upper level of the Fun Room runs up into the Flack Room, a 100-by-150-foot room with breakdown along the west side.

The Flack Room has several major exits plus numerous minor routes. The Subway leads from the west side of the Flack Room for 400 ft downhill to the junction of Salle de Martel and the Fun Room, coming out under a large breakdown block. The Flack–Handley Turnpike is a narrow ledge along the south wall near the ceiling that provides the only connection with the upper-level Handley Room. Floyd Collins Avenue is a 60-foot-wide walking passage heading north for 600 ft before ending in breakdown at the junction with Rat Alley Crawl, a low, occasionally muddy crawl east to the Bone Room.

The Flack–Handley area can also be entered by proceeding north from the Throne Room 350 ft and bearing left at each opportunity until reaching the Sand Room. On the way two junctions are passed, each of the other passages also go to the Sand Room, but via a more difficult way. The eastern side lead at the first junction gets to the Sand Room via Sally's Waterfall. The northern stream route proceeds up the Upper Stream Passage through a tight section that makes

Fig. 13.13 Bruce Boss in the T.T. Crawl

travel difficult before opening as a canyon in the floor of the Sand Room. The Upper Stream Passage continues north from Sand Room into the North Upper Level Organ Section.

In addition to these three passages, there are two others which intersect the Sand Room. The Nine Foot Bat Passage heads southwest from the Sand Room for 250 ft into a multilevel maze, and Ascending Way goes west toward the Handley Room. The western passage branches 75 ft from the Sand Room with the northern branch heading into the base of the Flack Room and, by another route, into the Sarver Room. The western branch continues west for another 75 ft before bending sharply north at a junction with a high trickle stream coming in from the south (a miserable passage, never pushed to its end). The main passage goes north for about 400 ft before gradually becoming too small; however, after 125 ft of large passage there is a narrow opening in the west wall that leads up the steeply ascending Handley Silo into the Handley Room.

The Handley Room is a 300-foot-long, 40-foot-wide room with an arm, the Left Hand Passage, trending west from its northern end. The 20-by-20-foot Left Hand Passage breaks up in about 250 ft into a series of large fingers going north for 300 ft before ending. The Earthquake Passage is the first of these fingers, and it leads north to the Fossil Room by way of T.T. Crawl (Fig. 13.13).

13.3.10 Big Canyon Section

The Big Canyon section consists of 2.7 miles of tall stream passage at base level fed by the Hedricks Stream resurgence at its north end (at the bottom of Bowen Room Drop). It proceeds downstream and south as a gravel-floored passage for 4,000 ft to the Big Canyon sump. Beyond the 270-foot-long sump, the passage (known as Wilson's Watery Wonderland) again assumes its tall canyon character and proceeds downstream and south another 6,500 ft before finally ending at David's Dungeon in an impenetrable sump. An unexplored stream passage from the north intersects it via a waterfall about half way between the Big Canyon sump and David's Dungeon.

Big Canyon can be entered by rappelling the 130-foot flowstone-covered Bowen Room Drop. A stream from a side lead forms a waterfall over the center of the drop; however, it is possible to stay out of the water except at the bottom of the rappel. Once off rappel, cavers can step to a dry area to remove their vertical gear. At the base of the drop a large passage heads north a short distance before ending under the Waterfall Room in a breakdown pile that provides an easier route than the Bowen Room Drop. This route was opened up after the original map was published.

The Hedricks Stream resurges off to one side of the bottom of the drop, pools, and then flows south down Big Canyon. After wading the first 3-foot-deep pool and passing the first of many flowstone cascades, cavers can walk down the gravel-floored shallow stream in passage 10 ft wide and 25 ft high for the next 4,000 ft interrupted only occasionally by breakdown and pools no more than 3 ft deep. About 1,500 ft downstream from the drop, a stream enters from the east over a flowstone cascade marking the route into the East Loop. Another 2,000 ft downstream Big Canyon intersects the Meatgrinder passage entering high on the west wall and the character of Big Canyon begins to change. Silt banks become evident and the ceiling drops until one must crawl a short distance in mud before the passage opens into a room where the stream sumps. Just upstream of the Big Canyon sump, Dave's Red Wagon Dig was dug through silt for 200 ft before ending at a rock constriction, but the passage continues to blow air.

The 270-foot-long Big Canyon sump is three ft high, but has been blocked by silt banks on occasion. It opens into Wilson's Watery Wonderland, large canyon passage much like northern Big Canyon. Just downstream of the sump, the stream is ponded by a breakdown pile and thereafter it continues as a low-gradient stream. About 2,000 ft downstream, Harold's Falls enters from the north marking a virgin lead not entered. After another 2,000 ft of similar high canyon stream passage, the passage widens and fills with breakdown at the Peanut Room with Cowart's Crawl coming in from the south. The breakdown clears as one heads downstream, but reappears 500 ft later for several hundred feet above Seekin's Falls, a series of short cascades. In the next 1,000 ft the canyon narrows and enters the Scarlet Pimpernel section with its red tinted speleothems before dividing into two separate passages. The passages rejoin at a large breakdown-floored room, David's Dungeon, with the stream ponding and disappearing under the breakdown.

13.3.11 Jones Canyon Section

The 1.9 miles of passage in the Jones Canyon section is bounded on the north by the Lipps Maze and the Flack–Handley sections, which provide its tributary streams. They join to form a large breakdown-floored stream passage which proceeds downstream and south until interrupted by a breakdown pile at the First Waterfall where it joins the upper-level Belfry stream passage. The main passage continues at the higher level until it abruptly ends and the caver is forced to descend to the lower stream passage. The stream passage gradually becomes a belly crawl until intersecting the top of a 25-foot dome at the Second Waterfall. The Jones Canyon Maze stream passage also intersects the top of the dome. The two streams join, but quickly flow over the Third Waterfall into the silt-floored Canfield Room and down to the Rectal Sump. The Jones Canyon Maze is a series of upper-level rooms connected by a maze of passages. It provides a route from Jones Canyon to the Big Canyon via the Meat Grinder Passage.

The Breezeway provides the narrow exit from the spacious Left Hand Passage into Jones Canyon. While initially walking passage, it soon becomes a short belly crawl followed by a 200-foot dry stoopway until it intersects the Treasure Passage. The Pendulum, a rock hung from the ceiling which swings when set in motion, marks the intersection. The Treasure Passage goes for about 800 ft, the last several hundred of which were dug open. Its current end is in the maze at the end of the Nine Foot Bat Passage, but it does not connect.

Continuing south from the Pendulum, the passage gradually opens up and after 900 ft intersects Skid Row, a stream passage exiting Lipps Maze. Downstream Jones Canyon continues as a 20-by-20-foot breakdown-floored stream passage with large conspicuous gravel banks along the way. Its character changes dramatically after 600 ft where it is blocked by breakdown. Ascending through the breakdown, the caver enters a large upper level with the Belfry Passage stream entering near the ceiling and forming the First Waterfall. The passage continues south; a large, dry route with occasional canyons across the floor to the stream below. The upper level ends about 1,000 ft south of the First Waterfall in a muddy passage. About 150 ft north of the end, one can duck under the east wall along the floor, traverse over the Three Inch Ledge, and gradually descend to the stream below. While the route initially allows walking in a shallow stream, the ceiling gradually descends and suddenly one must turn sharply west and proceed through a very narrow four-foot-high passage for 15 ft before regaining the wider stream passage.

The four-foot-wide passage proceeds south gradually forcing cavers to belly crawl in the shallow stream before opening up just before the top of the Second Waterfall. The passage continues downstream below the 20-foot drop at the Second Waterfall as a high canyon, but is quickly interrupted by another 20-foot drop at the Third Waterfall. The stream passage continues lined with silt banks for 200 ft before ending in Rectal Sump.

Standing in the Jones Canyon stream at the top of the Second Waterfall, one can see a second stream passage entering the top of the dome from the west at the same level. Crossing the dome is precarious, but once across, walking up this 2-foot-deep stream is the route to Jones Canyon Maze. By ascending through a hole in the ceiling, one enters the large room, Grand Central Station. If instead one stays in the stream passage and proceeds upstream, the route becomes a belly crawl in a shallow stream eventually opening up and providing access to the Lost Lamp Room, Romper Room, and House of Cards.

High leads going north from Grand Central Station lead to the Rutherford Room and Big Mother Pit. At the south end of Grand Central Station is a very short tight squeeze into the Meat Grinder Passage which proceeds as low, dry passage for 1,200 ft to its junction with Big Canyon.

13.3.12 Main Lipps Section

The Main Lipps section has 3.5 miles of passage consisting of a mile-long stream passage with supporting tributaries. It is bounded on the north by the Lipps and Deems Entrances and on the south by its connection with the Humphreys Stream. The large, dry, high-level Lunar Way passage runs east–west between the Lipps and Humphreys Stream Passages and intersects both of them at their ceilings. Near the middle of the Lipps Stream Passage the entrance stream is pirated at Lipps Piracy; however, just south a new stream enters and occupies the passage. The middle section of Lipps Stream Passage is noted for its high, narrow canyon (Hell's Fissure) which requires cavers to repeatedly ascend and descend in order to proceed. Farther south, the Lunch Room marks the intersection of three large passages: Lipps, Lunar Way, and Handley's Lost Sandwich Passage. Handley's Lost Sandwich Passage is a large breakdown-floored stream passage which is abruptly terminated at its north end by a breakdown pile. It was in this breakdown that Bob Handley lost his paper bag which contained a sandwich.

The Lipps and Deems entrances are located at the base of very small sinkholes. The passages leading south from the two entrances combine and form the Lipps Stream Passage, about 600 ft downstream from the Lipps Entrance. The Deems Entrance looks more like a rabbit hole and is rarely used. The Lipps Entrance is 2 by 3 ft and quickly opens into an entrance room with a short drop that can easily be climbed down to a stream. The passage downstream is littered by breakdown, but walking is easy until the stream turns sharply and enters a low breakdown-clogged passage to the west which serves as the route to Lipps Maze. However, a dry crawl continues straight ahead to a climbable drop into the Deems Stream just upstream of its junction with the Lipps Stream.

The Lipps Stream continues south from the junction in easy walking passage for 350 ft before passing a high lead on the west marking the entrance to T.H.E. Crawl, a dry passage requiring over 3,000 ft of hands and knees crawl before reaching the top of a 90-foot drop into the Waiting Room. Along the way there are plenty of opportunities to descend back into the Lipps Stream Passage which parallels T.H.E. Crawl in a deeply incised canyon as much as 40 ft below. About 600 ft from the entrance to T.H.E. Crawl, the Bone passage takes off to the west and splits into a myriad of muddy crawls. Another 300 ft south and the passage is almost blocked by a well casing from a well dug in the late 1970s.

Downstream 900 ft from its junction with T.H.E. Crawl, the Lipps Stream Passage has degraded into a narrow 6-foot-high canyon. Suddenly, the Lipps Stream disappears into Lipps Piracy Passage through one of many inconspicuous potholes. The Lipps Piracy Passage is a low crawl over a gravel floor which proceeds for 250 ft to the southeast before the ceiling and gravel come too close to allow further pursuit. However, the paleo-Lipps Stream canyon continues as before, but gradually becomes the very tall, narrow Hell's Fissure interrupted by numerous water-filled potholes separated by passage constrictions which force the caver to constantly change elevation. A new stream enters the fissure, and 400 ft downstream it cascades over a 21-foot drop from the base of Hell's Fissure into the Waiting Room.

The Waiting Room is a high dome with a 20-by-50-foot base. Both the 21-foot and 90-foot drops into the room require use of a rope or ladder. Proceeding 300 ft south in dry, breakdown-littered passage, one encounters the 50-foot-diameter Lunch Room with its large breakdown collapse pile at the convergence of five major passages.

The largest of the passages entering the Lunch Room is Handley's Lost Sandwich Passage which meanders 2,300 ft to the north as a breakdown-floored stream passage, typically 40 ft wide and 20 ft high. A considerable amount of silt covers the walls, floor and breakdown. This character is interrupted about 1500 ft from the Lunch Room by a collapse dome which completely fills the floor of the passage and forces the caver to ascend into the shale at the top of the dome 90 ft above. Coming back down to stream level the passage continues 40 by 20 ft until it abruptly ends in a breakdown collapse. Several routes have been found under the breakdown, but none have been found that lead to the hopefully large passage beyond.

To the west of the Lunch Room, Lunar Way runs 1,300 ft as a dry 20-by-20 foot silt-floored walking passage to Humphreys Stream Passage. The entrance to Lunar Way from the Lunch Room is on a high shelf along the west wall. Unfortunately, there is a short tight constriction before one can enter the large passage beyond. Lunar Way appears to continue to the east of the Lunch Room for 700 ft as Lunar Way Extended, low passage with several low wide rooms. Again, silt is present and may be responsible for closing Lunar Way Extended for each of its branches ends in silt.

The Lipps Stream flows south from the Lunch Room for 2,300 ft as breakdown-floored stream passage, sometimes as large as 50 ft wide by 25 ft high. Large silt banks line the route in sections.

13.3.13 Lipps Maze Section

The Lipps Maze section is a complex maze of 1.6 miles of predominately low, dry passages located between the Main Lipps Stream and the upstream end of Jones Canyon. To enter the maze from the Lipps Entrance, one only has to follow the stream discovered just inside the entrance for about 400 ft and then leave it for the first lead on the left (northeast). After 600 ft of stoopway and some crawling, the passage breaks into the west edge of a maze at the tight Rehajo Connector. It is not difficult to take the wrong turn in the mile of predominantly low, dry passages that in broken spider web fashion converge in an east–west central area.

The eastern edge of the maze is marked by the wide Flat Room from which eight short leads radiate. The southernmost lead off the Flat Room quickly becomes the western arm of Skid Row, a high narrow canyon that contains an intermittent stream and winds 300 ft south to its junction with Jones Canyon. About 100 ft downstream from the Flat Room, Skid Row splits in two. The right fork meanders, crosses the left passage, and then intersects with the eastern arm of Skid Row. The eastern arm of Skid Row can be followed 400 ft upstream in a 20-by-20-foot trunk passage to a breakdown collapse, or 150 ft downstream through a narrow stream fissure to its junction with the western arm of Skid Row.

13.3.14 Erwins Section

The Erwins section is bounded on the north by the Erwins entrance and on the south by Lunar Way. The water in its 0.4 miles of narrow stream passage is often deep, and three drops are encountered before the passage deteriorates into a belly crawl (often a sump) just before connecting with the Lunar Way. The Erwins section is the least attractive part of the entire cave system and is rarely visited.

The 2 foot high by 6 foot wide Erwins entrance captures the intermittent wet weather stream of a small valley. The entrance passage is a low, dry crawl for 200 ft before turning south into a stream canyon with a stream entering from the northeast. The next 800 ft of passage is low narrow stream passage with many pools requiring cavers to become totally wet before arriving at the 25-foot First Waterfall drop. After another 250 ft of similar passage a 19-foot drop is encountered at the Second Waterfall followed quickly by the 8-foot Third Waterfall drop. The ceiling gradually lowers until 150 ft later the caver is crawling in mud entering a 25-foot-long belly crawl through a seasonal sump. Once through the sump area, it is only 200 ft of dry cobble crawl to the junction with Lunar Way, a large dry trunk passage that intersects the Humphreys Stream Passage 450 ft to the west. The Erwins stream disappears in the cobbles and has not been traced further.

13.3.15 The Humphreys Section

The Humphreys entrance, Lunar Way and the Lipps Stream Passage provide access to the 5.1 miles of passage in the Humphreys section of the cave. The entrance passage is a low stream crawl initially over gravel, then scalloped rock, until the First Drop (23 foot) is reached. A low side lead to the west above the first drop provides a bypass; however, it also encounters a drop before rejoining the main passage. The combined passages continue low over a gravel floor until the 31-foot Second Drop enters The Highway, a large stream passage at base level with a breakdown floor. The Highway drains a maze of passages in North Humphreys which lie below the entrance passage. The Highway proceeds downstream and south as a large passage past the junction with the Lunar Way before it narrows and enters breakdown. It opens into large passage again at its junction with the Lipps Stream and continues south past a high side passage which leads to the Stadium and "S" rooms and the maze of passages along the west edge of the system. The combined Lipps and Humphreys Streams continue downstream and south for almost a mile in a wide passage with large breakdown. The Lower Lipps Stream Passage gradually narrows into a canyon and becomes smaller with a gravel floor before ending at the Lipps sump. The streams from Four Domes and the "S" Room form the Pythagorean Room waterfall and flow south into the Lipps Stream just before its sump.

The Humphreys entrance is in the same narrow valley as the Erwins entrance and downstream from it. It captures the overflow of the wet weather stream that normally enters Erwins and is separated from Erwins by a series of natural bridges. The Humphreys entrance is 6 by 6 ft, but quickly narrows due to a collapse pile from a skylight just inside the entrance. Immediately after the pile, the cave becomes 200 ft of gravel-floored stoopway with occasional shallow pools and surface debris, ending in a short drop into walkable stream canyon passage. In the early 1970s the same passage was a belly crawl up to the French Connection. The French Connection, a 1 foot wide by 2 foot high slit near the ceiling, was the normal route in preference to continuing through a short stoopway in 3-foot-deep water. Hurricane Camile cleaned out much of the gravel, and the 3-foot pool in the bypass to the French Connection has not been observed since. In fact, the original WVACS explorers said they did not use this entrance because it changed after each major flood event.

Once in the walkable stream canyon beyond the French Connection, it is 250 ft to the First Drop. It is easy to rig a rope or cable ladder out of the 23-foot waterfall which cascades into the shallow pool at the base of a high room. The passage continues as a crawlway, much of which is on gravel that filled the passage after Hurricane Camile, for 700 ft to the Second Drop. While the First Drop is along a wall and dry, the Second Drop is 31 ft long with the last half overhung and partially in the spray of the waterfall. The drop enters the ceiling of The Highway and deposits the caver on a large breakdown pile.

The 40-by-40-foot Highway runs upstream to the north for 1,200 ft to a sudden breakdown terminus. While most of the way is littered by breakdown, the gravel floor of Elysian Fields is especially pleasant. Just north of Elysian Fields are the large breakdown blocks of the Wombat Room with several high leads off to the east.

From the Second Drop it is 1,300 ft south to the junction of the Lipps and Humphreys Streams. While the first 500 ft after descending the breakdown pile is easy walking beyond the high junction with Lunar Way, the passage continues clogged with breakdown. The route requires a lot of scrambling, and dead end forays are common. The breakdown finally departs at the Lipps–Humphreys Junction, a flat area with a gravel floor (Fig. 13.9). Five hundred feet downstream from the junction is a flowstone cascade marking the entrance to the Rimstone Passage, and 500 ft further the cave floor is exposed showing polished rock with a blue tint. The gigantic breakdown block just to the south lies across from the overhung entrance of Purgatory Way.

Rimstone Passage includes several hundred feet of stream cascading down rimstone pools, the largest of which is 3 ft deep. The entrance to the Stadium Room is via a small hole near the ceiling just beyond this deep rimstone pool. Along the south side of the Stadium Room the passage drops into a stream way which winds south for 500 ft as stoopway to the 28-foot drop into the "S" Room. The "S" Room is more a section of walking passage that winds back upon itself, than a room. At its far end, a side lead takes off 1,000 ft to the north as a narrow stream passage to a series of Four Domes. A rope was left hanging in the fourth dome to allow ascending to the domes beyond. Just downstream from the "S" Room is the 27-foot waterfall and drop into Pythagorean Room where Purgatory Way joins at the floor. The passage continues downstream for 600 ft before dividing into upper and lower routes; the dry upper Angel Passage being much preferable. The crawl through Angel Passage joins the Lower Lipps Stream Passage 500 ft later just upstream from Lipps sump.

Returning to the section of exposed blue tinted bedrock in the Lower Lipps Stream Passage, the ceiling channel above enters the suspended Purgatory Way implying it was the paleo-water route. To enter Purgatory Way from this side requires traversing a 1-foot-wide overhung shelf along the west wall near the ceiling from the nearest breakdown block to the entrance, a very precarious and exposed route. Once in Purgatory Way, it continues west as a dry stoopway for 500 ft to a series of quick turns in tight passage before intersecting a north–south stream way. While the north stream passage soon breaks up and becomes small, the downstream route continues for 700 ft to the base of Pythagorean Room. But the downstream route is narrow, torturous canyon requiring constant changes in elevation to squirm through the constricted route.

The Lower Lipps Stream Passage south of the junction with Purgatory Way is a large two-level stream passage in which the correct route is sometimes the high level and sometimes the lower one. Inevitably, you will chose the wrong route and end up blocked by breakdown in the lower level, or suspended up high with no way to proceed. After 2,500 ft of trying multiple levels and scrambling over breakdown, the stream passage narrows and leaves most of the breakdown behind. After straddling a 3-foot-deep pool, it is an easy walk to the short stretch of breakdown marking the junction with Angel Passage.

Five hundred feet downstream from the junction with Angel Passage, the Lower Lipps Stream sumps in gravel. This section is pleasant walking passage with a gravel floor and gravel banks along the walls. A tributary stream enters from the southeast just before the sump, but its passage is too small to push.

13.3.16 Foxhole Cave

Although not physically connected, Foxhole Cave overlies the central portion of Organ Cave. Multiple generations of cavers have tried to connect Foxhole to Organ Cave by pushing tight leads and doing multiple digs to no avail.

The uppermost entrance is developed in a headwall at the downstream end of a broad sink. Thirty feet inside this entrance the cave passage is interrupted by a steep-sided trash-filled sinkhole. Foxhole Cave is developed in the

Hillsdale Limestone along the axis of the Caldwell Syncline. The cave is 2.75 miles in length.

Foxhole is a classic dendritic cave system; all three of the cave's major passages converge in the largest room in the cave, Desolation Row, located in the south. To the east of this room, a dry crawlway, True Grit, leads 1100 ft to several tight pinches. To the northwest, an intermittent stream way is developed 1300 ft to the Entrance Maze area and the two entrances. To the north, the major stream passage leads 1300 ft first through the Talcum Powder Room, then to the two-level Aztec Two Step, and Litsey's Leap. North and below Litsey's Leap is Handley's No-Go, a short, damp gravel crawlway that leads to the February Extension, a northern continuation of the Foxhole Main Stream. Developed along the Extension are the Debevoid, Hensley's Dome, Borden's Abortion, and the Jewel Box, a low formation area 900 ft north of the No-Go.

To the southeast of Desolation Row is a three-dimensional maze and to the west is the second-largest room in the cave, the Mynock Room. This room and the Debevoid are the only portions of the cave west of the Foxhole Main Stream and the Caldwell Syncline. Vertical and overturned strata can be observed above and west of the Debevoid.

South of Desolation Row a large walkway is developed south for 450 ft. This passage quickly becomes a stoopway, then a hands and knees crawl, and finally continues as a belly crawl. Its southern terminus is a clay plug. Immediately south of Desolation Row is a large pit with fluted walls. The pit's drain leads north to the Southern Maze, and an exposed lead under the pit accesses a north–south breakdown crawl that traverses the entire length of the higher larger room.

The stream from Hensley's Dome has been dye-traced into Desolation Row where it becomes lost in breakdown, later reappearing at Southern Maze and then exits the cave. It reappears in the Organ Cave System at Crowbar Crawl for a short distance before disappearing again, later flowing out of the base of the Thierry Room, down past PD-60 to Hedricks Sump and hence to Big Canyon. The Foxhole Stream has also been dye-traced to the east waterfall of the Waterfall Room.

13.4 The Geologic Framework of Organ Cave

Organ Cave is one of the best studied geologically of the caves in the Greenbrier Karst. The sections that follow (Sects. 13.4–13.8) are drawn from Deike (1988), courtesy of the West Virginia Speleological Survey.

13.4.1 Exposed Stratigraphy in Organ Cave

The cave contains extensive exposures of bedrock. Solutionally weathered walls reveal bed thickness and chert and shaly layers quite well. However, fractured surfaces, which are very common, often conceal even the presence of chert. Mud and silt obscure the details in many areas. The Maccrady–Hillsdale contact can be observed in many areas such as in the Organ Main Stream Passage along the tourist route, at the Fun Room and Salle de Martel, in the Sarver Room, in Hedricks Maze, and in Lower Lipps Stream Passage (Fig. 13.14).

Fig. 13.14 Maccrady shale and base of Hillsdale Limestone exposed at the *bottom* of the Salle de Martel. The dip of the beds increases rapidly to near vertical about 100 ft to the *right* of the area shown in this photograph. Photograph by W.K. Jones

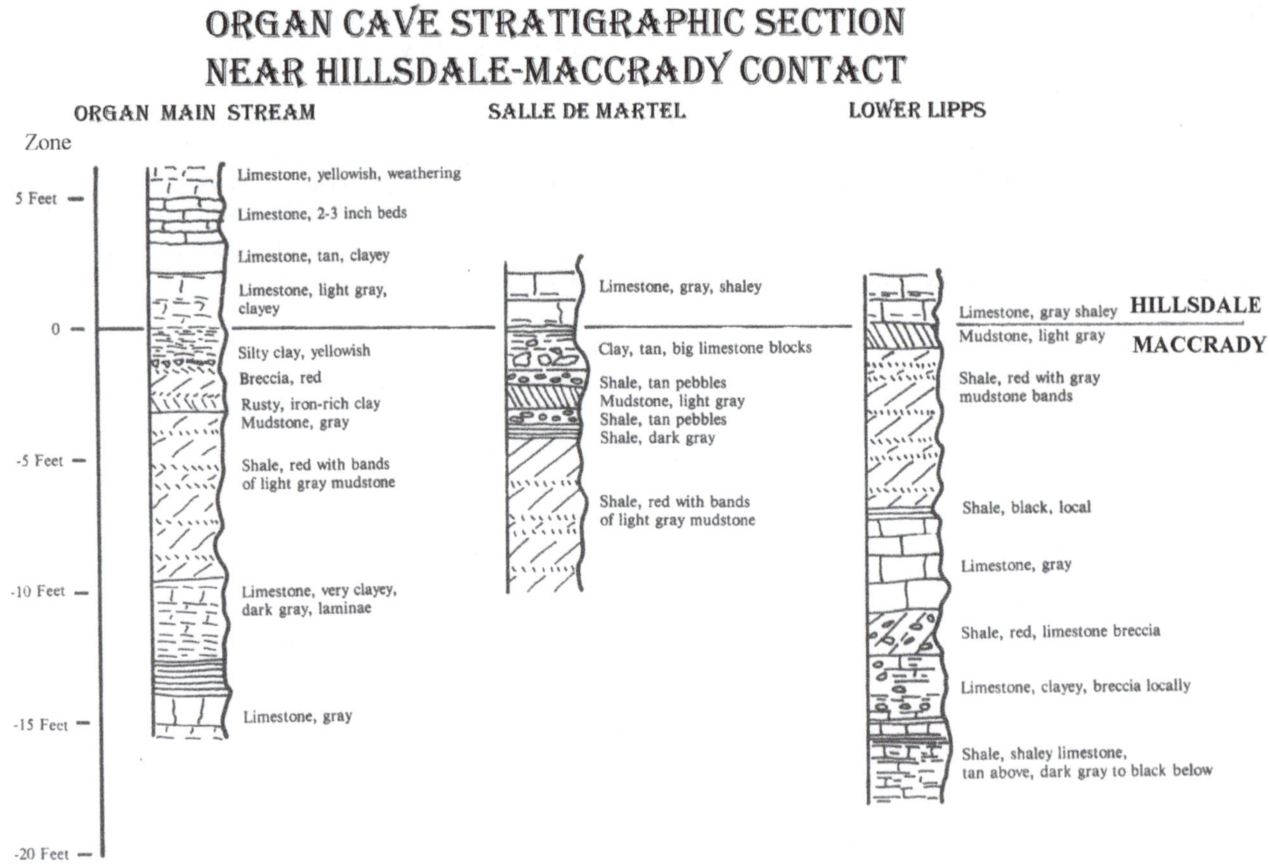

Fig. 13.15 Stratigraphic section in Organ Cave at the Maccrady–Hillsdale contact

Figure 13.15 presents several sections of the contact zone. In general, some thin to shaly beds occur in the Hillsdale Limestone above the Maccrady shale contact. A very shaly bed 3–4 ft above the contact is exposed above the waterfall in Organ Main Stream Passage. A gray or tan mudstone 8 in. to 2 ft thick often marks the top of the Maccrady Formation. Locally, thin tan or even white beds occur in the red shales. In Lower Lipps and in Floyd Collins Avenue, a limestone is exposed 8–10 ft below the top of the Maccrady, and in Lower Lipps another calcareous bed is about 16 ft below the contact. Heller (1980) reports frequent discontinuous layers of yellow to tan argillaceous limestones and calcareous shale in the uppermost Maccrady.

At most exposures the contact seems gradational, but some sections, including those near Salle de Martel, show tan shale pebbles in the shale matrix in the top foot or two of the Maccrady, the "reworked" Maccrady of Rutherford and Handley (1976). In Lower Lipps red shale breccia is exposed 10 ft below the Maccrady contact.

In most parts of the cave no special solution features were noted at the shale contact which seems to have been reached by solution and erosion in passages which originated 10–15 ft above the contact in the limestone beds. This situation is different from that reported in Ludington Cave by Palmer (1974), where major passages originate by solution of the basal Hillsdale beds at the Maccrady shale contact and were enlarged primarily by erosion of the underlying shale. The seeps which have deposited considerable flowstone in Floyd Collins Avenue are mainly found 3 ft above the contact, but some seeps are 2–3 ft down in the shales. The lower Hillsdale and the upper Maccrady are a productive aquifer for wells in Monroe and Greenbrier Counties. On the basis of chemical differences between well water and water in cave–spring systems, Heller (1980) concludes that the well water is drawn from a confined diffuse-flow aquifer. Fracturing which produced blocky weathering in the upper Maccrady might explain its being permeable to water, and dolomitic rock reported in lower Hillsdale (Heller 1980) may also be porous. Probably much of the well water is from the coarse-grained horizons in the lower Hillsdale, where dissolution is also very common. The cherty zone 15–25 ft above the contact in the Organ Cave System is typical of this permeable and easily weathered zone.

The Greenbrier Group rocks exposed in the east and central parts of Organ Cave System (east from Flack Room) are illustrated in Fig. 13.16. The section has been labeled

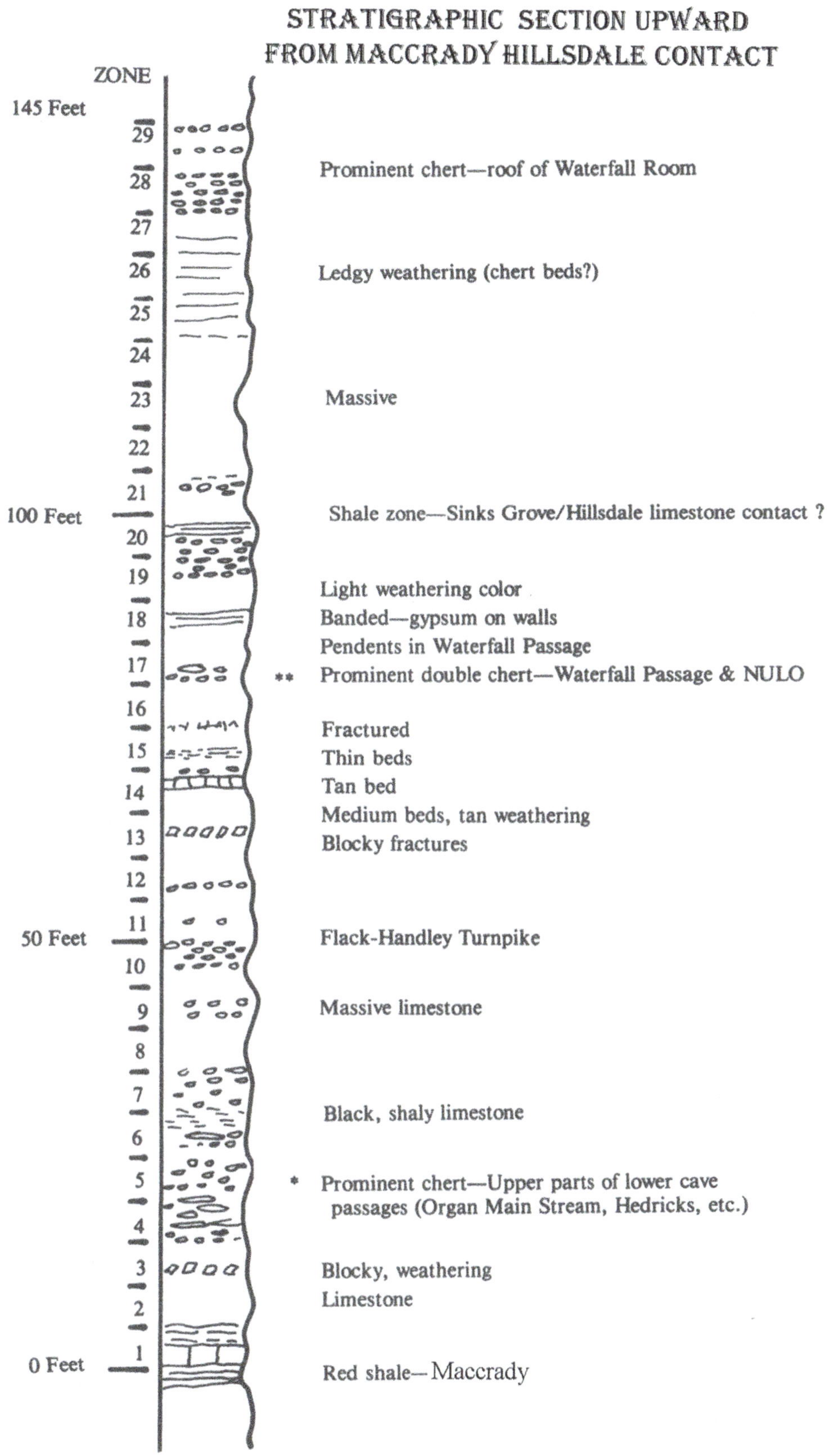

Fig. 13.16 A 145-foot stratigraphic section upward from the Maccrady–Hillsdale contact. East side of cave system

"zones 1–29" in 5-foot intervals upward from the Maccrady shale contact. The lowest 25 ft (zones 1–5) is widely exposed in the Organ Main Stream and Hedricks Stream Passages, in Lower Lipps, and in numerous other passages at or immediately above the Maccrady contact. Zones 14–19 are exposed in the Upper Stream Passage and adjacent passages. The beds between zones 7–13 are best seen in the USP–Hedricks Connector, the chimney that connects Old Saltpetre Route near Hedricks Stream Passage with the overlying Upper Stream Passage, where the section seems to be undisturbed. Other exposures of this section between Organ–Hedricks and Upper Level Organ are in areas of folding and/or faulting.

The rocks exposed in the cave system in and west of Flack Room and Handley Room are shown in Fig. 13.17. The lower part of this section including the top of the Maccrady shale is exposed in Lower Lipps. The section up to zone 16 is visible in Humphreys entrance passages, where it is unusually cherty. The section can be followed upward from its base to the prominent double chert in zone 18 in Hells Fissure. This chert horizon is widely exposed near the roofs of passages from Lipps through Jones Canyon and Left Hand Passage to Handley Room. Rocks above this chert are seen in the Belfry above Jones Canyon.

The section shown in Fig. 13.17 is separated from that shown in Fig. 13.16 by faulting in the Flack Room–Handley Room area. As far as observed it differs somewhat in detail because of lateral variations in stratigraphy. For example, the cherty zone 4 is present, but the nodular cherts of zone 5 are not obvious in Lipps. Zones 6 and 7 are very cherty in Humphreys, but show only isolated chert beds in upper Lipps or Organ. The double chert seen in Waterfall Passage and NULO is about 82 ft above the Maccrady shale in Fig. 13.16. The prominent double chert in passage roofs west from Handley Room is about 85 ft above the Maccrady shale in Fig. 13.17. Thus, this zone 17 chert (Fig. 13.16) is considered to correlate with the zone 18 chert of Fig. 13.17. In the outcrop above Humphreys entrance there are thick shaly beds about 90 ft above the base of the limestone. This is quite possibly the shale used by Ogden (1976) to mark the top of the Hillsdale, but they do not correlate easily with the thinner shaly horizons seen in the east side of the cave. Careful examination of the Humphreys entrance area, and of beds exposed in the high walls of the Waterfall Room, may well clarify the relationships. One particularly interesting bed is zone 6, which collapses to form piles of black shaly limestone chips in some areas. Examples are in the south end of 1812 Room (Fig. 13.18), in Octopus Alley, and notably in the "Shale Rooms" along Hedricks Stream upstream from Sarver Room.

A fault isolated section occurs in the Revak Room and passages to the east toward Raccoon Room, Fig. 13.19. The paired cherts in this section may correlate with cherts of zone 17 in Fig. 13.17. Correlation of the section in the cave with published reports is problematic. Wells (1950) reports abundant chert in the lower Hillsdale, but Ogden (1976) does not. Ogden reported chert as common in the overlying Sinks Grove, but only scattered chert is usually mentioned in these beds. In Organ Cave System there is fairly abundant chert throughout most of the section. The position of the Hillsdale–Sinks Grove (or Denmar) contact has not been fixed. The top 10–15-foot shaly bed Ogden used to mark the top of the Hillsdale was not identified. The highest thin-bedded or shaly zone identified in the cave (Fig. 13.16) is in the middle of zone 20, 97 or 98 ft above the base of the Hillsdale formation. The thicker shaly beds exposed above Humphreys entrance (Fig. 13.17) are about 90 ft above the base of the Hillsdale formation. This is conveniently near the reported thickness of the Hillsdale in the area, but only more detailed work might locate the contact with certainty.

13.4.2 Stratigraphy of Foxhole Cave

Foxhole Cave lies along the axis of the Caldwell Syncline 150–200 ft above the horizon of the Maccrady shale. The entrance maze lies in and below a coarse limestone with fossils and some chert. Immediately above is a thick cherty zone. About 18 ft below the fossiliferous limestone is a sequence of very cherty beds about 17 ft thick. This is followed downward by 22 ft of limestone which is massive above and thin-bedded below. There is some chert in the beds just below these limestones.

The observed section, some 60 ft thick, is probably all in the Sinks Grove Limestone. No definite correlation with the rocks exposed in Organ Cave System has been established. The extensive chert seen in the roof of the Waterfall Room is 140 ft above the base of the Greenbrier Group and are the only rocks in Organ Cave System that might correlate with any of the cherty horizons in Foxhole Cave.

13.4.3 Relation of the Cave to Geologic Structure

The Organ Cave Plateau is developed in the center of the Caldwell Syncline, whose axis is shown in the county geologic map (Price, 1939) as 0.25 mile west of the Organ Entrance. In fact, the Hedricks Stream Passage in the cave is rather close to this axis. The fold has a 2–3° plunge to the southwest. Regional strike and trend of the fold axis is about N30°E. Complex details of the structure of the syncline are revealed in the cave. The axis of Brown's Mountain Anticline lies about four miles away to the southeast, while that of the Sinks Grove Anticline is about 3.5 miles northwest. The Caldwell Syncline is bounded nearby by minor

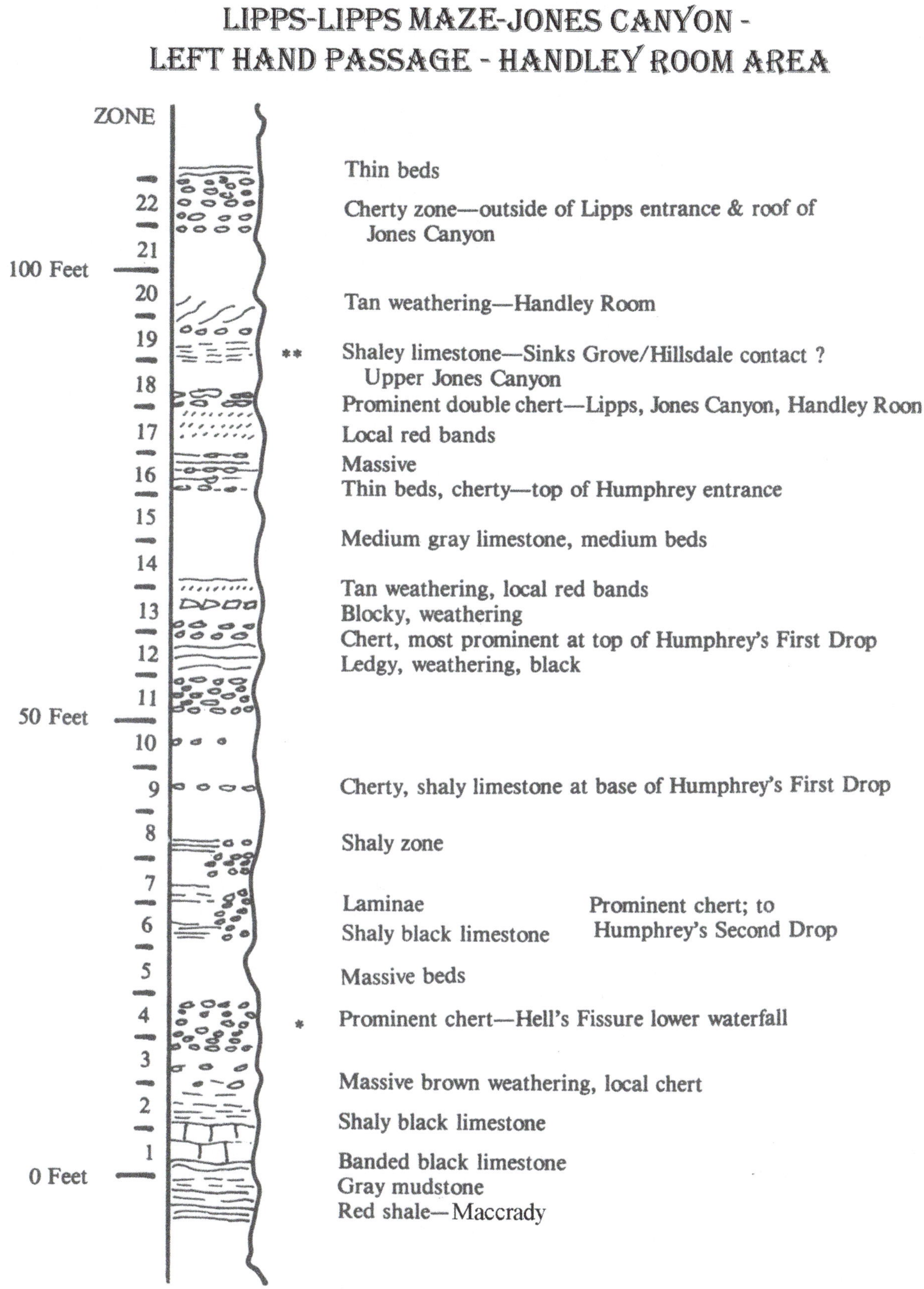

Fig. 13.17 Stratigraphic section in Lipps–Lipps Maze–Jones Canyon–Left Hand Passage–Handley Room area

Fig. 13.18 Shale outcrop and breakdown pile in 1812 Room. Photograph by W.K. Jones. Used with permission

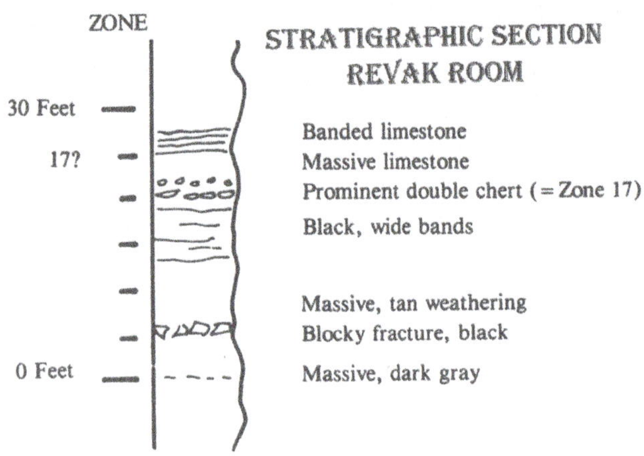

Fig. 13.19 Stratigraphic section in the Revak Room

anticlines, the Maple Grove Anticline about 3500 ft southeast of the synclinal axis, and an unnamed anticline about 9,000 west. The cave does not extend to either of these. Ogden (1976) says he agrees with Price (1939) on the location of these fold axes where they cross the county line at Second Creek. Actually, Ogden's detailed map seems to place the synclinal axis about 600 ft further west than Price's map. Ogden's location would place the axis about in line with Wilson's Watery Wonderland, the downstream continuation of Organ–Hedricks Stream beyond Hedricks and Bowen sumps.

Ogden reports that Winslow and Baroody (1976) mapped minor folds and many small faults in the Pocono Group near the county line. Ogden found only a few reverse faults in the Greenbrier Group, mostly with displacement estimated at 20 ft or so. Some beds of the Greenbrier also exhibit swarms of vertical clay-filled partings interpreted as solution pressure cleavage. The solution cleavages may account for considerable compression of the Greenbrier rocks. Fairly numerous small reverse faults, with displacements usually less than 2 ft may be observed. These sometimes splay from bedding planes into overlying beds. In addition, breakdown often reveals slickensides on inclined rock surfaces where the displacement seems small, but cannot be determined. Slickensides are also found on bedding in areas of deformation where the dip is steep. Finally, there is reverse faulting, with known displacement in one case of 20 ft, associated with the abrupt monoclinal folds that border the axis of the Caldwell Syncline in the cave. The lower Greenbrier has undergone considerable complex deformation, and this has had notable effects on the development of the Organ Cave System.

Jointing is not nearly as well developed as in the folded Appalachians, but numerous joints with rather irregular spacing and often limited extent are present. Ogden (1976) measured 873 joints in Monroe County, finding the most common set at around N37°E, parallel to regional fold axes. Diagonal joints at about N27°W and N85°W were fairly common, as were cross joints at about N58°W. Jointing has had an effect on orientation in various passages in Organ Cave System.

13.5 Passage Development and Stratigraphic Relations in the Basal Hillsdale

The passages in the Organ Cave System show a great deal of stratigraphic control. Extensive parts of the system are developed in zones 1–5, the basal 25 ft of the Greenbrier Group (Fig. 13.20). The major passages are the main stream passages (Hedricks, Organ Main Stream, Lower Lipps), with free-surface streams which have eroded into the underlying Maccrady in many places, sometimes to a depth of more than 15 ft. Weathering and erosion of the Maccrady shales has permitted widespread collapse of the Greenbrier limestone. There does not appear to be much evidence of initial solutional development in the contact zone.

Most of the passages extend upward into zones 4 and 5, a coarse bioclastic limestone with chert beds (particularly in zone 4) and nodules and rods (mostly in zone 5). The passages generally have ceiling channels and joint crevices in these beds. There is some anastomosing of the channels, and some lead away from the present stream passages, for example the Discovery Passage (Fig. 13.21). Many of these diverging passages are blocked by sediment. They tend to be tubular (sometimes with incised canyons) or fissures showing joint control. It is likely that the passages had their beginnings in a network of water-filled tubes which followed zones 4 and 5. They are often of considerable size and probably became part of a loosely organized conduit flow system before they were abandoned. The lower and larger parts of the passages are much modified by breakdown, but appear to be the result of solution and erosion by free-surface streams such as those that now occupy them.

The Hedricks Stream Passage extends thousands of feet down the plunge of the Caldwell Syncline, so that there is an elevation difference of more than 400 ft from the upstream (blind valley) to downstream (sump) end of the passage. For any point along this route there would have been a time when the passage downstream was water-filled, while upstream the passage contained a free-surface stream. The water-filled parts would still be particularly sensitive to stratigraphy and structure and apparently composed of a somewhat complex series of tubes. Heller found that well water, usually from a diffuse-flow aquifer, was chemically different from cave and spring water from conduit flow systems. Presumably, the water in these passages in Organ Cave System was not carried into a diffuse-flow aquifer. Spring outlets must have been established before the openings could enlarge significantly, and conduit flow was then quickly established.

Upstream (and up-plunge) the free-surface streams selected a single route with very few loops, dissolved and eroded to the base of the Greenbrier Group, and have continued to erode into the Maccrady shale. The process of concentration to a single flow route probably began in the water-filled passages in response to high-flow conditions and might be expected to have been nearly completed before the passages became air-filled, but there is little evidence of an enlarged water-filled conduit visible in the stream passages today. Careful observations by divers might clarify the picture, since the downstream ends of Organ Cave System passages are still water-filled.

13.5.1 The Organ Main Stream Passage

Organ Main Stream Passage presents a complex case (Fig. 13.21). It lies on the east flank of the Caldwell Syncline. The passage trends down-dip in most of the Organ Cave tourist section, with development in zones 4 and 5, and downcutting well into the Maccrady in many places. Open passages in beds 4 and 5 join at Organ Main Stream Passage and Fat Man's Misery, and there are some sediment-filled openings in these horizons as well.

Some 200 ft downstream from Fat Man's Misery the Organ Main Stream Passage becomes strike-oriented, and the original, upper part of the passage continues 800 ft along the strike toward the 1812 Room above the waterfall that enters Organ Main Stream here. The 1812 Route is in zones 2 and 3, and any continuations of the development in the cherty 4 and 5 zones are lost until the Rotunda Room. From here through the 1812 Room the Old Saltpetre Route is in zones 4 and 3, again showing some sign of initial development in the cherty zone 4, particularly south of the 1812 Room. The trend here is diagonally down-dip.

The Lover's Leap Passage leads from the 1812 Room in zones 5 and 6, and its dissolutional walls and irregular roof profile suggest phreatic development. Toward its silt-filled termination it has risen through the beds into zones 6–9, but its elevation is about 50 ft below the strike-orientated reach near Rotunda Room. Its elevation suggests it continues as part of Revak Room, but it is separated from those passages by a major fold and faulting. Lover's Leap was probably the earliest flow route used by the ancestral Organ Main Stream south of Organ Entrance.

A route the full length of the 1812 Room was next to develop. Beyond the south end of the 1812 Room, flow then continued south a short way to enter Lover's Leap Passage. Flow paths which developed from here followed cherty zones 4 and 5 down-dip, dropping 110 ft in elevation along

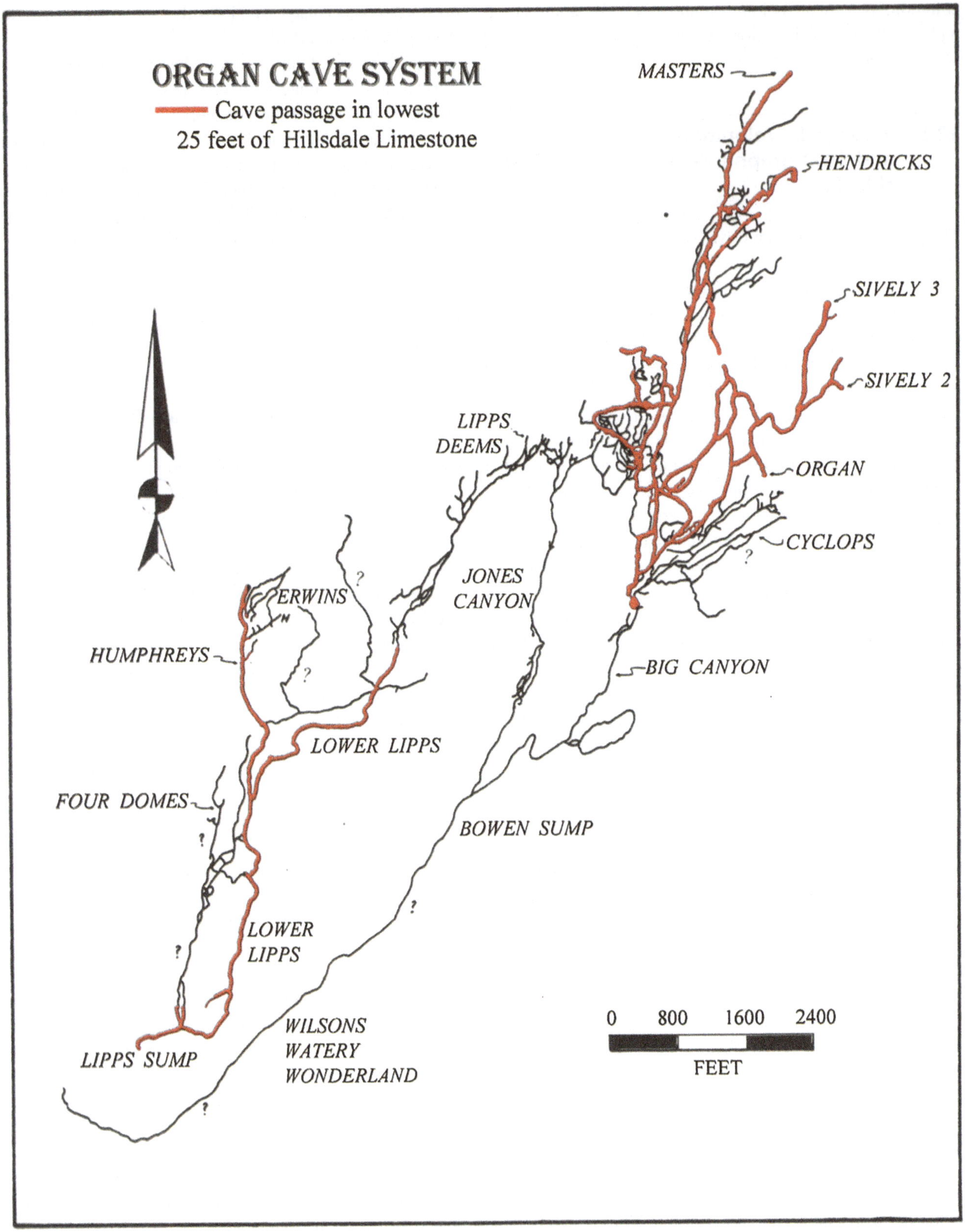

Fig. 13.20 Cave passage in the lowest 25 ft of the Hillsdale Limestone

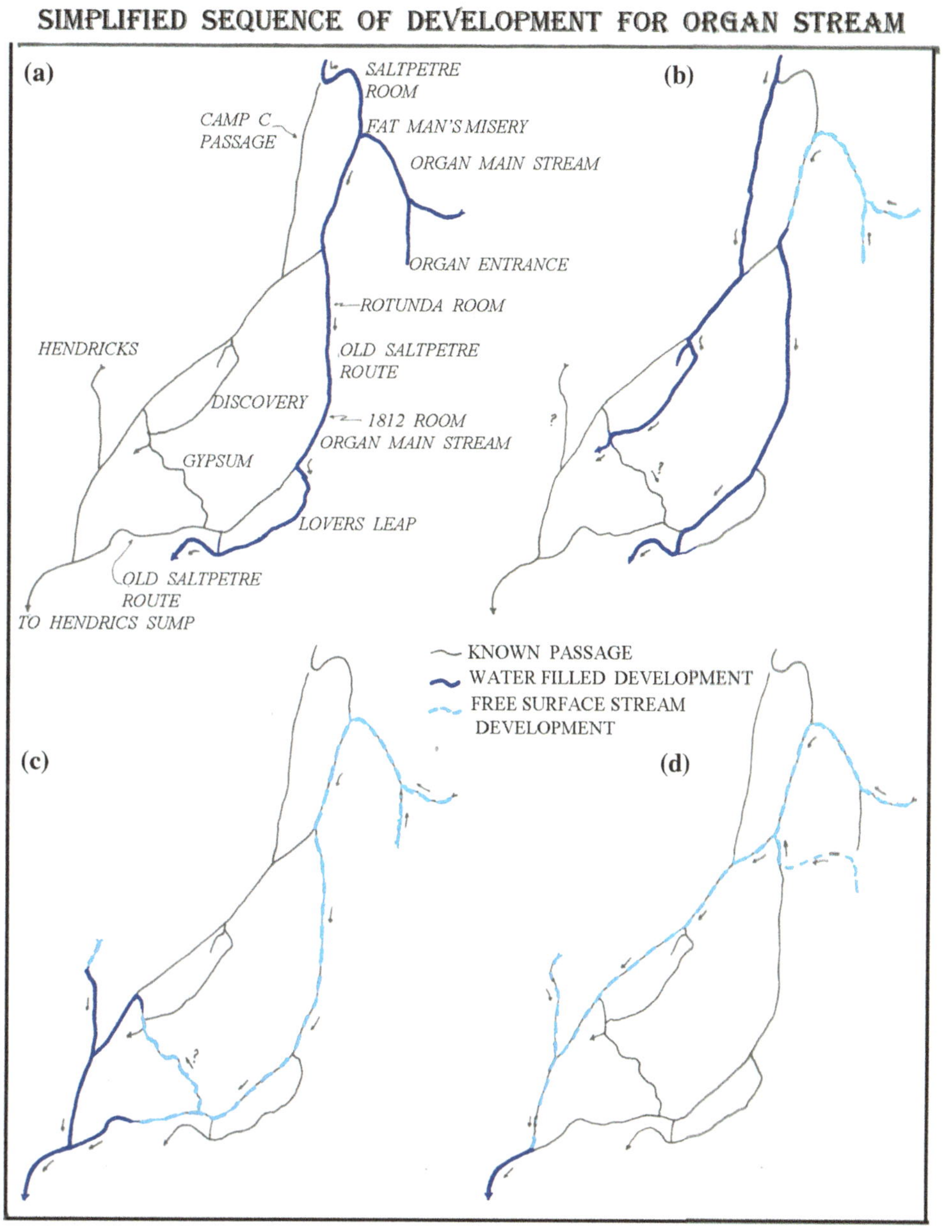

Fig. 13.21 Simplified sequence of cave development in main organ area

the Old Saltpetre Route to the vicinity of the present Organ–Hedricks Stream. Two routes down the steepening dip were developed under phreatic conditions. The Old Saltpetre Route (and Gypsum Passage) may have originally developed to carry water from evolving Hedricks Stream upward to join early Organ Main Stream at this point, as described later. The Organ Main Stream was diverted from this route by development of the present route beyond the waterfall. The 1812 Room–Old Saltpetre Route continues to be enlarged by occasional flow from the Organ Entrance Stream.

From the waterfall to its junction with the Hedricks Stream, the Organ Main Stream Passage is oriented diagonally down-dip at varying angles. In this reach there are extensive openings in zones 4 and sometimes 5: Camp C Passage, Discovery Passage, and Gypsum Passage (which extends all the way to Old Saltpetre Route). These are continuous with various ceiling channels in the Organ Main Stream Passage. The Organ Main Stream seems to have used these preexisting openings to develop a new route down to the Hedricks Stream Passage. Some of these passages may already have been carrying water from the Hedricks Stream route, either from Paleo Saltpetre Passage via Camp C Passage, and/or through the ceiling tubes just upstream from Organ–Hedricks Junction. The first flow from the waterfall area of Organ Main Stream could conceivably have followed Gypsum Passage to Old Saltpetre Route and Lover's Leap. It might also have temporarily gone up to Throne Room in Upper Stream Passage, which is about 20 ft lower in elevation than Lover's Leap. These must have been very temporary routes, because they consist of small passages. The Organ Main Stream must have made the transition from the 1812 Room route to the present route in a rapid sequence of shifts. This makes it seem probable that the Hedricks Stream became a free-surface stream near the present Organ–Hedricks Junction soon after the new route developed.

A possible history (Fig. 13.21) is then: Organ Main Stream develops the strike-oriented route toward 1812 Room close the base level. At this time Hedricks Stream would have reached its sump far upstream, above Rimstone Ceiling Passage junction. Indeed, there is evidence that the water from the ancestral Hedricks Stream joined Organ Main Stream by way of the Paleo Saltpetre Passage and the Saltpetre Room as described later. While base level fell 150 ft and Hedricks opened and developed its down-plunge route near the synclinal axis, Organ Main Stream continued to traverse the 1812 Room, 800 ft to the east and up the flank of the fold, by now a free-surface stream. Development of Camp C Passage and Discovery Passage would date from this interval, the water derived from Hedricks. Hedricks water may even have followed the beds upward along Old Saltpetre Route to join Organ Main Stream at one time. Organ and Hedricks water first joined with minor flow through Gypsum Passage. Fairly early in this interval the Organ drainage began to use or open the Old Saltpetre Route down toward the synclinal axis. Finally, the Organ Main Stream developed a cutoff from the waterfall into the Camp C Passage–Discovery Passage openings and integrated a new route all the way to the present Organ–Hedricks Junction.

13.5.2 Organ Entrance Stream

The stream which presently drains the large blind valley into the Organ Entrance has had a complex history. During recent vadose development of this part of the cave the stream probably entered Cyclops Avenue for a long period of time, but finally abandoned Cyclops for Organ Entrance. Organ Entrance is 400 ft further downstream in the blind valley than Cyclops, diagonally down the dip of the top of the Maccrady shale. The elevation is about 40 ft lower than Cyclops. The phreatic ancestors of this drainage developed the high-level passages leading toward Revak Room from the area of the Raccoon Room. The same drainage later initiated the lower passages including Cyclops.

In Organ Entrance the stream went down-dip to enter the 1812 Room and followed Old Saltpetre Route down to Organ–Hedricks Stream. This route is still used in high water. But very recently bulldozer work in the creek bed 50 yards upstream from the entrance caused the creek to sink short of Organ Entrance at low water stages. The water now bypasses the 1812 Room and follows the passage north to the Waterfall, where it drops into Organ Main Stream.

13.5.3 Hedricks Stream

At the time Organ Main Stream was flowing to the 1812 Room, Hedricks Stream would have reached base level near the present junction with Rimstone Ceiling Passage or Paleo Saltpetre Passage, 2000–2500 ft upstream from the present Organ–Hedricks Junction. In that area of Hedricks there are several large passages lying east of the Hedricks Stream Passage (Fig. 13.22).

At Paleo Saltpetre Junction, the upper half of Hedricks Stream Passage, developed in zones 4–6, goes to the east to become Paleo Saltpetre Passage, while the lower half in beds from zone 3 down into the Maccrady shale continues as modern Hedricks Stream Passage. With a broad canyon as its floor, the Paleo Saltpetre Passage continues a little east of the synclinal axis to a junction with Rimstone Ceiling Passage. The canyon is lost in fill at the bend only 50 ft from the junction. From here it probably followed a different route, now filled, developed in zones 4 and 5, to join the present Hedricks Stream Passage route at or near Shale Room.

The tubular upper part of the Paleo Saltpetre Passage joins the upper part of Rimstone Ceiling Passage coming down-dip from the northeast. These flows combined to follow Paleo Saltpetre Passage to the south along the strike, in zone 5, probably to continue through the Saltpetre Room of Organ Cave, to join Organ Main Stream drainage and go down the 1812 Route thence on through 1812 Room. Most of this route was water-filled during development. The elevation difference along this strike-orientated route from Paleo Saltpetre Junction to Rotunda Room is about 60 ft. The next route opened was the upper tube of Rimstone Ceiling Passage south to its junction with modern Hedricks Stream Passage. This might be the first time Hedricks water used its present course from there south toward Hedricks sump.

Another possibility, suggested by the ceiling tubes and development in zones 4 and 5, is that Hedricks water got only as far as 100 ft above Organ–Hedricks Junction, then went east through "T"-Room and up Gypsum Passage to join Organ Main Stream beyond 1812 Room. Water following this route, perhaps including water from Organ Main Stream as it developed a route toward Organ–Hedricks Junction, may next have forced the route upward into Upper Stream Passage at Throne Room. The next development could have been the passage parallel and east of Hedricks Stream downstream from Organ–Hedricks Junction. The flow might have been upward along zones 4 through Old Saltpetre Route, and/or up the chimneys of the

Fig. 13.22 Simplified sequence of cave development for Hedricks Stream Passage

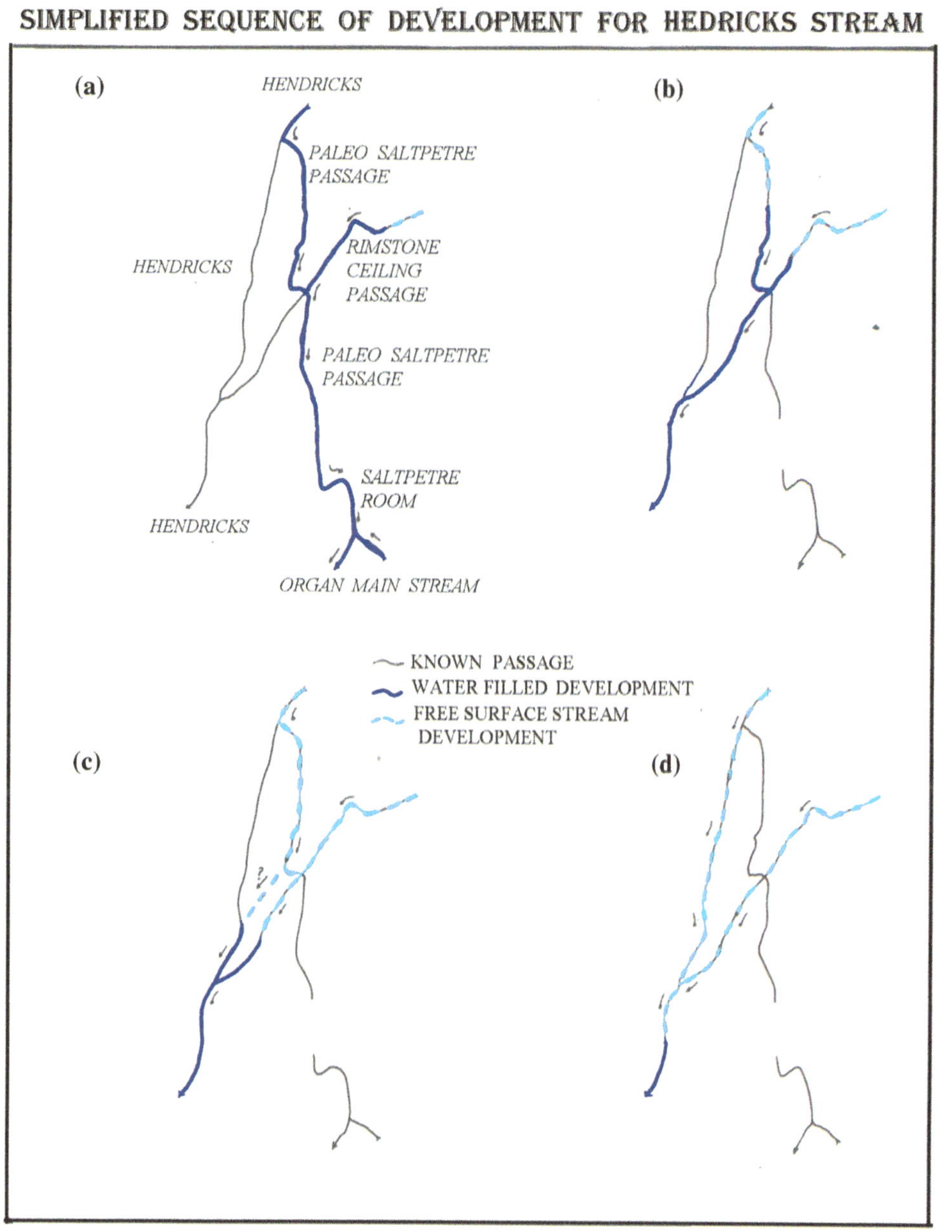

USP–Hedricks Connector to Upper Stream Passage. Last would be opening the route to Hedricks sump. After a short interval the flow developed the hypothesized route to Shale Room and another segment of modern Hedricks Stream, from there to Rimstone Ceiling junction. The water from upstream Rimstone Ceiling Passage continued to use that passage all the way south. Both streams incised canyons in their routes, both were now free-surface streams. Finally, the main flow forced a route from Paleo Saltpetre Passage junction to Shale Room, partly in zone 2 at first, in an area where the stream is now incised 10 ft into the Maccrady. Thus, the Hedricks Stream shifted in stages to follow the synclinal axis as base level fell.

13.5.4 Lower Lipps

Humphreys and Lower Lipps Streams are developed mostly in the base of the Hillsdale. The Humphreys entrance is high in the stratigraphic section, about 75 ft above the Maccrady. The First Drop is over cherty beds, and its base is at a shaly horizon in zone 9. Second Drop is over very cherty beds of zones 6 and 7. Some cherty horizons (zones 4 and 5, zone 17 in eastern Organ) seem susceptible to initial phreatic development, but the chert horizons in Humphreys entrance passage are holding up downcutting by the free-surface streams. Humphreys and Lipps Streams have initially developed in cherty zone 4 and above, much like the situation seen in Organ–Hedricks Stream, and both are downcut to or below the Maccrady shale contact. This situation continues nearly 1000 ft downstream from the Lipps–Humphreys Junction, but at Rimstone Passage and finally at Purgatory Way, the openings in zone 4 trend off to the west into a complex of passages that were probably the earlier route of this drainage, lying further up-dip.

Approaching Purgatory Way, the passage floor is cut ever deeper into Maccrady rocks, revealing a bluish limestone bed 9 ft below the contact. At Purgatory Way the development in zone 4 leaves the present Lipps Stream and follows Purgatory Way to the west. Just below Purgatory Way the Lower Lipps Stream Passage is entirely in Maccrady: the upper tube in shale, the lower canyon in limestone and clayey limestone below. Some 600 ft downstream the passage is developed in a few feet of basal Hillsdale, but cut 17 ft into shale, limestone, and limestone breccia. The dip of bedrock and passage gradient increase for about 400 ft with similar rocks exposed, then the gradient flattens and in 300 ft the shale passes beneath the floor and cherty beds of zone 4 appear again in the roof. This part of Lower Lipps is the best case observed for flow beginning a new route at the contact and even in the Maccrady shales. The reason why the water was able to initiate development of part of the route below Purgatory Way in these rocks is not known in detail, but obviously they were permeable, apparently locally more so than the lowest Hillsdale beds. This shaly contact zone is locally a principal aquifer for area water wells.

From Skyline Caverns Crawl to the junction with Angel Passage the roof of Lower Lipps Stream Passage is usually in zone 4, and the floor either bedrock- or gravel-covered is several feet above the Maccrady shale. In the last 400 ft of this section the passage turns west, nearly parallel to the strike, no longer running obliquely down-dip.

In the final 500 ft to the Lipps sump the whole Lower Lipps Stream Passage rises through the stratigraphic section, and the roof finally reaches the top of zone 6, about 15 ft above cherty zone 4 and 35 ft above the Maccrady shale. The passage follows the local dip direction as it climbs through the bedding. In part, this dip is near west, away from the Caldwell Syncline, and there must be a local wrinkle in the structure, almost a small anticline, to account for this. At the sump the dip steepens to 8° in the direction of the plunging folds. The original tube was not graded and gained elevation in the downstream direction in part of this area. It is now downcut by a free-surface stream.

The elevation of the sump is about 30 ft above the elevation of the Organ resurgence, which is about 3500 ft distance at S75°W. This would place the Maccrady–Hillsdale contact at Lipps sump at the same elevation as at the resurgence. David's Dungeon, the final Organ sump, is 1,500 ft in the same direction. Its stratigraphic horizon is unknown, but it is likely well above the Maccrady. Near the resurgence the dip is reported as 15° toward the synclinal axis. This is enough that the beds may rise to bring the contact to the surface at or near the resurgence. But it is also possible that the resurgence is from beds 100 ft to 200 ft up into the Greenbrier Group.

13.5.5 Fun Room–Flack Room–Floyd Collins Avenue

There is extensive development of passage in the lowest 25 ft of the Hillsdale in the Fun Room–Flack Room–Hedricks Maze area west of the synclinal axis. These passages carry water from a blind valley west of US 219, 3,000 ft north of its intersection with WV 63. The large passages trend down-dip to the Hedricks Stream Passage near the fold axis.

Floyd Collins Avenue (Fig. 13.23) has a roof in zones 2 and 3, and the floor usually extends into Maccrady shale. Zone 4 cherts are exposed in the room at its north end, but development of solution passage in this bed at this location is a continuation of strike-oriented Rat Alley Crawl, which is a tubular conduit in zone 4, not obviously related to the development of the Floyd Collins Avenue.

Down-dip Floyd Collins Avenue opens into Flack Room, where the cherty zones 4 and 5 are exposed in the upper walls and ceiling. At the southeast end of the Flack Room the dip steepens from about 10° to as much as 40° for nearly 200 ft before flattening out nearer the synclinal axis. This monoclinal fold shows some faulting. Several passages follow the beds down the fold to reach Slate Creek Passage at or near the Fun Room (Fig. 13.23).

The Subway is developed in zones 3, 2 and 1, and leaves Flack Room behind breakdown to the southwest. Locally eroded into Maccrady, it reaches Slate Creek at lowest floor level. The route directly to Fun Room is a group of wide interconnected openings following cherty zones 4 and 5 down the monocline. These become a single wide tube on top of the false floor of the Fun Room, where they are joined from the northeast by strike-oriented Octopus Alley, also in zones 4 and 5. Fun Room joins Slate Creek as the upper part of that passage, whose floor is in Maccrady shale. There is a lower level to Fun Room, in zones 1–3, eroded into the Maccrady. This level carries some modern drainage, probably from Floyd Collins Avenue. Under the east wall of Flack Room another passage begins in zones 6 and 7, but soon drops to zones 4 and 5 as it drops steeply down the dip to intersect Octopus Alley and continue to Slate Creek. A canyon in its floor also crosses Octopus Alley. Directly above this passage is a wide opening in zone 17, at ceiling level of Flack Room, with a canyon cut below it. This passage leads by two routes to the Centrifugal and, hence, to Sand Room on the Upper Stream Passage level of the cave, all in the same stratigraphic horizons. These two exits from

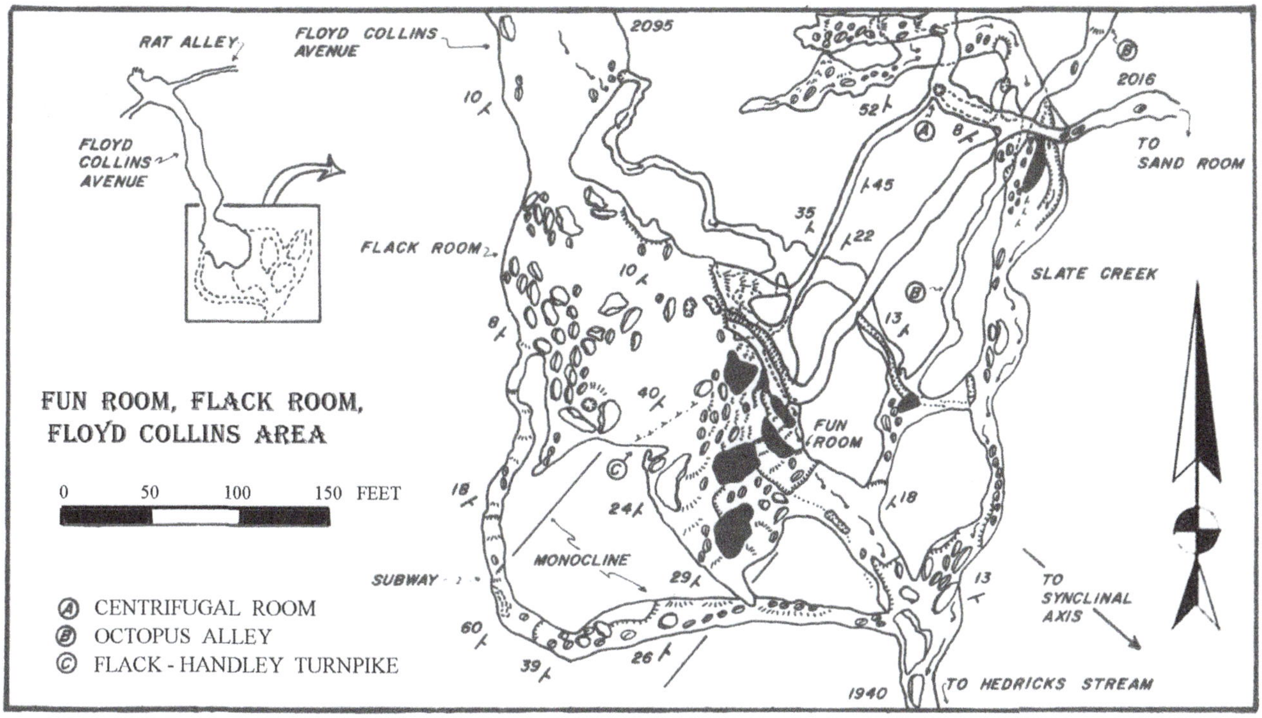

Fig. 13.23 Fun Room–Flack Room–Floyd Collins Avenue area

Flack Room are perhaps 30 ft apart vertically, but in beds 50 ft apart stratigraphically: Faulting accompanies the monoclinal fold here. A final passage from this end of the Flack Room leads upward through the stratigraphic section above Flack–Handley Turnpike, to enter the overlying Handley Room. A possible history of Floyd Collins Avenue and its downstream branches takes as a starting point that the route has developed specifically to carry water from the surface drainage ancestral to the present blind valley. No evidence suggests its evolution from earlier passages. Development did not begin in the cherty 4–5 zones, but closer to the Maccrady contact, quickly opening a single route down-dip as far as the Flack Room. If the route once terminated at, and was tributary to, Rat Alley Crawl, this is not evident—probably it did not and base level was as low as about 2075 ft elevation, the roof of Flack Room, when this new route from the surface to the main cave conduits developed. This is approximately the same elevation as the horizontal strike-oriented development in the 1812 Room.

From the Flack Room the routes, except for Subway, all exhibit wide phreatic solution of particular beds. The highest route exits upward along faulting to Handley Room. Other routes lead to Octopus Alley and the upper parts of Slate Creek. These passages in zones 4 and 5 would have been as much as 100 ft below water level at this time. The water would have flowed to the water-filled Hedricks Stream Passage route, which was developing in zones 4 and 5. There is no known development out of Flack Room between the high route to Centrifugal and the much lower outlets to Slate Creek. At some time Flack Room was filled almost to its roof with gravel and fine sediment, burying breakdown blocks. This might have been at a time (or times) when a large opening admitted the surface stream, which deposited the sediment in deep water standing in Flack Room. Flood water may have backed up into the Handley Room, which is part of the earlier history of the cave.

Then followed a period of lowering base level when water entering Flack Room used various exits to lower levels and removed large amounts of sediment as well. Earliest was the route to Sand Room following zone 17. Starting on a sediment floor a small free-surface stream ran down the steep bedding, gradually carving a 15-foot canyon at its exit from Flack Room. It cut shallow canyons in passages beyond, but probably entered water-filled passages before reaching Sand Room. Later outlets coursed down-dip directly toward Slate Creek. The northernmost one was in use long enough for a canyon to be eroded 10 ft deep crossing Octopus Alley to reach Slate Creek, which became the main stream way. The route through Fun Room was not incised by any stream, but a lower level was opened which still carries water. Subway is a tributary to the bottom of Slate Creek, just above the Maccrady, and it seems it must have had a late origin after Slate Creek was enlarged downward well below zones 4 and 5. Except at its lower end it seems to be the work of a free-surface stream, without obvious phreatic beginnings. It may have been a temporary route at a time when the other routes toward Fun Room became sediment-choked.

13.5.6 The Hedricks Maze

The Hedricks Maze is developed on the west flank of the Caldwell Syncline, and many of the passages are developed in the steeply dipping rocks of the same monoclinal fold seen between Flack Room and Fun Room. From the fold axis near the Hedricks Stream to Sarver Room the dip increases to 30°, and the same deformation is seen here as is seen in the stream passage 200 ft north of Sarver Room. Beyond Sarver Room the dip flattens to as little as 10°, then steepens again to at least 34° entering the Bone Room. Further north this second steepening of the dip reaches 45° in Hedricks Maze itself, and beds are locally vertical close to Handley's Climb. Where Room Without A Name crosses this structure, 550 ft north of Bone Room, the dip increases to as much as 54°.

Most passages in this section have developed in cherty zones 4 and 5, with development extending below into the Maccrady shales at various places. Breakdown has extended the Bone Room roof into zone 10. Free-surface streams presently traverse Bone Room–Sarver Room and Room Without A Name to reach the Hedricks Stream. At least in part these routes are of later origin than the maze.

Of particular interest is the Hedricks Maze itself, a series of interconnected phreatic tubes with minor free-surface stream modification that are developed in zones 4 and 5 which here are dipping at 20°–45°. The tilted maze occupies a vertical range of 70 ft and is filled with deposits of large, rounded cobbles. Water left the maze through the tube of Octopus Alley, which crosses the Sarver Room at its roof and extends to the Fun Room. For some 250 ft south of Sarver Room the passage is immediately adjacent to the place where the beds make their second fold to steeper dips toward the synclinal axis. This fold dies out toward Fun Room, and from there the water must have followed Slate Creek and crossed over to the Organ–Hedricks Junction area.

When the maze was water-filled, the Hedricks drainage would have been using the Paleo Saltpetre Passage and 1812 Room, based upon their elevations. Thus, water from Hedricks Maze may have first developed the Organ–Hedricks Junction area, which was later utilized by Hedricks Stream, and joined by Organ Main Stream later still. Alternatively, the water may have followed the beds upward from Fun Room into the location of Flack Room, and left by a now lost route, or even followed faulting on up into Handley Room, which could still have been active at that time.

13.6 Passage Development and Stratigraphic Relations in the Upper Hillsdale

13.6.1 The Upper Stream Passage–Sand Room Area

Only a few passages, Lover's Leap Passage for example, are known in the rocks of zones 7–12. The extensive Upper Stream Passage is developed in zones 13–21, from 65–100 ft above the base of the Hillsdale. This passage contains a stream perched 60–70 ft above Hedricks Stream near the axis of the Caldwell Syncline. For much of its length south of East Passage junction, Upper Stream Passage has a nearly straight alignment at N15°E, as does Hedricks Stream below, pointing to probable fracture control.

In several reaches of Upper Stream Passage between East Passage and the Waterfall Room the ceiling is at a chert bed in zone 17. This chert bed is often double, with a thin chert bed 6–10 in. above the major bed. In these areas there are often rounded solution forms in the limestone just below, and sometimes above, the chert, which might mark where solution of the passage began. But the roof is as much as 7 ft below the chert bed in much of NULO and rises 5 ft into overlying zone 18 close to Sand Room. The floor is an incised canyon cut by a free-surface stream and also rises about 12 ft in stratigraphic horizon. It is in zone 12 at East Passage, but is most commonly in 15 or upper 14 overall.

South of Sand Room the roof of Upper Stream Passage is commonly at or above the chert in zone 17, with local tubular side loops and ceiling channels in zone 18 south of Throne Room. The floor remains in zone 14 much of the way to the Revak Room, beyond which the passage is mostly in zone 16–18, with prominent pendants in 18 close to the Waterfall Room (Fig. 13.24).

There is some evidence that the coarse sparry limestone containing chert in zone 17 was the bed initially dissolved, and that the stream downcut to zone 14 before becoming perched on that horizon. The upper part of 14 is dark fine shaly or thin-bedded limestone and may indeed be impeding further deepening of the passage, but upstream at East Passage erosion has reached 7 ft below zone 14, only 35 ft above the roof of Hedricks Stream. Apparently no continuous route through joints exists that would let the water reach that lower passage.

Sand Room is crossed by the Upper Stream Passage canyon and predates it (Fig. 13.25). A ceiling tube 2 ft higher than Sand Room's roof came from NULO and is silt-filled to the south, above the canyon. Sally's Waterfall Passage is in zone 18, Sand Room in 18 and upper 17, and together they may form a strike-oriented tube looping around the synclinal axis. Nine Foot Bat Passage leads south out of

Fig. 13.24 Passage south of Sand Room. Keyhole passage in North Upper Level Organ (NULO) just south of sand room. The zone 17 chert is at the level of the person's feet. Photograph by W.K. Jones

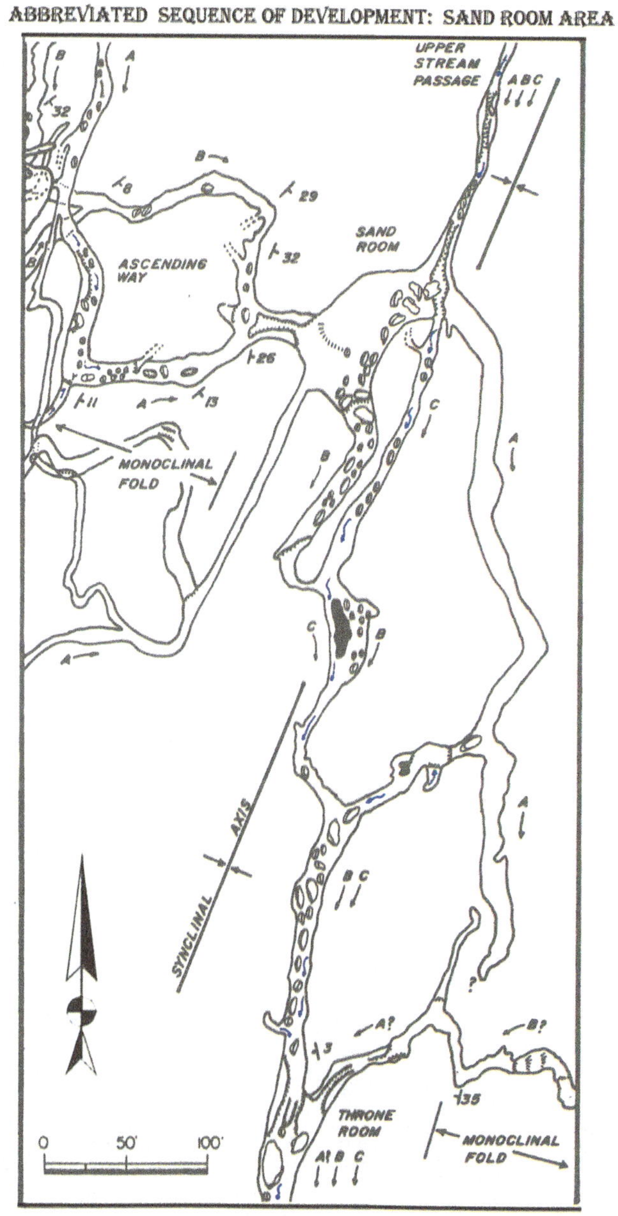

Fig. 13.25 Abbreviated sequence of development of Sand Room area

Sand Room in the same horizon, but eventually turns west into a maze that turns upward into the steeply folded monocline mentioned in describing the Flack Room area. This maze lies beneath the end of Treasure Passage and may have derived water from the far west side of the Organ Cave System.

The passage west from Sand Room forks 50 ft from the room in an area where the dip quickly increases to 30°. The north fork begins in zones 15–17, and although the floor varies, the chert horizon in 17 can be followed to the Centrifugal, and also up the dip which steepens again to the high route out of Flack Room already described. All of these passages near Sand Room were considerably influenced by solution along the zone 17 chert. These passages were water-filled at the time of development, and probably during part of the time when water from Flack Room flowed through some of them. There may also have been water entering from the Upper Stream Passage and from Nine Foot Bat Passage. Indeed, Sand Room may be in part the original phreatic extension, 50 ft under water, of the free-surface drainage coming down Floyd Collins Avenue to Flack Room. Flow into zone 17 at Flack Room was along faulting, and this route along soluble horizons was much shallower (had less phreatic head) than flow down the monocline in zones 4 and 5, a more prominent soluble horizon.

The left part of the west passage from the Sand Room passes into a zone of lower dip and rises through the stratigraphic section to enter a northeast trending passage developed in zones 19–21, where there is another cherty horizon. This solution canyon may always have been tributary to Sand Room and still contains a small stream. A side passage follows bedding west to the area of the monocline, but here the beds quickly become vertical to slightly overturned in the Handley Silo, which leads up the Handley Room. This tributary carried water from points unknown under the Handley Room, excavating sediments out of the Handley Silo from below. Water in Sand Room originally exited by Sally's Waterfall Passage, which probably rejoined the present Upper Stream Passage at Throne Room.

Upper Stream Passage south of Sand Room is interpreted as a more recent development fed by NULO water, water from the left fork of the west passage and perhaps originally also from the right fork and Flack Room. Sally's Waterfall Passage was not used by these free-surface streams. There is evidence of an early water-filled stage in parts of the passage, including local use of preexisting tubes in zone 18 south of Throne Room. These may be remnants of the earlier Sally's Waterfall route. Pendants suggest a period of standing water at about 1950 ft elevation near the south end of the passage close to the Waterfall Room (Fig. 13.26).

Fig. 13.26 Ray Garton at the pendants just north of the Waterfall Room

13.6.2 Handley Room, Upper Lipps, and Other High-Level Passages

From Handley Room through Left Hand Passage, Jones Canyon, Lipps Maze and Upper Lipps, the roof of most passages is in a light limestone with two prominent beds of chert, 6–10 in. apart. Although they cannot be followed continuously across the faulting between Handley Room and Flack Room below, these cherts (zone 18 in Fig. 13.14) are considered correlative with those of zone 17 (Fig. 13.16), based upon stratigraphic position. Solution channels in this horizon suggests that over a wide area solution of the passage began in zone 18. The dip of the beds is 6°–9° over most of this area.

The Handley Room (Fig. 13.27) is a large trunk passage fragment. The roof and upper walls are in gently dipping rocks including zone 18 chert, which dips 12° southeast. Lower walls, the Silo, and passage down to Flack–Handley Turnpike and Flack Room are in steeply dipping rocks separated by faulting from those above (Fig. 13.28). The room tends at an angle of about 40° to the strike of the beds in its roof and adjacent Left Hand Passage, but nearly parallel to the strike of the steeply dipping beds below. Handley Room is probably a major early stream route toward Second Creek, originating in zone 18, and downcut by a free-surface stream in the manner of Hedricks Stream, 160 ft lower in elevation. The room is blocked in both directions by gravel fill, and the floor is breakdown and sediments with no obvious sign of a former vadose stream course.

The Left Hand Passage and its large branches extend northwest up-dip from Handley Room about 600 ft, rising almost 100 ft in elevation following zone 18 (=17) and beds just below it. Anastomosing ceiling channels are developed in zone 18 in several areas, but quite a number of the passage roofs are in zone 17 (=16), a massive bed just below the chert layers. In the Breezeway area are several joint passages in zones 18 and 19 that cross and probably predate major development of Left Hand Passage.

While the initial openings, often in zone 18, were water-filled, the Left Hand Passage was probably soon transformed into a large branching system carrying free-surface streams from blind valleys ancestral to those presently feeding Floyd Collins Avenue, and perhaps Jones Canyon as well. The water entered the trunk passage at Handley Room. These passages were abandoned when Floyd Collins Avenue and Jones Canyon developed, but small streams, some still active, incised a complex of canyons below their floors. The abandoned canyon closest to Handley Room is cut 27 ft below zone 18 through zone 13 which is very cherty, but does not intersect the faulting, although it encounters steepening dip just as it reaches Handley Room.

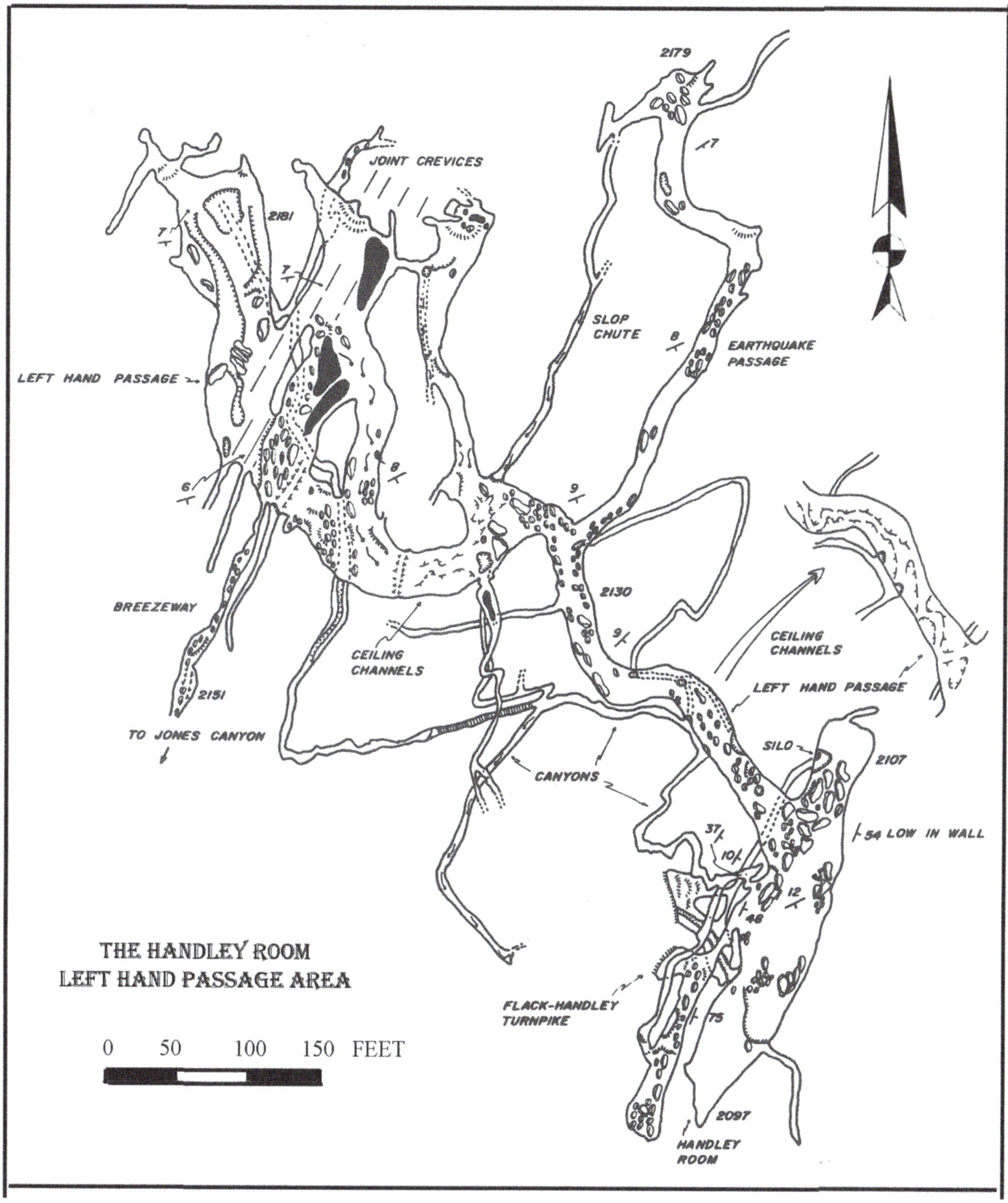

Fig. 13.27 The Handley Room–Left Hand Passage area

Lipps Maze consists of mostly small tubes, some joint controlled, developed in zones 17–19, with zone 18 and its prominent chert. The area of maze development extends from near the Lipps entrance 1,000 ft east along the strike and ranges through more than 70 vertical feet at least from 2160 to 2230 ft elevation. It is quite likely that the head of Jones Canyon, Breezeway, Treasure Passage and the solution of tubes and open joints of upper Left Hand Passage are

Fig. 13.28 View looking down Handley Silo. Note vertical orientation of beds exposed in the roof. The beds below this passage level out to a dip of about 20°. Beds 40 ft higher in Handley Room are nearly horizontal above an intervening fault. Photograph by W.K. Jones

part of this development extending 1,500 ft from the Lipps Entrance. The maze is phreatic in origin with some later canyon development by small free-surface streams. Treasure Passage follows the beds down-dip below any other part of these passages and apparently took water from the head of Jones Canyon before that route was opened.

Jones Canyon trends diagonally down-dip. Development was begun close to zone 18, now exposed in the roof. Subsequent development was by the free-surface stream that still follows the canyon. At the Belfry there are solution tubes 15–20 ft above zone 18; their relationship to Jones Canyon is probably accidental. Jones Canyon has not been examined south of Belfry.

Jones Canyon encountered the maze passages and made minor use of them in developing the present route. The present drainage from Lipps Maze has cut the canyon of Skid Row keeping nearly at grade with Jones Canyon. The water that now follows Jones Canyon may formerly have gone to Left Hand Passage.

In upper Jones Canyon, near Skid Row, red bands can be seen in zone 17 and low in zone 15. These zones are at least 75 ft above the horizon of the Maccrady shales. In upper Lipps Stream, development in zone 18 continues down-dip at a low angle to the strike all the way through Hells Fissure to Lipps Drop. Various tubes and branches at roof level are in these beds, while the Hells Fissure stream canyon cuts deeper into the underlying beds until below the drop it is about 85 ft below zone 18, at the top of the Maccrady shale. The passage remains at roughly the same horizon from the base of the drop to the junction with Humphreys, where the floor is about 5 ft above the base of the Hillsdale formation. Lipps Stream above the Lunch Room has probably used both some preexisting solution tubes, and developed some, opening the route to larger passages at the Lunch Room. Doline development over parts of the maze of tubes has expanded to develop the small drainage basin that feeds Lipps Stream, which has always been a small stream compared to that in Jones Canyon, for example. The whole development of upper Lipps Stream is probably a late event in the cave history.

13.6.3 The Revak Room

The Revak section is developed in about 30 ft of rock represented in Fig. 13.19. These rocks have gentle dip and are separated by a fault from the steeply dipping rocks which are continuous with those of the Upper Stream Passage below. The situation is like that at Handley Room where the rocks of Handley Room–Lipps, etc., meet those of Flack Room, but it is on the east side of the synclinal axis. Correlation with other stratigraphic sections is tentative; the double chert shown may be identical to that in zone 18 on the west side of the cave system, and to that in zone 17 in Upper Stream Passage.

The highest passages above Revak Room are developed in the chert zone and the zone above in this section, while Revak Room, the lead to Revak Silo, the passage to Cathedral Room, and the loop through Walker's Ramble are mostly in the 10 ft of rock below the chert. Thus, development of these mostly tubular passages is above or below the chert, sometimes with the chert forming a bridge between levels. The passages are oriented diagonally down-dip. They show shallow canyons and other free-surface stream modification, and gravel bars testify to the competence of the water which once flowed here. All are ancient flow routes from the Organ Entrance blind valley, but none ever seems to have concentrated the drainage in a single path. Such a single path developed under vadose conditions through Cyclops Avenue at a later time in the cave's history.

Passages in Foxhole Cave lie 180 ft and more above Hedricks Stream, close to the synclinal axis. Development is in limestone with prominent cherty horizons, and, as in the Organ Cave System, the initial development was often in the cherty beds, now seen close to roof level of the passages.

13.7 Structure Controls in the Organ Cave System

13.7.1 Joints and Minor Faults

Joint control of passage orientation is clearly visible in some passages in the Organ Cave System. Examples are found in Lipps Maze, Breezeway, Nine Foot Bat Passage, Discovery Passage, Walker's Ramble area, and various other places. These straight reaches with joints exposed are most common in passages developed underwater. Even then, perhaps only 20% of the total length shows clear joint control, although the map suggests some joint influence on perhaps 50% of the passages. Most of the observed joint control is exercised by the joint set at N15–30°E, roughly parallel to the strike. Joints in other directions were used occasionally as seen in Gypsum Passage.

The northeast orientation of many passages, especially the major stream passages, suggests that the major joint set was used where convenient in the development of these passages, but direct evidence in the passages is scant. Most of the ceiling tubes show only local joint influence. There are exceptions. Organ Main Stream has a ceiling channel for 250 ft above Organ–Hedricks Junction area that is oriented N25° E. Joint control is quite evident at the acute bend 1,300 ft upstream from Big Canyon sump.

Minor faulting is seen at many places in the cave. Faulting often involves only a few beds, presumably passing into slip along bedding planes, and speaks of considerable compressional movement within the limestone. Slickensides on breakdown blocks attest to the considerable influence fault planes have on rockfall (Fig. 13.29). Very large fallen blocks in the Slickenside Room separated from the ceiling along a fault plane that dips N70°W at 35°. Movement was just about east, at a 20° angle to the dip of the fault plane.

Several faults can be seen in Upper Stream Passage south of Throne Room. Just south of Throne Room a reverse fault dipping 19° east offsets the chert bed at ceiling level about 20 in. The ceiling tube here developed on the fault plane.

About 300 ft south of Throne Room, at the USP–Hedricks Connector, there is a reverse fault dipping 24° west, displacing the chert bed 24 in. A silt-filled ceiling tube seems to follow this fault, and it may have directed the original flow in the larger passages in beds above the chert. Its influence cannot be followed very far along the passage.

Fig. 13.29 Slickenside Room. The large breakdown block is parted along the fault plane of a reverse fault. Slickensides on hanging wall of fault are visible on the roof. Photograph by W.K. Jones

Joints and reverse faults of small displacement both directed dissolution and free-surface stream erosion to some extent where they were conveniently oriented with regard to the potentiometric gradient. Jointing had the greater influence. Minor faulting had more visible influence on breakdown than on dissolution.

Faults with displacement of 20 ft or more were noted in a few places in the cave. Most notable is the faulting between Flack and Handley Rooms, which had considerable effect on local water movement, as discussed above. Faulting near Revak Room probably involved considerable displacement, and its influence on water movement has been mentioned.

There is a fault with apparent large displacement in Foxhole Cave at the pit in Desolation Row that leads into the Southern Maze. If the cherts at the base of the pit are the same as those in Desolation Row, vertical displacement is about 25 ft, and total displacement about 70 ft. Minor features of rooms and passages are influenced by this fault, but it has no major impact on passage trend or elevation. Nearly all of Foxhole Cave is in rocks above the fault plane.

13.7.2 The Caldwell Syncline: The Influence of Strike and Dip

Several sections of the cave system have been mentioned as strike-oriented. The Paleo Saltpetre Passage and part of the 1812 Route, Octopus Alley, and Rat Alley Crawl were strike-oriented passages. The Lipps Maze complex also extends along strike. These are routes developed under phreatic conditions.

Passages developed directly down-dip include the Left Hand Passage and underlying Floyd Collins Avenue, Room Without A Name and some of Organ Main Stream. These passages are primarily of vadose origin. Smaller dip-oriented passages include the Old Saltpetre Route and routes from the Flack Room to the Fun Room. Some of these smaller passages are phreatic in origin and development. Many of the major passages are oriented diagonally down-dip, as seen in Upper Lipps, part of Lower Lipps, Jones Canyon, Organ Main Stream, and Cyclops.

The longest passages in the northeastern part of the cave, Hedricks Stream Passage and Upper Stream Passage, trend down the plunge of the Caldwell Syncline, usually quite close to the axis of the fold. Figure 13.30 presents a structure contour map of the rocks in which Organ Cave System is developed. Contour is on top of the Maccrady shale. Solid contours represent good control from observation in the cave, although it should be remembered that the elevations are derived from cave surveys of variable accuracy. Dashed contours are approximate. In Fig. 13.30, the axis of the Caldwell Syncline is clearly defined as far south as the Waterfall Room. The plunge to the southwest is about 350 ft in 5,000 ft, averaging about 4°. In the cave the plunge is seen to vary from 0° to 6°. South of the Waterfall Room the location of the axis and the plunge is uncertain. Figure 13.30 shows the axis as located by Price and Heck (1939) on the county geologic map.

The elevation of the top of the Maccrady shale where the fold axis crosses Second Creek was approximated as follows: Ogden (1976) reports that Dickson Spring is at the base of the Patton Limestone, which is 300 ft stratigraphically above the Maccrady. The elevation of Dickson Spring

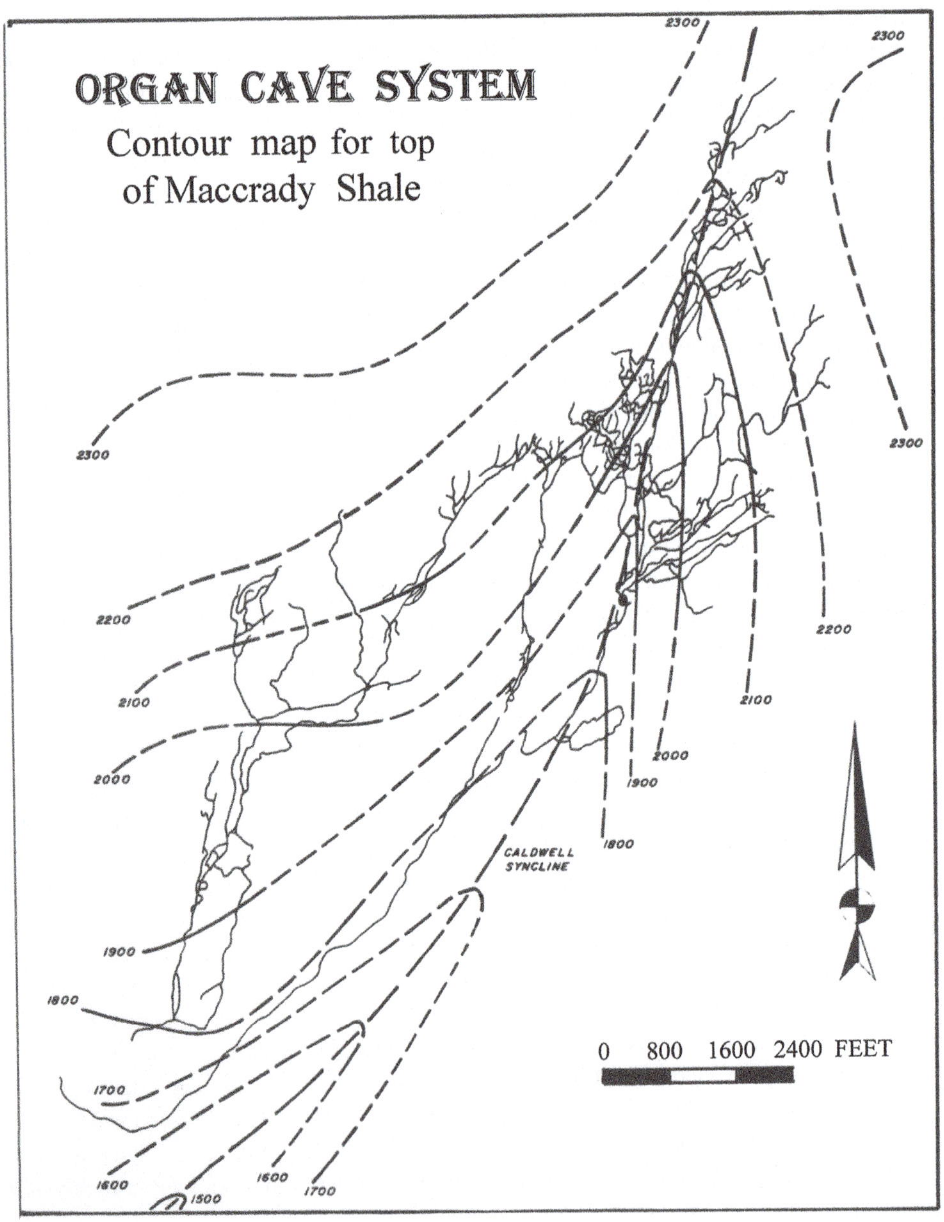

Fig. 13.30 Contour map of the top of the Maccrady shale

is about 1800 ft, so the top of the Maccrady is at about 1500 ft. Ogden reports the dip at Dickson Spring as 13° east, and the synclinal axis is more than 1,000 ft further east on the county map. Thus, the Maccrady should be at least 100 ft lower, 1400 ft, at the axis. Hedricks sump is in zone 3 at elevation 1825, so the Maccrady shale is about 1810 at the sump, or approximately 1800 at the base of Bowen Room Drop. The result is a plunge from there to Second Creek of about 400 in 11,000 ft, or less than 2°. Dashed contours were drawn on this basis.

Figure 13.30 makes it clear that Big Canyon and Wilson's Watery Wonderland probably do not follow the axis of the Caldwell Syncline very far south of Bowen Room Drop. The more nearly the axis is followed, the more these low-gradient passages must rise through the beds, possibly several hundred feet above the Maccrady shale in this case. Some passages do rise (or fall) stratigraphically, but most show considerable stratigraphic control. Thus, low-gradient passages, like Big Canyon and Wilson's Watery Wonderland, tend to follow the strike. It is most likely that these passages are more strike-oriented than they seem in Fig. 13.30 and remain in the Hillsdale at least until they pass the adjacent Lower Lipps Stream Passage. At the stratigraphic horizon of the cave the synclinal axis may well be further east than shown in Fig. 13.30.

The 1800 foot contour in Lower Lipps follows what must be a minor fold. Wilson's Watery Wonderland turns northwest to David's Dungeon in a manner suggesting it is following the strike around the same fold. At the Lipps sump, the Maccrady shale is already below 1800 ft elevation. The contours imply it would be more than 100 ft lower at David's Dungeon, which would then be well up in the section. The actual relations of the area beyond Big Canyon sump to stratigraphy and structure are unknown.

The complex northeastern part of the Organ Cave System is developed in an area where the syncline has a steeper plunge and is more tightly folded, than elsewhere in the cave system. Some passages parallel the strike around the syncline. Most of the many passages east of the axis run obliquely down-dip toward or to the synclinal axis. Passages on the west limb of the fold are less numerous and usually trend more directly down-dip toward the synclinal axis.

13.7.3 Deformation Near the Synclinal Axis

The monoclinal fold and associated fault seen on the west limb of the Caldwell Syncline between Flack Room and Room Without A Name has not been seen further north. It is reported in Debevoid in Foxhole Cave, which lies above the passages of Hedricks Maze. To the south the structure can be traced as far as the Nine Foot Bat Maze about 300 ft south from Flack Room. It might be found in the House of Cards area down in Jones Canyon, which was not examined.

The similar fold seen in Revak Room on the east limb of the syncline can be seen 250 ft south in the lead to Cyclops Hall, in Old Saltpetre Route 300 ft north at a lower level, and crossing Organ Main Stream 900 ft north, about 350 ft upstream from Organ–Hedricks Junction, where it is subdued (Fig. 13.31). No evidence of it can be seen 2,000 ft further north where the Rimstone Ceiling Passage and other passages reach far up the east limb.

Figure 13.32 shows a section through the monocline on the east limb of the Caldwell Syncline, following Old Saltpetre Route and Revak Silo, which is almost directly above Upper Stream Passage. The Hedricks Stream is approximately 100 ft east of the synclinal axis here. Old Saltpetre Route follows the cherty beds 4 and 5 and beds just below, conforming to the irregular dip, which reaches more than 40° at the steepest part on the monocline. The passage generally does not reach down to the Maccrady shale, and the Hedricks Stream is 5 ft above the shale here.

The Upper Stream Passage–Hedricks Connector is shown in Fig. 13.32, although its connection above is not shown as it is 90 ft out of the line of section. This is the route that reveals a seemingly continuous stratigraphic section up to zone 17, the horizon with chert beds seen in Upper Stream Passage and NULO. Upper Stream Passage is further from the fold axis, and the rocks dip 8° to the west. Revak Silo follows the beds below cherty zone 19 up the monocline. Dips reach 58°, and the beds are not strictly parallel to those in Old Saltpetre Route below.

At the top of the Revak Silo the cherty beds in the Revak Silo roof dip 30° west and are overlain by the series of beds which dip 8° west in the high passages around Revak Room.

Fig. 13.31 View looking upstream showing the Organ–Hedricks Stream Junction. The Hedricks Stream enters from the *left*, and the Organ Main Stream enters from the *right*. Low-angle thrust faults in the center of the photograph appear to be truncated against the ceiling. The contact of the Maccrady shale is about one foot above floor level. Chert of zone 4 is exposed in the ceiling. Photograph by W.K. Jones

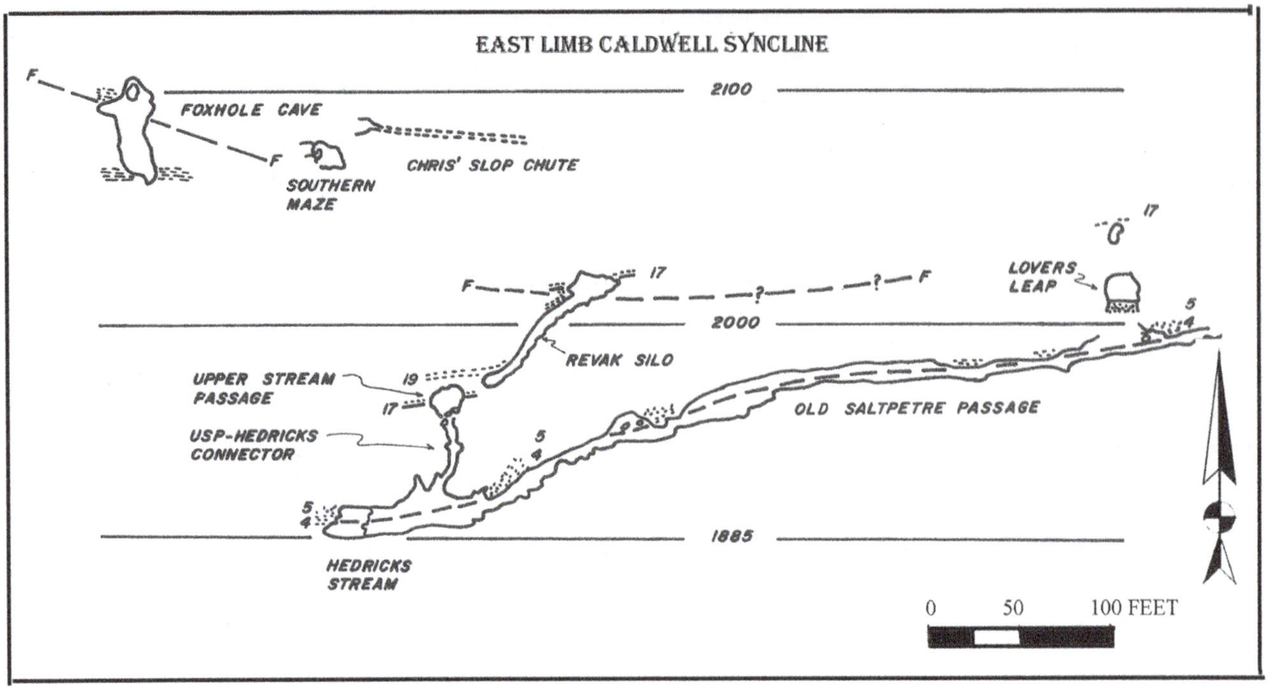

Fig. 13.32 Section along the east limb of the Caldwell Syncline

A similar disharmony of the rocks at the top of the monocline is seen in Revak Room, 190 ft to the south. The maximum dip there is about 40° and continues upward at about 25° to run into the overlying series with a dip of 6°. Just above the Revak Silo there is a wedge of cherty beds dipping 18° east between the two rock series which dip west toward the synclinal axis. The rocks series are evidently separated by faults. A probable reconstruction of the fault plane is shown in Fig. 13.32, showing it as a very low angle reverse fault that passes into the bedding to the east. This makes it possible that the chert beds in Revak series correlate with zone 17 in the rocks below as postulated elsewhere. The separation of zone 4 and zone 17 at Upper Stream Passage–Hedricks Connector is the same as the separation of zone 4 and the Revak chert beds near Lover's Leap, within the accuracy of the surveys.

Faulting or monoclinal folding of the Revak series of rocks has not been seen, but could lie just west of Revak Room. Beds 20 ft thick below the Revak cherts, observed in passages to the east, are absent in the west wall of Revak Room and above the Revak Silo, so the fault is evidently cutting up through the section to the west. The fault shown in Foxhole Cave lies about 240 ft west of and 80 ft above Revak Room. It is related to, if not identical with, the fault in Revak Room.

Figure 13.33 shows a section of the monoclinal fold developed on the west limb of the Caldwell Syncline as seen at Room Without A Name. In this figure, the elevation of NULO, derived from survey data, appears to be too great considering the stratigraphic horizon in which it is developed. Debevoid in Foxhole Cave is drawn from the map, and its geologic structure added from notes on rough map sheets and hearsay. It was not visited during this study. Hedricks Stream is very close to or at the synclinal axis in this section. Two zones of steep dip are visible, the lower one projected from the drain from Room Without A Name, located 250 ft further southwest along the strike. Hedricks Stream flows on the Maccrady shale, and the Room Without A Name is cut into the Maccrady shale. The ceilings of both passages are in cherty zones 4 and/or 5. Minor faulting is suspected in the area of steep dip (up to 50°), but no major displacement is obvious. More severe deformation is known 200 ft southwest, where the dip increases to vertical at Handley's Climb, and in Debevoid in Foxhole Cave, which lies about 150 ft higher in the stratigraphic section.

The section shown in Fig. 13.34 is through the Bone and Sarver Rooms and higher-level Fossil Room which are developed 500–600 ft southwest of Room Without A Name. Here Hedricks Stream is slightly west of the synclinal axis, and dip is 3°. The geometry of the double monocline is different here, with an abrupt lower fold with more relief than seen to the north. The lower passages have floors cut into Maccrady shale, except for the Hedricks Stream Passage, where the floor is about 5 ft above the contact. The roofs of the passages are often developed in cherty zones 4 and 5, and major connecting tubes leading to Hedricks Maze, including Octopus Alley, are in these zones.

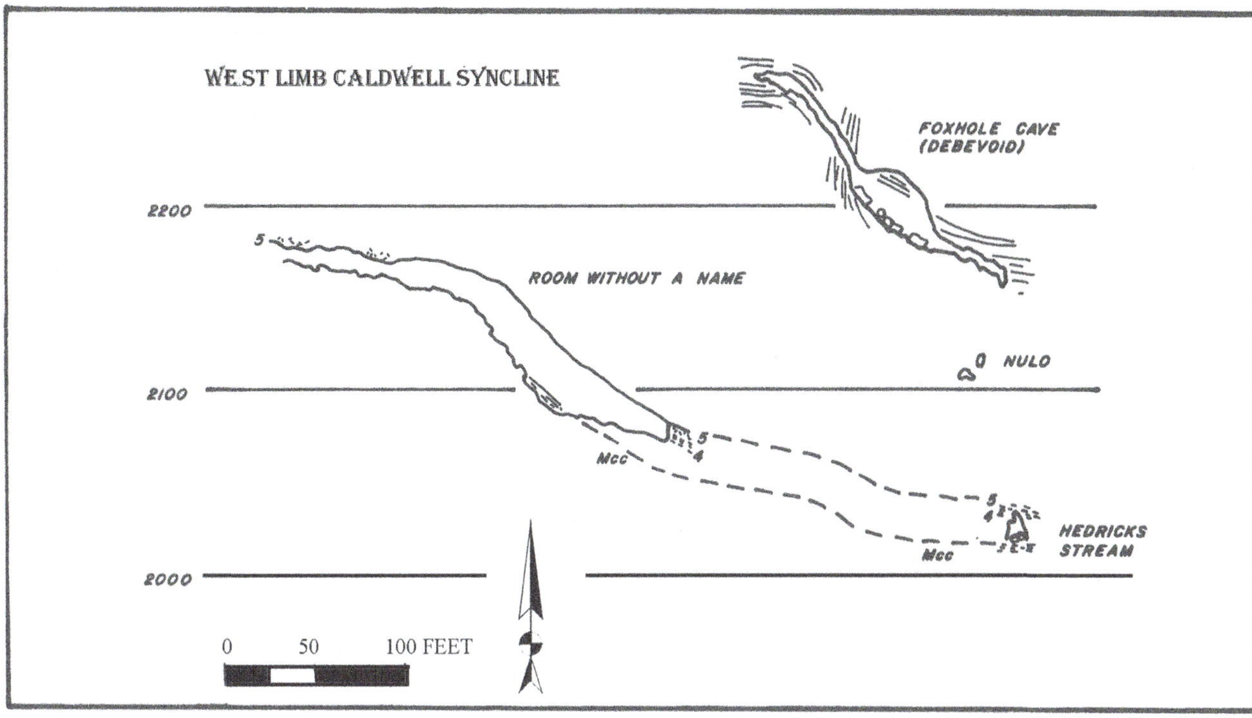

Fig. 13.33 Section along the west limb of the Caldwell Syncline

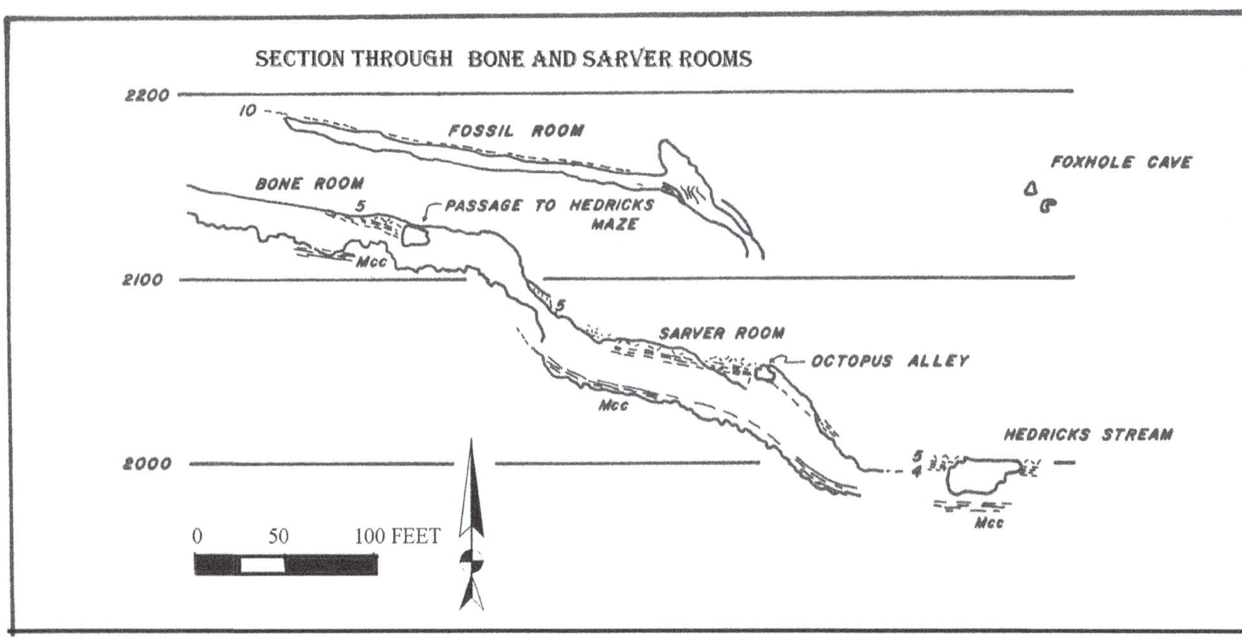

Fig. 13.34 Section through the Bone and Sarver Rooms

No major faulting has been noted, although a zone of prominent fractures can be seen at the bottom of the steep climb up to the Bone Room. Good exposures in Hedricks Maze 100 ft northeast show no faulting, but there is a local area of 45° dip, steeper than that observed in the Bone Room. Minor features of structure have considerable local variations in short distances.

Fossil Room overlies Bone and Sarver Rooms, only 60 ft away at its northwest end, 100 ft at the southeast end. Its roof lies just at and beneath the zone 18 chert as shown in

Fig. 13.17. The separation of this horizon and zones 4 and 5 in Bone Room is about what is expected; it has not been altered by faulting.

Beds in Fossil Room dip 7° southeast to the south end of the room, where they abruptly turn down to dip as steeply as 72° in as little as 15 ft distance. Some beds are intensely fractured perpendicular to bedding in the tight folds. There is apparent drag-folding and disharmony between folded beds and those above. The nature of the structure in Fossil Room is not clear, but it involves severe deformation and probably faulting not seen in Bone and Sarver Rooms.

Figure 13.35 shows a section through the north end of Handley Room at the Handley Silo, extending to Sand Room. All passages shown are in zones 14–21, on the NULO level of the cave, except for Handley Room which is in zones 14–17 in the western section (Fig. 13.17). The Handley Silo is 350 ft southwest along strike from Fossil Room. Here a double structure is visible in these horizons 80 ft above the Maccrady shale. The eastern fold is a monocline with dips observed as steep as 32°. The western fold abruptly becomes vertical to slightly overturned in the Handley Silo, and faulting is in evidence. Beyond Centrifugal there is a room only 40 ft north of the Handley Silo, in rocks just below zone 17, where the exposed structure includes 60° dips and intersects very deformed rocks 60 ft west and 20 ft below Handley Room. Thus, the folds can be interpreted as drag folds beneath a low-angle reverse fault, as shown in Fig. 13.35. Handley Room is thus located on the fault.

The relationship of this structure to the upper sequence of beds exposed in Handley Room is striking. Fifteen feet of beds in the west wall of Handley Room directly above the Handley Silo are undisturbed except for slight folding at their bases. Handley Room ceiling, in zone 17 (Fig. 13.17) five ft below zone 18 chert, dips uniformly east at 8°. In the east wall less than 2 ft of undisturbed rock lies below the roof horizon, then a few feet of fractured rock, and the lower 4 ft of the 8-foot wall is a very cherty bed that dips 50° east. The fault apparently passes upward into this wall. It may cut the zone 18 chert a short way further east, but this relationship is not clear in Fossil Room where the beds below zone 18 can be seen to fold down steeply, but zone 18 involvement was not observed. No major folding or faulting was observed in Foxhole Cave, about 450 ft east of Handley Room and in part at about the same elevation.

Figure 13.36 depicts a section through the cave at Flack and Handley Rooms about 150 ft southeast of Fig. 13.35. The Hedricks Stream is a little distance east of the synclinal axis, and the dip is about 3° west. In the lower part of the section just above the Maccrady shale the monocline interrupts the gentle dip seen in Floyd Collins Avenue and Flack Room. Dip in Flack Room roof increases abruptly to 40° and prominent cherty zones 4 and 5 appear to be offset about 20 ft by reverse faulting. No faulting is seen in the Maccrady contact in the Subway just south of this section, and the fault may pass into bedding plane slip to the west, possibly on the contact zone. The fault seems to be next to the wall south of Flack–Handley Turnpike and at floor level on the north side of Flack Room just north of the high-level outlet where zone 17 is exposed. The ceiling above the high-level outlet route is extensively fractured with zone 17 on the floor of the bedding opening, but zones 4 and 5 in Flack Room roof are very close by.

The fault seems to pass upward above the Turnpike, and the route from there up to Handley Room follows it. The

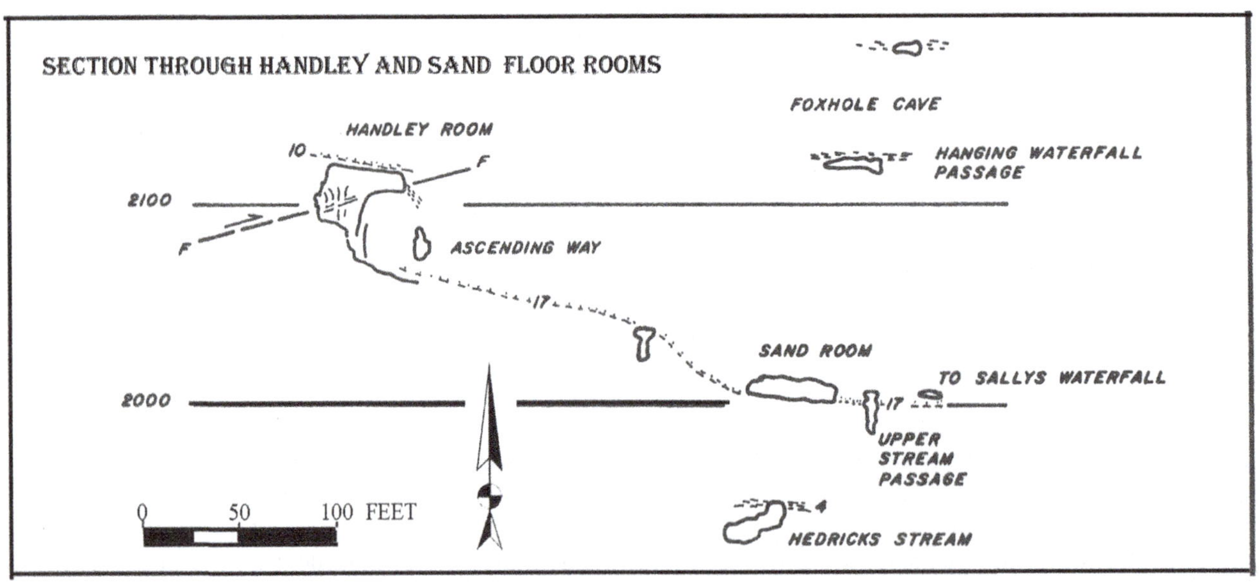

Fig. 13.35 Section through the Handley and Sand Floor Rooms

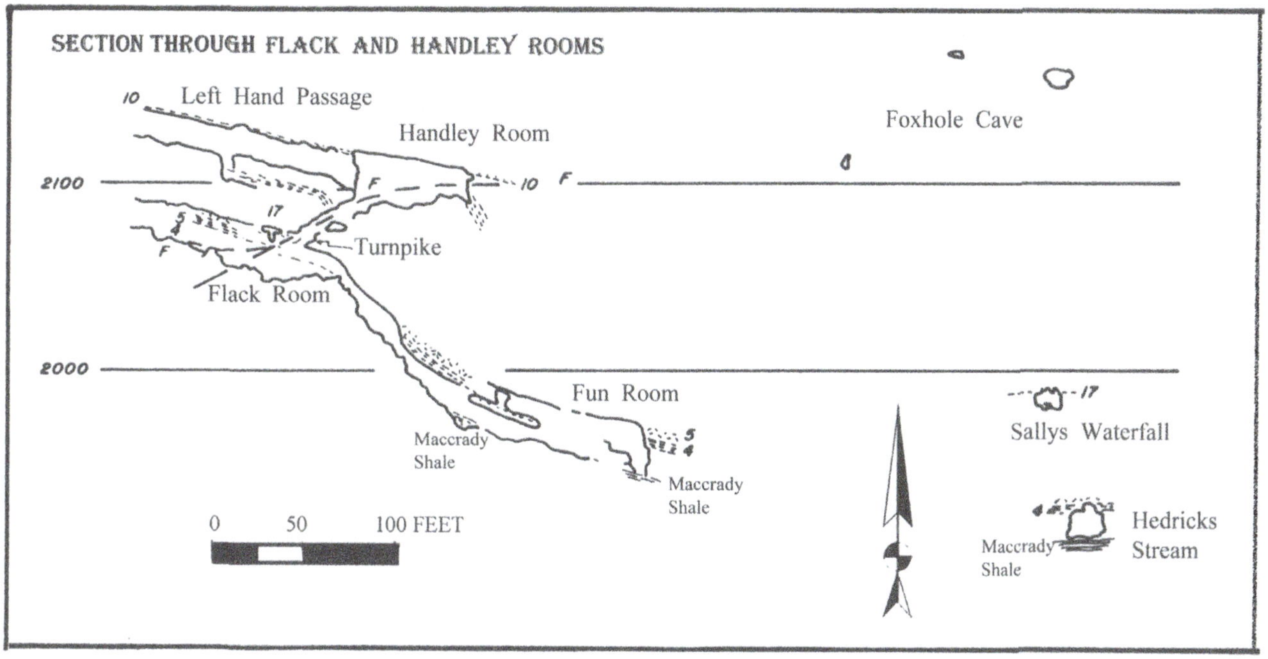

Fig. 13.36 Section through the Flack and Handley Rooms

relationship observed on the east wall of Handley Room is the same as that at Handley Silo, and this is presumably the same faulting seen at Handley Silo.

Above the fault lie the little disturbed rocks of zones 12–18 (west section). In the canyon extending from under Left Hand Passage to Handley Room the cherty zone 13 can be seen folded sharply to dip 37° east above the fault. A zone of fracturing in the west wall of Handley Room above Flack–Handley Turnpike is about 13–20 ft below zone 18 chert. This should be above the fault plane and implies multiple fault planes or brecciation in the area of drag folds.

The separation between zone 18 chert and the Maccrady shale is about 80 ft just west of Handley Room, but the maps suggest it decreases to about 55 ft at the upstream end of Floyd Collins Avenue. This change could be due to horizontal stratigraphic changes or could be a result of loss or repeat of beds due to faulting. The first is unlikely, and although faulting nearly parallel to the bedding is possible, it is not demonstrated. The difference may be an artifact of computer fitting of diverse surveys.

Figure 13.36 is a composite sketch of the structure of the synclinal axis area. Offset beds do not seem to line up on single faults, but this is probably due to including sketches far out of any single line of section.

As noted with regard to Flack Room and Subway, the Maccrady shale contact is involved in folding, but perhaps is not usually offset by faulting. Faults in the rock above may pass into bedding plane slip in the uppermost shales of the Maccrady. Some of the fragments in the "reworked Maccrady" seen in various outcrops might be breccia resulting from slip along the top of the formation.

The upper sections with chert (18 in Lipps and 17 in Revak Room) have not been observed offset by faulting, so that there is a visible disharmony with underlying folded and faulted beds. But on both flanks of the syncline the faults may cut these beds a little closer to the fold axis, where no exposures have been examined. The reported structure in Debevoid suggests that the faulting on the west limb of the fold cuts on upward through the section. Certainly, the pattern of faulting is complex and deserves further investigation.

There is a similarity between the passages leading diagonally down-dip to Revak Room area and those leading more directly down-dip to Handley Room. In both cases the dip-oriented passages reach toward the synclinal axis only as far as the fault zone at the top of the monoclinal folding. Large passages (Handley Room and Revak Room) then develop parallel to the fold, where the beds which were followed to this point encounter the fault plane. Revak Room is 1100 ft from Handley Room, with its sediment floor about 100 ft lower in elevation. Revak Room is probably more recent in origin. Base-level elevation may have played a role in the location of this room, but more important was the restriction of movement further down-dip by the structure and the availability of the fault as an avenue of water movement.

Eventually passages following the steeply dipping beds below the fault at Revak Room developed to lower levels. From Handley Room the passages followed the fault down in the direction opposite to the dip to reach steeply dipping zones 4 and 5. Most of these steep passages have been used by free-surface streams, but are largely phreatic in development. While they can be interpreted as routes where the water found its way to lower levels, in each case it is quite possible that the flow was originally upward.

Time below base level probably consisted of several types of passage, all largely stratigraphically controlled. There are segments (often with at least 70 ft of vertical relief) that parallel the strike, probably developed in the shallow phreatic zone. Other segments lead down-dip at angles, perhaps to depths of 100–150 ft. Finally, there are segments that lead back to shallower depths, sometimes by following beds up-dip, usually in areas of steep dip, sometimes by following faults or otherwise cutting across beds to horizons higher in the stratigraphic section.

Later in the cave history free-surface streams developed a system of routes that follow dip and plunge more closely using parts of the preexisting system, and adding new passages as necessary to carry the drainage. Parts of these new routes had phreatic beginnings, but soon were converted to free-surface flow. Even the Hedricks Stream had tubular phreatic beginnings, still evident in its roof, and does not strictly follow the axis of the Caldwell Syncline down the plunge, although it is fairly close to it as far as Bowen Room Drop.

13.8 Sediments

13.8.1 Clastic Sediment

The most common sedimentary deposits in the Organ Cave System are silty and sandy gravels. The gravel is often only partially rounded by stream transport. Particle sizes from 1/2 to 4 in. are common, and individual deposits differ in grain size.

These sediments often form terraces in both large and small passages, although only occasional terrace surfaces have survived later erosion. The deposits are found nearly to the ceiling of many passages, and at most if not all elevations in the system. Some of the breakdown common in large passages is or has been buried in these deposits. Thus, at some time in the history of most passages there was water flow with competence to move coarse sediment, and a supply of sediment sufficient to nearly fill the passages. It is possible that much of the sediment was carried through the parts of the cave that were in the vadose zone to be dumped wherever the water entered the phreatic zone, building bars and deltas into the standing water, and grading terraces upstream from those points. Some coarse sediment would be carried on into the phreatic zone, where water volume and passage size produced sufficient velocity to move it, as in times of high water. As base level fell, existing deposits would be eroded by free-surface streams, and the material carried further downstream to be redeposited. It is not necessary to postulate enough sediment to nearly fill the whole cave system at one time.

There are some deposits of silt. These are often found on top of other sediments and on top of breakdown. They probably speak of backwater from high water levels in floods. There are also passages with deposits of limestone slabs barely worn by the water that moved them, as in Left Hand Passage. These are the beds of free-surface streams with open access to surface sinks and probably were continuations of large sinking streams.

13.8.2 Breakdown

Breakdown is extremely common in large passages in the cave and makes travel slow and difficult in many of the stream passages in the basal Hillsdale Limestone (Fig. 13.37). It may be that the shaly zone 6 is responsible for much of this rockfall, but this was not investigated. Certainly, breakdown has extended above zone 6 in some areas and occurs in passages like Jones Canyon that are not in the basal Hillsdale. Slickensides are visible on breakdown at various places, and fault planes, where present, are evidently important planes of weakness leading to collapse.

Fig. 13.37 Overflow Passage just south of the 1812 Room. Zones 4 and 5 cherts are visible in the upper walls and ceiling channel. Local joint control influenced passage development. The terraced gravel and limestone slab fill is well displayed. Photograph by W.K. Jones

13.9 Hydrology

The discussion of hydrology is from Jones (1988), courtesy of the West Virginia Speleological Survey.

13.9.1 The Hydrogeologic Setting

The Organ Cave System is developed in a synclinal plateau which has the lower part of the Greenbrier limestone exposed to the surface in a southwest trending "trough" which is bounded to the east and west by the underlying Maccrady shale. Small surface streams drain from the shale into blind valleys and dolines near the limestone contact. The underground streams have downcut 10–30 ft into the shale at a few points within the cave, but the shale generally acts as a relatively impermeable boundary. The drainage divides are based on the position of surveyed cave passages, results of dye-tracer tests, topographic divides, geologic contacts, and water budget analysis of the known springs. The positions of the divides are not exact and are least certain on the western and southern margins of the Organ Cave basin. The mature karst developed on the plateau has obliterated all traces of any original surface stream channels. There is no evidence that the present catchment area has changed very much from the time just prior to the onset of karstification. Recharge to (and runoff from) the basin may have been much higher at earlier stages in the development of the basin. Large stream gravels in some of the fill deposits in the higher-level canyons of Organ Cave System seem to suggest periods of very high discharge (velocity) through the system.

Lewis Cave, situated to the north of the Organ Cave basin, drains north to Soloman Springs (elevation 1950 ft) which discharges in a steep valley tributary to the Greenbrier River. Cricket, Hellems, and Destitute caves all drain to a spring on the north side of Second Creek 600 ft upstream of the US 219 Bridge. Several additional small springs emerge on the north side of Second Creek between US 219 and Patton. No tracer tests have been conducted to these springs, but they presumably account for water sinking near the bluff above Second Creek and sinkholes immediately south of the Organ Cave basin. Storage, which maintains all of the main cave streams under base flow conditions, may be mostly in the epikarst which overlies the cave passages (Williams 1983).

A water budget analysis of the Organ Cave resurgence suggests that the spring should account for all the drainage from this system. The Organ Cave Plateau receives about 37 in. of precipitation annually which is evenly distributed throughout the year. The mean temperature is 52 °F, and potential adjusted evapotranspiration is 27 in. (Thornthwaite equation). Since all runoff from the Organ Cave basin is subsurface, the predicted spring discharge is 10 in. or a mean annual discharge of 3.11 cubic feet per second (cfs) from a 3.1 square mile catchment area. The closest karst basin for which actual discharge data are available is the Davis Spring basin which drains 74 square miles and is situated about five miles to the northwest of Organ Cave. The Davis Spring basin is in a similar hydrologic and climatic setting and has a mean discharge of 140 cfs or about 15 in. of runoff. This suggests that precipitation can rapidly infiltrate into the karst drainage system and a soil moisture deficiency exists during most of the summer months (Jones 1973). Based upon a linear comparison of the Organ Cave resurgence discharge per square mile ratio with that for Davis Spring, the base flow discharge at the Organ Cave resurgence should be about 0.3 cfs, mean annual discharge about 7 cfs, and peak discharge over 100 cfs. This implies that over half of the annual discharge comes from a few storm events and that most of the year stream flow is minimal. This may help to explain the very underfit streams which flow in the large canyon passages of the cave.

The Organ Cave resurgence is situated at base level on Second Creek and appears to be the only outlet from the cave. Under low-flow conditions the water appears to rise from several openings in the bottom of the spring pool. A direct accurate measurement of discharge is not possible because the spring is always in backwater from Second Creek. Base flow was roughly measured using tracer dilution techniques (September 21, 1985) at 0.4 cfs.

The stream passages are generally large with gradients averaging over 400 ft per mile in the upper reaches of the system. The cave stream levels respond very rapidly to storms.

The maximum relief on the plateau is about 180 ft, and no major (or permanent) surface streams drain onto the limestone. Much of the recharge to the cave system is from rainfall, but full passage flooding is rare except immediately above the permanent sumps. The lower canyons (Lipps and Big) have a lower gradient (Lower Lipps averages 254 ft per mile, and Big Canyon averages only 47 ft per mile), and the two terminal sumps (elevation 1800 ft) are very near base level for the Organ Cave resurgence on Second Creek (elevation 1770 ft). The average elevation on the plateau is 2300 ft, so there is over 400 ft of relief between base level and the upper parts of the cave system. An upward flexure of the Maccrady shale in Lower Lipps canyon is exposed for about 1000 ft parallel to the stratigraphic strike without any apparent change in the passage gradient. The upper levels of the cave were developed before the establishment of present base level and lithologic and structural influences predominate. The relative gradients of the main stream passages parallel to the strike are shown in Fig. 13.38, and the cross section in Fig. 13.39 shows the relative positions of main passages plotted perpendicular to the strike.

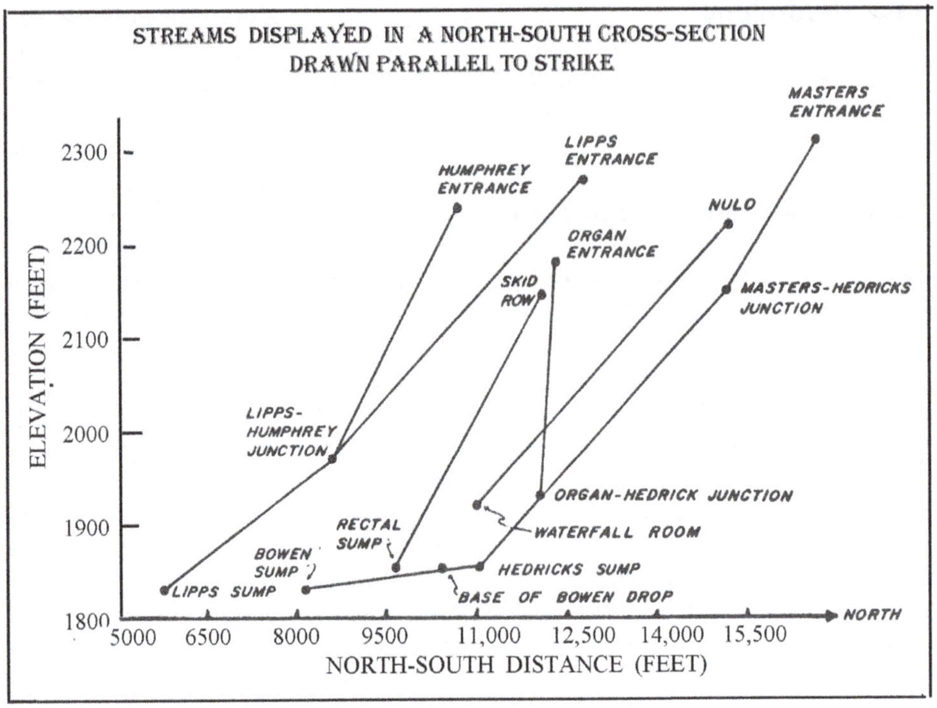

Fig. 13.38 Streams displayed in a north–south cross section drawn parallel to strike. Observed stream gradients in the upper (northern) reaches of the system are greater than those for streams in the downstream (southern) part of the cave (15X vertical exaggeration)

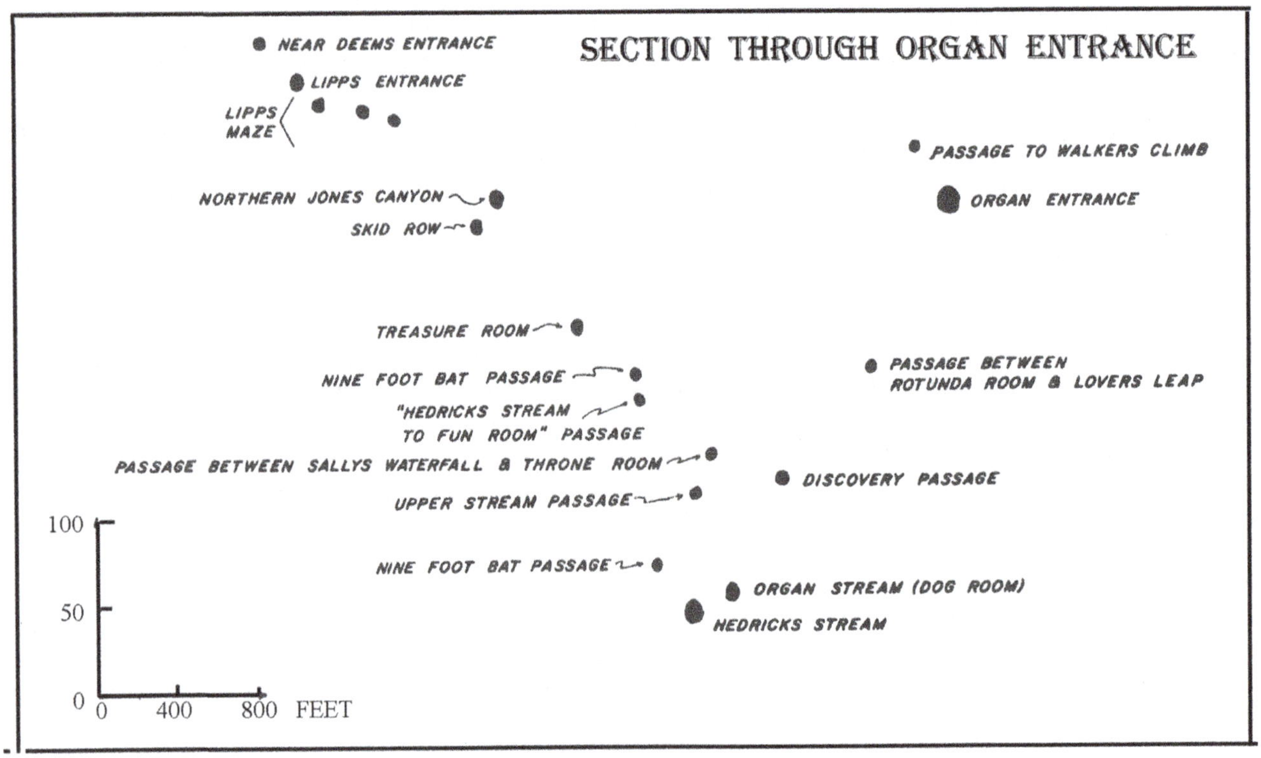

Fig. 13.39 East–west cave passages drawn in cross section perpendicular to strike through the Organ Cave Entrance. Observe influence of Caldwell Syncline. The Hedricks Stream is close to the axis of the syncline (8X vertical exaggeration)

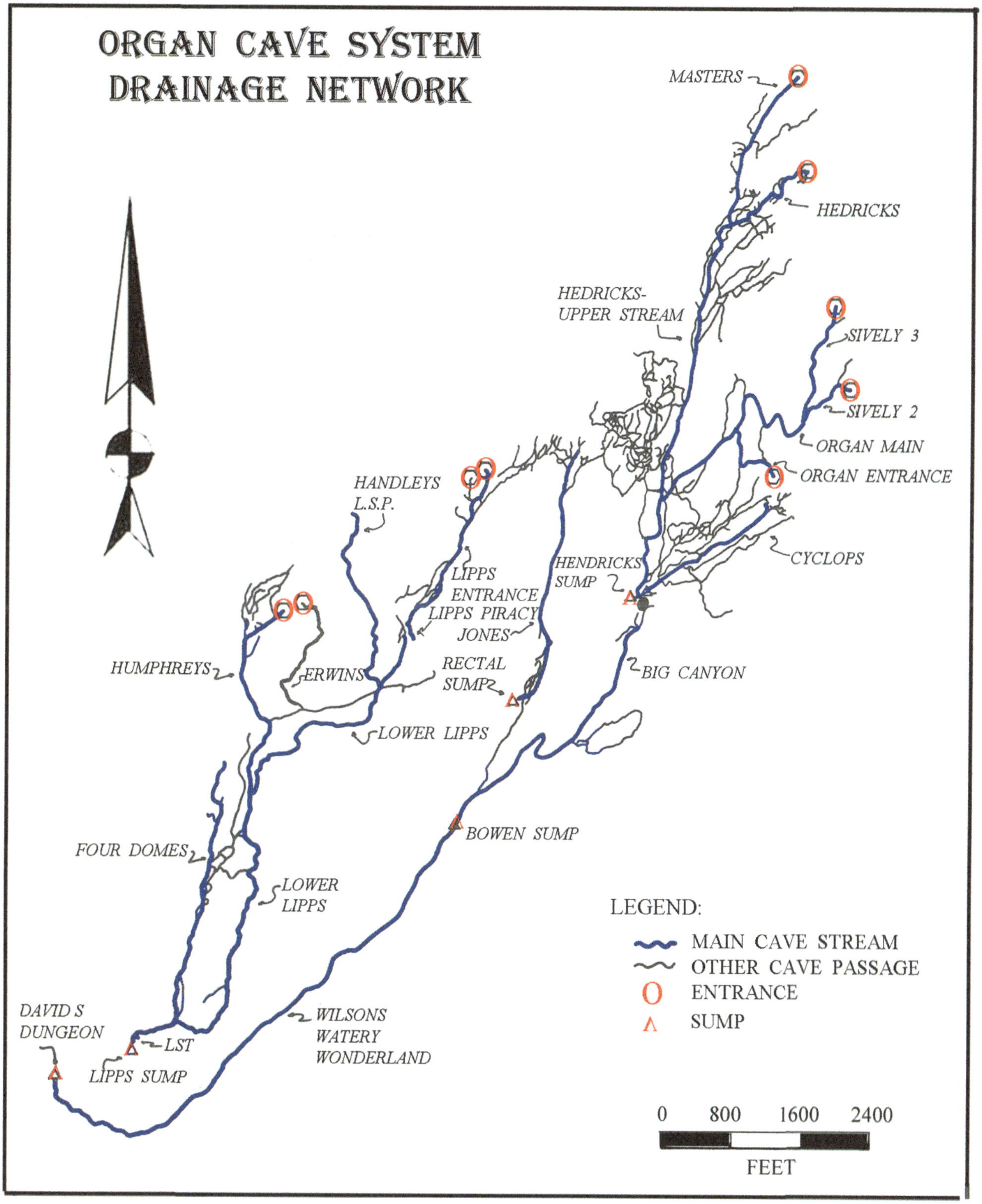

Fig. 13.40 Organ Cave System drainage network

13.9.2 Contemporary Drainage Patterns

The cave is presently drained by two master streams which trend parallel to the strike and probably merge before their resurgence on Second Creek. The Organ–Hedricks–Big Canyon stream closely follows the trough of the Caldwell Syncline, at least until it reaches the Bowen sump. At this point the stream appears to veer up-dip to the west (possibly perched on the Maccrady shale) and has been mapped to within 2,700 ft of the resurgence. The final sump at "David's Dungeon" is also directly in line and below the Lower Lipps canyon which is the second major stream canyon to the west of, and subparallel to, the Organ–Hedricks–Big Canyon passage. Intra-cave dye-tracing has shown that all of the tributary and distributary passages are accounted for at or before reaching the Bowen or Lipps sumps (Fig. 13.40). The overall drainage pattern of the system is an anastomatic dendritic network of passages in the upper (northeast) part of the cave. All converge prior to reaching the Organ Cave resurgence. The three-dimensional nature of the Organ Cave System drainage network is illustrated by the Upper Stream Passage stream which is perched on a chert layer and flows directly above the Hedricks Stream Passage for 4,500 ft (Fig. 13.39).

Extensive stream tracing work was done in conjunction with the survey of the cave, and the results are summarized in Fig. 13.40 and in Jones (1988). The principal tracer was fluorescein sodium, although a few tests were conducted using Rhodamine WT. Dye injection amounts were usually 50–100 g, and passive detectors of activated charcoal were used to recover the dye. For several of the tests, the charcoal packets remained in the cave for over 1 year before they were retrieved and tested. Clear positive tests for fluorescein were obtained, but high background fluorescence in the 580 nm range produced ambiguous results for the Rhodamine analyses.

Several distributary routes were found within the Organ Cave System. One is the stream passage which can be followed from the Lipps entrance down to the piracy passage where all the water (at normal flow conditions) disappears to the east in a gravel-floored passage which gradually becomes too low to follow. This stream does not reappear in the Jones Canyon stream, but reemerges just upstream from Lipps sump along with the water diverted from Jones Canyon at the Rectal Sump. This suggests that another as yet undiscovered stream passage must lie between Jones Canyon and Lower Lipps Canyon.

Another branching of flow exists under high water conditions when water flowing into the Organ tourist entrance is pirated from the west side of the large entrance room to the Rotunda Room where some of the water flows north down the Waterfall Passage and some flows south through the 1812 Room to Hedricks Stream.

Fig. 13.41 Paul Stevens negotiates the French Connection near the Humphreys entrance

Acknowledgements The following persons have made major contributions to the Organ Cave System project. These include Paul J. Stevens, George R. Dasher, Dr. George H. Deike for Geology, Frederick V. Grady for Fossils, Dr. John R. Holsinger for Biology, and William K. Jones for Hydrology. The WVSS Bulletin from which much of this chapter was drawn would not have been possible without the relentless work of Paul Stevens and his wife Lee. The project co-chair and our friend Paul died of amyotrophic lateral sclerosis (ALS) in 2007 (Fig. 13.41).

References

Cole, J.R. 1917. *History of Greenbrier County*. J.R: Cole publisher.

Davies, W.E. 1958. *Caverns of West Virginia*, vol. 19A, 2nd ed, 330. Morgantown, WV: West Virginia Geological and Economic Survey.

Deike, G.H., III. 1988. Geology. In *Caves of the organ cave plateau*, ed. P.J. Stevens, vol. 9, 13–39. Barrickville, WV: West Virginia Speleological Survey Bulletin.

Faust, B. 1964. *Saltpetre caves and Virginia history*. Falls Church VA: Econo Print.

Flack, R. 1949. Charleston Grotto group performs second spectacular stunt of the current season. *The NSS News* 7 (7): 4.

Hartline, D. 1964. Greenbrier Caverns (Organ-Hedricks System). *The Massachusetts Caver* 3 (4).

Heller, S.A. 1980. A hydrogeologic study of the Greenbrier Limestone karst of central Greenbrier County, 167pp. Ph.D. Thesis, West Virginia University, West Virginia.

Humphreys, B. 1926. A History of organ cave. Unpublished paper, Greenbrier Historical Society, Lewisburg WV.

Jones, W.K. 1973. *Hydrology of limestone karst*. Morgantown, WV: West Virginia Geological and Economic Survey Bulletin 36.

Jones, W.K. 1988. Hydrology. In *Caves of the organ cave plateau*, ed. P.J. Stevens, vol. 9, 40–48. Barrickville, WV: West Virginia Speleological Survey Bulletin.

Ogden, A.E. 1976. The hydrogeology of the central Monroe County karst, 263pp. Ph.D. Thesis, West Virginia University, West Virginia.

Palmer, A.N. 1974. Geologic influence upon cave passage orientation in Ludington Cave, Greenbrier County, West Virginia. In *Proceedings Fourth conference on karst geology and hydrology*. ed. H.W. Rauch and E. Werner, 33–40. Morgantown, WV: West Virginia Geological and Economic Survey.

Prolix, Pergerine (P.H. Nicklin). 1837. *Letters descriptive of the Virginia springs*, 2nd ed. Philadelphia, PA: H.S. Tanner Publisher.

Price, P.H., and E.T. Heck. 1939. *Greenbrier County*, 846. Morgantown, WV: West Virginia Geological Survey.

Rice, O.K. 1970. *The Allegheny Frontier, West Virginia Beginnings 1730–1830*. Lexington, KY: University Press of Kentucky.

Rutherford, J.W., and R.H. Handley. 1976. The Greenbrier Caverns. *National Speleological Society Bulletin* 38 (3): 41–52.

Stevens, P. 1982. The history of organ cave: The Moretti years. *Capital Area Cavers Bulletin No. 1*, published by D.C. Grotto and Potomac Speleological Club.

Stevens, P. 1988. *Caves of the organ cave plateau*, vol. 9. Barrickville, WV: West Virginia Speological Survey Bulletin.

Stafford, R. 1961. Cavers threaten cave. *The NSS News* 19 (8): 99.

Staurt, Colonel. 1971. *Memoir of Indian wars and other occurrences*. Parsons, WV: McLain Printing Co.

Wells, D. 1950. Lower middle Mississippian of Southeastern West Virginia. *American Association of Petroleum Geologists Bulletin* 34 (5): 882–922.

Williams, P.W. 1983. The role of the subcutaneous zone in karst hydrology. *Journal of Hydrology* 61: 45–67.

Winslow, M., and R. Baroody. 1976. Structure map of Greenbrier County, West Virginia. Unpublished.

Caves of the Dickson Spring Basin, Monroe County

14

William K. Jones

Abstract

Dickson Spring drains about 26 mi^2 to Second Creek and is the largest catchment in Monroe County. Water emerges at the base of a cliff in the lower Patton Limestone. Eight significant caves are reported within this basin, and all of the passages are developed in the Patton, Sinks Grove, and Hillsdale units. Several of the entrances receive small streams and are at the contact of the Hillsdale Limestone and the Maccrady Shale. The caves, with the exception of Haynes Cave, are all in the headwaters (southern) part of the drainage basin and are complex from a geologic and hydrologic viewpoint. Most of these caves have received little attention in the last forty years because of access limitations.

14.1 Introduction

The Dickson (Dixon) Spring basin is elongate (Fig. 14.1) with the northwest flank of Little Mountain near Gap Mills forming the southern boundary and Dickson Spring on Second Creek at the northern end. Water flow is from south to north. The longest water trace in the basin is from Burnside Branch Cave with the water flowing about 8 miles to the north to resurge at Dickson Spring (Jones 1997). The "Monitor Lineament" cuts perpendicularly across the basin about 2 miles (3 km) south of the spring (Werner 1975). This obvious surface feature starting near Bickett Knob and terminating at Second Creek is a straight line of dolines that trend N76E for about 5 miles. The geologic control of this lineament is not understood but may be along a fracture zone. The lineament is not associated with any mapped faulting and cuts across the structural grain at an odd angle. No significant caves are found along the lineament, and subsurface flow appears to pass under the lineament no observable deflection from the path to Dickson Spring. With the exception of Canterbury Cave, the cave passages appear to be controlled by jointing, but the orientation of the main passages tends to be almost north-south while the regional strike is about N30°E.

Caves within the basin are described by Davies (1965), Hempel (1975), and Bray et al. (2012a). There are seven significant caves totaling about 17 miles (27 km) of passage within the southern (upstream) half of the basin. Haynes Cave is the only cave described in this chapter north of the Monitor Lineament, and this cave is included only because of its proximity to Dickson Spring. Most of the cave entrances in this basin are at sink points along the Greenbrier Limestone–Maccrady Shale contact. The caves tend to have intricate multilevel patterns and a good bit of vertical relief. The caves are deep enough that the gradient from the sumps to Dickson Spring is less than 0.5°. There may be some hydrological connection between different caves although this has only been established with the water in Rehoboth Church Cave reappearing to the north in Hurricane Ridge Cave. All of the caves except Canterbury and Haynes are subject to flooding.

Electronic supplementary material
The online version of this chapter (doi:10.1007/978-3-319-65801-8_14) contains supplementary material, which is available to authorized users.

W.K. Jones (✉)
PO Box 356, Warm Springs, VA 24484, USA
e-mail: wkj30@hotmail.com

W.B. White (ed.), *Caves and Karst of the Greenbrier Valley in West Virginia*,
Cave and Karst Systems of the World, https://doi.org/10.1007/978-3-319-65801-8_14

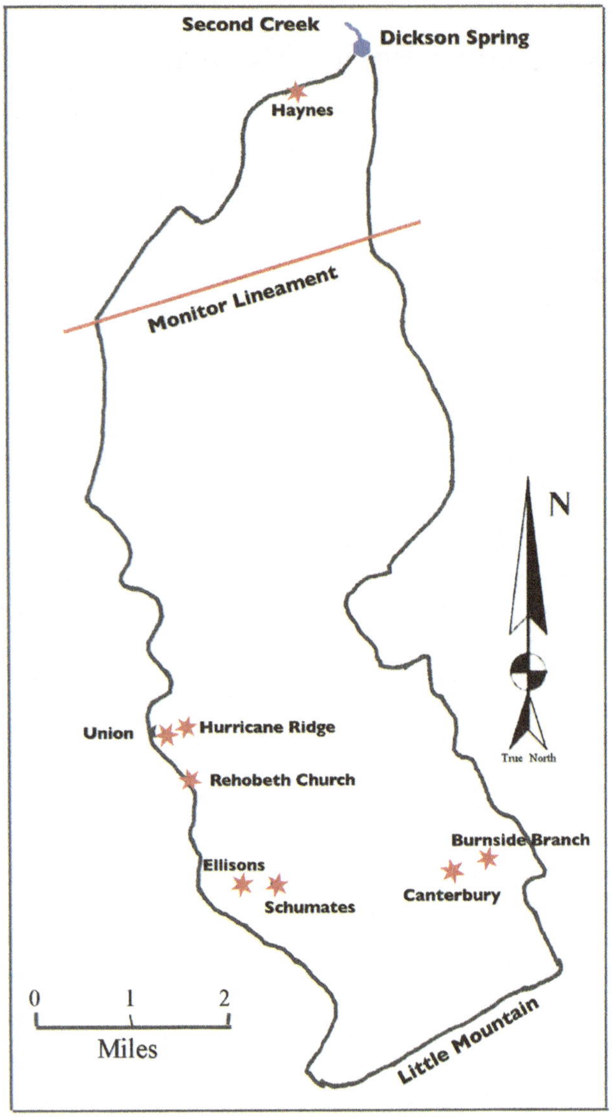

Fig. 14.1 Sketch map showing relative position of principal caves within the Dickson Spring drainage basin

14.2 Caves of the Dickson Spring Basin

The caves in this chapter are described from south to north (Fig. 14.1). These caves are all on private land, and it may be difficult to obtain permission to enter them. Union and Hurricane Ridge Caves have ongoing surveys. Surveys and maps for the other caves date back to the early 1970s, and much of the early data has been lost. The surveyed lengths listed for the caves are in some cases estimates from work done forty years ago. These caves are very interesting geologically and will hopefully receive additional study in the future.

14.2.1 Canterbury Cave

Canterbury Cave is an unusual (at least in the area covered by this book) three-dimensional maze. The cave has 1.51 miles (2.44 km) of surveyed passage but at no point in the cave is more than 350 ft from the entrance (Hempel 1975). The listed depth of 30 ft is probably low. The passages are very intertwined and sinuous and do not follow any particular orientation. The cave is on or near the axis of the overturned Canterbury Cave Syncline (Ogden 1975) and is probably in the Sinks Grove Limestone. The cave formed under closed-channel flow conditions, probably by water rising under pressure from some depth. If so, this may be the only fully hypogenic cave in the Greenbrier Karst. Thermal water rises at several springs in Ordovician rocks 13 miles to the east-north-east but no thermal springs are known from the Greenbrier Group.

The entrance is in a doline in a field southwest of the old Canterbury farmhouse. Canterbury Cave is perched about 120 ft higher in elevation than nearby Burnside Branch Cave, but a connection between the two caves is unlikely (there are no active streams in Canterbury Cave). Several crawlways lead away from the entrance, and one leads to a 30-foot drop (Fig. 14.2). The room at the base of the drop has several passages radiating from it, and these passages usually have more passages and so on. Several rooms and dome pits are found in the lower sections of the cave. There are relatively few junctions or interconnections between passages that pass very close to one another. The cave was partially surveyed by members of the Pittsburgh Grotto in the early 1970s. The only available "map" of the cave is a three-dimensional wire model (Fig. 14.3) built by Charles F. Williams in 1975. The passages simply appear to wind and loop on multiple levels without any obvious control by the rock structure (Fig. 14.4). This cave is well worth additional study if and when access can be obtained.

14.2.2 Burnside Branch Cave

Burnside Branch (Woodland School) Cave has a surveyed length of 1.99 miles (3.20 km). It has five entrances and captures most of the flow from the upper part of Burnside Branch. The entrances are about 2400 ft east of Canterbury Cave and line up in a north-northwest direction along the west side of Burnside Branch (Fig. 14.5). Most of the passages appear to be phreatic and formed under closed channel flow conditions. This cave has a very different pattern, more controlled by the geologic structure, than Canterbury Cave. Burnside Branch Cave is a multilevel maze on three distinct

Fig. 14.2 Photograph of Charles F. Williams on the 30-foot climb to the entrance passage to the central room in Canterbury Cave. Photograph by the author

levels and may be a good example of a "flood-water maze cave" (Palmer 1975, 2011). The cave was described by Davies (1965) as Woodland School Cave and Hempel (1975) as Burnside Branch Cave.

The lower entrance passages trend northwest 500 ft and intercept south-southwest trending passages and may follow the local strike. The main entrance passage as described by Hempel (1975) drops rapidly through a series of cascades and becomes a complex maze near the intersection of the strike oriented passages. The lower stream passage ends in a sump about 1000 ft south of the maze and 254 ft below the entrance. Small connector climbs provide access to the upper and lower levels of the cave. The cave is in the Sinks Grove Limestone, and chert nodules are exposed in many of the passages (Fig. 14.6).

14.2.3 Schumates Icehouse Cave

Schumates (Shumates) Icehouse Cave is in the Burnside Branch valley near Gates. The original entrance was through a trap door in the floor of a small building used as natural refrigerator. The cave is briefly described by Hempel (1975) and was resurveyed by members of the VPI caving club in the 1980s. The cave is said to have 2.48 miles (4 km) of passage and reach a depth of 174 ft. Unfortunately the map and survey data have been lost so little information is available on this interesting cave.

Hempel (1975) describes the entrance passage as a tight pinch to a small drop in a maze. After about 600 ft of mostly crawlway, an inconspicuous climb leads to an upper level with about 2000 ft of walkway. The cave is in the Sinks Grove Limestone, and the stream in the cave may flow to Ellison's Cave half a mile to the west.

14.2.4 Ellison's Cave

The entrance to Ellison's Cave is in a doline on the west side of the Gates Road. The entrance passage trends north for about 400 ft to a climb down of about 20 ft to a small room. A tight low fissure passage with many chert nodules to impede progress leads about 600 ft to an intersection with a large canyon passage with several branches. In wet weather, there is a stream flowing in this passage and another (separate?) steam in a parallel passage about 100 ft to the west. These streams may flood and several sumps are present during higher flow conditions. This is a surprisingly complex cave developed in the Sinks Grove Limestone.

A brief description of Ellison's Cave was presented by Hempel (1975). The cave is said to have 3.03 miles (4.9 km) of passage and reaches a depth of 177 ft. The water from nearby Schmates Icehouse Cave may flow through Ellison's Cave. The cave was surveyed during the 1980s by members of the VPI cave club but the map and survey notes have been lost.

14.2.5 Rehoboth Church Cave

Rehoboth Church Cave was described by Hempel (1975). The entrance is a stream sink at the western end of a doline behind the historic Rehoboth Church. The entrance passage descends to a low stream passage, and several good-sized rooms are reached by following the stream. The steam

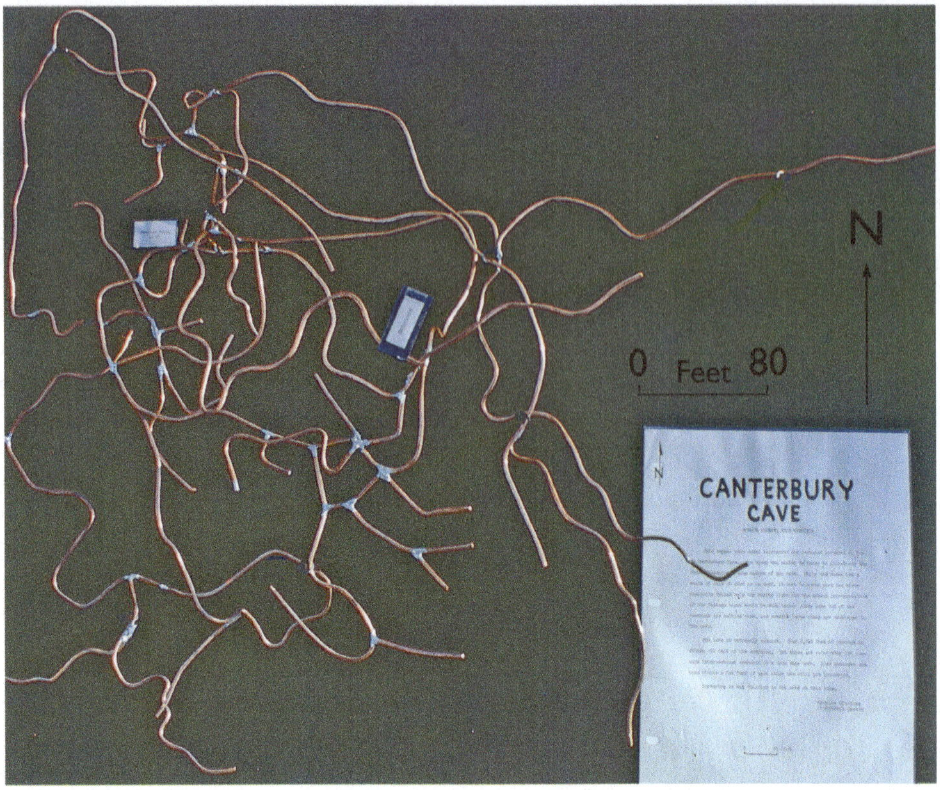

Fig. 14.3 Map of Canterbury Cave constructed by Charles F. Williams using copper wire to present a 3-D model of the cave circa 1975. Note how the passages are intertwined with relatively few nodes or interconnections between the passages

Fig. 14.4 One of several passages connecting climbs in Canterbury Cave. Photograph by the author

passage ends at a sump about 500 ft west of the entrance. Several upper-level rooms that trend north-south branch off from the stream passage. The main passages are oriented north-south with infeeders oriented southwest. Rehoboth Church Cave is developed in the Hillsdale Limestone, and the water has been traced one mile north to Hurricane Ridge Cave and from there to Dickson Spring. The cave has about 1.34 miles (2.16 km) of passage and a depth of 125 ft. A map was published by Hempel (1975).

14.2.6 Union Cave

Union (Caperton) Cave was described by Davies (1965), Hempel (1975), and Bray et al. (2012b). The part of the cave originally described by Davies (1965) and Hempel (1975) was about 1500 ft long and ended at "the big room." The group from the West Virginia University Student Grotto that was exploring the nearby Hurricane Ridge Cave pushed a series of tight climbs starting at the base of the Big Room and entered a large stream passage that connected to several side passages on different levels (Fig. 14.7). They finished with 4.96 miles (7.98 km) of passage and a depth of 326 ft. Much of the cave consists of slot canyons on different levels. The streams in the southern part of the cave flow north (Ecstasy River, Fig. 14.8, and the UnPaleo Trunk) and sump on the west side of the main passage. Water in the main northern passage (Echo Hall Figs. 14.9, 14.10 and 14.11)

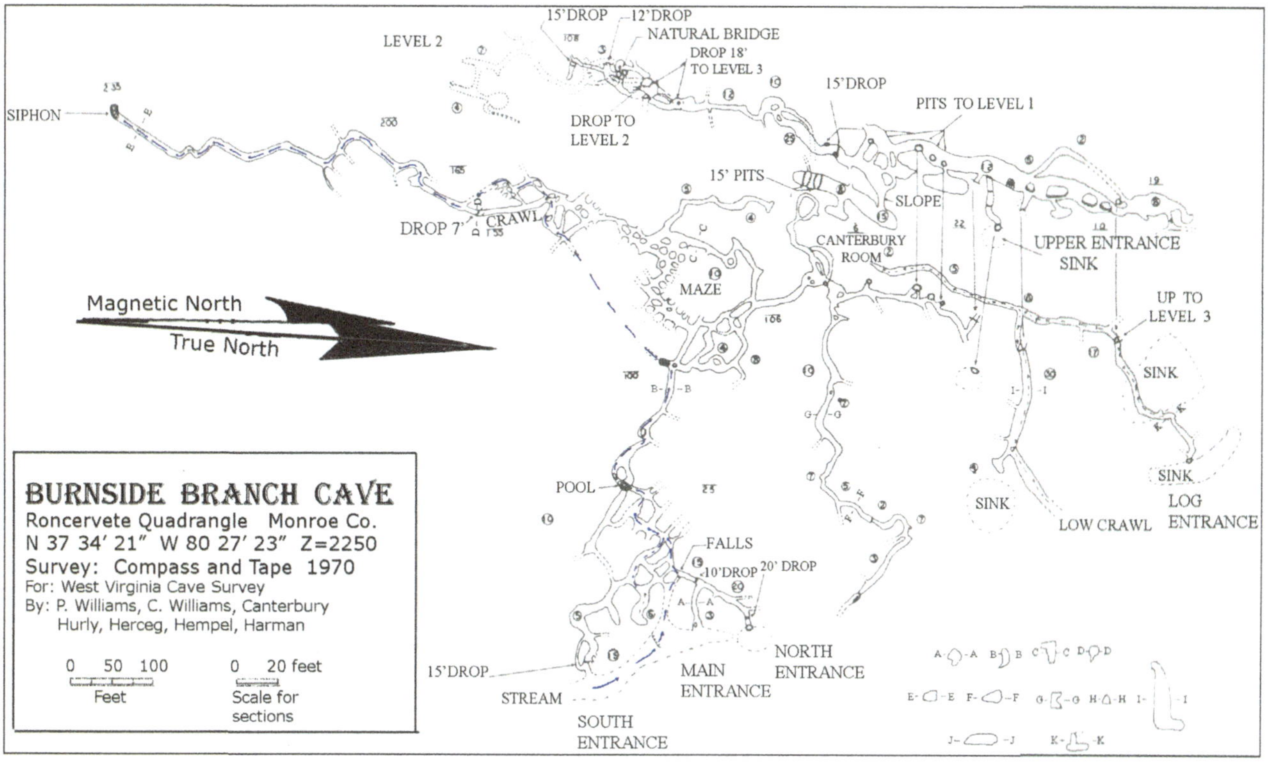

Fig. 14.5 Map of Burnside Branch Cave

Fig. 14.6 Low infeeder passage in Burnside Branch Cave showing exposed chert nodules in the walls. Photograph by the author

flows to the south and turns west at about the midpoint of the cave at the UnPaleo sump. There is a drainage divide at the extreme northern end of the cave with water flowing north. All of the water eventually resurges at Dickson Spring. CPG Pit is just to the east of a dome in the cave and may provide a second entrance with some digging.

The passages are oriented in a north-south direction and trend a little north of the regional stratigraphic strike. Union and Hurricane Ridge caves are east of the Hurricane Ridge syncline but near north-south oriented anticlines and synclines mapped by Ogden (1976). The cave is developed in the Patton Limestone

14.2.7 Hurricane Ridge Cave

Hurricane Ridge (Peddlers) Cave trends parallel to Union Cave less than one-half mile to the west (Fig. 14.7). The cave was described by Bray et al. (2012b), and exploration and mapping is ongoing. At this writing, the cave contains 2.41 miles (3.89 km) of passage and reaches a depth of 251 ft. A connection with Union Cave is unlikely because Hurricane Ridge Cave is developed stratigraphically lower and none of the passages mapped in either cave show promise of a connection. The entrance is a steam sink at the base of a cliff at the western end of a large doline. The entrance is at the contact of the Maccrady Shale and the Hillsdale Limestone and was dug open in 1996 by Jeff Bray, Tom Malabad and Kristen Matak. This cave contains a stream passage (Fig. 14.12) with several low spots and can flood quickly. The first 600 ft of passage is wet and often tight with chert nodules to make traversing unpleasant. Numerous side passages, rooms, and different stream passages lead away from the entrance series (Figs. 14.13, 14.14 and 14.15). The stream from Rehoboth Church Cave is encountered at the downstream infeeder to Cripple Creek, and this water flows north to a sump. This is a complex cave on different levels developed in the Hillsdale Limestone.

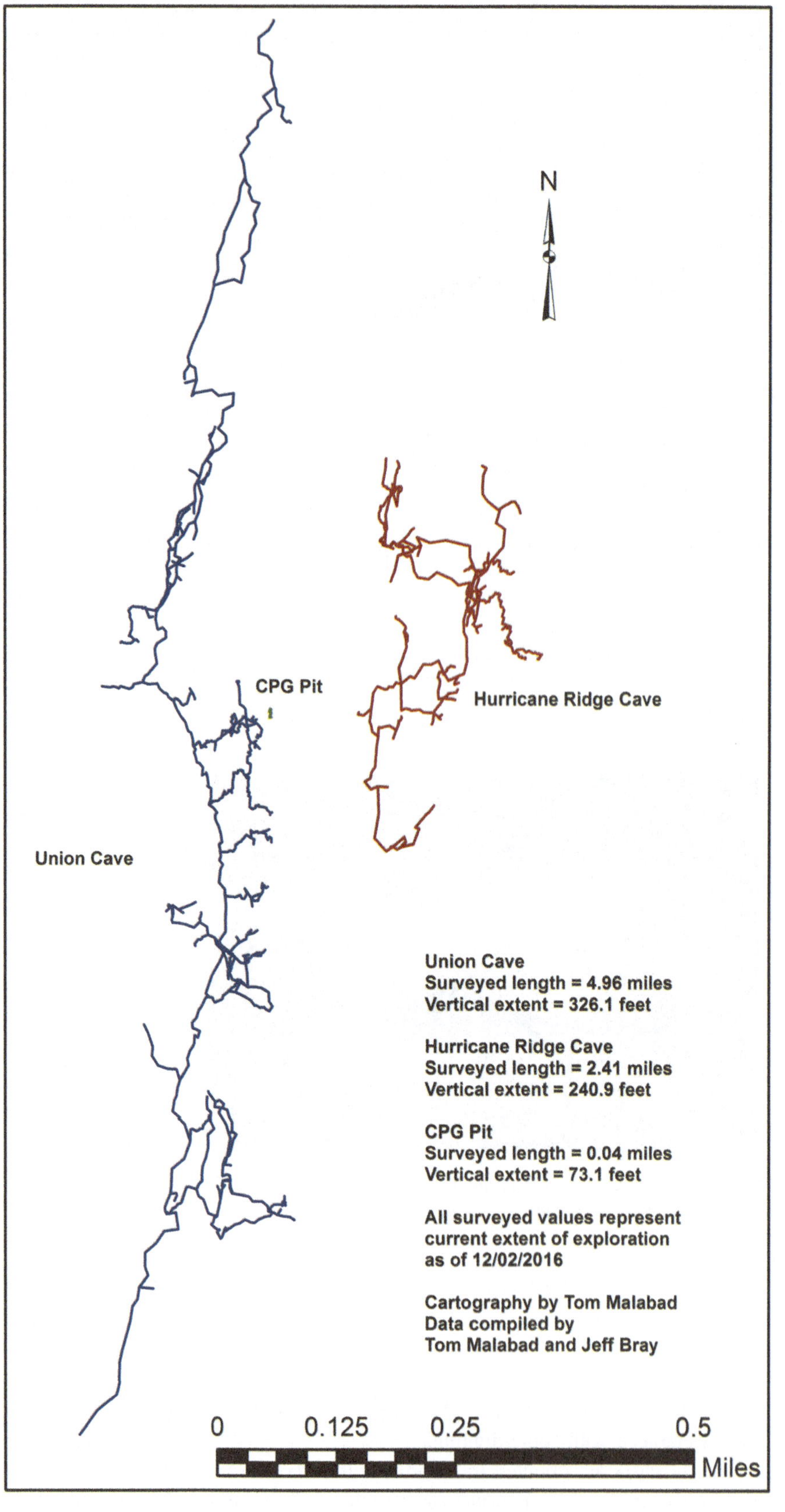

Fig. 14.7 Line maps showing passages in Union and Hurricane Ridge Caves. Map by Tom Malabad and used with permission

Fig. 14.8 Ecstasy River passage just south of junction with main entrance stream in Union Cave. Photograph by Ed McCarthy and used with permission

Fig. 14.9 View looking north into the Planetarium just south of the lake in Echo Hall in Union Cave. Photograph by Ed McCarthy and used with permission

14.2.8 Haynes Cave

Haynes (Burwell) Cave is apparently in the Dickson Spring catchment, but there is no active stream passage. Except for a ceiling drip and flowstone mound 100 ft from the entrance, this cave is very dry and dusty. Haynes Cave is now owned and managed by the West Virginia Cave Conservancy. The cave is gated (Fig. 14.16) to protect artifacts left from saltpeter mining dating back to the late 1790s and the civil war. Numerous relics from the mining operations are well preserved in the cave including wooden bridges and a windless (Figs. 14.17 and 14.18). A wooden saltpeter trough was left on top of a flowstone mound just inside the entrance at the end of the civil war. The trough had accumulated a 1-in. thick coating of calcite in about 100 years (Fig. 14.19).

The cave is developed on two levels with the main passage trending generally south from the entrance. This upper-level passage contains numerous artifacts from the saltpeter mining operations and ends in "the signature room" where the walls are covered with names and dates of visitors. A register has been kept here for many years. There are numerous gypsum flowers on the walls. A lower level passage extends for 1000 ft from near the windless along the strike direction of N30°E. The cave is in the Patton Limestone (electronic map M-14.1).

Fig. 14.10 Echo Hall just past climb up into Upper Echo Hall, Union Cave. Photograph by Ed McCarthy and used with permission

Fig. 14.11 Tom's bolt climb into Graham's Dome at the southern end of upper Echo Hall in Union Cave. Photograph by Ed McCarthy and used with permission

Fig. 14.12 Mainstream passage in Hurricane Ridge Cave downstream of OJ Junction. This is a tall canyon named "Cripple Creek" for the numerous pot holes. Photograph by Ed McCarthy and used with permission

Fig. 14.13 "Budway," a large upper-level passage in Hurricane Ridge Cave. Photograph by Ed McCarthy and used with permission

Fig. 14.14 "Fantastica" in the southern part of Hurricane Ridge Cave. Photograph by Ed McCarthy and used with permission

Fig. 14.15 Large stalagmite in "Fantastica." The speleothem was named "Hale-Bop" after the comet visible in the night sky on the day of the discovery trip. Photograph by Ed McCarthy and used with permission

Fig. 14.16 Gated entrance to Haynes Cave. Photograph by the author

Fig. 14.17 One of the bridges spanning a lower level canyon in Haynes Cave. These bridges date to the saltpeter mining operations in the cave during the civil war. Photograph by the author

Fig. 14.18 A civil-war era windless in Haynes Cave. Photograph by the author

Fig. 14.19 Calcite encrusted trough near the entrance to Haynes Cave in December, 1964. The trough was destroyed by vandals within a month of when this picture was taken. Photograph by Robert Bergad. Used with permission

Acknowledgements Special thanks are due to Mike Futrell, Thomas Malabad, and Ed McCarthy for help with the cave descriptions and photographs.

References

Bray, J., T. Malabad, and K. Matak. 2012a. Hurricane ridge cave. In *MayaCon 2012, NSS Convention Guidebook*, ed. M. Adler, 239–241. Huntsville, AL: National Speleological Society.

Bray, J., T. Malabad, T., and K. Matak. 2012b. Union Cave. 249–251. In *MayaCon 2012, NSS Convention Guidebook*, ed. M. Adler. Huntsville, AL: National Speleological Society.

Davies, W.E. 1965. *Caverns of West Virginia*, vol. 19A. Morgantown, WV: Geological and Economic Survey.

Hempel, J.C. 1975. *Caves and Karst of Monroe County, W.Va*, vol. 4, 139pp. Barrickville, WV: West Virginia Speleological Survey Bulletin.

Jones, W.K. 1997. *Karst Hydrology Atlas of West Virginia*, vol. 4, 111pp. Leesburg, VA: Karst Waters Institute Special Publication.

Ogden, A.E. 1976. The Hydrogeology of the Central Monroe County Karst, 263pp. Ph.D. Thesis, West Virginia University.

Palmer, A.N. 1975. The origin of maze caves. *National Speleological Society Bulletin* 37 (3): 57–76.

Palmer, A.N. 2011. Distinction between epigenic and hypogenic maze caves. *Geomorphology* 134 (1–2): 9–22.

Werner, E. 1975. Long lineaments in southeastern West Virigina. *West Virginia Academy of Sciences Proceedings* 47: 113–118.

Windy Mouth Cave

15

David A. Shank, Megan Curry Fucci and Ira D. Sasowsky

Abstract

Windy Mouth Cave is a mostly abandoned paleospring conduit that drained water from the Big Levels surface northward to the Greenbrier River. The formation of the cave was controlled by a combination of structural and hydrologic factors. Geologic structure provided a pre-solutional network of faults, joints, and bedding planes in the bedrock that was later exploited by groundwater flux. The cave is situated on the western limb of the Sinks Grove Anticline. Beds dip gently to the northwest and strike generally N40–50°E in the cave. Conduits are primarily oriented along strike, while a smaller component is oriented sub-parallel to dip. In plan view, Windy Mouth Cave appears as a branchwork system with a minor anastomotic morphological element overprinted on the dominant dendritic pattern. There are three levels in the cave. Upper levels are phreatic tubes that are connected to small vadose canyons at their origin, and contain some active water. The middle level comprises mainly large phreatic passages, but with substantial clastic fills. The lower levels are well-developed canyon passages that run down dip and crosscut and incise below the main level of the cave. Conduit cross-section morphology is complicated by interbedded chert and shale layers in the Hillsdale Limestone host rock. The impermeable layers form resistant ledges that split individual conduits into multiple levels. Fluvial sediment deposits that are suitable for paleomagnetic analyses were not found in the upper levels, although the position and hydrologic genesis of the system suggest that the upper levels were formed first and are thus the oldest portion of the cave. A magnetically reversed sample was found near the base of one section that was presumably deposited during the reversal of the geomagnetic field which ended at 788 ka. This sets the minimum age of the cave. The modern-day hydrology is markedly different from the past. The drainage basin area is much smaller (~ 2 km^2), and the resident streams have considerably less discharge. Much of the drainage has been pirated to the Scott Hollow drainage basin located south and west of Windy Mouth Cave.

15.1 Introduction

Windy Mouth Cave is one of three large cave systems (along with Organ and Scott Hollow Caves) known from the portion of the Big Levels limestone plateau lying south of the Greenbrier River in Monroe and Greenbrier Counties. The cave's reported length and depth are 29 km (18 miles) and 61 m (200 ft), respectively (Gulden 2017), ranking it as 39th on the US long cave list. Although it has a history of exploration going back over 60 years (see Chap. 5), difficulties in accessing the entrance and traversing the cave have

Electronic supplementary material
The online version of this chapter (doi:10.1007/978-3-319-65801-8_15) contains supplementary material, which is available to authorized users.

D.A. Shank
Strategic Mining Solutions, 473 Brockway Road, Frankfort, NY 13340, USA

M.C. Fucci · I.D. Sasowsky (✉)
Department of Geosciences, University of Akron, Akron, OH 44325-4101, USA
e-mail: ids@uakron.edu

W.B. White (ed.), *Caves and Karst of the Greenbrier Valley in West Virginia*,
Cave and Karst Systems of the World, https://doi.org/10.1007/978-3-319-65801-8_15

Fig. 15.1 Entrance passage crawl. Photograph by I.D. Sasowsky

precluded the completion of a comprehensive map for the system, or even a delineation of its full extent and length. The cave is usually accessed by traversing along the south bank of the Greenbrier River, and when water levels are high, or weather cold, this is challenging. A caver died by drowning in the river in 1998 after exiting the cave. Additionally, the front sections of the cave involve very long (Fig. 15.1), and frequently low, crawls. When we first considered working in the cave, we spoke with Jim Hixson who had surveyed in the cave earlier. His advice was "Wear double kneepads," which we interpreted as a joke. Our subsequent initial trip into the cave convinced us that he was correct, and on later trips we wore double kneepads, and even affixed padding into the palms of our gloves. Nevertheless, the cave has its appeal (Fig. 15.2) and many walking passages beyond the entrance series (Fig. 15.3).

Brief descriptions of the cave, some with associated maps (discussed below), have appeared over the years (Davies 1958, p. 132; Peters and Davis 1960; Davies 1965, pp. 26–27; Whittemore 1971; Dore 1995; Dasher 2002). The most

Fig. 15.2 Speleothems in trunk passage south of the waterfall room. Photograph by I.D. Sasowsky

Fig. 15.3 Walking passage beyond entrance crawls. Photograph by E. McCarthy. Used with permission

complete written description, running just over two single-spaced typewritten pages (Davis and Peters 1960), is unpublished as far as we know and was shared with us from the personal files of Jim Hixson. The present chapter makes use of these early works (including maps discussed below), our personal experiences in the cave, and the scientific studies of Shank (2002) and Curry (2002), to discuss the geologic and hydrologic conditions and history of this major system.

15.2 Setting

The area near the cave is within the transition zone, referred to as the "Folded Plateau," adjacent to the boundary between the Valley and Ridge and the Appalachian Plateaus physiographic provinces. Specifically, this area is situated between the Allegheny Structural Front to the west and the Intraplateau Structural Front to the east (Kulander and Dean 1986). The region is characterized by gently dipping northeast trending folds in Middle-Late Paleozoic sedimentary rocks (Heller 1980; Ogden 1976). Scott Hollow Cave (Chap. 16) lies directly south of Windy Mouth Cave. This position, along with the general alignment of the trunk passages in the two caves, has led to the hypothesis that they may consist of one more extensive system, as yet unconnected by explorers.

15.2.1 Stratigraphy and Structure

The Greenbrier Group carbonates and siliciclastics within which the cave is formed are exposed in a northeast trending belt that outcrops along the eastern margin of West Virginia. Locally, the outcrop belt forms a broad sinkhole plain that ranges in elevation between 560 and 690 m (1820–2250 ft) and is referred to as the "Big Levels." Hills that surround the sinkhole plain are capped by clastic rocks of the Mississippian Mauch Chunk Group that conformably overly the Greenbrier Group (Heller 1980). In the study area, the sinkhole plain is deeply dissected by the Greenbrier River, a major tributary to the New River in the Ohio River drainage basin. Numerous karst features such as blind valleys, sinkholes, and disappearing streams dominate the scenic landscape of the plain. There is also a notable lack of surface drainage on the Big Levels. Groundwater enters the system on the karst upland and discharges through springs and seeps at river level, 150 m (490 ft) below.

The Greenbrier Group is approximately 240 m (900 ft) thick (Fig. 15.4) in southern Greenbrier County (Wells 1950) and is bounded at its upper and lower contacts by Mississippian Shales (Ogden 1976). The upper contact is made with The Lillydale Shale of the Mauch Chunk Group (Bluefield Formation). The Lillydale is only found at the highest elevations in the study area. The Maccrady Shale bounds the lowest unit of the Greenbrier Group and plays a major role in the hydrogeology and karst development in the area (Palmer 1974). The cave is confined to the Hillsdale Limestone (Fig. 15.5) with one possible exception, in the First Creek dome, where it may extend vertically into the Sinks Grove Limestone. In an overview of the stratigraphy in nearby Organ Cave, Deike (1988) mentions that the contact may be observed in a large room that extends high into the overlying strata. Organ and Windy Mouth Caves are similar in that they are both confined to the lower Greenbrier Group strata, as are many other caves in the region.

Several minor folds, mapped and named by Heller (1980), exist between major structures in the study area (Fig. 15.6). These include the Fort Spring Anticline and Syncline as well and another unnamed fold pair that lies just west of the study area. An additional unnamed thrust fault and fold pair lies just to the east. Heller (1980) attributes these minor structures to splay faulting through the relatively competent Greenbrier Group carbonate sequence from underlying Lower Paleozoic strata.

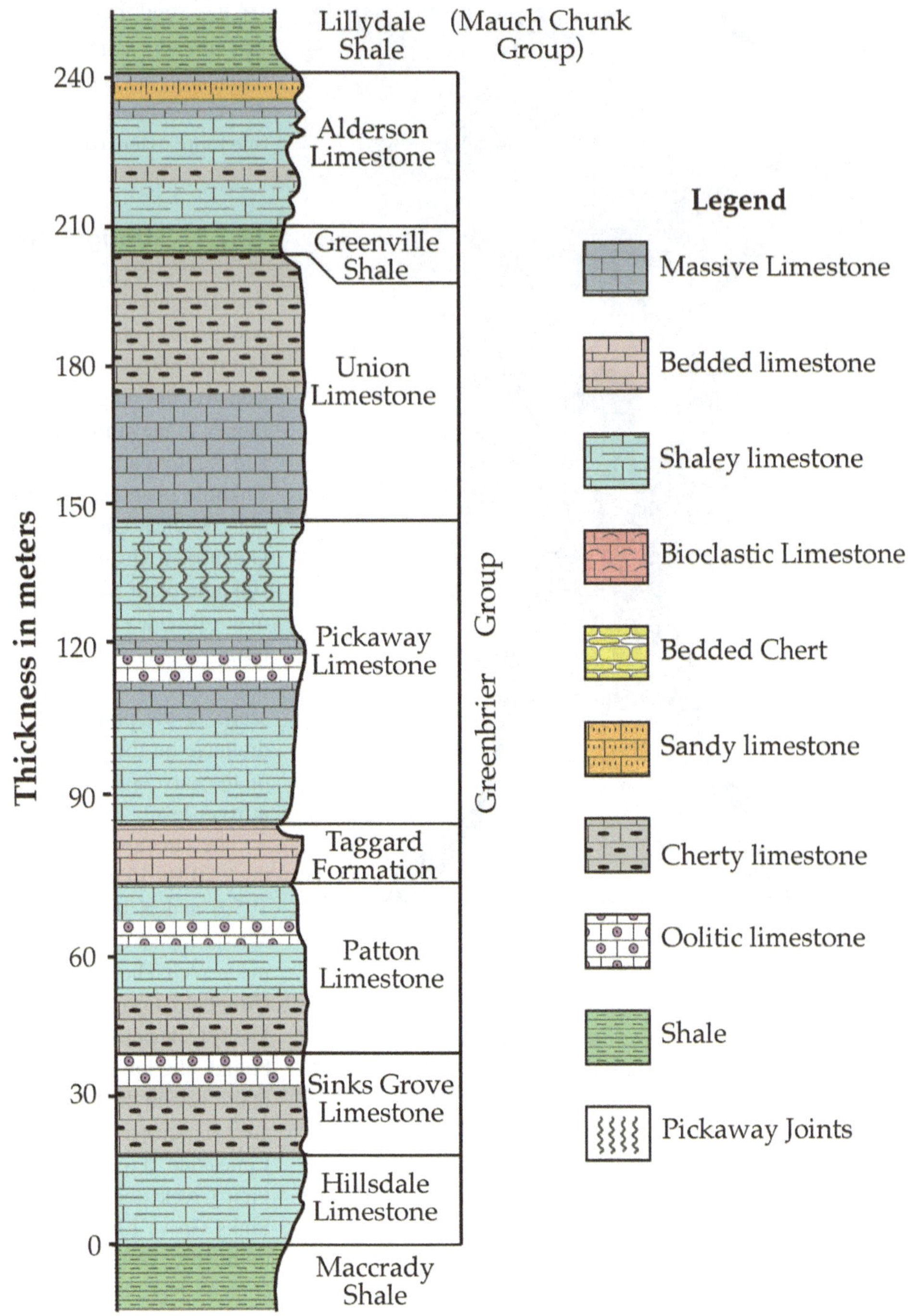

Fig. 15.4 Generalized stratigraphic column of the Greenbrier Group carbonates in southern Greenbrier County. *Source* Shank (2002), after Heller (1980)

15.2.2 Hydrology

The cave is considered to lie within the Scott Hollow groundwater basin, although no dye tracing has confirmed this. The Scott Hollow drainage basin is bounded to the west and south by two topographic highs: Flat Top Mountain and Swoopes Knobs (Fig. 15.7). These form the highest points within the study area and are capped by clastic rocks of the Mauch Chunk Group (Fig. 15.4; Heller 1980). The Greenbrier River and Second Creek form the northern and eastern boundaries. The Greenbrier River serves as regional base level.

It is likely that drainage through the cave currently serves as direct flow to the Greenbrier River. Groundwater flow in most of the mature karst of southeast West Virginia is predominantly through large diameter solution conduits (Jones

Fig. 15.5 Stratigraphic column of the Hillsdale Limestone of the Greenbrier Group in Organ Cave. *Source* Shank (2002), after Deike (1988)

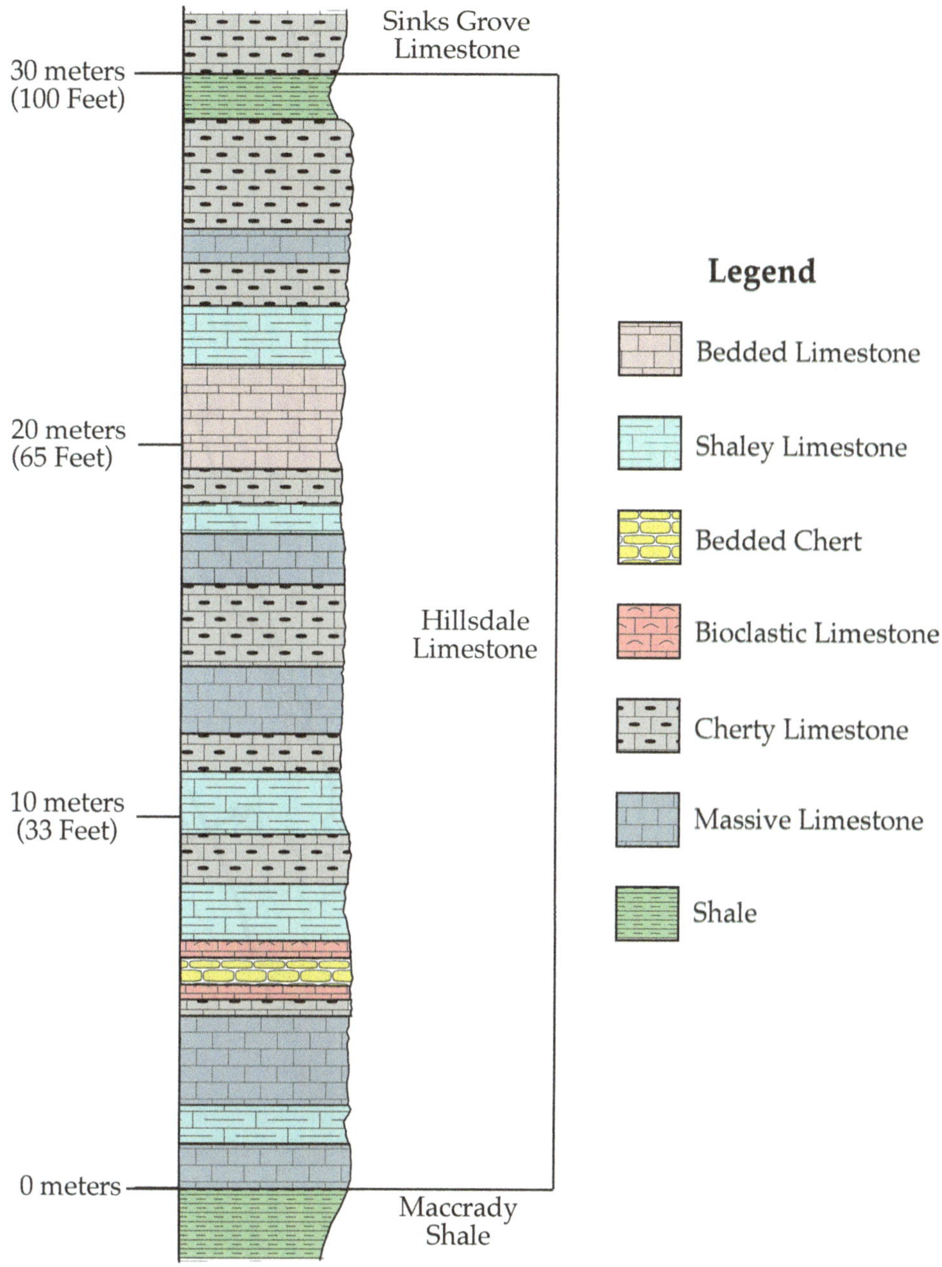

1997). Surface streams are commonly short-circuited through insurgences on the Big Levels. Subsurface stream piracy complicates the delineation of drainage basin boundaries. Dye trace tests reported by Jones (1997) determined that the Scott Hollow drainage basin covers approximately 47 km^2 (18 mi^2), 90% of which is carbonate outcrop.

It is known that water in Scott Hollow Cave currently enters that cave system on the karst upland surface as far south as the town of Sinks Grove (see Chap. 16). It flows into several smaller subsurface tributaries to the Mystic River in Scott Hollow Cave. That north-flowing water eventually resurges at several points along the Greenbrier River, downstream of the entrance to Windy Mouth Cave. These waters are underneath Windy Mouth Cave, which contains limited higher-level vadose flow at the present time. It appears the waters currently flowing in Windy Mouth Cave originate in closed depressions further north, closer to the river.

Average annual runoff for the Kanawha/New River basin, of which the Greenbrier is a tributary, is reported by Jones (1997) to be 43.2 cm (17 in.). The stream gradient for the Greenbrier River in the area of Windy Mouth Cave is 0.0014 (~7 ft/mile). Annual mean discharge for the Greenbrier River over the decade 1991–2001 was 58 m^3/s (2050 ft^3/s; USGS 2001). The present-day entrance to Windy Mouth Cave is at the apex of a meander in the southern wall of the Greenbrier River Gorge, and it appears to be an abandoned

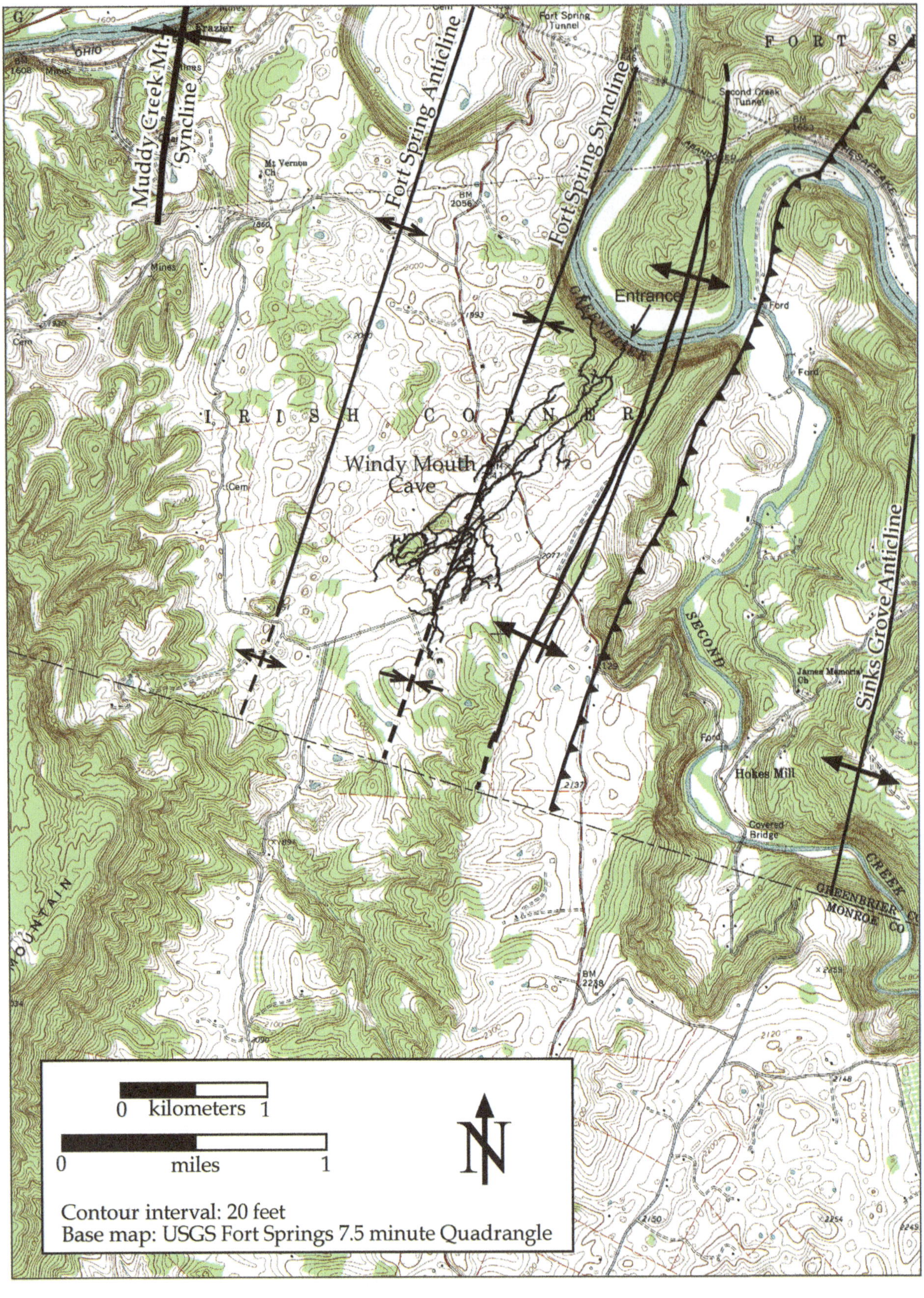

Fig. 15.6 Structural map of southern Greenbrier County with Windy Mouth Cave and first-order folds. After Heller (1980)

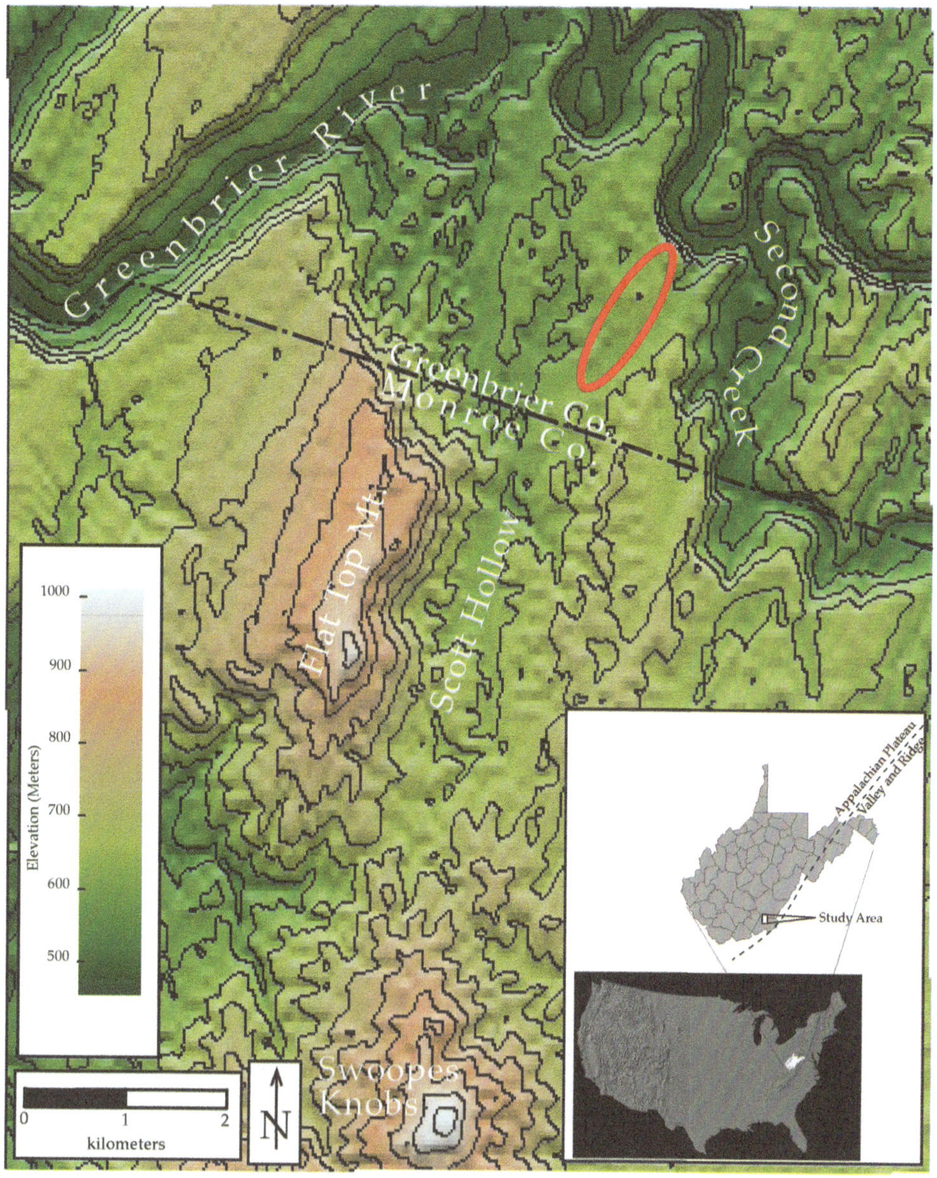

Fig. 15.7 Shaded topography of the study area showing major surface features and general extent of Windy Mouth Cave (*red ellipse*). Base created from USGS Fort Springs digital elevation (DEM) data

spring mouth that once conveyed water from the Big Levels to the Greenbrier River. The initially reported entrance elevation (Davies 1958) is 1750 ft amsl. Our own calculations place it at circa 100 ft. above river level, or 1710 ft, (521 m) amsl. The cave trends southward from the entrance, underneath the karstified limestone upland surface.

15.3 Maps

There are four known versions of maps existing for the cave, and another version is currently being prepared during ongoing survey (see Chap. 5). The earliest known map ("Version 1," Fig. 15.8) is a hand-drawn outline on an 8.5 × 11-in. sheet, dated March 5, 1960, and labeled "Rough sketch, not to scale." It is annotated with many useful descriptions, and it presents estimated distances between major junctions. The next map, Version 2 (Fig. 15.9), is a cleaner page-sized version of that map, published in a caving club newsletter. The third map (Version 3) is a 4 × 6-in. outline map ("sketch survey") and it appears in the supplement to the compendium *Caverns of West Virginia*, attributed to the West Virginia Association for Cave Studies (Fig. 15.10). The final existing map, Version 4 (electronic map M-15.1), is a line plot prepared by Shank (2002) using the software *Compass* (Fountainware, Denver, Colorado) with survey data collected by Jim Hixson and colleagues. Each of these maps contains useful information on the position, length, navigability, and hydrology of the cave.

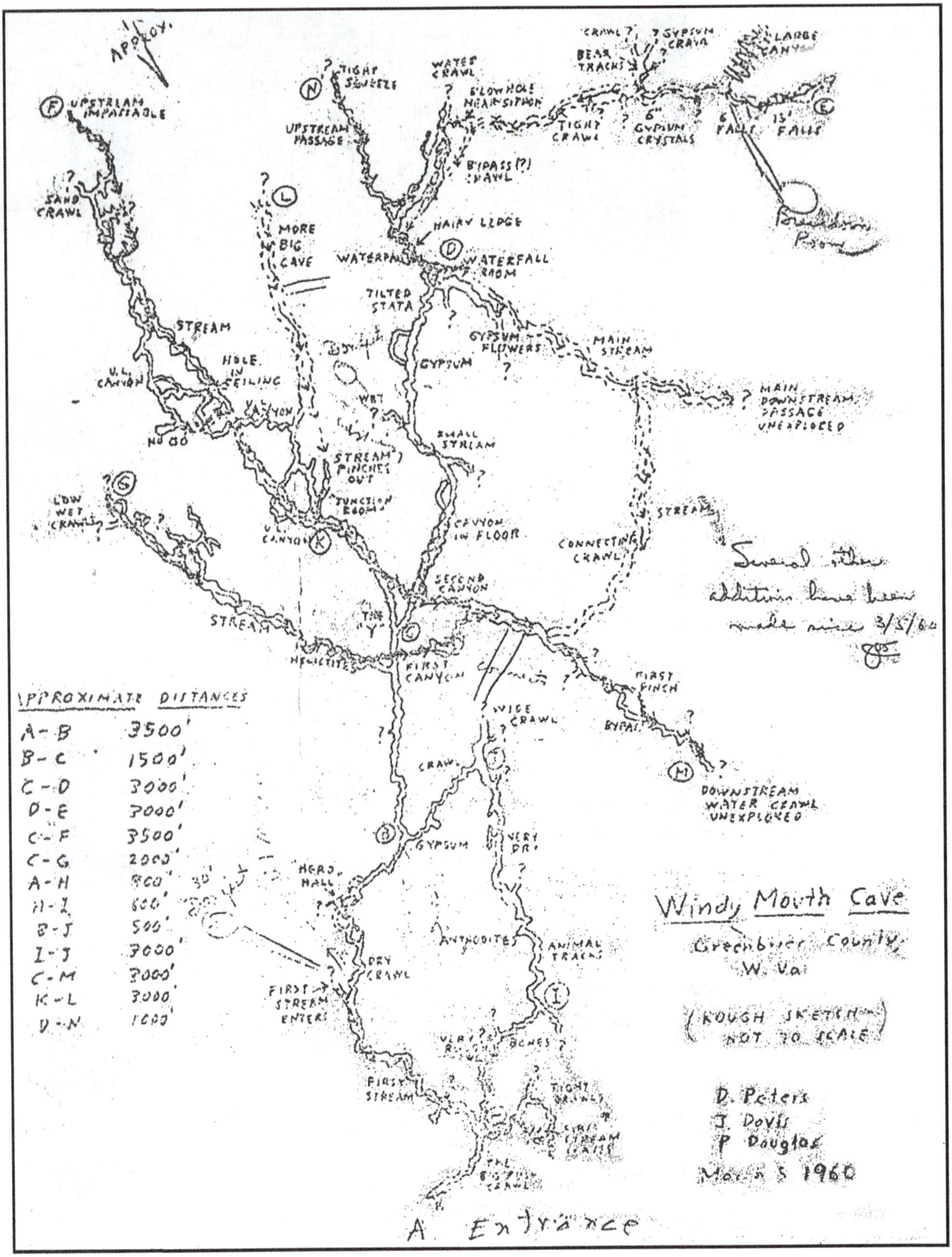

Fig. 15.8 Windy Mouth Cave map version 1. This is a reproduction of the March 5, 1960 map by D. Peters, J. Davis, and P. Douglas, which was further annotated by Jim Hixson. Not to scale, but see approximate distances indicted in table at *bottom left*

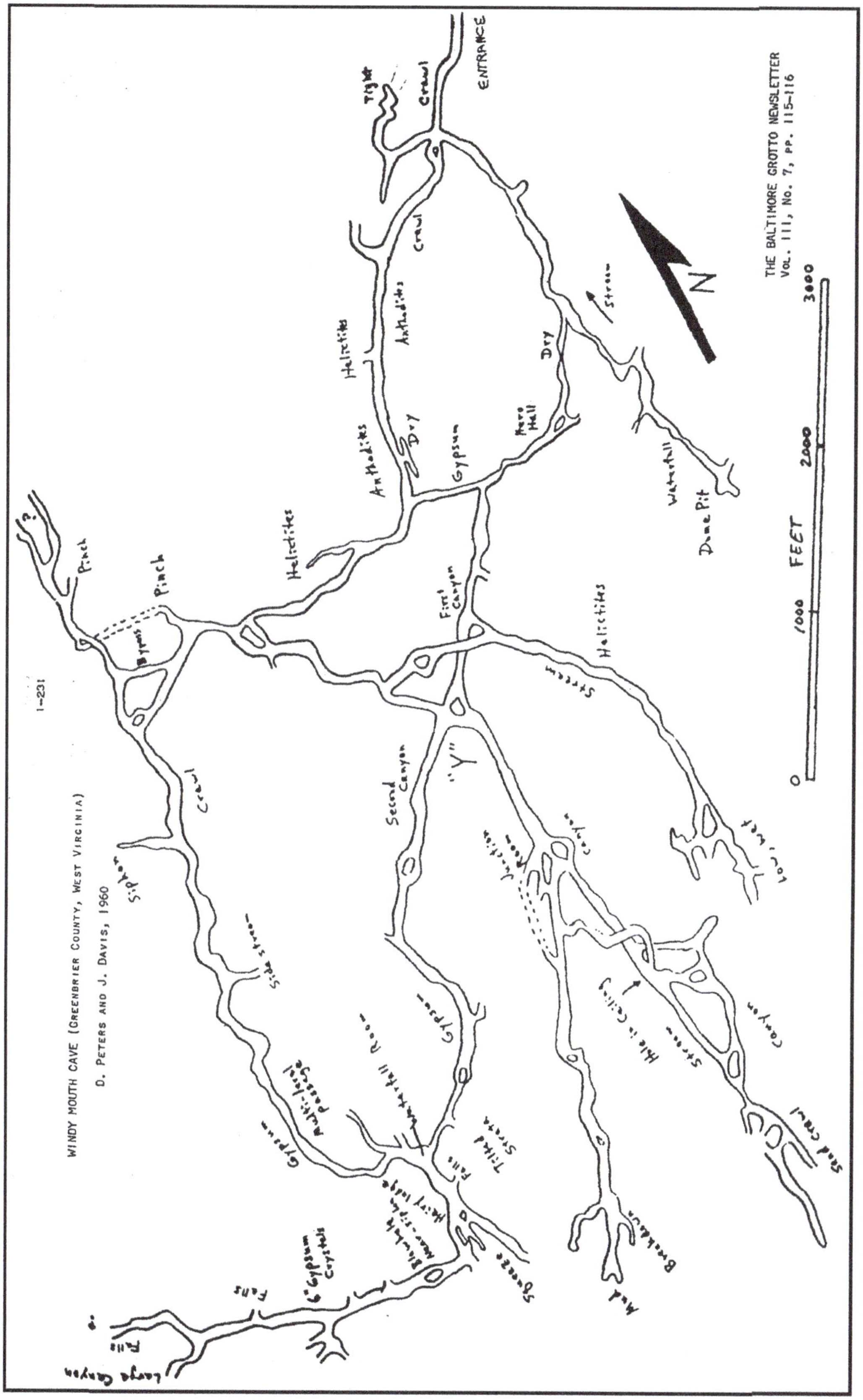

Fig. 15.9 Windy Mouth Cave map version 2 (Peters and Davis 1960)

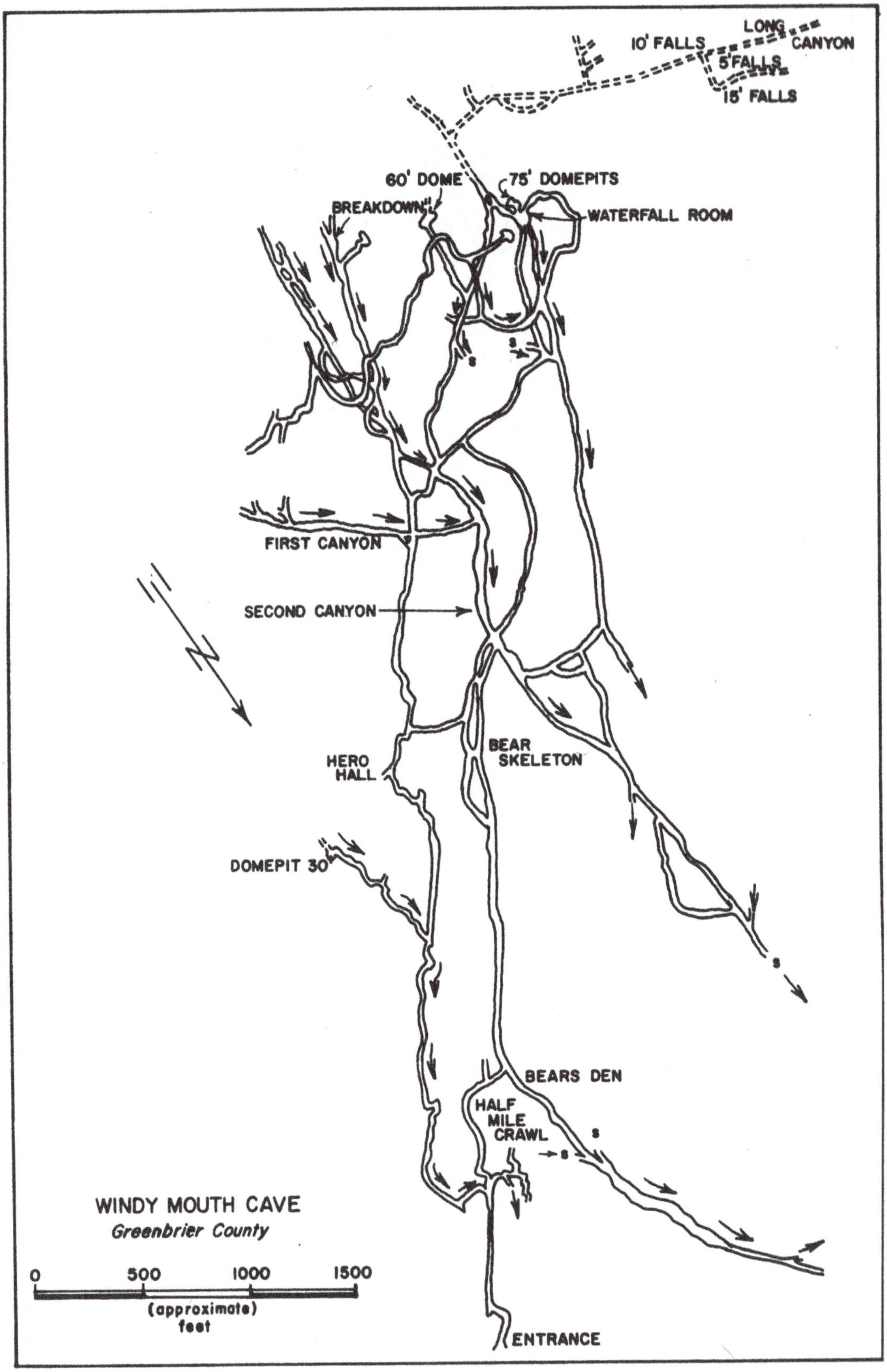

Fig. 15.10 Windy Mouth Cave map version 3, from Davies (1965). Attributed to West Virginia Association for Cave Studies

15.4 General Form and Structural Influences

Cave passages in this region have been interpreted to originate under phreatic conditions in the Hillsdale Limestone and to gradually become vadose as they mechanically incise into the relatively insoluble Maccrady Shale (Palmer 1974). Incision into the Maccrady Shale is usually localized within conduit systems where active streams form canyons. The situation with Windy Mouth Cave is complex due to its length and polygenetic nature as will be described below.

15.4.1 General Form and Position

Examination of the cave map in relation to the landscape clearly shows that at least some portions represent a paleospring conduit that drained to the Greenbrier River (Fig. 15.11). A lower passage to the west currently carries water which is interpreted to feed some active spring on the Greenbrier River (local base level), but this has not been traced. The cave generally trends southward from the entrance. The southernmost part of the cave aligns with the northernmost portion of Scott Hollow Cave, suggesting that the caves might be part of the same conduit system. This may be possible, but exploration to date (see Chap. 5) has not revealed a connection. Furthermore, although the caves are only separated by about 600 horizontal meters (2000 ft), the southern portions of Windy Mouth are likely circa 30 m (100 ft) higher than the nearest reaches of those in Scott Hollow Cave. This suggests that Scott Hollow Cave evolved at a later phase than Windy Mouth Cave, and may be isolated from that drainage.

The overall plan view morphology of Windy Mouth Cave fits the general classification by Palmer (1991) of a "branchwork cave," where several lower-order conduits converge into fewer higher-order conduits in a dendritic pattern (Fig. 15.10). The cave also exhibits some characteristics of an anastomotic pattern. The hybrid morphology is the result of a combination of recharge types. The dominant branchwork pattern results from numerous sinkhole inputs, whereas the anastomotic element originates from the influence of a single, relatively high discharge, sinking stream input. The cave is remarkably flat in profile, with vertical relief primarily coming from a few steep gradient infeeders and domes.

15.4.2 Structural Controls

Windy Mouth Cave is developed in gently dipping (5–7°) limestone strata on the western limb of the Sinks Grove Anticline. Trunk passages have formed just off bedding, oriented N30°E–N65°E (Fig. 15.12). Structural measurements taken at the surface near the study area show that the dominant strike direction of bedding is between N40°E–N50°E (Ogden 1976) and N20°E–N30°E (Heller 1980). Ogden (1976) found that the dominant joint set in the area is oriented N31°E–N41° E. Many passages observed in the cave display at least some degree of joint control. There are several sets of joints that were observed and measured in the cave, although they are not the predominant control on conduit formation.

First and Second Canyons are two major canyon passages whose general orientation is different from that of the rest of the cave. They are deep, narrow fissures that display drawdown passage morphology in cross section. Both canyon passages contain active streams that flow almost directly north–south, oblique to northwest dip, and the main northeast orientation of the cave. Canyon passages are incised 3–4 m (10–12 ft) below the floor of the trunk passage where they intersect. Canyon development is evidence of recent hydrologic activity that contrasts the paleohydrology of the cave.

Some passages in the cave terminate at faults associated with large rooms or domes and active streams. Other Greenbrier County caves such as Taylor Falls and Culverson Creek Caves also contain major passages that abruptly terminate in similar fault structures (Heller 1980). Faulting and active streams are associated with some of the large dome rooms that were observed in Windy Mouth Cave.

There are at least three "significant" faults in Windy Mouth Cave that have influenced the formation of conduits and/or the (paleo) hydrology of the cave. Two significant thrust faults and two minor reverse faults were observed in the southwest section of the cave, in and around the Waterfall Room (Fig. 15.13). A third thrust fault is located approximately 200 m southeast of the Waterfall Room faults.

The Waterfall Room fault is the main fault in a fault zone that occupies the southwest section of the cave. The main fault is a thrust, oriented N41°E 27°SE, with at least 2 m (7 ft) of displacement. Drag associated with the movement of the fault complicates the displacement estimate. Small-scale anticline and syncline folds are observed immediately adjacent to the fault plane.

Three large domes, the first of which lies just 3 m to the south of the Waterfall Room, contain abundant evidence of a fault zone associated with the Waterfall Room fault. Overturned beds, slickensides (on breakdown blocks), and brecciated zones can be observed in these domes.

In summary, the morphology of Windy Mouth Cave is dictated largely by structural factors that have been exploited by groundwater flux. The cave has developed primarily along strike of gently dipping bedrock in the lower Hillsdale Limestone. The resulting "branchwork" pattern of the cave is analogous to a dendritic drainage basin on the surface. Minor anastomoses formed from recharge through a single,

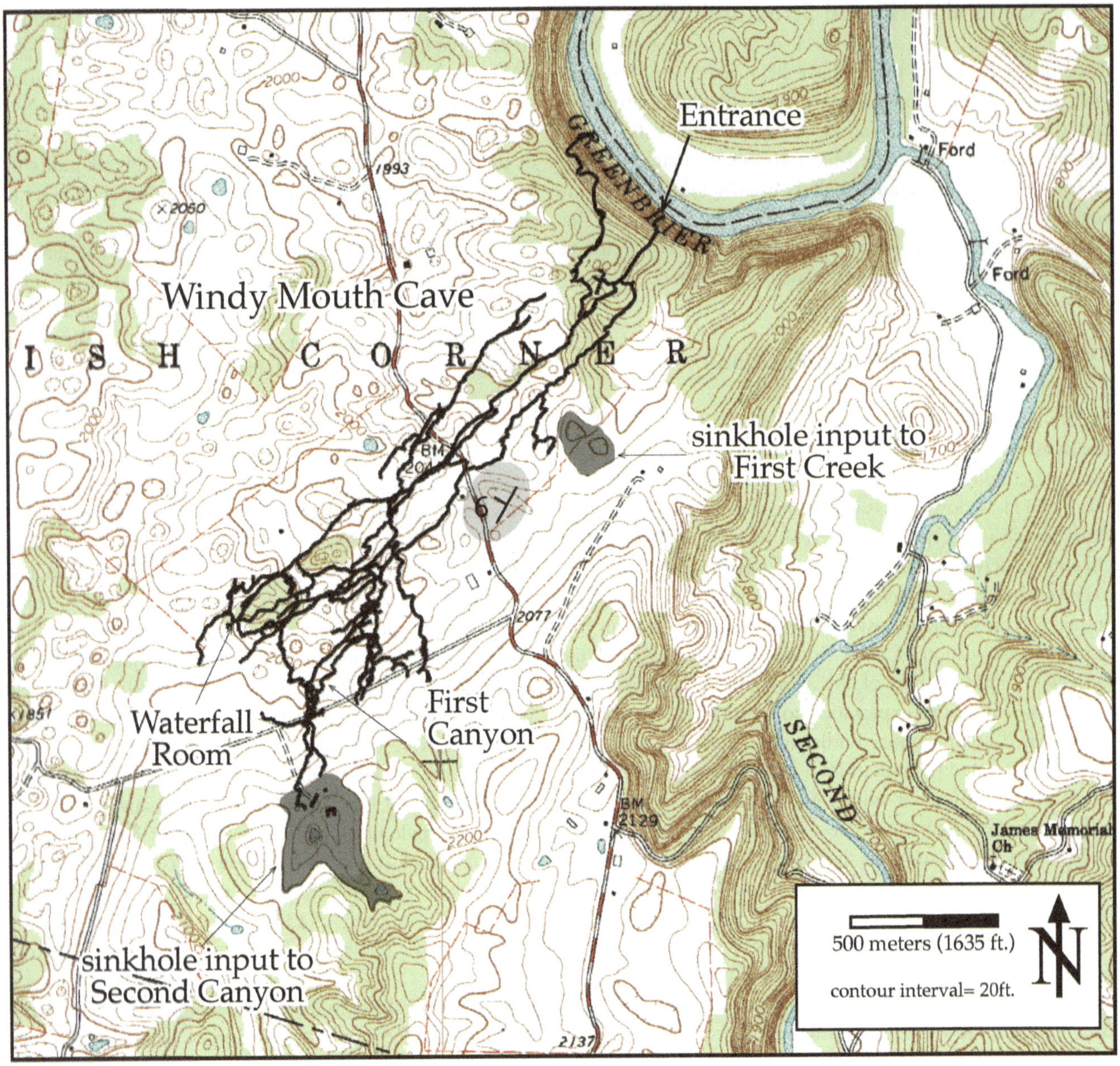

Fig. 15.11 Topographic map showing sinkhole insurgences into Windy Mouth Cave. Base USGS Fort Springs 7.5-min topographic quadrangle

large input point (sinking stream) in the cave have been overprinted by the more dominant and numerous recharge-type sinkholes.

Bedding planes provided the most important pre-solutional openings for conduit development. Conduits that have developed along the long axis of the cave are primarily strike-oriented along bedding planes. Closed loops and angulate passage elements are generally formed along joints. One major dome at the end of First Creek Passage is also formed as a result of jointing. Faults physically and hydrologically connect the upper and main levels of the cave in the Waterfall Room. The three large dome rooms in that area are also formed as a result of faulting associated with a fault zone.

15.5 Paleohydrology and Geomorphology

Windy Mouth Cave is a multi-level cavern system that contains abundant sediment deposits. These, along with other paleohydrologic evidence from the shape and distribution of conduits in the system as a whole and individually,

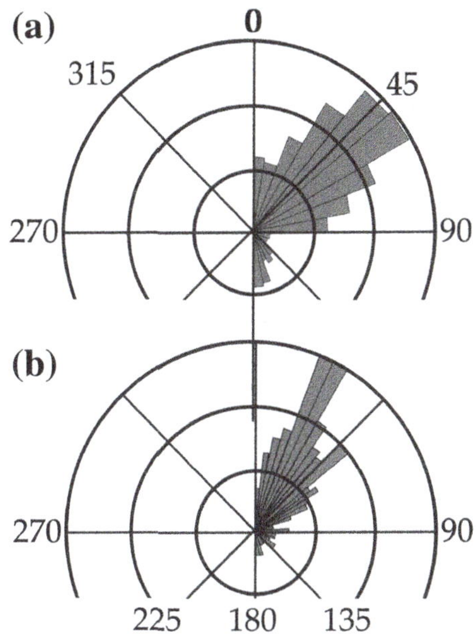

Fig. 15.12 Rose diagrams comparing cave passage orientation to stratigraphic strike (Shank 2002)

allow interpretation of the former hydrologic function of the cave, as well as its age. The cave has three distinct levels. Each level consists of a network of interconnected, hydrologically related conduits that are at similar stratigraphic elevations and share characteristics of morphology and general trend of orientation. The three levels are referred to as the upper levels, main or "trunk" level, and the canyons, with the canyons forming the lowest levels in the cave.

15.5.1 Passage Characteristics

The morphology of Windy Mouth Cave in plan view is characterized by several low-order conduits converging into higher-order conduits that flow into the main trunk passage in a dendritic stream pattern. The plan view pattern emulates that of Palmer's (1991) "branchwork"-type cave. In terms of conduit orientation, there are two distinct trends. The dominant orientation trend follows the northeast strike of the beds. A minor trend exists sub-parallel to dip, roughly north–south.

Primary development of trunk passages in Windy Mouth Cave is almost exclusively phreatic. Tubular conduits are oriented along the northeast strike of the beds. Their cross sections are generally elliptical and show smooth, rounded ceilings that roughly parallel bedding. Some joint control on conduit formation is present although such passages account for only a small fraction of total conduit length. Bedding planes provided the most important pre-solutional pathways for phreatic tubes with vertical joints playing a less dominant role (Fig. 15.14). North–south conduits have formed sub-parallel to local dip. These passages are primarily vadose canyons and shafts that show more pronounced vertical cross-section morphology.

Variation in conduit orientation between the vadose canyons and the phreatic trunk is attributed to the hydrologic differences in their origin. Gravity-driven flow, through vertical fractures or down dip in the vadose zone, formed vertical conduits (dome pits) and north–south oriented canyons. During the active development stage of the cave, these conduits collected recharge from discrete surface inputs (i.e., sinkholes and/or sinking streams) and transmitted water to the trunk conduits along solutional openings that originated along bedding planes or joints. Canyon passage formation resulted from vadose flow incision of the floor that occurred after the initial solutional opening enlarged enough to permit turbulent flow. A marked change in conduit orientation occurs where vadose conduits intersect the strike-oriented phreatic trunk passages.

Trunk passages are phreatic and developed along strike in response to groundwater flow near the water table. The strong correlation between trunk passage orientation and strike resulted from flow through the phreatic zone along bedding planes, toward the spring at base level. The extensive development of conduits at the main level suggests that base level remained static at this elevation for extended periods of time.

Upper levels are present in the southern portion of the cave. The upper-level conduits are comparatively shorter in length and less extensively developed as the other levels. Small active streams are present where the upper levels are connected to the main level at the southern end of the system near the Waterfall Room.

The upper levels of the cave are the most complex in terms of orientation and cross-section morphology. Smaller vadose canyons that are located upstream feed a less extensively developed phreatic conduit system that formed prior to the development of the main level below. Upper-level canyons run immediately above and parallel to the lower-level canyons, separated by only a few meters of bedrock between them. The vertical distribution and morphological similarities of the upper and lower levels in the upstream reaches of the cave suggest that the upper levels formed under similar hydrologic conditions and just prior to the lower levels.

Although minor streams are present in the upper level, the initial hydrologic activity that formed these passages while they were in the phreatic zone has ceased. The spatial distribution of the upper levels with respect to the remainder of the cave is interpreted to indicate that they were the first conduits to be hydrologically abandoned during the initial phase of base-level drop.

Trunk passages in the main level are largest in cross section and the longest in length of all cave elements. They are generally phreatic and contain numerous sediment

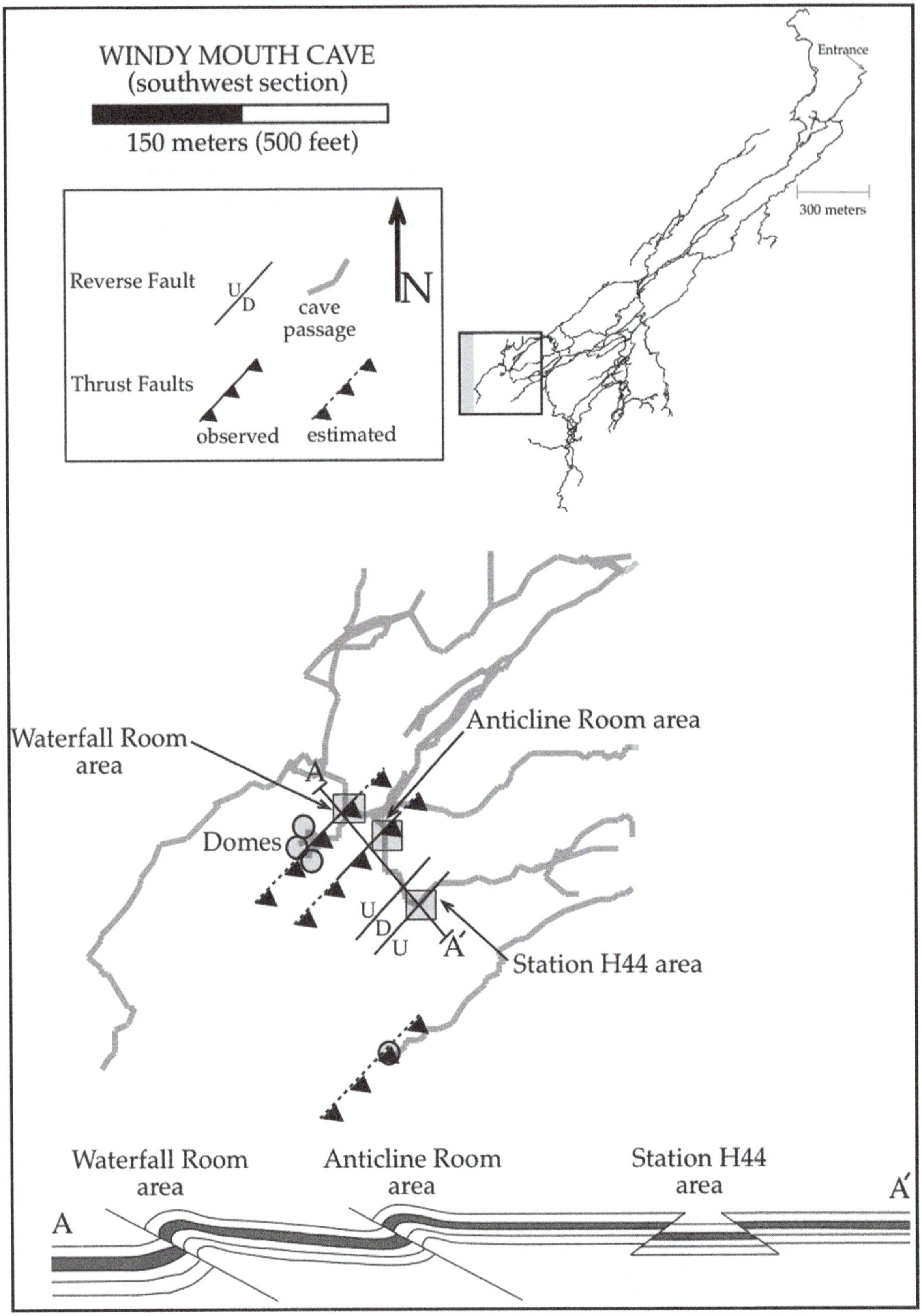

Fig. 15.13 Structural map and cross section illustrative of the southwest section of Windy Mouth Cave (Shank 2002)

Fig. 15.14 Tube/canyon formed with shaley limestone in lower portion. Photograph by E. McCarthy. Used with permission

deposits in a wide variety of forms including imbricated gravel bars and fine-grained, laminated sections. The main level is developed along local strike and is directly connected to the paleospring mouth (entrance) of the cave.

The main level of the cave consists primarily of a well-developed trunk passage. The trunk passage in Windy Mouth Cave received paleoinput from smaller feeder conduits that were located up dip and the trunk transmitted the input to the spring mouth. The base level must have remained relatively stationary at the elevation of the main level for an extended period of time to produce the well-defined trunk passage. The large cross section and extensive gravel deposits are interpreted to indicate that the main level of the cave was hydrologically active during a period where base level fluctuated around the elevation of the trunk passage.

The lowest elevations in the cave are vadose canyons. These conduits are relatively large, well-developed canyon passages that run down dip, crosscut, and are incised below the main level of the cave. They are long sinuous passages that are many times deeper than they are wide. Maximum height reaches approximately 12 m (40 ft), while width varies from little more than 3 m (10 ft) to less than a half of a meter (1 foot). The canyons contain active streams that run oblique to the general orientation of the cave, sub-parallel to local dip. Canyons are typically only navigable upstream from the junction with the trunk passage. The lower-level canyons originate in large deposits of breakdown at southern most, upstream, reaches of the passages.

The canyons contain modern streams and are thus still being developed. The north–south orientation of these lower conduits reflects the different hydrologic factors that governed their evolution in comparison with the overlying levels. Canyon passages are decoupled from the trunk passage where they intersect and incise beneath it and continue down dip. As base level dropped below the elevation of the trunk passage, conditions became vadose and subsequently, water flow was shifted down dip.

15.5.2 Clastic Sediments

As with most caves in the Appalachian Highlands, Windy Mouth Cave contains passages with evidence of being nearly completely blocked by clastic sediments in the past, which have been subsequently re-excavated. Numerous sediment deposits are located throughout the main level of Windy Mouth Cave. They record several stages of hydrologic activity including aggradation and subsequent excavation that occurred periodically throughout the active history of the cave. Sediment deposits range in character from well-sorted clays to unsorted gravels, sands, and silts, which show distinct grouping in a ternary plot (Fig. 15.15). Distinct marker beds are traceable throughout the main level of

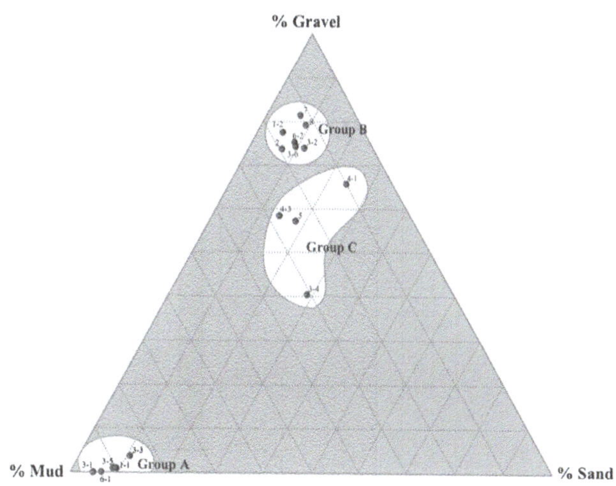

Fig. 15.15 Ternary plot of cave clastic sediment size (Curry 2002)

the cave. Sediments deposited from flooding of the Greenbrier River subsequent to hydrologic abandonment are also found in the floor of the entrance passage. Sediment samples for paleomagnetic analyses were collected from five locations in the cave (see Shank 2002 for details).

A key marker bed in the cave sediments is a fine-grained, approximately 0.5-m (1.5 ft)-thick, laminated clay layer that is found at or near the ceiling of the trunk passage. The clay unit is underlain by a ~1-m (3 ft)-thick, unsorted, gravel/sand/silt layer. Both layers can be traced almost continuously throughout the main level. A sharp, well-defined boundary exists between the two distinct units.

The extensive gravel deposits throughout the trunk passage indicate that high-velocity, turbulent flow conditions must have existed in the cave during this phase. Sedimentation must have occurred while substantial recharge was entering the system, perhaps via a sinking stream from the surface. Large sinkholes on the surface may have been the source of surface drainage. Imbrications within gravel bars indicate paleoflow direction toward the spring mouth in the main level. The bulk of gravel is chert (Figs. 15.16 and 15.17).

The fine-grained laminated clays located near the ceiling were perhaps deposited after the conduit was completely filled with water and sediment. Stream flow velocity would decrease dramatically if the space allowed for flow to occur becomes restricted and choked with aggrading sediments. The abrupt transition from turbulent to quiet water conditions is expressed in the sharp contact between the gravel and clay units.

The most extensive stratigraphic section of fluvial deposits in the cave is located in the Waterfall Room (Fig. 15.18). The 2-m section comprises alternating beds of laminated clay/silt units with unsorted, coarse-grained gravel, sand, and silt (Fig. 15.19). Abrupt and periodic facies changes are indicated in the section by sharp contacts between coarse and fine-grained units.

Upper-level sediment deposits are sparse. Where they are found, the deposits are discontinuous and untraceable. Most sediment deposits in the upper levels are autochthonous and occur as breakdown from the ceiling. Sediments that are suitable for paleomagnetic analyses were not found in the upper levels of the cave.

The lower levels contain active streams in canyons. Flow direction in the canyons is directly evident. The streams are currently downcutting, sub-parallel to dip. Sediments that were suitable for paleomagnetic sampling were found where the canyons intersect the trunk passage. The sediments are lodged into the upper portion of the canyon ceiling and the floor of the trunk passage. These deposits appear to be representative of when the canyon became decoupled from the trunk, hydrologically abandoning the trunk conduit.

Deposits from flooding of the Greenbrier River after the main phase of river incision are found in the 260-m (850 ft) entrance passage. An active stream that flows into the floor

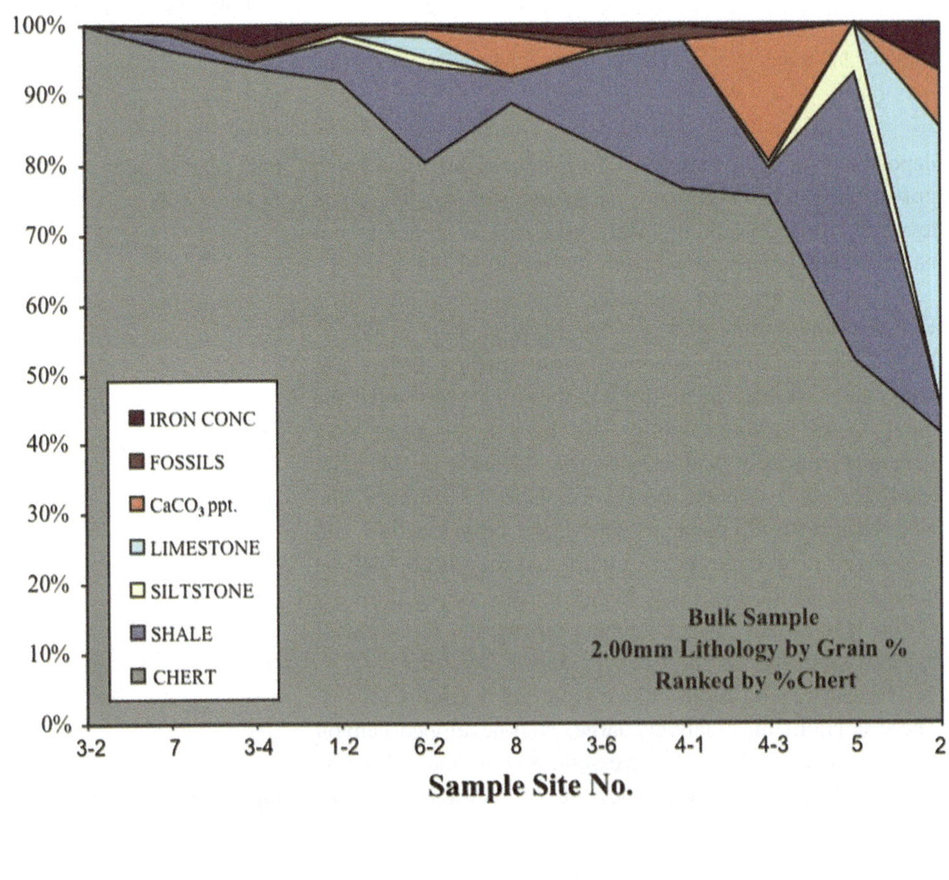

Fig. 15.16 Graph of coarse fragment lithology of cave sediments (Curry 2002)

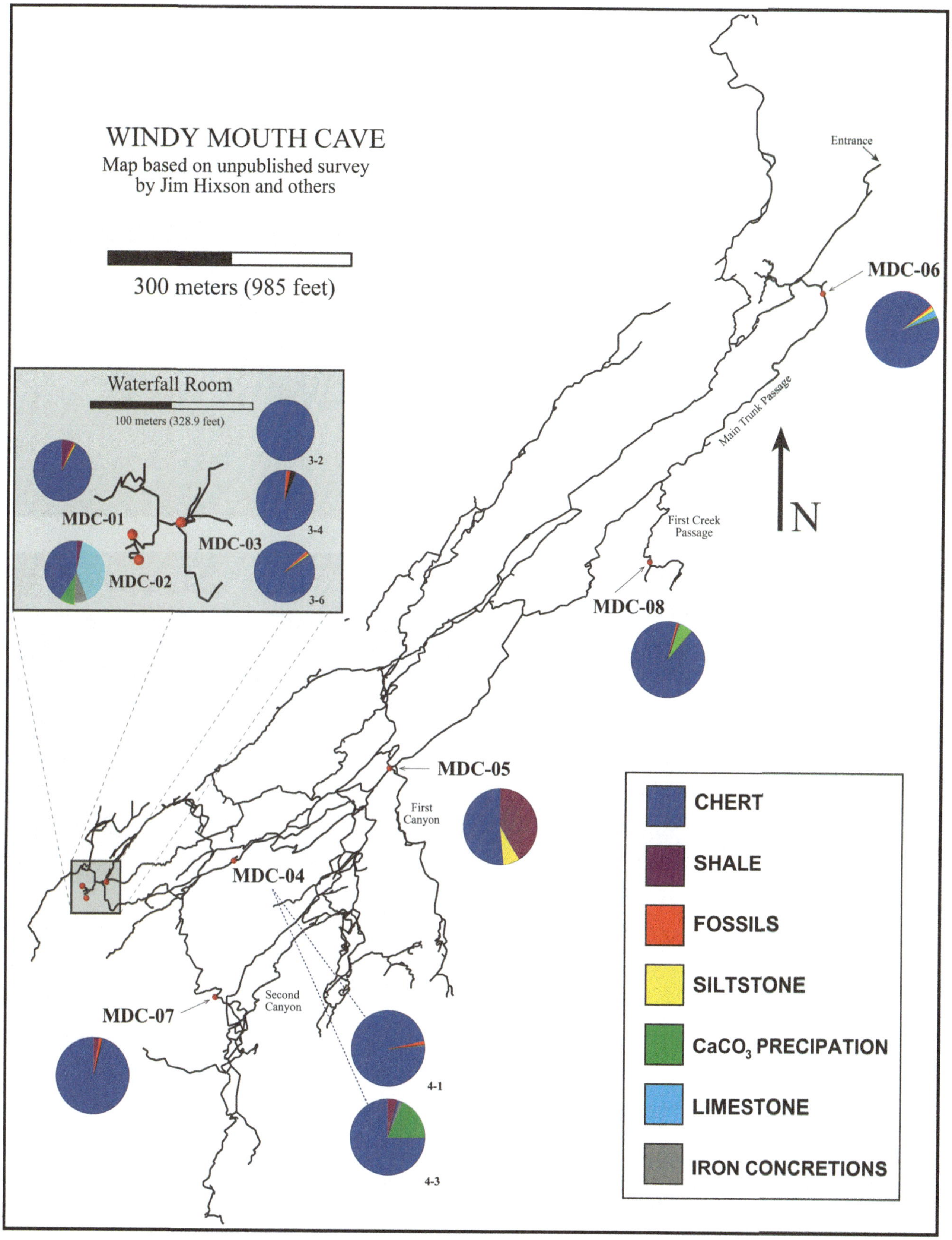

Fig. 15.17 Map showing distribution of coarse fragment lithologies throughout the cave (Curry 2002)

Fig. 15.18 Photographs of coarse sediment deposits in Windy Mouth Cave. **a** Sediment deposits in the trunk passage; note that erosion of the material has left a flowstone/column suspended above cave floor. **b** Gravel/clay deposits in the Waterfall Room. Photographs by I.D. Sasowsky

of the cave where the entrance passage ends has effectively removed flood sediments beyond the entrance passage. This implies that the absence of flood deposits in the remainder of the cave may represent postcave formation hydrologic activity and thus would not represent the period of active cave formation.

The fluvial history of Windy Mouth Cave is complex. The interbedded coarse and fine-grained deposits in the cave suggest episodic aggradation, flooding, and excavation. Aggradation of sediments indicates a net rise in base level or an increased sediment supply from the surface. When conduits became filled with sediment, flow was restricted, and

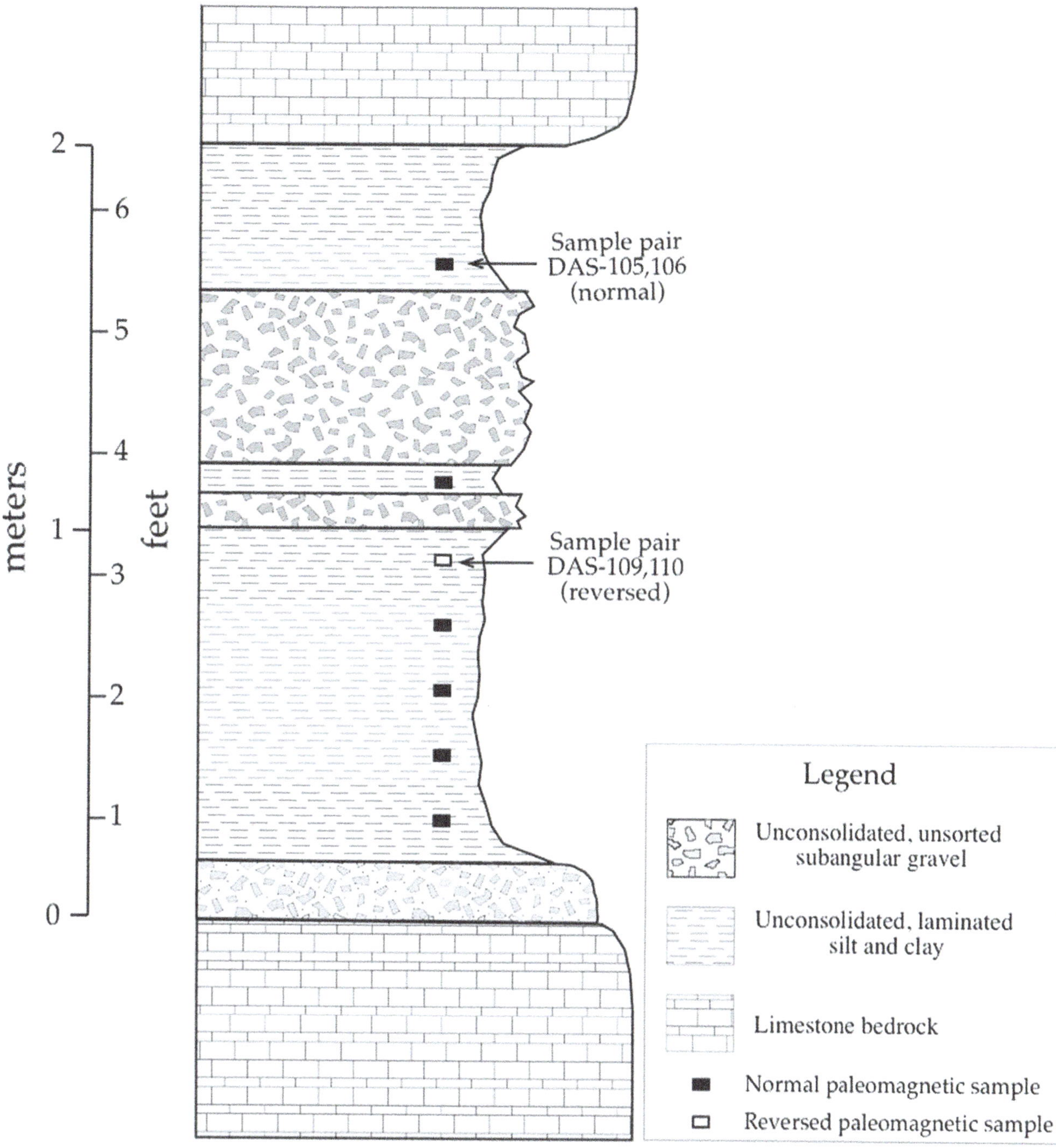

Fig. 15.19 Stratigraphic column of fluvial deposits in the Waterfall Room (Shank 2002)

fine-grained clasts settled out near the ceiling of the conduit. Subsequent lowering of base level, and thus an increase in hydraulic gradient through the conduits, generated high-velocity flow and excavation of aggraded deposits. Paleoflow indicators such as imbrications on gravel deposits are interpreted to show that flow direction was unidirectional, toward the spring mouth.

15.5.3 Paleomagnetism of Clastic Sediments

Paleomagnetic samples of fine-grained clastic cave sediments samples were taken from various locations throughout Windy Mouth Cave where the following criteria could be met: (1) sediments were relatively undisturbed or "in situ," (2) sediments were fluvial/clastic in nature (i.e., they were deposited by very low-velocity streams and do not represent insoluble residue weathered from the host limestone), and (3) they should be fine-grained (silts and clays) and preferably thinly bedded or laminated.

Twenty sample pairs (40 samples) were collected in spatially oriented plastic cubes according to methods described in Sasowsky et al. (1995). Stratigraphically equivalent pairs were taken to insure accuracy of analytical results. Details are found in Shank (2002), and results are discussed below.

The samples all provided paleomagnetic signals of high fidelity (Fig. 15.20). Best-line fitting shows good clustering of normal directions and consistency in the reversed sample directions (Fig. 15.21). The results support the idea that most of the cave was developed since the last major reversal of the earth's geomagnetic field since most samples are of normal polarity. A lone pair of magnetically reversed sediments (Fig. 15.19) was found near the bottom of the most extensive deposit of clastic sediments in the cave.

15.5.4 Landscape Evolution

The observed magnetic reversal implies that trunk passages in Windy Mouth Cave were active prior to 788 ka, thus close to the base level as it existed at that time. This would place the Greenbrier River 30 m (100 ft) above its present average river bed elevation of 490 m (1610 ft). Therefore, the river has incised 30 m (100 ft) below the present elevation of the spring mouth over the past 788,000 years. The average incision rate is then 39 mm/ka (39 M/Ma). This rate is comparable to others calculated for other similar karst terrains in the New River, Virginia, 27 mm/ka (Granger et al. 1997), the Obey River of Tennessee, 30 mm/ka (Anthony and Granger 2004), the Cheat River, West Virginia, 56–62 nm/ka (Springer et al. 1997), and the Green River, Kentucky, 30 mm/ka (Granger et al. 2001).

The river incision rate is based on two assumptions: (1) Windy Mouth Cave is a water table-type cave that was formed when base level (Greenbrier River) was at the same elevation as the cave entrance (spring mouth); (2) the paleomagnetic reversal located in the Waterfall Room sediments resulted from deposition of the clay unit during the Matuyama Reversed Chron, near the boundary with the Brunhes Normal Chron.

15.6 Present Hydrology and Hydrochemistry

The modern-day hydrology of Windy Mouth Cave contrasts sharply with that of the past. There is no indication that coarse clastic materials are currently being carried into the cave. Modern stream flows, although minor in comparison to paleostreams, continue to flow along pathways that began to form during the earlier stages of cave development.

Water samples were taken from the two most accessible streams in the cave. Sample locations were First Creek (sample# 200) and the Waterfall Room (sample# WR1). These samples were collected during the near-drought conditions in 2001 that created unusually low flow conditions in the cave streams. Active streams in the cave are typically found at First Creek, First Canyon, Second Canyon, and the Waterfall Room. The conditions that existed during this study reduced flow in the canyons to almost zero. Details on field and lab methods are given in Shank (2002).

Although the cave system is largely abandoned hydrologically, active streams that persist in the cave contain water that is drained from the Big Levels surface and therefore the cave system represents the nature of water in the drainage basin. Delineation of flow paths from the surface and through the cave is important to establish the physical boundaries of the drainage basin and potential contaminant transport pathways through the system.

Within the cave, First Creek is a relatively small stream, <30 L/s (<8 gallon/s) that can be traced upstream to a large dome room where it flows into the cave from the eastern side of the ceiling, >11 m (35 ft) above the floor. The existence of the dome is attributed to the stream flowing along a near vertical joint that strikes N45°E.

The dome was formed as dissolutionally aggressive water entered the groundwater system from the sinkhole located almost directly above the dome room (Fig. 15.11). A shaley, dolomitic unit forms the base of the cascade and provides a resistant bed along which the water flows into the main trunk passage. The stream flows north along First Creek Passage, the access passage to the dome room, until the passage terminates at a junction with the main trunk passage. First Creek then turns northeast and flows down the main trunk passage. The stream disappears into a non-navigable opening in the floor, just a few meters from the water sampling location.

Chemical analyses of the water in First Creek (Table 15.1) indicate a relatively short residence time and minimal natural filtration. The water is undersaturated with respect to calcite and dolomite. Given the low flow conditions that existed at the time of sampling, one would expect that these saturation values to be at their maximum. With increased flow, residence time would decrease and thus water would be even less saturated with calcite and dolomite. Negative saturation indices (SI values) are indicative of a short residence time and/or direct recharge for groundwater (White 1988).

Elevated levels of nitrate in First Creek are compared with those higher nitrate levels found in the study area by Ogden (1976), Heller (1980) and Davis (1999). High nitrate levels are indicative of the lack of natural filtration of animal waste and fertilizers from the agricultural land on the

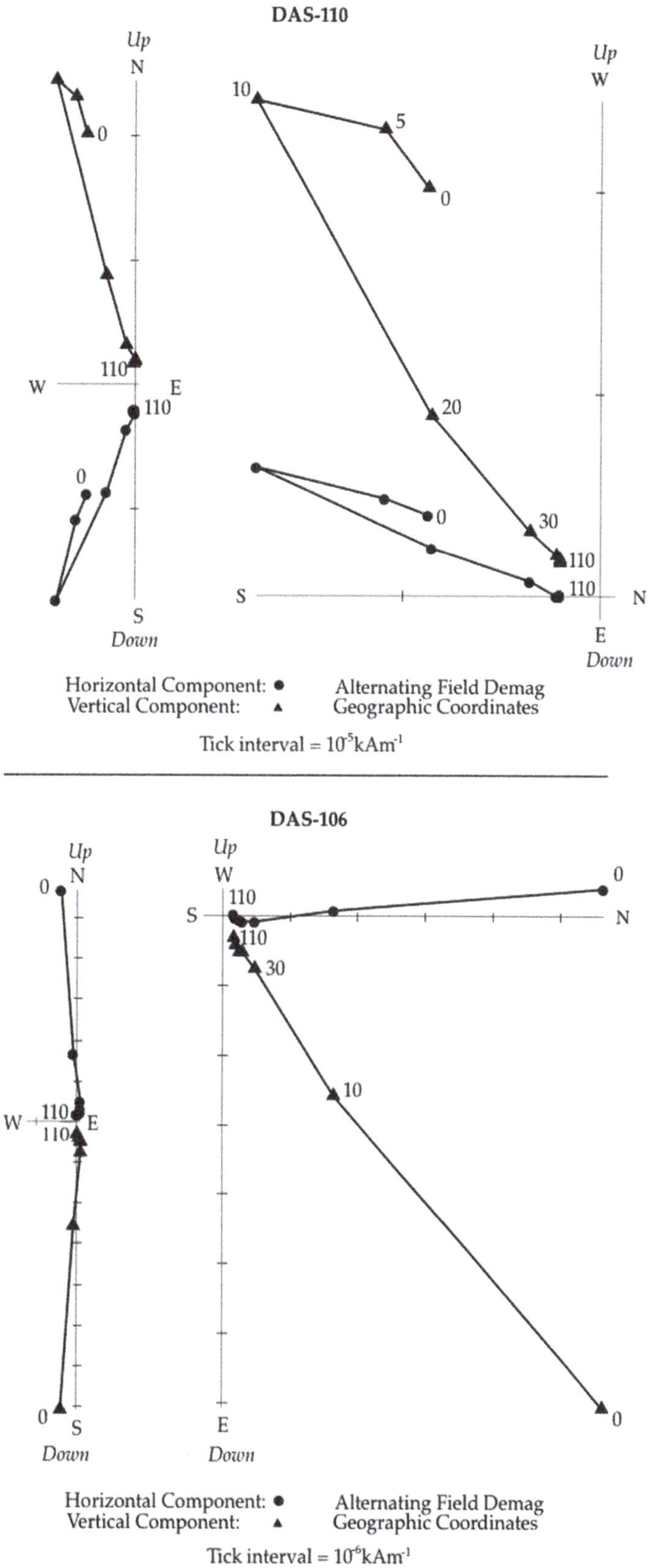

Fig. 15.20 Zijderveld diagrams showing typical reverse (*top*) and normal (*bottom*) polarity samples

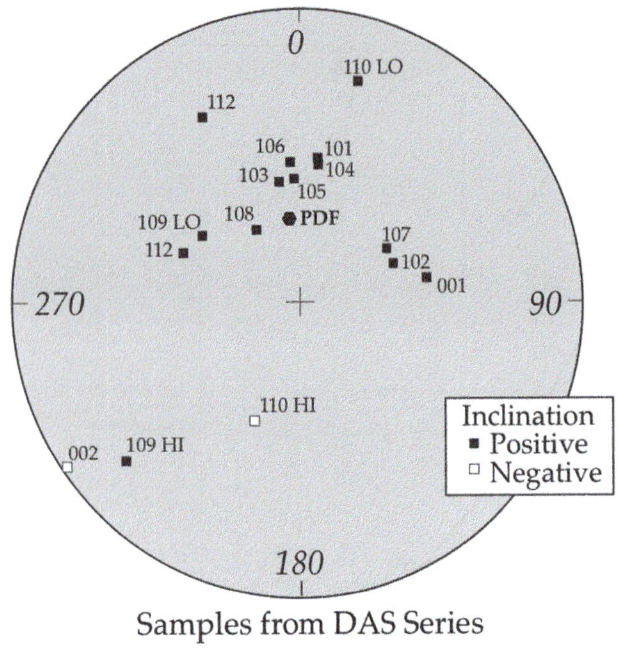

Fig. 15.21 Pole plot of paleomagnetic directions

surface. There also may be a concentrating effect in the waters due to the low flow conditions that existed during sampling.

The Waterfall Room stream is chemically very similar to First Creek. It had a slightly higher discharge during the time of sampling. Negative SI values and elevated nitrate levels in the Waterfall Stream also indicate short residence time and lack of natural filtration of waters (Table 15.1).

The falls are formed as the stream cascades over steeply dipping beds associated with the thrust fault in the Waterfall Room. Massive breakdown piles obscure the downstream reaches of the stream, but it appears to flow beneath the established main level of the cave. There are multiple sources to the stream, but the majority of flow comes from a stream in the upper levels of the cave that cannot be traced to its origin. However, there are several sinkholes located on the surface above the southern reaches of the cave where the stream appears to originate from (Fig. 15.11).

The stream in Second Canyon was not active during this investigation and therefore could not be sampled; however, it is known to flow when precipitation levels are considered normal. The source of Second Canyon stream is a large asymmetrical sinkhole located at the terminus of the passage on the surface (Fig. 15.11). The upstream passage terminates underground at a large impassable pile of breakdown. The Second Canyon stream flows north, sub-parallel to dip, but cannot be traced beyond the surveyed portion of the cave. It is inferred from that the stream continues to flow north, incising below the known levels of the cave, toward base level.

Drainage through Windy Mouth Cave is classified as a part of the Scott Hollow drainage basin (Jones 1997). However, the fact that it contains an active stream network with drainage from the surface suggests that Windy Mouth Cave is its own autonomous drainage basin. Calculations

Table 15.1 Aqueous geochemical results. Elemental and TDS values are in mg/L

	Waterfall room	First creek
HCO_3^-	208	202
pH	6.99	6.98
Temp (°C)	12.1	11.2
Cond. (μS)	403	420
Sat. index	−0.469	−0.550
Cl^{-1}	4.05	4.67
SO_4^{-2}	2.95	3.02
NO_3^-	22.62	16.25
PO_4^{-2}	0.197	0.126
Al^{+3}	0	0.01
Ca^{+2}	70.82	63.63
Fe^{+2}	0.01	0
Mg^{+2}	6.82	8.31
Mn^{+2}	0	0
SiO_2	9.14	9.25
K^+	1.25	1.34
Na^+	2.60	3.80
TDS	319.31	303.16
CBE (%)	+0.26	−0.05

yield a drainage basin area for Windy Mouth Cave of 1.95 km^2 (0.75 $mile^2$).

Water chemistry data reinforce the inferences of short residence time and minimal filtration. The stream in Windy Mouth Cave is undersaturated with respect to calcite and dolomite, indicating minimal exposure time to limestone and/or dolostone and continued cave development at lower levels in the bedrock. Elevated levels of nitrate indicate a lack of natural filtration that results from direct input of water into the cave.

Windy Mouth Cave is an extensive cave system that developed when the Scott Hollow drainage basin was different than it is today. Paleoinput and drainage basin size of the cave was substantially greater during its development as evidenced by the size and extent of conduits. Drainage was pirated from Windy Mouth Cave when the Greenbrier River incised beneath the current level of its entrance. Drainage was diverted to the west, into Scott Hollow and other caves that are located on the western limb of the Sinks Grove Anticline, as a result of base-level lowering. Modern discharge from the Scott Hollow drainage basin is from river level springs on the Greenbrier River downstream and west of Windy Mouth Cave.

15.7 Summary

The formation of Windy Mouth Cave was controlled by a combination of structural and hydrologic factors. Stratigraphy also had a role in cave development, albeit to a lesser extent. The cave system is almost completely abandoned hydrologically; however, extensive fluvial sediment deposits in conduits indicate that it was an important route for groundwater from the Scott Hollow drainage basin in the past.

Geologic structure provided a pre-solutional network of faults, joints, and bedding planes in the bedrock that was later exploited by groundwater flux. The cave is situated on the western limb of the Sinks Grove Anticline. Beds dip gently to the northwest and strike generally N40–50°E in the cave. Conduits are primarily oriented along strike, while a smaller component is oriented sub-parallel to dip. Bedding planes provided the most important pre-dissolutional fissures, although closed loops are formed along joints.

Faults are locally important to cave system hydrology. A fault zone located in the southern reaches of Windy Mouth Cave connects the upper and main levels (Fig. 15.13). The fault plane breaches an impermeable layer of bedded chert, allowing water to descend through the Waterfall Room and into breakdown on the floor. The faulting and related fracturing extend upward an estimated 30 m (100 ft) into the overlying strata, creating an opening for water to enter the system from the surface. At least three such domes exist in the cave in the Waterfall Room, First Creek, and an unnamed dome in the upper levels.

In plan view, Windy Mouth Cave appears as a branchwork system similar to that described by Palmer (1991). A minor anastomotic morphological element is overprinted on the dominant dendritic pattern. The pattern is suggestive of primary recharge originating from discrete input points such as sinkholes, while in the past focused recharge may have entered the cave from a sinking stream. Imbricated gravel bars in the trunk passage of the cave support the interpretation that considerable direct recharge was received via a sinking stream.

Trunk conduits originated as phreatic tubes that directly connected to the spring mouth (entrance). Tributary conduits were vadose canyons that ran down dip until they reached the phreatic trunk where they altered course and ran along strike.

There are three levels in the cave (Fig. 15.22). Upper levels are phreatic tubes that are connected to small vadose canyons at their origin. Conduit cross-section morphology is complicated by interbedded chert and shale layers in the Hillsdale Limestone host rock. The impermeable layers form resistant ledges that split individual conduits into multiple levels. Fluvial sediment deposits that are suitable for paleomagnetic analyses were not found in the upper levels, although the position and hydrologic genesis of the system suggest that the upper levels were formed first and are thus the oldest portion of the cave.

The main levels (trunk conduits) are the most extensively developed and comprise the majority of conduits in the cave. They are primarily phreatic tubes that have been reshaped by a transition to vadose conditions that resulted from base-level lowering.

Lower-order infeeders are connected to the higher-order trunks to form a dendritic drainage pattern in the main levels. The infeeders are well-developed vadose canyons that intersect and are incised below the trunk conduits. The junctions between First and Second Canyons with the trunk are important because they illustrate the hydrologic decoupling between the trunk conduit and vadose canyons. Originally, the vadose canyons originating in the south acted as infeeder streams that descended down dip into the phreatic zone where they intersected the trunk conduit. As base level lowered, the vadose conduits incised below the trunk and consequently became hydrologically decoupled by diverting drainage into the newly forming conduits beneath. Currently, the canyons are occupied by small streams that discharge through springs at river level.

Fluvial sediment deposits in the main levels consist of gravel, sand, and clay. Coarse deposits are generally unsorted and occasionally imbricated. They represent high discharge flow regimes that probably originated from a sinking stream input source. Fine-grained deposits are

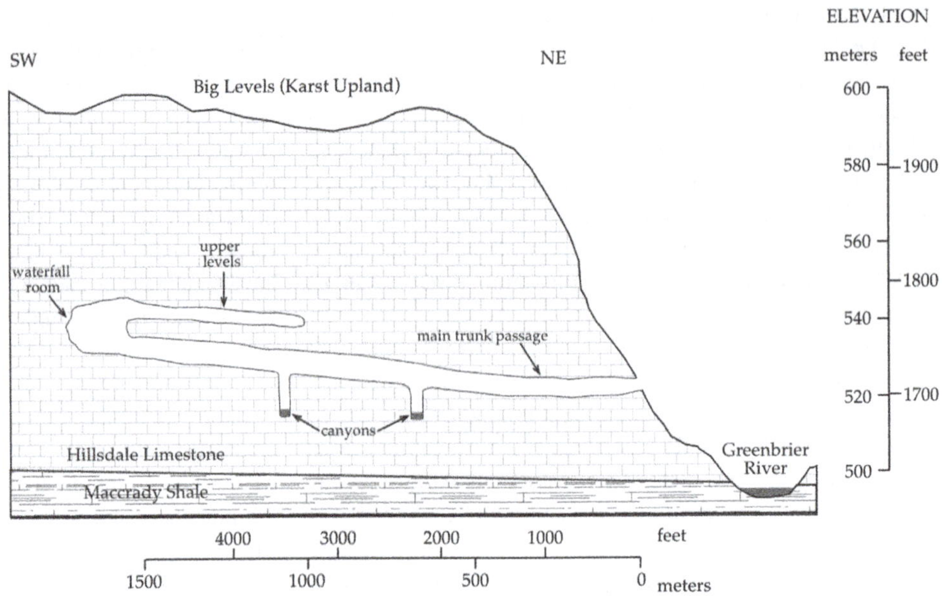

Fig. 15.22 Simplified longitudinal cross section of Windy Mouth Cave that illustrates the distribution of different levels within the cave and relation to Greenbrier River (Shank 2002)

typically found at or near the ceiling, which indicates sediment aggradation until the conduits were choked with sediment, restricting the flow to quiet water conditions. The final stage of cave sediment evolution was a high discharge stream that excavated most of the sediments. This probably resulted from a high hydraulic gradient caused by base-level lowering that excavated most of the deposits.

A magnetically reversed sample found near the base one section was presumably deposited during the reversal of the geomagnetic field which ended at 788 ka. This sets the minimum age of the cave. Because Windy Mouth Cave is a water table cave, formed when local base level was near the trunk conduits, the assumption is made that the reversed sediments were deposited when the river bed was 30 m (100 ft) above its current elevation. Conversely, the river has incised 30 m (100 ft) since those reversed sediments were deposited. This yields an average incision rate for the Greenbrier River of 0.04 m/ka (1.6 in./ka).

The modern-day hydrology is markedly different than in the past. The drainage basin area is much smaller (~ 2 km^2), and the resident streams have considerably less discharge. Much of the drainage has been pirated to the Scott Hollow drainage basin located south and west of Windy Mouth Cave. Current stream sources are sinkholes located on the surface above the origins of active cave streams (Fig. 15.11). Chemical analyses of cave water indicate a short residence time and minimal filtration evidenced by negative SI (calcite) and slightly elevated levels of nitrate.

Acknowledgements The following people provided exceptional assistance during the field work and other aspects of the study, which is greatly appreciated: Mike & Pat Dore, Sarah Benjamin, Verne Friberg, David McConnell, Bill Harbert, Natasha Demrovsky, Yvonne Droms, Laura Hudnall, Mike McFall, Steve Rhodes, Dave Seslar, Jenn Ulmer, Jim Hixson, and Julie Bennett. Financial support was provided by the Cave Conservancy Foundation (Virginia) and Richmond Area Speleological Society. We thank Ed McCarthy for allowing use of his photographs.

References

Anthony, D.M., and D.E. Granger. 2004. A Late Tertiary origin for multilevel caves along the western escarpment of the Cumberland Plateau, Tennessee and Kentucky established by cosmogenic ^{26}Al and ^{10}Be. *Journal of Cave and Karst Studies* 66 (2): 46–55.

Curry, M.D. 2002. Sediments from an abandoned karst groundwater conduit: Windy Mouth Cave, Greenbrier County, West Virginia: University of Akron (Ohio) undergraduate student project in geology, 23 p. + appendices.

Dasher, G. 2002. *The Caves and Karst of West Virginia*. Barrackville, WV: West Virginia Association for Cave Studies Bulletin 19, 264pp.

Davies, W.E. 1958. *Caverns of West Virginia*, vol. 19(A), 330pp. Morgantown, WV: Geological and Economic Survey.

Davies, W.E. 1965. *Caverns of West Virginia, Supplement to Volume 19(A)*, 72pp. Morgantown. WV: West Virginia Geological and Economic Survey.

Davis, S.B. 1999. Aquifer development in a folded and fractured limestone: The Scott Hollow drainage basin, Monroe County, West Virginia, 210pp. Masters Thesis, University of Akron.

Davis, J., and D. Peters. 1960. Windy Mouth Cave. Unpublished manuscript describing the cave by the surveyors, 3 single-spaced typed pages, dated April, 1960.

Deike, G.H. 1988. Geology. In *Caves of the organ cave plateau, Greenbrier County, West Virginia*, ed. P.J. Stevens, vol. 9, 13–39. Barrackville, WV: West Virginia Speleological Survey Bulletin.

Dore, M. 1995. Windy Mouth Cave. In *Underground in the Appalachians, A guidebook for the 1995 NSS convention*, ed. C. Zokaites, 92–94. Huntsville, AL: National Speleological Society.

Granger, D.E., J.W. Kirchner, and R.C. Finkel. 1997. Quaternary downcutting rate of the New River, Virginia, measured from differential decay of cosmogenic ^{26}Al and ^{10}Be in cave-deposited alluvium. *Geology* 25 (2): 107–110.

Granger, D.E., D. Fabel, and A.N. Palmer. 2001. Pliocene-Pleistocene incision of the Green River, Kentucky, determined from ^{26}Al and ^{10}Be in Mammoth Cave sediments. *Geological Society of America Bulletin* 113 (7): 825–836.

Gulden, B. 2017. *USA Longest Caves* (list as of Nov. 28, 2016). http://www.caverbob.com/usalong.htm. Accessed May 6 2017.

Heller, S.A. 1980. A hydrogeologic study of the Greenbrier Limestone karst of central Greenbrier County, West Virginia, 167pp. PhD. dissertation, West Virginia University.

Jones, W.K. 1997. *Karst Hydrology Atlas of West Virginia*, 96pp. Charles Town, WV: Karst Waters Institute Special Publication No.4.

Kulander, B.R., and S.L. Dean. 1986. Structure and tectonics of Central and Southern Appalachian Valley and Ridge and Plateau Provinces, West Virginia and Virginia. *American Association of Petroleum Geologists Bulletin* 70 (11): 1674–1684.

Ogden, A.E. 1976. The hydrogeology of the central Monroe County karst, West Virginia, 263pp. Ph.D. Dissertation, West Virginia University.

Palmer, A.N. 1974. Geologic influence upon cave passage orientation in Ludington Cave, Greenbrier County, West Virginia. In *Proceedings of the fourth conference on Karst geology and hydrology*, H.W. Rauch, and E. Werner, 33–40. Morgantown, WV: West Virginia Geological and Economic Survey.

Palmer, A.N. 1991. Origin and morphology of limestone caves. *Geological Society of America Bulletin* 103: 1–21.

Peters, D., and J. Davis. 1960. (Map of) Windy Mouth Cave. *Baltimore Grotto Newsletter (National Speleological Society)* 3 (7): 115–116. Reprinted in *Speleo Digest 1962*, ed. A.P. Haarr, and W.T. Plummer, 1–231. Pittsburgh, PA: Pittsburgh Grotto of the National Speleological Society.

Sasowsky, I.D., W.B. White, and V.A. Schmidt. 1995. Determination of stream incision rate in the Appalachian Plateaus by using cave sediment magnetostratigraphy. *Geology* 23 (5): 415–418.

Shank, D.A. 2002. Hydrologic and structural controls on the evolution of a karst aquifer, Windy Mouth Cave, West Virginia, 121pp. M.S. Thesis, University of Akron.

Springer, G.S., J.S. Kite, and V.A. Schmidt. 1997. Cave sedimentation, genesis, and erosional history in the Cheat River Canyon, West Virginia. *Geological Society of America Bulletin* 109 (5): 524–532.

Wells, D. 1950. Lower middle Mississippian of Southeastern West Virginia. *American Association of Petroleum Geologists Bulletin* 34 (5): 882–922.

White, W.B. 1988. *Geomorphology and hydrology of Karst Terrains*, 464. New York, NY: Oxford University Press.

Whittemore, R.E. 1971. Windy Mouth Cave. *The Region Record, (Virginia Region of the National Speleological Society)* 1 (4): 114–115.

Scott Hollow Cave

16

Stacey B. Lane, Melisa R. Bishop, Michael J. Dore and Ira D. Sasowsky

Abstract

The Scott Hollow Drainage basin is a mature karst system that serves as a northward flowing subterranean tributary to the Greenbrier River. A well-developed conduit system transports waters through this upland basin via a series of branchwork passages. The main underground stream, the Mystic River, flows 10.4 km (6.5 miles) and has both eastern and western tributaries, fed by a combination of concentrated and diffuse recharge. Upper-level passages are mostly disconnected from the active lower ones and seem related to an earlier phase of speleogenesis. The complex flow system is impacted by surface land use in this rural basin, and well drilling has resulted in intersection with the cave in at least two places. Aquifer development within this basin follows structurally controlled initial porosity and "perching" on an underlying shale unit. The northward flowing Mystic River is controlled by regional flexures resulting from Alleghenian compressive stress and an unmapped thrust fault. Tributaries joining the Mystic River from the east generally trend down dip along bedding planes, while western tributaries follow steeper, more complex paths, at least partially controlled by a thrust fault. This has resulted in chemical differences between the waters constituting these tributaries, although all are fed by sinking streams from the upland surface. Sediments within the cave include chert gravels in the active streams (many of which are manganese oxide coated), along with relict diamicts that are currently undergoing excavation.

16.1 Introduction

Scott Hollow Cave is one of the three long cave systems (along with Organ and Windy Mouth Caves) known from the portion of the Big Levels limestone plateau lying south of the Greenbrier River in Monroe and Greenbrier Counties. The cave's currently mapped length and depth are 47.5 km (29.5 miles) and 174 m (571 ft), respectively (Gulden 2017), ranking it as 17th on the USA long cave list. The cave had no known natural entrance, but its existence was postulated by Mike Dore, who dug open the entrance in 1984 (see Chap. 5). There are now four entrances (all dug). Mapping and exploration are ongoing, but are challenging because all of the entrances are near the middle of the system. Exploration trips downstream require multiple scuba dives. Trips upstream require hours of traversing stream passage and large breakdown. An upstream camp provides for logistical support in that area.

The first broadly published description of the cave, with a detailed map of 3 km (2 miles) of the central ("tourist") portion (Fig. 16.1), appeared in the 1995 NSS Convention Guidebook (Dore 1995). It includes several pages of description, photographs, and also a line plot with 35 km (21.5 miles) of passage. Dasher's (2012) comprehensive volume on the caves and karst of West Virginia contains a wealth of information on the cave, including Pleistocene fossils recovered from the cave sediments. The present chapter makes use of these early works (including maps discussed below), our personal experiences in the cave, and

S. B. Lane
5473 Beechmont Ave Apt. 208, Cincinnati, OH 45230, USA

M. R. Bishop · I. D. Sasowsky (✉)
Department of Geosciences, University of Akron, Akron, OH 44325-4101, USA
e-mail: ids@uakron.edu

M. J. Dore
P.O. Box 7, Sinks Grove, WV 24976, USA

W.B. White (ed.), *Caves and Karst of the Greenbrier Valley in West Virginia*,
Cave and Karst Systems of the World, https://doi.org/10.1007/978-3-319-65801-8_16

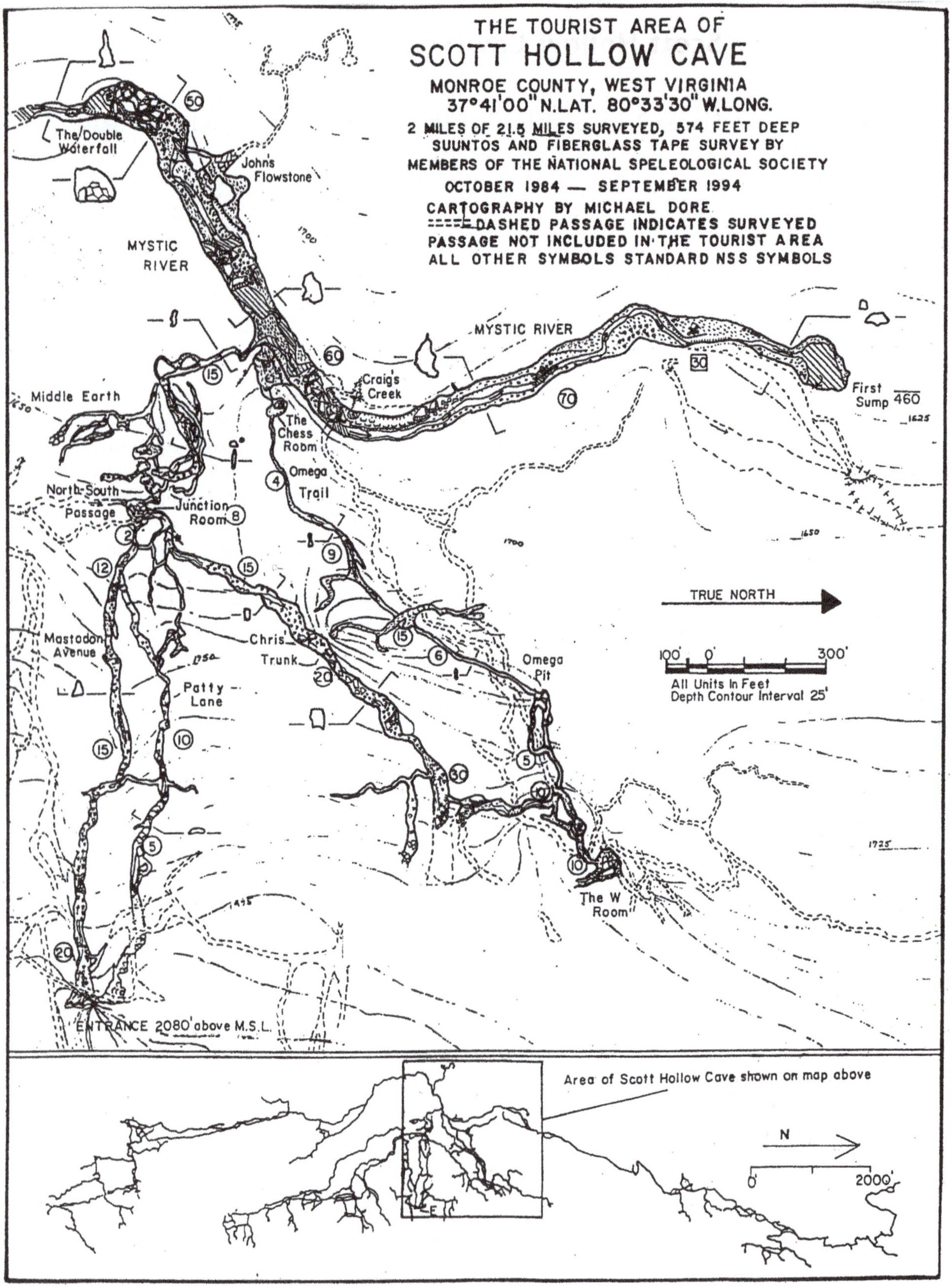

Fig. 16.1 Tourist area map of Scott Hollow Cave (Dore 1995) showing Mystic River and several tributaries, as well as overlying passages

the scientific studies of Davis (1999) and Bishop (2010), to discuss the geologic and hydrologic conditions and history of this major system.

16.2 Geologic and Geomorphic Setting

16.2.1 Geomorphic Setting

The area near the cave is within the transition zone, referred to as the "Folded Plateau," adjacent to the boundary between the Valley and Ridge and the Appalachian Plateaus physiographic provinces. Specifically, this area is situated between the Allegheny Structural Front to the west and the Intraplateau Structural Front to the east (Kulander and Dean 1986). The region is characterized by gently dipping, northeast trending folds in Middle-Late Paleozoic sedimentary rocks (Heller 1980; Ogden 1976).

Scott Hollow, the surface valley for which the cave is named, is a dry karst valley on the Big Levels surface, located 5 km (3 miles) south of the Greenbrier River at Fort Spring, Greenbrier County (Fig. 16.2). The valley itself extends an additional 5 km (3 miles) to the south, into and mostly in Monroe County. It is bounded to the west by Flat Top Mountain, to the south by the mountain-like Swoopes Knobs, and to the east by a low ridge congruent with the Sinks Grove Anticline. The Greenbrier Limestone, the underlying Maccrady Shale, and the overlying Mauch Chunk sandstones all dip west from the Sinks Grove Anticline. All of these beds outcrop in Scott Hollow with the shales along the east portion of the hollow, the limestones through the center and up the lower half of Flat Top Mountain, and the sandstones along the remaining upper half of Flat Top Mountain. The various shale layers, below and within the Greenbrier series, played an important role in the development and discovery of Scott Hollow Cave.

It was obvious to most cavers that there should be a large cave system underneath Scott Hollow, but there were few present-day cave entrances. Exploration in nearby Scott Cave in 1979 revealed the deep nature of the drainage; that cave was mapped to a depth of 130 m (426 ft) without reaching a main drain. Scott Cave is developed at the bottom of the Greenbrier series, riding the Maccrady Shale its entire 0.8 km (0.5 miles) length. This was a key to the discovery of Scott Hollow Cave. The original entrance dig is located in a small eastern blind valley, the stream of which flows west on the Maccrady Shale until contacting the bottom of the Greenbrier series at the dig site, and then continues underground sandwiched between the bottom of the Greenbrier and top of the Maccrady Shale. Bedding location was the primary reason for this dig site. Dowsing with a forked apple tree branch gave the exact spot to dig, which happened to be precisely on a significant fracture heading down dip into the cave. The present entrance, a piece of corrugated pipe beneath a trap door (Fig. 16.3), belies the fantastic cave system that lies below.

Windy Mouth Cave (Chap. 15) lies directly north of Scott Hollow Cave. This position, along with the general alignment of the trunk passages in the two caves, has led some to hypothesize that they may consist of a single more extensive system, as yet unconnected by explorers.

16.2.2 Stratigraphy and Structure

The Greenbrier Group carbonates and siliciclastics within which the cave is formed are exposed in a northeast trending belt that outcrops along the eastern margin of West Virginia. Locally, the outcrop belt forms a broad sinkhole plain that ranges in elevation between 560 and 690 m (1820–2250 ft) and is referred to as the "Big Levels." Hills that surround the sinkhole plain are capped by clastic rocks of the Mississippian Mauch Chunk Group that conformably overly the Greenbrier Group (Heller 1980). In the study area, the sinkhole plain is deeply dissected by the Greenbrier River, a major tributary to the New River in the Ohio River drainage basin. Numerous karst features such as blind valleys, sinkholes, and disappearing streams dominate the scenic landscape of the plain. There is also a notable lack of surface drainage on the Big Levels. Groundwater enters the system on the karst upland and discharges through springs and seeps at river level, 150 m (490 ft) below.

The Greenbrier Group is approximately 240 m (900 ft) thick (see Fig. 15.4) in southern Greenbrier County (Wells 1950) and is bounded at its upper and lower contacts by Mississippian Shales (Ogden 1976). The Maccrady Shale bounds the lowest unit of the Greenbrier Group and plays a major role in the hydrogeology and karst development in the area (Palmer 1974). Cave development is primarily in the Hillsdale Limestone (see Fig. 15.5) although stream incision in some passages cuts down into the upper few meters of the Maccrady Shale. There are many distinctive layers present even within the Hillsdale Limestone. Work for an undergraduate student project at Washington and Lee University identified seven distinct layers, including some very resistant bedded cherts (Kimmel et al. 2002).

Understanding of the stratigraphy has been a key to ongoing exploration, as it guides both in-cave and exterior digs. In 1993, shale position was again a key. Many of the lower Maccrady contact passages are confined to the bottom 10 m (35 ft) of Greenbrier limestone, but a tight, infeeder stream led to a fault that provided access to higher beds. A tough dolomitic shale, named the DT Shale, not only confines those lower passages, but also acts as the base layer for passages above, named Scoop City for the propensity of Scott Hollow cavers to scoop before survey. Scoop City ignores the patterns of the lower cave and is thus like an

Fig. 16.2 Terrain visualization of the Scott Hollow area showing the two major caves, Greenbrier River, and spring. Base map from USGS terrain data. Cave data from Dore and Hixson

entirely different cave. In 2003, a second entrance was dug into a remote section of Scoop City, and a few years later, Cleveland Grotto cavers finished a third entrance dig, also into Scoop City.

Mapping of lithologic boundaries and structural features is very difficult in this region due to limited exposures. Original mapping on the county level (Reger and Price 1926) identified only one fault in the county. Several minor folds, mapped later and named by Heller (1980), exist between major structures in the study area (see Fig. 15.6). These include the Fort Spring Anticline and Syncline as well and another unnamed fold pair that lies just west of the study area. An additional unnamed thrust fault and fold pair lies just to the east. Heller (1980) attributes these minor structures to splay faulting through the relatively competent Greenbrier Group carbonate sequence from underlying Lower Paleozoic strata.

One of the biggest questions remaining is the role of the Taggard Shale, a major shale layer that divides the Greenbrier series into what is locally referred to as the upper and lower limestones, with Scott Hollow Cave being wholly within the lower limestones. An interesting feature of the narrow Taggard outcrop in Scott Hollow is that it very closely aligns with Mystic River, 100–130 m (300–400 ft) below the surface (Fig. 16.4). One hypothesis suggests this was a natural location of the ancient surface creek draining the area, and the underlying saturated limestone became the first location for cave development in Scott Hollow, hence

Fig. 16.3 Present-day entrance to Scott Hollow Cave. Photo by I. D. Sasowsky

the location and size of Mystic River. Another interesting Taggard feature is that it appears to perch the drainage flowing off Flat Top Mountain. Significant upper limestones capture this drainage from the mountain, and there might be an entirely different cave system above the Taggard Shale.

16.2.3 Hydrology

Details on the general hydrology and meteorology of the region are given in Chap. 15. The Scott Hollow surface water drainage basin is bounded by previously discussed topographic highs. The Greenbrier River and Second Creek form low northern and eastern boundaries, and the former serves as regional base level. Surface water courses are generally absent in the basin, with the exception of short channels on clastic substrates that disappear when they contact the limestone.

The cave itself lies within the Scott Hollow groundwater basin, and initial dye tracing in 1985 and 1988 (summarized in Jones 1997) determined that the basin covers approximately 47 km^2 (18 mi^2), 90% of which is carbonate outcrop. This included inputs from Boyd Spring and Marcus Wiley Cave. Later work (Bishop 2010) refined and extended

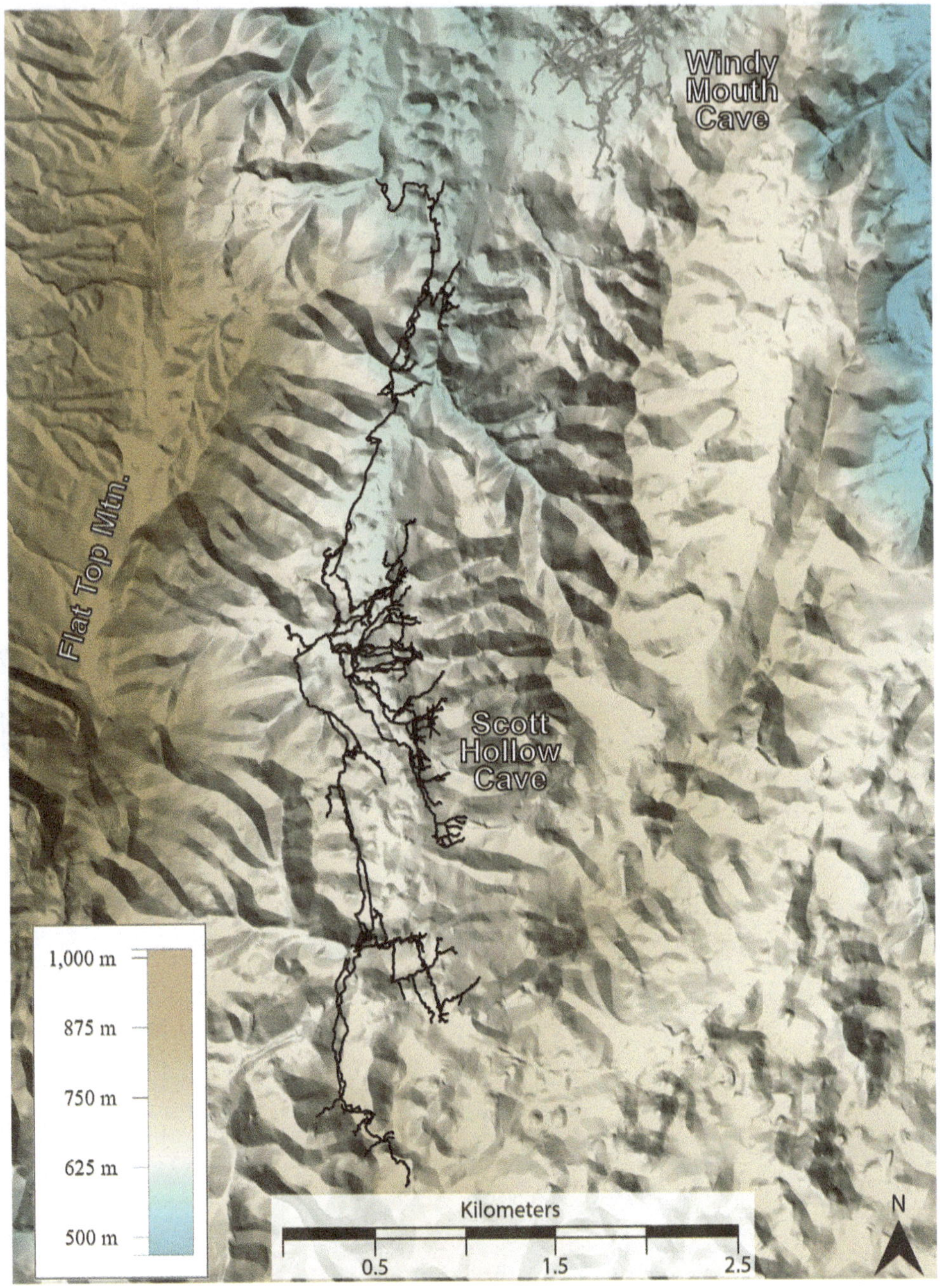

Fig. 16.4 Terrain visualization closeup of the Scott Hollow Cave area. Base map from USGS terrain data. Cave data from Dore and Hixson

the boundaries of the basin. She found connections between Bland's Sink, Anderson Sink, and Rolling Rock Cave to the east, as well as Erskine Cave and Lilly's Sink to the west (Fig. 16.5).

The cave lower levels are currently very hydrologically active. The main conduit, named Mystic River, flows generally northward along strike and has been traced to a resurgence (Gloria's Spring) on the Greenbrier River (Fig. 16.5). Mystic River (Fig. 16.6) is fed by numerous subsurface tributaries (Fig. 16.5) coming in from both the east (Chris's Trunk, Mastodon Ave., Patty Lane, North–South passage) and west (Craig's Creek, John's Flowstone). The eastern tributaries generally follow dip and begin at sinking streams coming off the Maccrady Shale. The western tributaries are quite steep and probably follow along a thrust fault that may daylight at the base of Flat Top Mountain, draining waters from that region. Gradients range from a low of 0.0088 (Mystic River) to 3.2 (Craig's Creek) (Table 16.1). Streams are generally perennial and respond to seasonal precipitation, occasionally making travel through the cave very difficult.

In order to evaluate flow in the central part of the cave, a flow measurement program was conducted (Fig. 16.7). Table 16.2 shows the discharge data for each stream

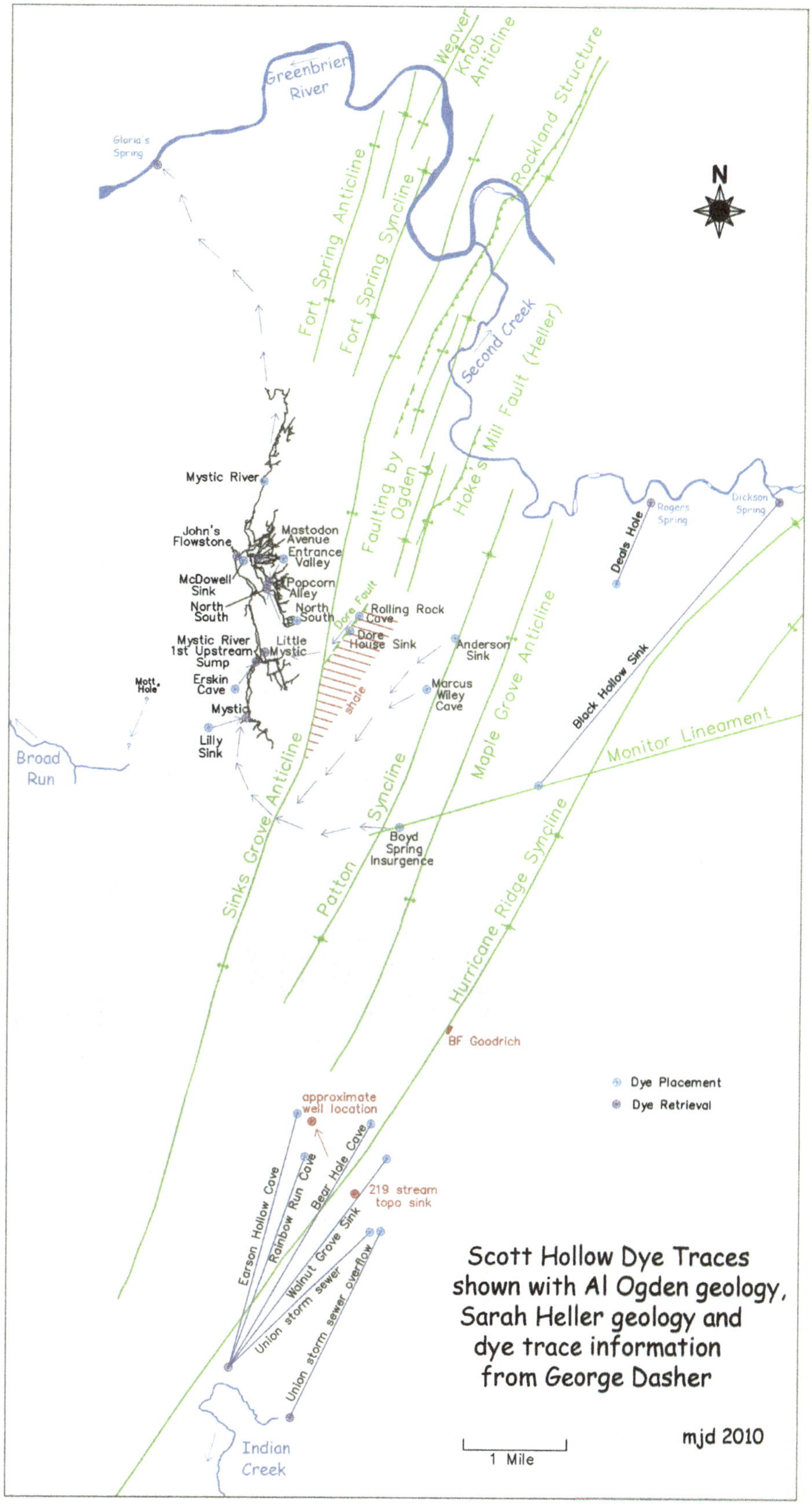

Fig. 16.5 Summary map showing dye traces in vicinity of Scott Hollow Cave (Dore 2010, unpublished)

Fig. 16.6 Downstream view of Mystic River. Photo by Ed McCarthy. Used with permission

measured during three sampling events. These three sets of data allow for preliminary comparisons and show indications of a seasonal effect in the groundwater flow for the study area. First to be considered are the western tributaries, John's Flowstone and Craig's Creek. Both have a similar pattern of increased flow from July to February and decreased flow from February to May. The smallest discharge values for both John's Flowstone and Craig's Creek are seen in July (0.021 cfs and 0.14 cfs, respectively). The greatest discharges are seen in February (0.55 cfs and 0.92 cfs, respectively).

Next to be considered are the eastern tributaries as a whole. These streams show a consistent increase in discharge from July through May. An increase is seen from July (0.70

Table 16.1 Gradients of selected streams in Scott Hollow Cave (Davis 1999)

Stream	Map length (ft)	Elevation change (ft)	Gradient
Mystic River	34,000	300	0.0088
John's Flowstone	600	450	0.75
Craig's Creek	125	400	3.2
North–South	4,500	400	0.089
Root Canal	2,300	400	0.17
Mastodon	1,250	400	0.32
Chris's Trunk	1.305	400	0.31

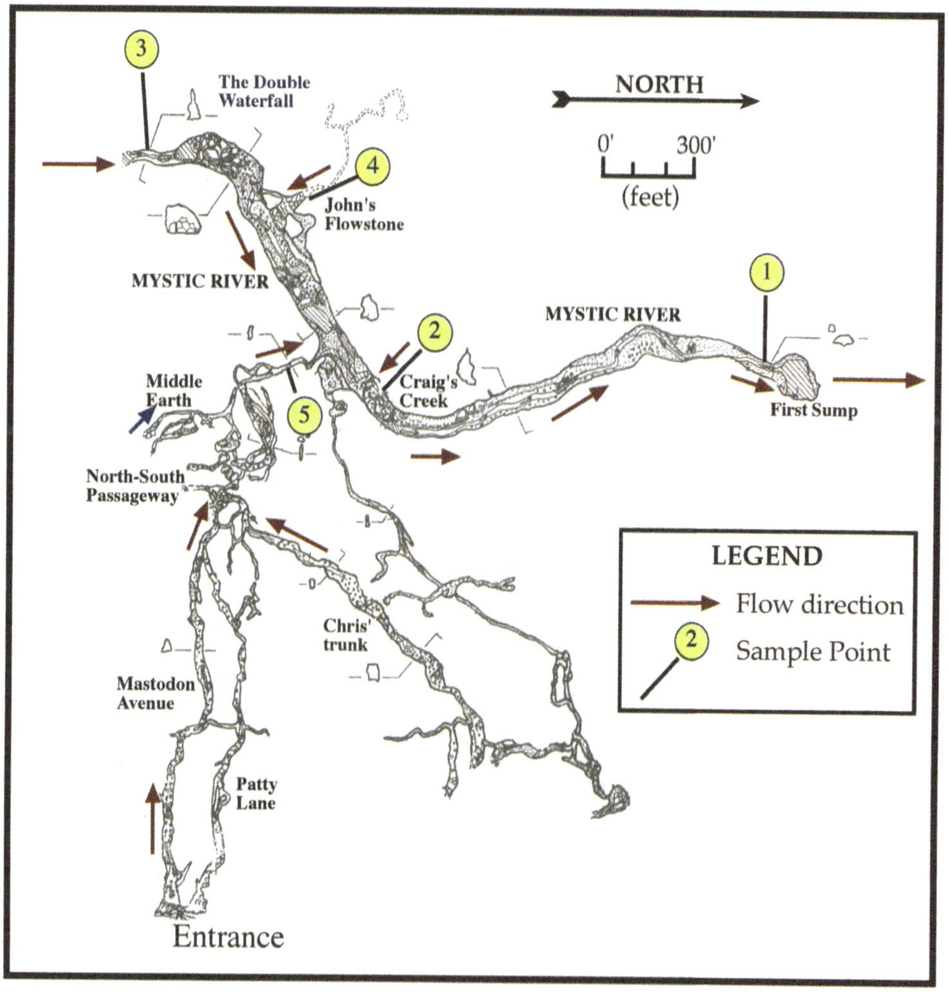

Fig. 16.7 Map showing central part of Scott Hollow Cave and locations for flow and chemistry sampling. Base map prepared by Dore, modified by Davis (1999)

cfs [cubic feet per second]) to February (1.53 cfs) and again to May (3.45 cfs). Discharge is more than doubled from one measurement to the next, for all three sampling events. This is a markedly different pattern from the western tributaries, which exhibit their highest discharges in February.

Last to be considered is the Mystic River itself. Both the upstream and downstream discharges show the same pattern of consistent increase seen in the eastern tributaries. Upstream and downstream Mystic River shows a slight increase from July (3.30 and 1.66 cfs, respectively) to February (4.90 and 1.91 cfs). A considerable increase in discharge is seen in May (24.4 and 8.89 cfs) where values are nearly five times greater than those for February.

All the streams show an increase from July to February, with July's discharges being the lowest of the three sampling events. This reflects a lack of precipitation in summer or increased precipitation in late winter. The study area received almost 15 cm (6 in.) of moisture during the weeks preceding measurement in February. The study area experienced mild weather with frequent rainy periods during the month of February. July only received 3 cm (1 in.) of precipitation prior to measuring.

Table 16.2 Stream discharge at several sampling locations in Scott Hollow Cave during the course of a year (Davis 1999)

Stream	July 9, 1997	February 21, 1998	May 17, 1998	Comments
Upstream Mystic River	3.30 cfs (±1.05 cfs)	4.90 cfs (±1.57 cfs)	24.4 cfs (±7.8 cfs)	Small increase summer to winter, higher in spring
John's Flowstone	0.021 cfs (±0.007 cfs)	0.55 cfs (±0.18 cfs)	0.29 cfs (±0.09 cfs)	Highest in winter, lowest in summer
Craig's Creek	0.14 cfs (±0.05 cfs)	0.92 cfs (±0.29 cfs)	0.65 cfs (±0.21 cfs)	Highest in winter, lowest in summer
Eastern Tributaries	0.70 cfs (±0.22 cfs)	1.53 cfs (±0.50 cfs)	3.45 cfs (±1.10 cfs)	Small increase summer to winter, higher in spring
Downstream Mystic River	1.66 cfs (±0.53 cfs)	1.91 cfs (±0.61 cfs)	8.89 cfs (±2.84 cfs)	Small increase summer to winter, higher in spring

cfs Cubic feet per second

A curiosity arises when evaluating the water balance between upstream and downstream Mystic River's discharges. It was expected that the downstream discharge would equal the upstream discharge plus tributary inputs, but this is not the case. To examine the water balance, the upstream and all tributary discharges were added together to obtain a calculated downstream discharge. This calculated discharge was then compared with the actual measured discharge for the downstream sump (Table 16.3). Measured downstream discharges were found to be smaller than calculated, showing an overall loss taking place within the system. Percentage losses are similar, though it is unclear where the loss occurs or is going. One possibility is that these waters are regained farther downstream, past the first sump.

The hydrology of the cave has also been impacted by anthropogenic effects. Two deep water wells have intersected the cave. One is exposed in the upper-level Scoop City passages (Fig. 16.8). The other intersected small voids adjacent to the North–South passage, apparently breached a confining unit, and has allowed water to flow uninterrupted into the vadose stream passage for over a decade. This feature, called "The Spigot," has been the subject of two Master of Science theses (Demrovsky 2003; Check 2007), and the flow from the deep confined aquifer (likely a fractured shale) accounts for the bulk of North–South discharge during many parts of the year.

16.3 General Form and Structural Influences

As previously mentioned, passage development in Scott Hollow Cave is complex, with many overlying passages that are non-concordant with active underlying ones. Nevertheless, study within the cave has revealed some major controls.

16.3.1 Form and Position

Examination of the cave map in relation to the landscape clearly shows a linkage of the presently active passages such as Mastodon Avenue, Mystic River, and Craig's Creek to the surface stream system. The master conduit of the system is Mystic River, which flows northward to discharge at the Greenbrier River. The elevation of Mystic River near the sump is barely above the local base level (Greenbrier River). The cave generally trends southward from the entrance. The northernmost part of the cave aligns with the southernmost portion of Windy Mouth Cave, suggesting that the caves might be part of the same conduit system. This may be possible, but exploration to date (Chap. 5) has not revealed a connection. Furthermore, although the caves are only separated by about 600 horizontal meters (2000 ft), the southern portions of Windy Mouth are likely circa 30 m (100 ft) higher than the nearest reaches of those in Scott Hollow Cave. This suggests that Scott Hollow Cave evolved at a

Table 16.3 Water balance comparison for the study area in Scott Hollow Cave showing overall loss occurring between the upstream and downstream Mystic River locations, regardless of the season (Davis 1999)

Month	Measured upstream mystic R. (cfs)	Measured tributary inputs (cfs)	Calculated downstream Mystic R. (cfs)	Measured downstream Mystic R. (cfs)	Total loss (cfs)	Percentage loss (%)
July	3.30	0.86	4.16	1.66	2.50	60.1
February	4.90	3.00	7.90	1.91	5.99	75.8
May	24.4	4.39	28.79	8.89	19.90	69.1

cfs Cubic feet per second

Fig. 16.8 Steel well casing from domestic well intersecting passage in Scoop City section of the cave. Photo by I.D. Sasowsky

later phase than Windy Mouth Cave and may be isolated from that drainage.

The overall plan-view morphology of Scott Hollow Cave fits the classification by Palmer (1991) of a "branchwork cave," where several lower-order conduits converge into fewer higher-order conduits in a dendritic pattern. The cave also exhibits some characteristics of an anastomotic pattern in the upper levels. The hybrid morphology is likely the result of different phases of speleogenesis, and the levels only connect in a few places throughout the system.

16.3.2 Structural Controls

Along its path, the main Mystic River conduit generally follows strike, though at the first sump it makes a rather sharp "S," which is also quite deep. Tributary passages from the east are mostly stratigraphically controlled, following dip, although the Root Canal passage follows a high-angle low-displacement fault near its confluence with the North–South Passage. Important intersections such as the Junction Room and the confluence of Middle Earth Stream with

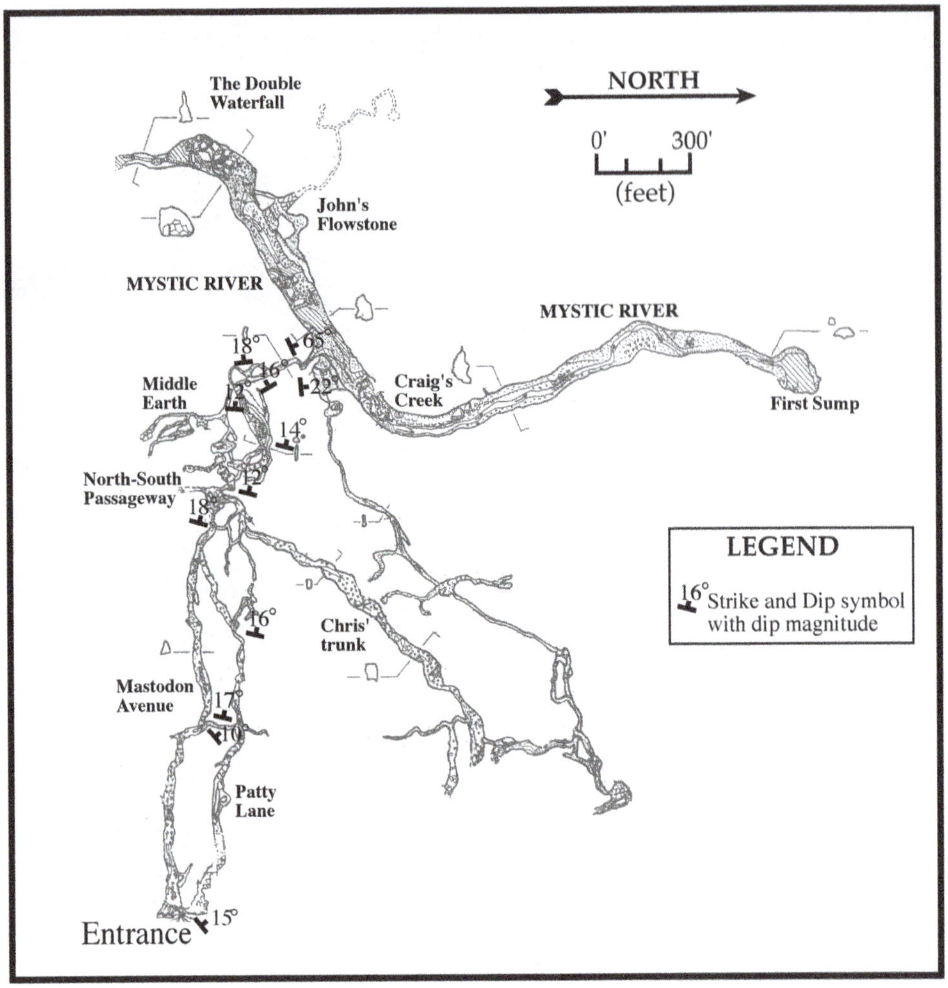

Fig. 16.9 Map showing central part of Scott Hollow Cave with strike and dip measurements. Base map prepared by Dore, modified by Davis (1999)

Mystic River are likely controlled by faults. Mapping with Brunton compass shows how dips increase closer to Mystic River (Fig. 16.9). The segment of Mystic River that runs SW–NE in the area of the Middle Earth confluence (Fig. 16.10) is controlled by a thrust fault (Strike N65°E, dip 15°SE) that is apparent in the cave, but has not been mapped on the land surface. The steeply ascending John's flowstone passage actually follows the ramp of the thrust upward to the NW (Fig. 16.11), and the ceiling of this passage is in part crusted with thick secondary calcite which has formed in the fault plane. Chert beds in the hanging wall are vertical or slightly overturned.

16.4 Speleothems and Clastic Sediments

Certain portions of the cave are known for their impressive speleothems. Massive flowstone deposits have formed from the western Craig's Creek (Fig. 16.12) and John's Flowstone passages (Fig. 16.13). Areas such as the Root Canal (Fig. 16.14) and Ambers Garden (Fig. 16.15) show some of the fantastic secondary mineralization found in the cave.

However, from a caver's perspective, one of the most memorable aspects of the passages is the pervasive breakdown. Whether in one of the entrance passages or in the main trunk (Fig. 16.16) this is a common characteristic, and one seems to always be scrambling over (or under) large slippery blocks. Beyond these autogenic materials, there are two classes of allogenic sediments present. The first are paleofills which include diamicts that have nearly filled some of the eastern tributaries. These are currently undergoing a phase of natural excavation, probably related to climate changes. Separately, though perhaps sometimes related are present-day alluvial sediments. The alluvium is primarily chert gravels, which are frequently found to be coated with thick manganese oxide deposits. In upstream Mystic River, and also in Mastodon Avenue, fossil bones and teeth from Pleistocene fauna have been recovered from finer-grained alluvium. The input points (sinking streams) that form the headwaters of the cave tributaries are all presently blocked.

Fig. 16.10 Downstream view of Mystic River showing relationship of joint and bedding orientation to conduit walls. This is looking in direction of Craig's Creek from junction or Middle Earth stream. Photo by W.K. Jones. Used with permission

However, this was not likely the case before European settlement of the area. Excavations into "natural" fill in the valley of the main entrance to the cave found a hewn log buried several meters deep. Likewise, we have observed antique canning lids in stream sediments in the North–South Stream that had to have washed in from an originally open sinkhole entrance. These occurrences indicate that the transport of sediment through the system, along with the opening of entrances, is episodic.

16.5 Hydrochemistry

Given that the numerous tributaries within Scott Hollow Cave are fed by different sinking streams, there is an expectation for great spatial and temporal variability of water quality. Davis (1999) used the five sampling sites where flow measurements were made (Fig. 16.7) to characterize the geochemical characteristics of groundwater in the cave. Three of these sites are tributaries to the Mystic River, the

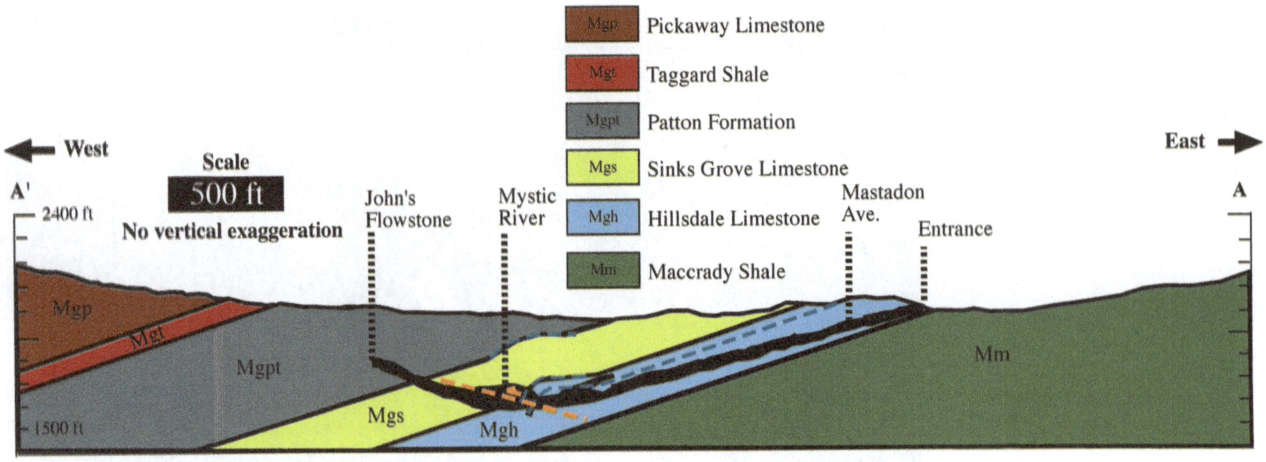

Fig. 16.11 Cross section along entrance passages of Scott Hollow Cave, across Mystic River, and up John's Flowstone. Based on the section prepared by surveyor M. Dore

Fig. 16.12 Rimstone dam at Craig's Creek confluence during high discharge. Photo by I.D. Sasowsky

basin's primary drainage stream. One site was located on the Mystic River, upstream of the tributaries, and another was also located on the Mystic River, downstream of the tributary inputs. Waters were analyzed for major cations and anions using conventional laboratory techniques, and the data were entered into the WATEQ4F program to determine, in particular, saturation indices for calcite ($SI_{calcite}$).

Western tributaries showed a slight variation in $SI_{calcite}$ values, while eastern tributaries had more exaggerated shifts in their $SI_{calcite}$ values. This suggests a combination of diffuse and concentrated recharge for waters in the western tributaries and dominantly concentrated recharge directly from surface sources for the eastern tributaries. The Mystic River has consistently undersaturated $SI_{calcite}$ values typical of waters flowing through large conduits.

$SI_{calcite}$ values were consistently less undersaturated at the downstream Mystic River location, suggesting a possible effect of mixing with the tributary waters. Mixing calculations were performed and found to account for only part of the decrease in $SI_{calcite}$ values from upstream to downstream. Differences in the calculated and measured decreases are attributed to buffering of waters after mixing, due to flow

Fig. 16.13 Upstream view of Mystic River, with John's Flowstone on right side. Photo by Ed McCarthy. Used with permission

being predominantly over limestone breakdown. This indicates dissolution of calcite is taking place, and thus, the cave is experiencing growth.

The Scott Hollow basin is home to agricultural activities including livestock grazing and crop growth. Nitrate levels were consistently above the 10 mg/L MCL set by USEPA for drinking water and showed a definite correlation to surface land usage. The greatest fluctuations can be seen in the tributaries, particularly the eastern ones, compared with relatively consistent values for the Mystic River. Phosphate and fluoride showed no distinctive patterns.

Fig. 16.14 Speleothems in Root Canal Passage. Photo by I.D. Sasowsky

16.6 Summary

Scott Hollow Cave is an extensive branchwork conduit system that has formed at several stratigraphic levels, all in the lower Greenbrier Limestone. Upper-level passages tend to be phreatic in style and are mostly dry at the present time. Lower-level passages contain active streams and are fed by sinking streams of various sizes from the east and west. The cave stream resurges on the Greenbrier River at Gloria's Spring, as shown by dye tracing.

Aquifer development and conduit formation within the Scott Hollow drainage basin are the result of a complex relationship involving stratigraphy, hydrogeology, geochemistry, and structural geology. Despite the complexities, a definite pattern can be observed, as drainage conduits have developed along structural features in the area. The entrance to the subsurface conduit system, and most of the study area with the possible exception of portions of the western tributaries, is developed within the Hillsdale Limestone, the basal unit of the Mississippian age Greenbrier Group. This conduit system provides the only known drainage for the upland Scott Hollow basin, transporting waters under pipe-like conditions northward, to the Greenbrier River.

Fig. 16.15 Speleothems in Amber's Garden. Photo by Ed McCarthy. Used with permission

A significant amount of the water transported through the Scott Hollow basin enters from sources to the south. These waters form the Mystic River and join with eastern and western tributaries fed by either precipitation or a mixture of diffuse flow and surface runoff, respectively. An overall loss of water takes place, despite tributary stream inputs, as the river moves northward through the study area, suggesting the presence of a highly complex flow system. Response to storm events is similar to, though not as rapid as, that of a surface basin.

Fig. 16.16 Large breakdown piles in the Kansas section of Scott Hollow Cave. Photo by Ed McCarthy. Used with permission

Overall, the eastern tributaries exhibit more variable water chemistry than the western tributaries. This supports the idea that eastern tributaries are fed by precipitation and runoff, while western tributaries are fed predominantly by diffuse flow. A decrease in $SI_{calcite}$ values from upstream to downstream Mystic River is caused by mixing with tributary waters and is partly due to the dissolution of limestone from the streambed, indicating that the cave is experiencing growth. Elevated nitrate levels emphasize the close relationship of surface land use with groundwater quality for karst basins and show a seasonal effect of concentrated nitrate levels during drier months when flushing of surface soils occurs less frequently.

The main conduit, as well as all tributaries, shows evidence for structural involvement in flowpath locations and orientations. Eastern tributary conduits descend along joint orientations that either parallel or are oblique to the dip of bedding (NW). Locations of western tributaries are influenced by joint orientation and the presence of a localized thrust fault. To the North, the main conduit follows the strike of bedding and large-scale joints. To the South, flowpath orientation for the main conduit seems to be oblique to the strike of bedding and it follows the orientations of joints and the localized thrust fault.

Clastic sediments are common throughout the lower passages and include materials from the surface as diverse as mastodon bones and canning jar covers. This implies a complex and episodic process of sediment transport that reflects climatic as well as anthropogenic influences in the basin.

Exploration and scientific studies have been constrained by sumps to the north and great distances to the south. If additional entrances beyond the existing four are ever found/dug, this will greatly assist in revealing the extent and contents of the cave.

Acknowledgements The following people provided exceptional assistance during the field work and other aspects of the study, which is greatly appreciated: Pat Dore, Sarah Benjamin, Verne Friberg, David McConnell, Natasha (née Demrovsky) Papuga, Mike McFall, Rebecca Bixby, and Dave Shank. We thank Ed McCarthy and Bill Jones for allowing use of their photos. Joshua Novello prepared the terrain figures.

References

Bishop, M.R. 2010. A detailed hydrologic study of the Scott Hollow Groundwater Basin, Greenbrier and Monroe Counties, West Virginia, 216pp. Master of Science Thesis, University of Akron.

Check, D. 2007. High resolution monitoring of anthropogenic groundwater flow modification: Monroe County, West Virginia, 331pp. Master of Science Thesis, University of Akron.

Dasher, G. 2012. *The Caves and Karst of West Virginia*, 264pp. Barrackville, WV: West Virginia Association for Cave Studies Bulletin 19.

Davis, S.B. 1999. Aquifer development in a folded and fractured limestone: The Scott Hollow drainage basin, Monroe County, West Virginia, 210pp. Master of Science Thesis, University of Akron.

Demrovsky, N. 2003. The effects of well drilling on karst aquifers in Monroe and Greenbrier County, West Virginia, 198pp. Master of Science Thesis, University of Akron.

Dore, M. 1995. Scott Hollow Cave. In *Underground in the Appalachians, A guidebook for the 1995 NSS Convention*, ed. C. Zokaites, 102–105. Huntsville, AL: National Speleological Society.

Gulden, B. 2017. *USA Longest Caves* (list as of Nov. 28, 2016). http://www.caverbob.com/usalong.htm. Accessed May 6, 2017.

Heller, S.A. 1980. A hydrogeologic study of the Greenbrier Limestone karst of central Greenbrier County, West Virginia, 167pp. PhD. dissertation, West Virginia University.

Jones, W.K. 1997. *Karst Hydrology Atlas of West Virginia*. Charles Town, WV: Karst Waters Institute Special Publication No. 4, 96pp.

Kimmel, S., N. Weber, R. Burnside, R. Thren, and C. Connors. 2002. Bedrock geology of the Scott Hollow Cave, Monroe County, West Virginia. (Undergraduate research project, Washington & Lee University), 49pp.

Kulander, B.R., and S.L. Dean. 1986. Structure and tectonics of Central and Southern Appalachian Valley and Ridge and Plateau Provinces, West Virginia and Virginia. *American Association of Petroleum Geologists Bulletin* 70 (11): 1674–1684.

Ogden, A.E. 1976. The hydrogeology of the central Monroe County karst, West Virginia, 263pp. Ph.D. Dissertation, West Virginia University.

Palmer, A.N. 1974. Geologic influence upon cave passage orientation in Ludington Cave, Greenbrier County, West Virginia. In *Proceedings of the fourth conference on Karst geology and hydrology*, ed. H.W. Rauch, and E. Werner, 33–40. Morgantown, WV: West Virginia Geological and Economic Survey.

Palmer, A.N. 1991. Origin and morphology of limestone caves. *Geological Society of America Bulletin* 103: 1–21.

Reger, D.B., and P.H. Price. 1926. *Mercer, Monroe, and Summers Counties*, 963. Morgantown: West Virginia Geological Survey.

Wells, D. 1950. Lower middle Mississippian of Southeastern West Virginia. *American Association of Petroleum Geologists Bulletin* 34 (5): 882–922.

The Laurel Creek Basin and Greenville Saltpeter Cave

17

William B. White

Abstract

The Laurel Creek Basin is located in an isolated outcrop of Greenbrier Limestone to the southwest of the main Greenbrier Karst. It is drained by Laurel Creek, a tributary of Indian Creek and thus of the New River. The underground route of Laurel Creek passes through Laurel Creek Cave, Rimstone-Crossroads Cave, and Greenville Saltpeter Cave. The caves are in nearly flat bedding near the crest of the Abbs Valley Anticline.

17.1 Introduction

The Laurel Creek Basin represents the southwest limit of the Greenbrier Karst. It is also the one basin in the Greenbrier Karst that is not part of the Greenbrier River drainage. Laurel Creek flows west and southwest over the Mauch Chunk shales and sandstones onto the limestone and then southwest as karstic drainage to its confluence with Indian Creek, a west-flowing tributary that drains directly into the New River.

Laurel Creek flows from clastic rocks to limestone, to more clastic rocks on its route to the New River. The portion of the route across the limestone is mostly underground through Laurel Creek Cave, Rimstone-Crossroad Cave, and Greenville Saltpeter Cave although a well-defined segment of dry valley marks an old surface route.

Laurel Creek Cave was closed by the owners following a rescue of cavers trapped by flood waters in 1982 (Chap. 5). Greenville Saltpeter Cave has been closed to protect endangered bat species. This chapter, therefore, depends on older published reports and on visits to the caves made many years ago.

Electronic supplementary material
The online version of this chapter (doi:10.1007/978-3-319-65801-8_17) contains supplementary material, which is available to authorized users.

W.B. White (✉)
Department of Geosciences, Deike Building, University Park, PA 16802, USA
e-mail: wbw2@psu.edu

17.2 Geomorphic and Geologic Framework

17.2.1 Local Geology

The Laurel Creek Karst is an island of limestone brought to the surface by an undulation in the Abbs Valley Anticline (Fig. 17.1). The limestone is surrounded by the shales and sandstones of the Mauch Chunk Formation. Because of their location on the crest of an anticline, local dips are very low. The basin and the caves are formed in the upper members of the Greenbrier Limestone. According to Davies (1949), Laurel Creek, Crossroads, and Greenville Saltpeter Caves are all in the Union Limestone.

17.2.2 The Landscape

The Laurel Creek Karst takes on the rough form of a closed basin (Fig. 17.2) because the limestone surface is lowered faster by dissolution than the surrounding clastics. The resistant sandstones that surround the basin produce a pronounced escarpment on the north and east sides. Laurel Creek has cut a deep narrow valley through the escarpment where it enters the basin from the east. The western boundary is a line of hills at the base of which is the south-flowing Indian Draft, also a tributary of Indian Creek.

The central portion of the basin is a dissected upland with a hilltop elevation near 1800 ft. The upland is pocked with sinkholes, and there is no defined surface drainage although there are several dry valleys. Presumably precipitation in the

W.B. White (ed.), *Caves and Karst of the Greenbrier Valley in West Virginia*,
Cave and Karst Systems of the World, https://doi.org/10.1007/978-3-319-65801-8_17

Fig. 17.1 Geologic map of the Laurel Creek karst taken from Reger and Price (1926). The *darker, blue-green* (Cgr) is the Greenbrier Limestone. The hatched *lighter green* is the Bluefield member of the Mauch Chunk group. *Green lines* are structure contours. The *dashed green line* marks the location of the axis of the Abbs Valley Anticline

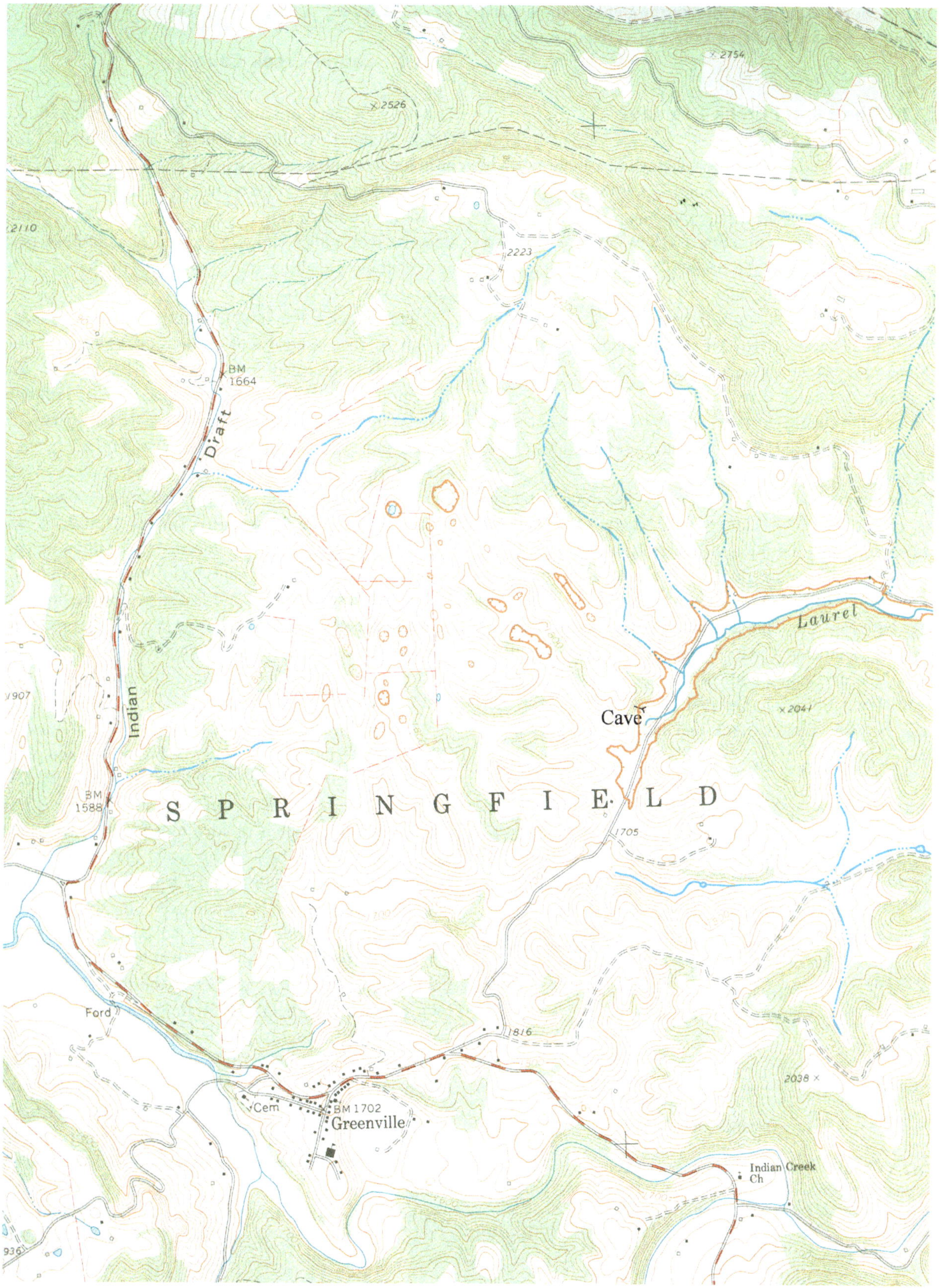

Fig. 17.2 Section of the U.S. Geological Survey Greenville 7.5 min quadrangle

central portion of the basin drains through sinkholes and eventually reaches Laurel Creek by underground routes.

17.2.3 The Laurel Creek Drainage

Laurel Creek has a large surface catchment east and north of the karst region. A profile of Laurel Creek through the karst was constructed from the topographic map (Fig. 17.3). The stream enters the basin through a valley cut in the escarpment on the east side of the basin. The stream contact with carbonate rock is near the edge of Fig. 17.2 where the slope of the stream channel abruptly flattens to the floor of a blind valley at an elevation of 1670 ft. At the southwestern end of the blind valley is the entrance to Laurel Creek Cave and the sink point of Laurel Creek (Fig. 17.4). Under low flow conditions, the stream channel is dry above the cave entrance (Fig. 17.5) but contains flowing water a few hundred yards upstream.

Downstream from the cave entrance, the floor of the dry valley rises over a low saddle to an elevation of 1705 ft and then descends to a flat floor at an elevation of 1690 ft. This flat sequence of valley floor is interrupted by the karst window that exposes a short surface exposure of Laurel Creek between the spring and the Water Entrance to Greenville Saltpeter Cave. Downstream from the karst window, the dry valley floor drops rapidly to the spring at the Mill Pond Entrance. The end of the profile is the confluence between Laurel Creek and Indian Creek at an elevation of 1580 ft.

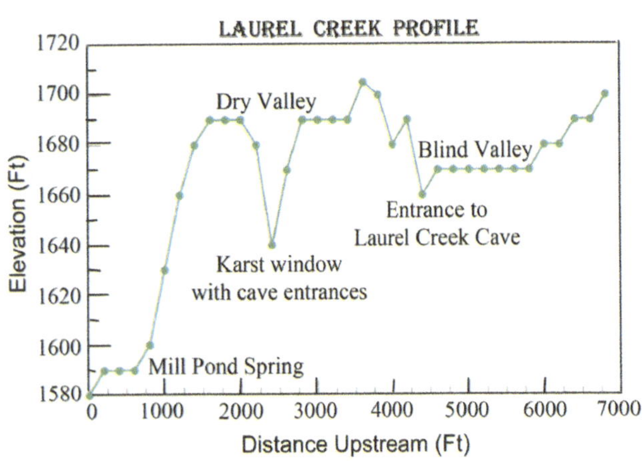

Fig. 17.3 Profile of the Laurel Creek dry valley measured upstream from the confluence with Indian Creek. Data taken from U.S. Geological Survey Greenville 7.5 min quadrangle

17.3 Caves

17.3.1 Laurel Creek Cave

The main entrance to Laurel Creek Cave is along the Laurel Creek Road 1.5 miles northeast of Greenville. It is a prominent opening at creek level 110 ft wide and 30 ft high. In recent years, the property owners have planted evergreens across the entrance (Fig. 17.6). Under low flow conditions, Laurel Creek either sinks in its bed or continues past the cave entrance to drain into a sinkhole in the streambed a short distance downstream.

Laurel Creek Cave, or at least its prominent entrance, was known to the earliest settlers in Monroe County. It was extensively described by Davies (1949). The Davies description was repeated by Hempel (1975). Following a caver entrapment by flood waters in 1982 (see Chap. 5), the owners have denied access to the cave (Fig. 17.7) and little new information has been gathered in recent decades. A map prepared in the 1940s shows the main passages in good detail and was reproduced by Davies (1949). A second survey by the VPI Grotto produced a map with more passage shown but with less passage detail that was published by Hempel (1975). The third and last survey by George Dasher and the Greenbrier Grotto in the late 1970s produced the map included in this chapter which shows more passage, more passage detail, and two more entrances (electronic map M-17.1).

Much of the description that follows is paraphrased from Davies (1949) and Hempel (1975). From the main entrance, a broad passage curves to the south and for more than 300 ft maintains the cross section of the entrance (Fig. 17.8). At this point, the passage is divided by a partition with the west section strewn with breakdown while the east section is an open passage 20 ft wide and 15 ft high. On the northwest edge of the breakdown is a small exit to the top of the hill. At the end of the divided section, the passage resembles the entrance section except that the width is about 50 ft. One thousand feet from the entrance is the theater room, 160 ft long with a ceiling height of 50 ft. From this room, the cave consists of two roughly parallel passages.

The upper level leads off from the southwest edge of the upper part of the room and is 10 ft above the lower passage. Three hundred feet from the theater room, the passages are connected by a broad corridor. After another 350 ft, the upper level is full of breakdown for 225 ft after which it changes direction to the south and crosses the lower level, continuing for 900 ft to where further advance is blocked by

Fig. 17.4 View of the blind valley looking upstream from the entrance to Laurel Creek Cave. Photo by the author

Fig. 17.5 A dry bed of Laurel Creek. The cave entrance is in the wood in the backgound. Photo by the author

silt and flowstone that reaches the ceiling. Throughout its entire length, the upper level is rather uniform in size averaging 20–3 ft wide and 6–15 ft high.

The lower level is less uniform. From the theater room, the passage is 25 ft wide and 15 ft high for 100 ft where a passage develops to the southeast. The passage turns abruptly to the west and for 420 ft is 20 ft wide and 4–7 ft high. Here the lower passage passes beneath the upper level and is connected to it by a pathway along the south wall. The passage is traversable for 250 ft where a lake is encountered

Fig. 17.6 Main entrance to Laurel Creek Cave. Laurel Creek flows past the entrance in the foreground. Photo by the author

Fig. 17.7 Warning sign along Laurel Creek Road at the cave entrance. Note wide marshy area that forms the alluviated floor of the blind valley. Photo by the author

which continues for 100 ft more at which point it narrows to a thin crevice. The lake can be reached from the upper level, 250 ft from the end. The lower level is floored with breakdown.

From the entrance to the theater room, the floor is covered with coarse gravels made up of cobbles from 4 to 10 in. in size in a matrix of finer sand and gravel. Banks of stratified clay reaching to the ceiling occur in this section in niches along the wall. The floor of the remainder of the cave is underlain by deep, fine-grained, olive-drab silt. On the upper level, it is dry but compact. On the lower level it is damp, and in some places the high clay content makes it sticky. Near the end of the upper level a slump pit exposes 15 ft of silt resting on rock, a thickness that is average for this level. On the west side of the theater room, the thickness reaches 40 ft, but in the remainder of the lower level the thickness ranges from a thin veneer to 6 or 7 ft (Fig. 17.9).

The underground route of Laurel Creek lies somewhere below the main cave passages although it is likely that the lakes in the rear of the cave connect with the active stream level. The surface stream does not enter the cave during periods of low flow, but during flood runoff, the stream overflows its banks and

Fig. 17.8 Entrance passage with gravel floor. Photo by George Dasher. Used with permission

Fig. 17.9 Stream channel cut into silt fill on floor of passage. Photo by Ed McCarthy. Used with permission

fills the cave passage, backing up water to form a lake in the blind valley above the entrance (Fig. 17.10).

17.3.2 Rimstone-Crossroad Cave

The Crossroad Entrance is in a low escarpment 25 ft east of the Laurel Creek Road (Fig. 17.11). The entrance is at the top of a breakdown slope that immediately drops down 15 ft to a main stream passage. The Rimstone Entrance is at the base of a 60-foot cliff, 400 ft north of the Water Entrance to Greenville Saltpeter Cave.

The Crossroad Entrance was reported by Davies (1949), but only a short segment of the stream passage was known. Members of the Pittsburgh Grotto explored the cave, pushing low air spaces and connected Crossroad Cave with Rimstone Cave. The map given here was surveyed in 1973 (Fig. 17.12).

Fig. 17.10 Laurel Creek Cave. Main entrance in flood. Photo by Debbie Kyle

Fig. 17.11 Crossroad Cave Entrance to the Rimstone-Crossroad Cave. Photo by the author

Rimstone-Crossroad Cave (spelled Cross Road by Davies 1949 but Crossroad on the map in Hempel 1975) is essentially a half-mile segment of stream passage. The stream rises from a sump at the eastern end, 75 ft upstream from the Crossroads Entrance. This segment of passage is keyhole-shaped with the stream flowing in a deep-cut trench. Downstream, the passage varies in width and height but except for a short segment just downstream from the Crossroads Entrance, contains a stream with depths ranging from 2 to 7 ft. The stream emerges from the Rimstone Entrance.

17.3.3 Greenville Saltpeter Cave

Greenville Saltpeter Cave (also called Head of the Mill Pond Cave) lies beneath a remnant of dissected upland immediately north of Greenville. The cave has four entrances. The southern or Mill Pond Entrance is near the resurgence of Laurel Creek. The three northern entrances are in a karst window 2000 ft to the north. From east to west, these are the Water Entrance, the Hilltop Entrance, and the Saltpeter Entrance. The cave is described by Davies (1949) but

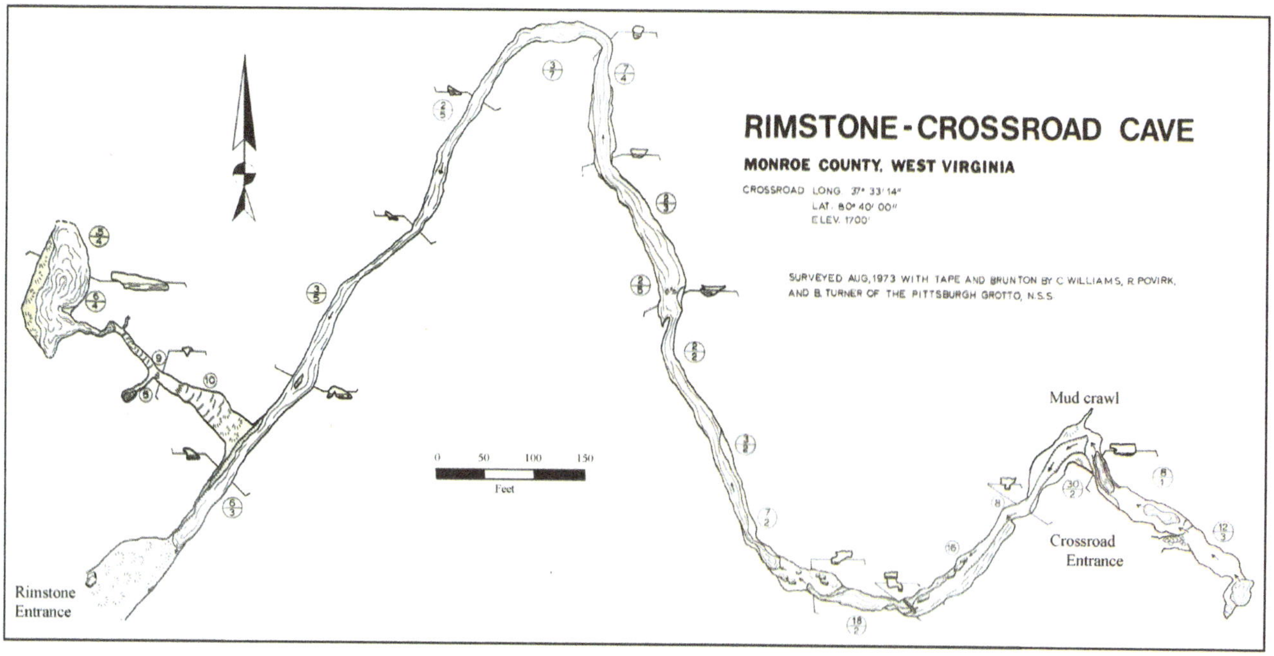

Fig. 17.12 Map of Rimstone-Crossroad Cave. Adapted from Hempel (1975)

without a map. The cave was first mapped by Charleston Grotto in 1948, and this map (Fig. 17.13) was published by Davies (1958). The most recent map was surveyed by Pittsburgh Grotto cavers in 1973 and was included in Hempel's (1975) report.

The spring run from the cave entrance was dammed just upstream from its confluence with Indian Creek to serve a grist mill which still survives as a historic monument. The Mill Pond is backed up to the cave entrance (Fig. 17.14). The Mill Pond Entrance is 5 ft high and 20 ft wide. It opens into a broad, sloping room on the north side of which is a low passage with a floor composed of large Rimstone pools. The main passage trends east for 1200 ft and then north, where it splits into multiple passages in the central portion of the cave.

The Water Entrance, 25 ft high and wide, carries the stream south along the east side of the cave through a passage that leads to a large breakdown room. It carries the main flow of Laurel Creek which rises from a spring on the north side of the karst window, flows across the floor of the window, and into the Water Entrance. The stream follows the passage and flows beneath the breakdown to a smaller breakdown passage for an additional 200 ft. The stream passages ends in a sump at the end of the passage. The remaining link between the sump and the Mill Pond Spring is through a deeper and probably water-filled passage. Passages from the breakdown room connect with a complex tangle of passages to the south which connect with the Mill Pond passage.

The Hilltop Entrance opens onto a talus slope that leads to a large chamber. To the south is a major passage which extends to a three-way junction between the Mill Pond passage and the complex passage from the Water Entrance.

The Saltpeter Entrance opens into an east–west trending passage that was the main locus of Saltpeter mining operations. A side passage to the south connects to a larger east–west trending passage which also contains the remnants of the Saltpeter mining. Cart ruts, burro tracks, mattock marks, and other relics are found in these passages. It appears that the Saltpeter mining was extensive, because artifacts are also found in most of the side passages in this area. The larger Saltpeter passage connects with the large chamber inside the Hilltop entrance.

Overall, the passage pattern of Greenville Saltpeter Cave is that of an anastomotic maze in the sense of Palmer (1975). Passage development is mainly along bedding plane partings so that passages curve and wander and interconnect at random points as would be expected from the initiation routes along bedding planes. Passage cross sections (where not modified by breakdown are irregular but smoothly sculptured. The hydrologic implications are that passages developed slowly along multiple pathways, by slowly moving water under completely phreatic conditions. This is consistent with the clastic sediments found in the cave. These are fine-grained clays and silts (except where the floors are covered with breakdown) as expected from material settled out of slowly moving water.

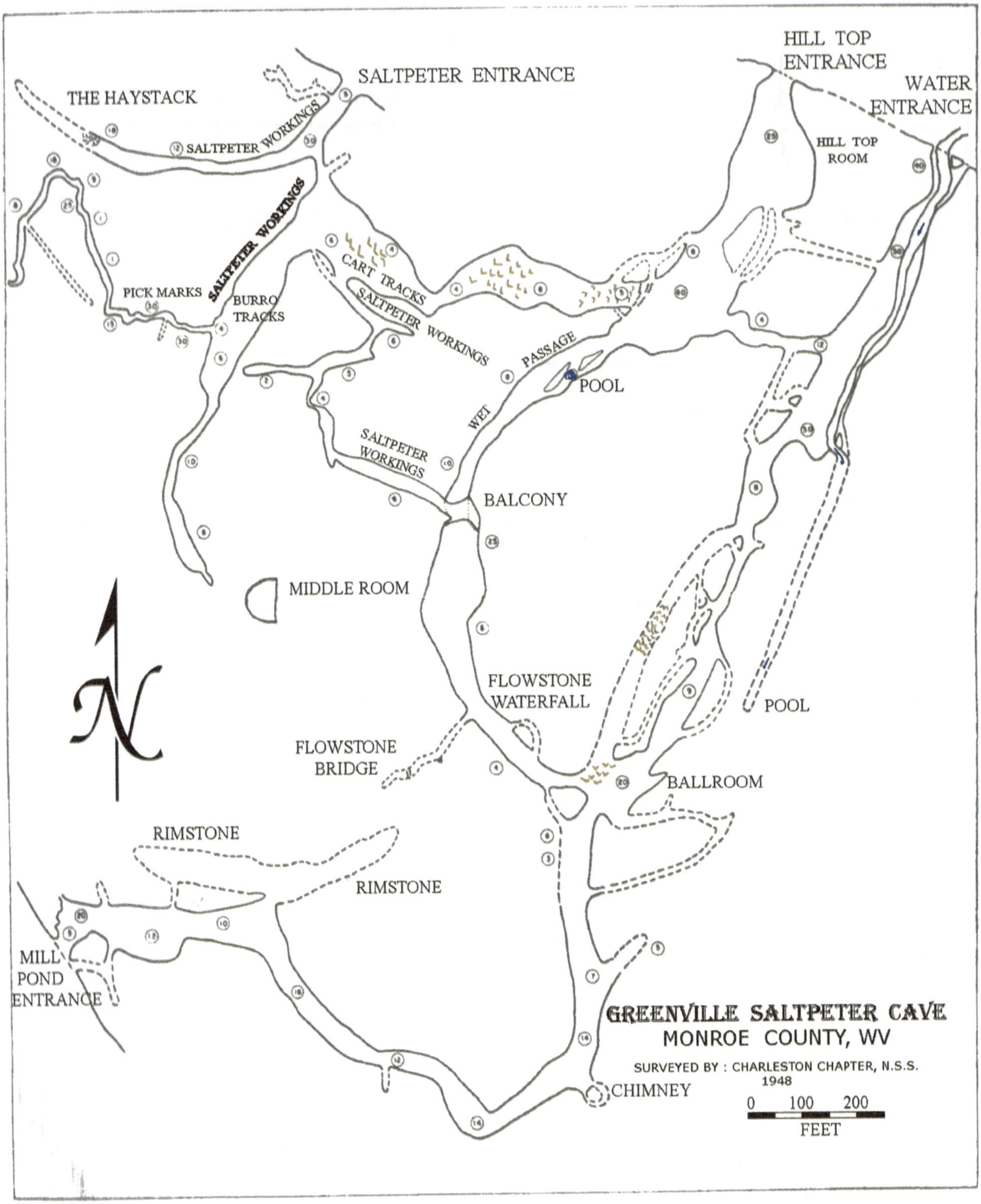

Fig. 17.13 Map of Greenville Saltpeter Cave taken from Davies (1958)

Fig. 17.14 Mill Pond at Greenville. The Mill Pond Entrance to Greenville Saltpeter Cave is in the woods at the head of the pond. Photo by the author

17.4 Water Chemistry

There is a single set of measurements of water chemistry along the Laurel Creek drainage (Groves 1992). Samples were collected on April 11 and 12, 1987, analyzed for the common aqueous species, and the saturation state of the water with respect to calcite was calculated. Groves' data were recalculated and are displayed in Fig. 17.15. The concentrations of Ca^{2+} and Mg^{2+} ions were combined and recalculated as Ca + Mg hardness, a useful measure of concentration of dissolved carbonate. Saturation indices were taken directly from the original paper.

As expected, the hardness is low where the stream is flowing on clastic rocks (BR) and increases rapidly when Laurel Creek flows onto the Greenbrier Limestone and continues to increase downstream. Groves assigned the increase in hardness to the kinetics of limestone dissolution as the water in the stream was exposed to more and more carbonate rock. However, an unknown factor is the amount of water from fractures and bedding plane partings leaking into the master conduit. These waters are derived from the epikarst and from runoff into sinkholes and are expected to have a higher content of dissolved carbonate.

All samples are highly undersaturated with respect to calcite and although the degree of saturation increases downstream. The final sample taken at the spring outlet of Crossroad Cave is only 10% of saturation. This is consistent with other observations that conduits fed by sinking streams usually traverse the conduit without coming into chemical equilibrium with the bedrock.

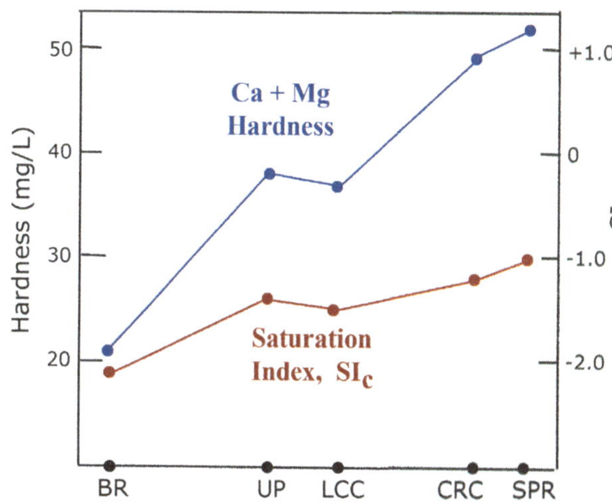

Fig. 17.15 Geochemical profile along Laurel Creek calculated from data of Groves (1992). The distance scale is approximate and based on estimated distances between sampling sites. Code for the sampling sites: *BR* Bridge across Laurel Creek Road at the eastern edge of the basin where the stream is flowing on shales and sandstones. *UP* Upstream from Laurel Creek Cave where the stream is flowing in an alluviated blind valley. *LCC* Entrance to Laurel Creek Cave. *CRF* Stream in Cross Road Cave. *SPR* Spring where Laurel Creek emerges into the karst window

17.5 Summary

The underground drainage of Laurel Creek is essentially a linear system now fragmented into three caves: Laurel Creek Cave, Rimstone-Crossroad Cave, and Greenville Saltpeter Cave. The system is low gradient and was formed before the dissection of the topography by Indian Creek and its tributaries. The surface tributaries of Laurel Creek provide most of the allogenic input; however, infiltration through the karst surface of the closed basin must also drain through the system. Also unknown is the contribution from Indian Draft, a loosing stream that flows south along the west side of the basin.

References

Davies, W.E. 1949. *Caverns of West Virginia*, vol. 19, 353pp. Morgantown, WV: West Virginia Geological Survey.

Davies, W.E. 1958. *Caverns of West Virginia*, vol. 19(A), 330pp. Morgantown, WV: West Virginia Geological Survey.

Groves, C.G. 1992. Geochemical and kinetic evolution of a karst flow system: Laurel Creek, West Virginia. *Ground Water* 30: 186–191.

Hempel, J.C. 1975. *Caves of Monroe County*, vol. 4, 149pp. Barrackville, WV: West Virginia Speleological Survey Bulletin.

Palmer, A.N. 1975. The origin of maze caves. *National Speleological Society Bulletin* 37 (3): 57–76.

Reger, D.B., and P.H. Price. 1926. *Mercer, Monroe, and Summers Counties*, 963. Morgantown, WV: West Virginia Geological Survey.

Terrestrial Fauna in the Greenbrier Karst

18

David C. Culver and Daniel W. Fong

Abstract

The West Virginia cave fauna includes species that leave caves periodically to forage (cave crickets and bats) as well as permanent inhabitants (both species that are limited to caves [troglobionts] and ones that occur elsewhere [troglophiles]). Troglobionts are the best studied, but troglophiles predominate in many caves. Major sources of food for the terrestrial cave communities are transitory organic matter brought in by bats and cave crickets, and riparian deposits along streams. A total of 40 troglobionts are known, 19 of which are endemic to the Greenbrier karst. The highest species richness is in the contact caves and Buckeye Creek drainage in Greenbrier County. Greenbrier County is a hotspot of troglobionts in the USA.

18.1 Introduction

The first summary of the West Virginia cave fauna was that of Reese (1934), who mostly emphasized the non-obligate cave-dwelling fauna. The next statewide study of invertebrates was that of Holsinger et al. (1976). They included both facultative and non-facultative cave dwellers, but with an emphasis of the obligate cave dwellers (terrestrial troglobionts and aquatic stygobionts). The most recent summary of the invertebrate cave fauna is that of Fong et al. (2007), and they included only troglobionts and stygobionts. Garton et al. (1993) summarized all records of vertebrates in West Virginia caves.

The Greenbrier karst is recognized as a center of troglobiotic and stygobiotic species richness (Barr 1967), and Culver et al. (2000) report that Greenbrier County ranks fifth among all US counties in number of troglobiotic species.

D.C. Culver (✉)
Department of Environmental Science, American University, 4400 Massachusetts Avenue NW, Washington, DC 20016, USA
e-mail: dculver@american.edu

D.W. Fong
Department of Biology, American University, 4400 Massachusetts Avenue NW, Washington, DC 20016, USA

W.B. White (ed.), *Caves and Karst of the Greenbrier Valley in West Virginia*,
Cave and Karst Systems of the World, https://doi.org/10.1007/978-3-319-65801-8_18

18.2 The Transitory Terrestrial Fauna–Overview

18.2.1 Transitory Species

Most of the attention of biologists studying the terrestrial cave fauna is directed toward the obligate cave-dwelling species (troglobionts), typically with tiny or no eyes and no pigment, and to the bats that use caves as hibernacula, maternity sites, or day roosts. However, there are many vertebrates and invertebrates that use caves in a variety of ways.

One category of these species is ones that blunder into caves or are carried in by air or water. Individuals in this situation have two fates—either they find their way back out or they become part of the food chain (see the next section on Energy Vectors). Garton et al. (1993) report six species of toads and frogs and two species of snakes from caves in the Greenbrier karst, and none of these species can reproduce in caves or survive for extended periods in caves.

Beyond the cases of accidental species, there are also birds that frequently nest in cave entrances. Nests of the eastern phoebe (*Sayornis phoebe*) are a frequent sight at cave entrances; Garton et al. (1993) list records from three caves in the Greenbrier karst, but it is surely much more common than this. Occasionally turkey vultures (*Cathartes aura*) nest in the twilight zone of caves. Such caves are easily recognized both by their smell, and the fluffy white newborn chicks.

There are also a number of arthropod species that are primarily summer visitors to caves, especially cave entrances. These summer visitors include many species of Diptera, and cavers are often greeted by a number of Diptera (including Tipulidae, Muscidae and others) flying around the cave entrance. Unfortunately, the fauna of cave entrances is little studied, both in the Greenbrier karst and elsewhere. This is unfortunate because it is clearly a highly structured community. A notable exception to the lack of study is a series of papers by Richard Graham in the now defunct journal *Caves and Karst*, on the entrance zone fauna of western caves (e.g., Graham 1968). Diptera and other arthropods that congregate in entrances in summer months probably do so for two primary reasons:

To avoid high temperatures on the surface
To reduce predation

While predation is certainly reduced in cave entrances, it is not eliminated. There are two common predators in cave entrances in the Greenbrier karst—the orb-weaving spider *Meta ovalis* and the cave salamander *Eurycea lucifugus*. *E. lucifugus* is really misnamed because it is more common in twilight zones and entrances than in aphotic parts of the cave (Camp and Jensen 2007). Holsinger et al. (1976) report that there are many sight records of *M. ovalis* "too numerous to mention."

Cave crickets (especially *Euhadenoecus fragilis* and *E. puteanus*) are frequently observed near cave entrances, although *E. fragilis* occurs throughout caves, including the dark zone (Holsinger and Culver 1988). The concentration of cave crickets is especially noticeable during summer months around dusk, just before the cave crickets exit the cave to forage (Fig. 18.1). Presumably, cave crickets utilize caves as day roosts and egg laying sites to minimize predation. Cave crickets are keystone species for the transitory organic matter terrestrial community (see below) because of the fecal material they leave in the cave. In the Mammoth Cave region (Lavoie et al. 2007), there are beetle predators that specialize on cricket eggs but this interaction has not been reported in the Greenbrier karst.

Another species that regularly leaves caves at night to forage is the Allegheny wood rat (*Neotoma magister*). Allegheny wood rats typically build conspicuous nests in the cave with nest material brought in from the outside. In West Virginia, they can nest in deep fissures and talus slopes as well as caves (www.wvdnr.gov/wildlife/RETspecies/asp). The species is declining, especially in the northern part of its range (Maryland and Pennsylvania due to habitat fragmentation, predation of great horned owls, and the raccoon parasite *Baylisascaris procyonis* (Balcome and Yahner

Fig. 18.1 Photograph of cave crickets on the ceiling of a cave. Photo by H.H. Hobbs III, used with permission

1996). The species is uncommon in caves in the Greenbrier karst, reported from only three caves, including Organ Cave (Garton et al. 1993). It is much more common in caves at higher elevations in the northern part of the state.

18.2.2 Bats

Of course, the best known and numerically most important vertebrates in caves of the Greenbrier karst are bats. Seven species of bats use caves in the Greenbrier karst for one or more of the following—hibernacula, autumn/late summer swarming, day roost, and maternity roosts (Table 18.1).

Of these, three are uncommon in the Greenbrier karst—*Corynorhinus townsendii virginianus, Myotis leibii*, and *M. septentrionalis* (Fig. 18.2). The endangered Virginia big-eared bat (*C. t. virginianus*) is much more common in the northern part of the state, where the colder temperatures in caves are more suitable as hibernacula for this species. Populations have been increasing since its listing as an endangered species (Harvey et al. 2011). *Myotis leibii* (eastern small-footed bat) are among the last bats to enter caves in autumn, and sometimes hibernate in cracks in the floor (Harvey et al. 2011). They are likely more common than the small number of records indicates, but population sizes are declining because of white-nose syndrome. White-nose syndrome is a fungal disease that is decimating many populations of eastern cave-dwelling bats (Moore and Kunz 2012). *Myotis septentrionalis* is occasionally found both in summer, using caves as night roosts, and in winter, where solitary individuals hibernate (Harvey et al. 2011). It has recently been listed as threatened by the US Fish and Wildlife Service as a result of the depredations of white-nose syndrome (www.fws.gov/endangered/).

Table 18.1 List of bats commonly occupying caves in the Greenbrier Karst and the ways they utilize caves

Scientific name	Common name	Hibernacula	Autumn/late summer swarming	Day roosts	Maternity roosts	WNS	Federal status
Perimyotis subflavus	Tricolored bat (eastern pipestrelle)	Y		Y		Y	
Corynorhinus townsendii virginianus	Virginia big-eared bat	Y		Y	Y		Endangered
Myotis leibii	Eastern small-footed bat	Y		Y		Y	
Myotis lucifugus	Little brown bat	Y				Y	
Myotis septentrionalis	Northern long-eared bat	Y		Y		Y	Threatened
Myotis sodalis	Indiana bat	Y	Y				Endangered

From Harvey et al. (2011). WNS refers to species for which white-nose syndrome has been detected (www.whitenosesyndrome.org)

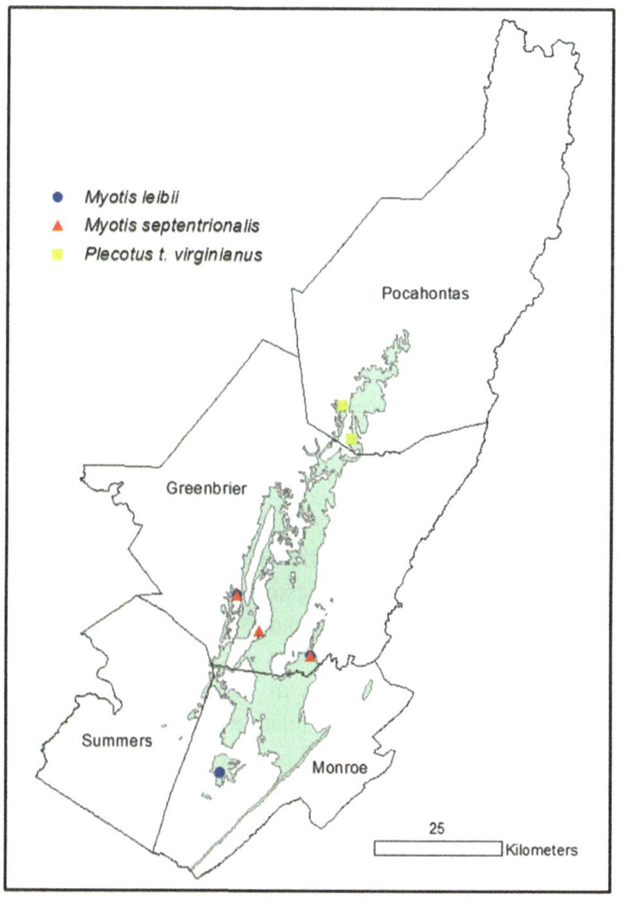

Fig. 18.2 Map of distribution of *P. townsendi virginianus*, *M. leibii*, and *M. septentrionalis*. Area in *gray* is the extent of the Greenbrier limestone. Base map from Weary and Doctor (2014)

Myotis lucifugus (little brown bat) is frequently found in caves in the Greenbrier karst during the winter, especially in the southern part of the area (Fig. 18.3). Little brown bats have severely declined in numbers due to white-nose syndrome, and it is likely it will be listed as threatened or endangered in the next several years. *Myotis sodalis* is also frequently found in caves in the Greenbrier karst, not with a more northerly distribution (Fig. 18.3). It is on the endangered species list and is especially vulnerable because 85% of the individuals hibernate in just 15 locations. Major hibernacula in West Virginia are north of the Greenbrier karst. They are occasionally found in caves in the summer as well.

The most frequently observed bat in caves in the Greenbrier karst is the tricolored bat, *Perimyotis subflavus*. Previously known as the eastern pipistrelle, *P. subflavus* is certainly much more common than the records indicate (Fig. 18.4). It is never present in large numbers in either summer or winter, and the combination of its small size and nearly ubiquitous presence results in an incomplete inventory of the caves it inhabits. It is susceptible to white-nose syndrome, but little is known about the overall impact of the disease on this species.

18.3 The Permanent Terrestrial Fauna–Overview

The permanent terrestrial fauna includes only species that complete their entire life cycle in the cave. Some of these species can also complete their life cycle in surface habitats, especially leaf litter, and others can only complete their life cycle in caves and other subterranean habitats. The first group is called troglophiles (sometimes eutroglophiles), and the second group is called troglobionts or troglobites (Sket 2008). Much more attention is paid to troglobionts compared to troglophiles because of the restricted ranges of troglobionts and their highly modified morphology, typically including eye and pigment loss, and appendage elongation (Culver and Pipan 2009). Indeed, most recent faunal inventories in the Greenbrier karst and elsewhere include only troglobionts (e.g., Fong et al. 2007).

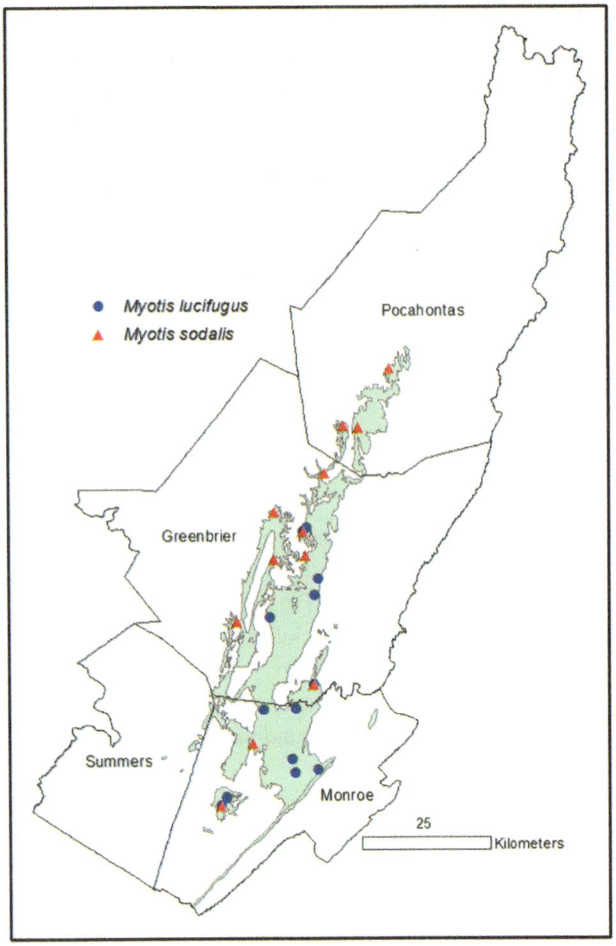

Fig. 18.3 Map of distribution of *M. sodalis* and *M. lucifugus*. Area in *gray* is the extent of the Greenbrier limestone. Base map from Weary and Doctor (2014)

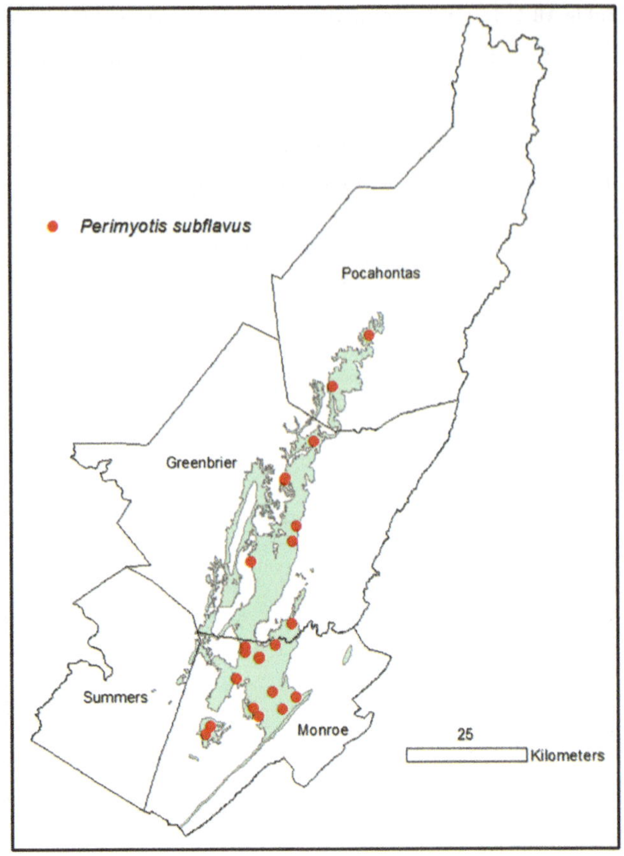

Fig. 18.4 Map of distribution of *P. subflavus*. Area in *gray* is the extent of the Greenbrier limestone. Base map from Weary and Doctor (2014)

An exception to the neglect of troglophiles is Schneider's (2009) study of the invertebrate fauna of 12 pits in a 2 km^2 area near Buckeye Creek Cave in Greenbrier County. She extirpated both the resources and fauna in all 12, and then added resources in the form of leaves to six and rat carcasses to six. The pits were small enough that (3–19 m deep and 5–26 m long) she was able to do a complete census every month for 23 months. Her study was and is unprecedented in its scale and repetition. Of special interest in the present context is her summary of the abundance and temporal occurrence of nearly 100 arthropod species. Since permanent residents should be present in all samples, those species occurring in all or nearly all months are most likely to be permanent residents. Species found in more than 20 of the 23 sampling months are listed in Table 18.2.

Of the ten species, only two are troglobionts—a springtail and a mite. The troglobiotic springtail (*Sinella hoffmani*) was the most common species, and a troglophilic millipede (*Pseudotremia hobbsi*) was the second most common, and certainly the most visible terrestrial resident of many caves. It is a measure of our lack of understanding of the dynamics of cave communities that there is no apparent concentration of troglophilic *Pseudotremia* in shallower regions of the cave closest to the entrance, relative to the much less common troglobioitc white millipede *Pseudotremia fulgida*. Except for the cave cricket *Euhadenoecus fragilis,* which leaves the cave at night in the summer to forage, the other species are highly likely to be permanent residents. While the number of troglophilic species may be high in part because the caves (pits) are small and include some entrance zone species, the pattern of a mix of troglophilic and troglobiotic species is the rule, rather than the exception, even in deep caves (Pipan and Culver 2012).

Collecting of cave fauna for taxonomic studies in the Greenbrier Karst had its start in the late 1920s, especially with the collections of J.M. Valentine, who subsequently published a number of descriptions of beetle species in the 1930s. His most famous discovery was the unique genus

Table 18.2 Common arthropods found in Schneider's study pits

Order	Species	Caves (of 12)	Months (of 23)	Individuals
Chordeumatida	*Pseudotremia hobbsi*	12	23	1486
Collembola	*Sinella* sp.	12	23	935
Orthoptera	*Euhadenoecus fragilis*	12	23	649
Collembola	*Tomocerus* sp.	12	22	129
Collembola	*Sinella hoffmani?**	12	22	2525
Coleoptera	Staphylinidae	12	22	360
Julida	*Ophyiulus pilosus*	11	22	165
Isopoda	Oniscoidea	6	22	137
Acari	*Rhagidia varia**	12	22	159
Diptera	Phoridae	12	21	478

All resources and organisms were removed, and colonization over a 23-month period was observed. Six pits were given leaf packs and six were given rat carcasses. All species found in more than 20 months are listed. Species marked with an asterisk are troglobionts. From Schneider (2009) and Schneider et al. (2011)
* Troglobionts

Horologion, described from Arbuckle Cave in Greenbrier County in 1932, and has never been seen since. By 1960, the cave fauna of the Greenbrier karst was becoming relatively well known and was featured prominently in a number of papers in the symposium on "Speciation and Raciation in North American Cavernicoles", organized by T.C. Barr. At present, there are 40 species of troglobionts known from Greenbrier karst caves (Table 18.3). There certainly are a number of undescribed and even undiscovered species. The undescribed species are discussed by Fong et al. (2007). We do not consider them further because they are undescribed and not all undescribed species actually turn out to be new species.

In terms of species, the fauna is dominated by the Collembola and Coleoptera, as well as the Arachnida taken as a whole (Fig. 18.5). Compared to hotspot terrestrial faunas of the French Pyrenees and the Dinaric karst, a number of beetle families are absent from the Greenbrier karst (Deharveng et al. 2012). The most speciose genera are the beetle genus *Pseudanophthalmus*, with 10 species and subspecies, and the Collembola genus *Pygmarrhopalites*, with five species. However, the geographic distribution of species in these two groups is quite different. Eight of 10 *Pseudanophthalmus* are endemic to the Greenbrier Karst, and none of the *Pygmarrhopalites*. Because of this, it is possible that the Greenbrier Karst is a center of radiation for cave beetles. Three other genera have three species—the pseudoscorpion genus *Kleptochthonius*, the Collembola genus *Pseudosinella*, and the millipede genus *Pseudotremia*.

Overall, 19 of the 40 species are endemic to the Greenbrier Karst (Appendix). Groups with high endemism are the pseudoscorpions (100%), *Pseudosinella* (67%), *Pseudotremia* (100%), and beetles (80%).

Although arachnids, beetles, and springtails have roughly the same number of species, their abundances are quite different. By far the most common (and smallest in size at around 1 mm) are Collembola. Collembola in the genera *Pseudosinella* and *Sinella* are nearly ubiquitous in the Greenbrier Karst and are often especially common near standing pools of water. The predaceous spiders and beetles

Table 18.3 Summary of the described troglobiotic species known from caves in the Greenbrier Karst

Class	Order	Family	Genera	Species	Records
Arachnida	Araneae	1	4	4	68
Arachnida	Opiliones	1	1	1	1
Arachnida	Pseudoscorpionida	2	2	4	24
Arachnida	Trombidiformes	1	2	3	30
Arachnida	TOTAL	5	9	12	123
Collembola	Collembola	3	4	10	139
Diplura	Diplura	1	1	1	26
Diplopoda	Chordeumatida	2	2	5	68
Insecta	Coleoptera	1	2	8	131
	Diptera	1	1	1	4
TOTAL		13	19	38	492

More detail is provided in Appendix

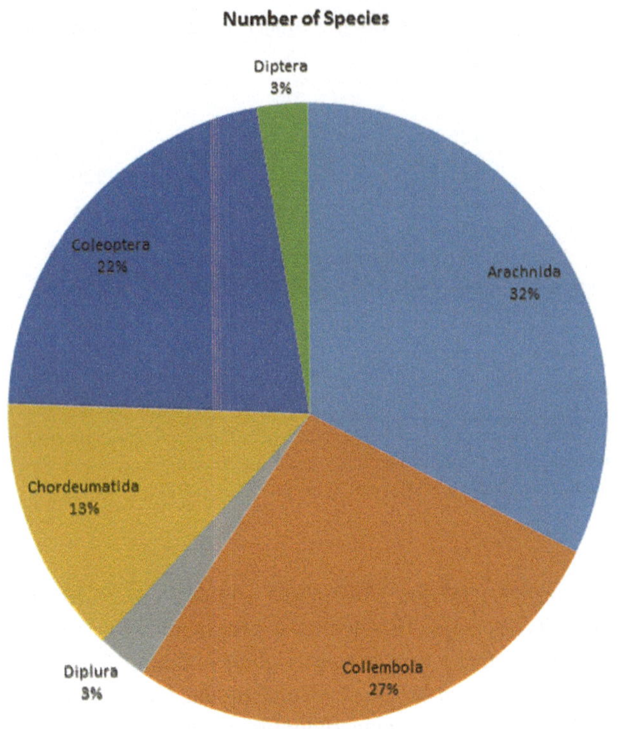

Fig. 18.5 Pie diagram of the number of species of troglobionts from different taxonomic groups

are much less common, although the relatively large (5 mm) *Pseudanophthalmus* beetles are relatively easy to spot. The most obvious troglobiotic terrestrial invertebrates in caves in the Greenbrier Karst are invertebrates, especially *Pseudotremia fulgida*, typically 25–35 mm long.

18.4 Energy Vectors

For any species to survive and reproduce in a cave, there must be a sufficient supply of organic carbon and nutrients. This is particularly critical in the aphotic environment of caves because there is no photosynthesis driving the production of organic carbon. Chemoautotrophy may supply organic carbon, but it has not been reported from any caves in the Greenbrier karst. Chemoautotrophy is typically found in redox environments, especially sulfidic caves (Kumaresan et al. 2015).

Fong (2011) conveniently divides the paths of allochthonous energy input into (1) transport by water; (2) transport by movement of animals; (3) passive transport by gravity or wind; and (4) tree roots. All the large cave systems in the Greenbrier Karst have active streams, and these streams carry organic matter and nutrients, utilized not only by the aquatic cave communities (see Chap. 19), but also for the terrestrial riparian community, that occurs along the banks of subterranean streams. Water percolating through epikarst also provides organic carbon, as demonstrated by Simon et al. (2007), not only for the aquatic epikarst community but for the very little known terrestrial epikarst community. The terrestrial epikarst community is largely known only as a by-product of sampling for the aquatic epikarst community. See Pipan and Culver (2005) for an example from Organ Cave. Bats and crickets are the primary transporters of organic carbon, especially their guano. Gravity and wind also bring organic carbon and nutrients into caves. In Schneider's (2009) study of shallow pit communities, this was the primary source of energy. It included not only leaf fall but also occasional dead animals that fell into the pits. Occasionally, roots may also be a source of nutrients, and roots are sometimes observed on cave ceilings, as in Sunnyside pit in Greenbrier County. Its importance as a food source is unknown.

Fong et al. (2007) divide terrestrial species into communities based on their energy source (Table 18.4). Terrestrial riparian communities, deriving organic carbon and nutrients from streams, are (1) transitory organic matter species, deriving organic carbon and nutrients from transport by animals and transport by wind and gravity and (2) terrestrial epikarst communities, deriving organic carbon and nutrients from water percolating through the soil.

Terrestrial cave communities may be carbon, nitrogen, or phosphorus limited. Simon and Benfield (2002) suggest cave stream communities are carbon-limited, but there are no data on terrestrial cave communities that bear on this. Barton (2015) points out that the widespread occurrence of nitrogen-fixing bacteria in caves suggests that nitrogen may be limiting. Schneider et al. (2010) found evidence for phosphorus limitation because both the average molar carbon–phosphorus ratio and the percent phosphorus were low in cave millipedes (*Pseudotremia fulgida*). Which factor (carbon, phosphorus, or nitrogen) is limiting is unclear, but all are in short supply.

18.5 Geographic Patterns of Species Richness

Records of troglobionts are known from 132 caves in the Greenbrier Karst. According to the records in Fong et al. (2007), an additional 10 caves were investigated but no troglobionts were found. All in all, this is one of the best sampled cave regions in the USA, but less than 15% of the known caves have been sampled. Those sampled include most of the large caves, and the geographic coverage is good (Fig. 18.6). Two of the caves in Fig. 18.6 do not fall on the mapped extent of the Greenbrier limestone (Indian Draft Cave in Monroe County and Sinks of the Run in Greenbrier County) and may actually be formed in the Reynolds Limestone of the Bluefield Formation (D.H. Doctor, pers. comm.) or the mapping of the Greenbrier series is inaccurate.

Table 18.4 Troglobionts associated with different energy fluxes, modified from Fong et al. (2007)

Terrestrial riparian	Transitory organic matter	Terrestrial epikarst
ACARI	ACARI	COLEOPTERA
Rhagidia varia	*Poecilophysis extraneostella*	*Horologion speokoites*
ARANEAE	*P. weyeri*	
Anthrobia coylei	ARANEAE	
Bathyphantes weyeri	*Anthrobia coylei*	
Phanetta subterranea	*Bathyphantes weyeri*	
Porrhomma cavernicola	*Phanetta subterranea*	
PSEUDOSCORPIONIDA	*Porrhomma cavernicola*	
Chitrella regina	PSEUDOSCORPIONIDA	
Kleoptochthonius hetricki	*Kleptochthonius henroti*	
COLLEMBOLA	*K. hetricki*	
Pygmarrhopalites clarus	*K. proserpinae*	
P. carolynae	COLLEMBOLA	
Pseudosinella gisini gisini	*Onychiurus janus*	
Sinella hoffmani	*Pseudosinella gisini gisini*	
DIPLURA	*Sinella hoffmani*	
Litocampa fieldingae	DIPLOPODA	
DIPLOPODA	*Zygonopus packardi*	
Pseudotremia fulgida	*Z. weyerensis*	
COLEOPTERA	COLEOPTERA	
Pseudanophthalmus fuscus fuscus	*P. grandis*	
P. fuscus constrictus	*P. grandis grandis*	
P. grandis	*P. grandis elevatus*	
P. grandis grandis	*P. hypertrichosis*	
P. grandis elevatus	*P. orthosculcatus*	
P. higginbothami	DIPTERA	
P. hypertrichosis	*Spelobia tenebrarum*	
P. lallemanti		
P. orthosulcatus		

See Appendix for species details. Note that some species utilize more than one type of energy flux

Culver et al. (2000) found that, of all the counties in the USA, Greenbrier County ranked fifth in the number of troglobionts. In a more detailed study of the southeastern USA, Christman et al. (2005) found the Greenbrier Karst to be a hotspot of terrestrial cave biodiversity. However, it is overshadowed by the species richness of northeast Alabama (Culver et al. 2006). Within the state of West Virginia, the Greenbrier Karst is the hotspot of terrestrial cave biodiversity, with Greenbrier, Pocahontas, and Monroe Counties ranking first, second, and third in numbers of troglobionts among West Virginia counties. None of these studies found the Greenbrier Karst to be a center of cave endemism.

Although not a center of endemism, there are number of species in the Greenbrier Karst known from a single cave (Fig. 18.7). They include one opilionid, one pseudoscorpion, two beetles, two Collembola, and two millipedes. In many cases, additional collecting may yield additional records, but it is likely all these species are restricted to a very small area. The distribution of the single-cave endemics does not correspond to any obvious geological barrier, nor are they clustered in any one area. None of the single-cave endemics occur on the isolated patch of Greenbrier limestone in the southwest part of the study area (near Greenville, West Virginia), and this patch of limestone has itself only one endemic species—the beetle *Pseudanophthalmus grandis* (Barr 2004).

The finest scale to assess species richness is that of individual caves. However, because most of the species richness is between caves, i.e., β-diversity predominates (Malard et al. 2009), it does not reflect diversity in an area, and because caves are very unevenly sampled, it provides an incomplete picture of the geography of species richness.

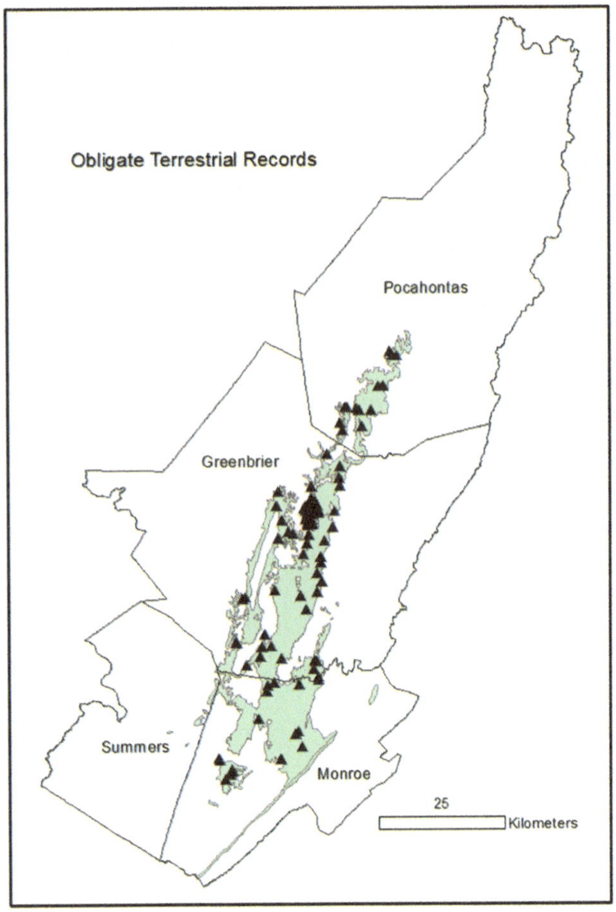

Fig. 18.6 Map of locations of collections of troglobionts. There are an additional 10 sites sampled that did not yield troglobionts that are not shown. The dense cluster of points is the result of Schneider and Culver's (2004) study of 65 caves in a 25 km^2 area. Area in *gray* is the extent of the Greenbrier limestone. Base map from Weary and Doctor (2014)

Caves with the highest number of troglobionts are shown in Fig. 18.8 and Table 18.5.

The species-rich caves are all in Greenbrier County and in the core area of the Greenbrier karst. Four of the nine caves are in the Buckeye Creek drainage (Buckeye Creek Cave, Fuell's Fruit Cave, Nellies Cave, and Rapps Cave). The Buckeye Creek basin is less karstified than other basins in the sense that there is surface flow, but it is dense in caves (Chap. 9, Jones 1973). Four of the remaining caves are large, so-called contact caves (Benedict Cave, McClung Cave, and Organ Cave, Chaps. 11 and 13). Higginbotham Cave is a large multi-km cave as well. Arbuckle is the outlier on the list—it is a small shallow cave.

The map of regional species richness at a scale of 10 km × 10 km is shown in Fig. 18.9. The pattern is similar to that of individual cave hotspots (Fig. 18.8). Species richness is highest in central Greenbrier County with a secondary center of species richness in southeast Greenbrier County, which is the Organ Cave Plateau.

18.6 Protection of the Terrestrial Cave Fauna

There are several aspects to the protection of cave fauna, especially (1) the nature of the threats and their remediation, and (2) prioritizing sites for protection. Threats can range from:

Destruction of the cave, e.g., quarrying.
Direct destruction of fauna, e.g., killing bats in hibernacula and maternity colonies.
Changes in surface land use that result in changes in the cave, e.g., habitat loss of energy vectors and subsequent reduction in spatial subsidies.
Groundwater contamination or reduction, e.g., cave stream pollution with accompanying pollution of riparian habitats.

In the Greenbrier Karst, the major land use change at present is the spread of home sites in the countryside with the accompanying changes to vegetation (reducing energy vectors), an increase in impervious surfaces (changing hydrology), and groundwater contamination as the result of the application of lawn chemicals. The area around Lewisburg is one of the fastest growing areas in the state, aside from the Washington suburbs in the eastern panhandle.

Countering this trend are protection efforts, especially those being undertaken by the West Virginia Cave Conservancy and The Nature Conservancy. While primarily dedicated to maintaining access for recreational cavers, the West Virginia Cave Conservancy protects entrances and the area surrounding them, maintaining them in a natural state. Their holdings include (wvcc.net) entrances and adjoining area for:

Benedicts Cave, Greenbrier Co.
Culverson Creek Cave, Greenbrier Co.
Haynes Cave, Monroe Co.
Lobelia Saltpetre Cave, Pocahontas Co. (managed for owner)
Maxwelton Sink Cave, Greenbrier Co.
McClung Cave, Greenbrier Co.
Rapps Cave, Greenbrier Co. (managed for owner)
Savannah Cave, Greenbrier Co. (managed for owner)

The Nature Conservancy owns three cave entrances and their immediate area:

General Davis Cave, Greenbrier Co.
Piercys Cave, Greenbrier Co.
Piercys Mill Cave, Greenbrier Co.

Overall, 22 of the 38 known troglobionts from the Greenbrier karst in found in one or more of the protected

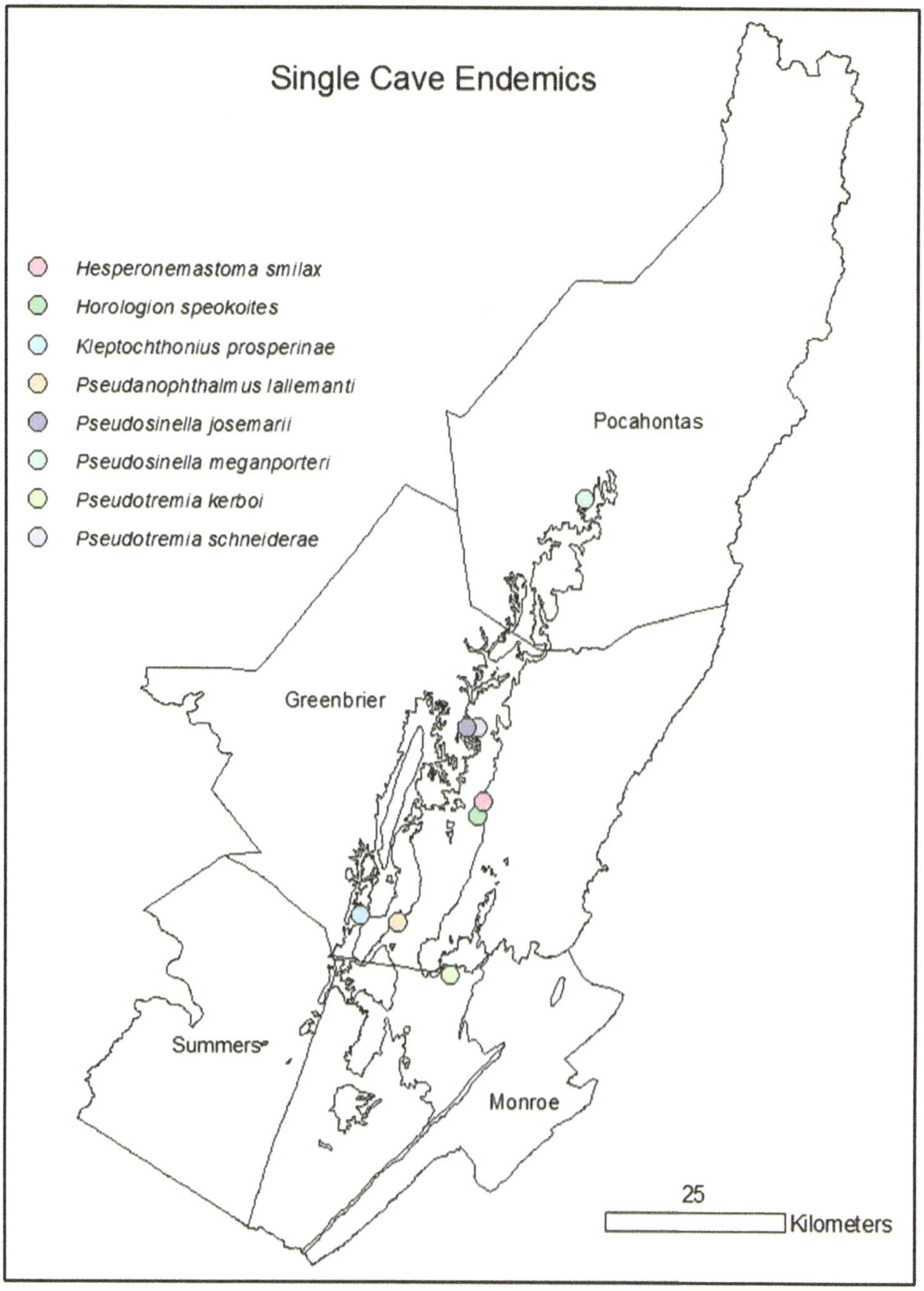

Fig. 18.7 Distribution of single-cave endemics in the Greenbrier karst. Area in *gray* is the extent of the Greenbrier limestone. Base map from Weary and Doctor (2014)

caves, a classic case of a glass half empty and half full. Important unprotected caves include (see Table 18.5)

Higginbotham Cave, Greenbrier Co.
Organ Cave, Greenbrier Co.
Arbuckle Cave, Greenbrier Co.
Nellies Cave, Greenbrier Co.
Buckeye Creek Cave, Greenbrier Co.
Fuell's Fruit Cave, Greenbrier Co.

In terms of bats, one additional way of setting priorities among caves for protection of the obligate terrestrial cave fauna is to give caves that are type localities priority. Both because it is a relative hotspot of subterranean terrestrial

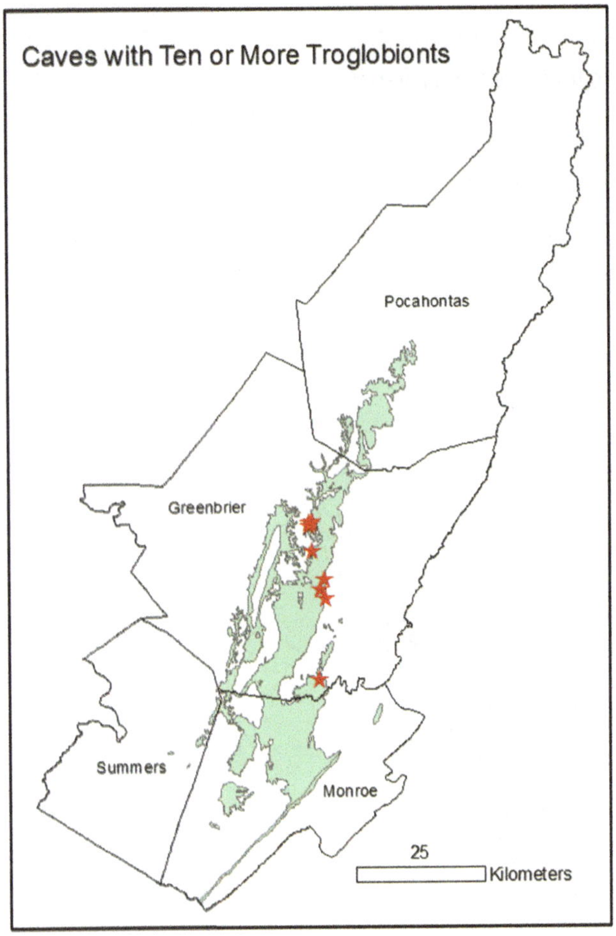

Fig. 18.8 Caves with ten or more troglobionts (see Table 18.5). Area in *gray* is the extent of the Greenbrier limestone. Base map from Weary and Doctor (2014)

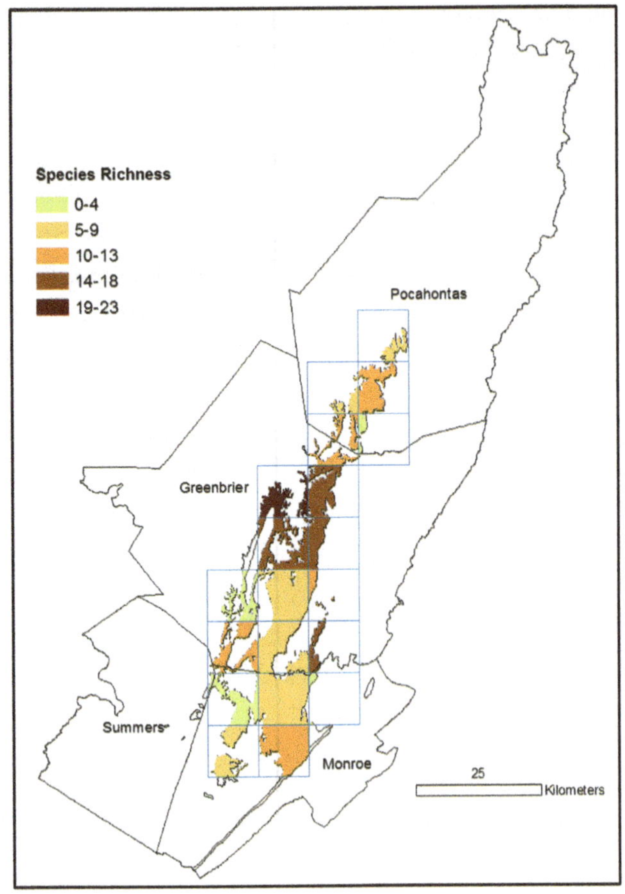

Fig. 18.9 Map of species richness using a grid size of 10 × 10 km. Smaller grid sizes result in gaps of coverage

Table 18.5 Caves with ten or more troglobionts

Cave	County	Number of troglobionts
Higginbotham's Cave	Greenbrier	15
Organ Cave	Greenbrier	13
Arbuckle Cave	Greenbrier	12
McClung Cave	Greenbrier	12
Nellies Cave	Greenbrier	12
Rapps Cave	Greenbrier	11
Benedicts Cave	Greenbrier	10
Buckeye Creek Cave	Greenbrier	10
Fuell's Fruit Cave	Greenbrier	10

biodiversity and there is a long history of biological collection, there are a large number of type locality caves—thirteen (Fig. 18.10).

Finally, the caves harboring bats must be an important part of any plan for the protection of biological resources. This is especially true because three species are on the US Threatened and Endangered Species list (Table 18.1), and because the fungal disease, white-nose syndrome (WNS), poses a serious threat to the continued survival of many bat species. The endangered Virginia big-eared bat is only known from two caves in the Greenbrier karst—Droop Mountain Cave and Martens Cave No. 2, both in Pocahontas County. The threatened northern long-eared bat is known from three caves—Organ Cave, Piercys Cave, and Sinks of the Run Cave. The endangered Indiana bat is known from 13 caves, but none of these are hibernacula. The TNC owned caves, General Davis Cave and Piercys Cave in Greenbrier County, are among these caves.

Priorities for the aquatic fauna are discussed in Chap. 19, as well as some overall comments about priority setting.

Appendix

See Table 18.6.

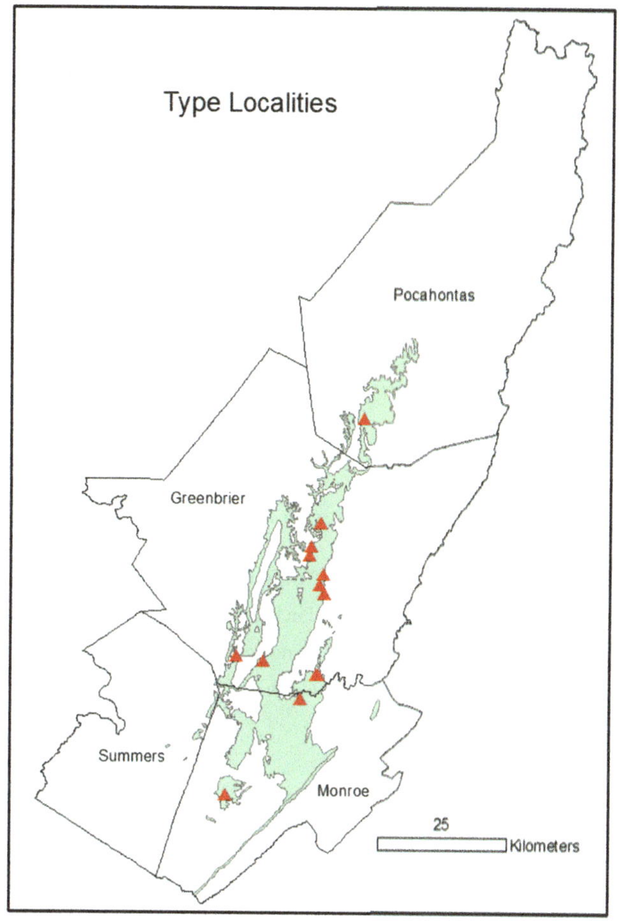

Fig. 18.10 Map of type locality caves of troglobionts

Table 18.6 Known troglobiotic species from the Greenbrier Karst

Class	Order	Family	Species	Taxonomic authority	Area Endemic?
Arachnida	Araneae	Linyphiidae	*Anthrobia coylei*	Miller 2005	
			Bathyphantes weyeri	Emerton 1875	
			Phanetta subterranea	Emerton 1875	
			Porhomma cavernicola	Keyserling 1886	
	Opiliones	Sabaconidae	*Hesperonemastoma smilax*	Shear 2010	Yes
	Pseudoscorpion	Chthoniidae	*Kleptochthonius henroti*	Vachon 1952	Yes
			Kleptochthonius hetricki	Muchmore 1974	Yes
			Kleptochthonius prosperinae	Muchmore 1965	Yes
		Syarinidae	*Chitrella regina*	Malcolm & Chamberlin 1960	Yes
	Trombidiformes	Rhagidiidae	*Poecilophysis extraneostella*	Zacharda 1985	
			Poecilophysis weyerensis	Packard 1888	
			Rhagidia varia	Zacharda 1985	

(continued)

Table 18.6 (continued)

Class	Order	Family	Species	Taxonomic authority	Area Endemic?
Collembola	Collembola	Arrhopalitidae	*Pygmarrhopalites carolynae*	Christiansen & Bellinger 1996	
			Pygmarrhopalites clarus	Christiansen 1966	
			Pygmarrhopalites commorus	Christiansen & Bellinger 1996	
			Pygmarrhopalites pavo	Christiansen & Bellinger 1996	
			Pygmarrhopalites sacer	Christiansen & Bellinger 1996	
		Entomobryidae	*Pseudosinella gisini gisini*	Christiansen 1960	
			Pseudosinella josemarii	Soto-Adames 2010	Yes
			Pseudosinella meganporteri	Soto-Adames 2010	Yes
			Sinella hoffmani	Wray 1952	
		Onychiuridae	*Onychiurus janus*	Christiansen & Bellinger 1980	
Diplopoda	Chordeumatida	Cleidogonidae	*Pseudotremia fulgida*	Loomis 1943	Yes
			Pseudotremia kerboi	Shear 2008	Yes
			Pseudotremia schneiderae	Shear 2008	Yes
	Chordeumatida	Trichopetalidae	*Zygonopus packardi*	Causey 1960	
		Trichopetalidae	*Zygonopus weyeriensis*	Causey 1960	
Diplura	Diplura	Campodeidae	*Litocampa fieldingae*	Conde 1949	
Insecta	Coleoptera	Carabidae	*Horologion speokoites*	Valentine 1932	Yes
			Pseudanophthalmus fuscus constrictus	Valentine 1931	Yes
			Pseudanophthalmus fuscus fuscus	Valentine 1931	
			Pseudanophthalmus grandis	Valentine 1931	Yes
			Pseudanophthalmus grandis elevatus	Valentine 1932	Yes
			Pseudanophthalmus grandis grandis	Valentine 1931	Yes
			Pseudanophthalmus henroti	Jeannel 1949	Yes
			Pseudanophthalmus higginbothami	Valentine 1931	Yes
			Pseudanophthalmus hypertrichosis	Valentine 1932	
			Pseudanophthalmus lallemanti	Jeannel 1949	Yes
			Pseudanophthalmus orthosulcatus	Valentine 1932	Yes
	Diptera	Sphaeroceridae	*Spelobia tenebrarum*	Aldrich 1897	

Data from various sources, including Fong et al. (2007)

References

Balcom, B.J., and R.H. Yahner. 1996. Microhabitat and landscape characteristics associated with the threatened Allegheny woodrat. *Conservation Biology* 10: 515–525.

Barr, T.C. 1967. Observations on the ecology of caves. *American Naturalist* 101: 475–491.

Barr, T.C. 2004. A classification and checklist of the genus *Pseudanophthalmus* Jeannel (Coleoptera: Carabidae: Trechinae), vol. 11, 52pp. *Virginia Museum of Natural History Special Publication*.

Barton, H.A. 2015. Starving artists: bacterial oligotrophic heterotrophy in caves. In *Microbial life of cave systems*, ed. A.S. Engel, 79–104. Berlin, Walter de Gruyter GmbH.

Camp, C.D., and J.B. Jensen. 2007. Use of twilight zones of caves by plethodontid salamanders. *Copeia* 2007: 594–604.

Christman, M.C., D.C. Culver, M. Madden, and D. White. 2005. Patterns of endemism of the eastern North American cave fauna. *Journal of Biogeography* 32: 1441–1452.

Culver, D.C., and T. Pipan. 2009. *The biology of caves and other subterranean habitats*. Oxford, UK: Oxford University Press.

Culver, D.C., L.L. Master, M.C. Christman, and H.H. Hobbs III. 2000. Obligate cave fauna of the 48 contiguous United States. *Conservation Biology* 14: 386–401.

Culver, D.C., L. Deharveng, A. Bedos, J.J. Lewis, M. Madden, J.R. Reddell, B. Sket, P. Trontelj, and D. White. 2006. The mid-latitude biodiversity ridge in terrestrial cave fauna. *Ecography* 29: 120–128.

Deharveng, L., J. Gibert, and D.C. Culver. 2012. Diversity patterns in Europe. In *Encyclopedia of caves*, ed. W.B. White, and D.C. Culver, 2nd ed., 219–238. Waltham, MA: Academic/Elsevier Press.

Fong, D.W. 2011. Management of subterranean fauna in Karst. In *Karst management*, ed. P.E. Van Beynen, 201–224. Dordrecht, Germany: Springer.

Fong, D.W., D.C. Culver, H.H. Hobbs III, and T. Pipan. 2007. In *The invertebrate cave fauna of West Virginia*, 2nd ed., vol. 16. Barrackville, WV: West Virginia Speleological Survey, Bulletin.

Garton, E.R., F. Grady, and S.D. Carey. 1993. *The vertebrate fauna of West Virginia caves*, vol. 11. Barrackville, WV: West Virginia Speleological Survey, Bulletin.

Graham, R.E. 1968. The twilight moth: *Triphosa haesitata* (Lepidoptera: Geometridae), from California and Nevada Caves. *Caves and Karst* 10: 41–48.

Harvey, M.J., J.S. Altenbach, and T.L. Best. 2011. *Bats of the United States and Canada*. Baltimore, MD: The Johns Hopkins University Press.

Holsinger, J.R., and D.C. Culver. 1988. The invertebrate cave fauna of Virginia and a part of East Tennessee: Zoogeography and ecology. *Brimleyana* 14: 1–162.

Holsinger, J.R., R.A. Baroody, and D.C. Culver. 1976. *The invertebrate cave fauna of West Virginia*, vol. 7. Barrackville, WV: West Virginia Speleological Survey, Bulletin.

Jones, W.K. 1973. *Hydrology of limestone karst in Greenbrier County, West Virginia*, vol. 36. Morgantown, WV: West Virginia Geologic and Economic Survey Bulletin.

Kumaresan, D., A.M. Hillebrand-Vioculescu, D. Wischer, J. Stephenson, and J.C. Murrell. 2015. Microbial life in unusual cave ecosystems sustained by chemosynthetic primary production. In *Microbial life of cave systems*, ed. A.S. Engel, 215–229. Berlin: Walter de Gruyter GmbH.

Lavoie, K.H., K.L. Helf, and T.L. Poulson. 2007. The biology and ecology of North American cave crickets. *Journal of Cave and Karst Studies* 69: 114–134.

Malard F., C. Boutin, A.I. Camacho, D. Ferreira, G. Michel, B. Sket, and F. Stoch. 2009. Diversity patterns of stygobiotic crustaceans across multiple spatial scales in Europe. Freshwater Biology 54(4): 756–776.

Moore, M.S., and T.H. Kunz. 2012. White nose syndrome; a fungal disease of North American hibernating bats. In *Encyclopedia of caves*, ed. W.B. White, and D.C. Culver, 2nd ed., 904–910. Waltham, MA: Academic/Elsevier Press.

Pipan, T., and D.C. Culver. 2005. Estimating biodiversity in the epikarstic zone of a West Virginia cave. *Journal of Cave and Karst Studies* 67: 103–109.

Pipan, T., and D.C. Culver. 2012. Convergence and divergence in the subterranean realm: a reassessment. *Biological Journal of the Linnean Society* 107: 1–14.

Reese, A.M. 1934. The fauna of West Virginia caves. *Proceedings of the West Virginia Academy of Science* 7: 39–53.

Schneider, K. 2009. *How the availability of nutrients and energy influences the biodiversity of cave ecosystems*. Ph.D. Dissertation, University of Maryland, College Park, MD.

Schneider, K., and D.C. Culver. 2004. Estimating subterranean species richness using intensive sampling and rarefaction curves in a high density cave region in West Virginia. *Journal of Cave and Karst Studies* 66: 39–45.

Schneider, K., A.D. Kay, and W.F. Fagan. 2010. Adaptation to a limiting environment: the phosphorus content of cave arthropods. *Ecological Research*.

Schneider, K., M.C. Christman, and W.F. Fagan. 2011. The influence of resource subsidies on cave invertebrates: results from an ecosystem-level manipulation experiment. *Ecology* 92: 765–776.

Simon, K.S., and E.F. Benfield. 2002. Ammonium retention and whole-stream metabolism. *Hydrobiologia* 482: 31–39.

Simon, K.S., T. Pipan, and D.C. Culver. 2007. A conceptual model of the flow and distribution of organic carbon in caves. *Journal of Cave and Karst Studies* 69: 279–284.

Sket, B. 2008. Can we agree on an ecological classification of subterranean animals? *Journal of Natural History* 42: 1549–1563.

Weary D.J., and D.H. Doctor 2014. *Karst in the United States: A digital map compilation and database*. USGS Open File Report 2014-1156. 2014. http://pubs.usgs.gov/of/2014/1156/.

The Subterranean Aquatic Fauna of the Greenbrier Karst

19

Daniel W. Fong and David C. Culver

Abstract

The Greenbrier Karst harbors 16 species of stygobionts known from 92 caves, and six of these caves are type localities of ten of the species. The fauna is dominated by crustaceans and especially amphipods of the genus *Stygobromus*, and they primarily occupy vadose streams and the epikarst, but are notably absent from phreatic waters. *Stygobromus spinaus* is the most widely distributed, found in 59 caves. The segmented worm *Trichodrilus culeri* and the salamander *Gyrinophilus subterraneus* are both endemic to single site. The amphipod *Gammarus minus* is the most intensely studied species, but the basic biology of the other species is little known. The Organ Cave system holds the record in species richness (8) and in terms of type locality (3 species) and deserves a coordinated effort for protection.

19.1 Overview of the Subterranean Aquatic Fauna

Most research on the subterranean aquatic fauna is focused on species that show troglomorphy, morphological features associated with the aphotic environment, such as elongated appendages and reduced or complete loss of eyes and body pigment compared to related surface species. The ecological category of stygobiont consists of species that complete their entire life cycles in subterranean waters with no part of their life cycles occurring in surface waters. Thus, species that are only mildly troglomorphic or even non-troglomorphic are considered to be stygobionts as long as they fit this ecological criterion. Parallel to the situation with the terrestrial fauna (see Chap. 18), both stygobionts and non-stygobionts commonly occur in subterranean waters of the Greenbrier Karst.

D.W. Fong (✉)
Department of Biology, American University, 4400 Massachusetts Avenue NW, Washington, DC 20016, USA
e-mail: dfong@american.edu

D.C. Culver
Department of Environmental Science, American University, 4400 Massachusetts Avenue NW, Washington, DC 20016, USA

19.2 Non-stygobionts

Non-stygobionts include a whole range of species, from those that passively wash into cave streams from surface waters from episodic events such as storms and rapid snow melt, to those that continually infiltrate cave streams from sinking streams and from the epikarst through ceiling drips, to those that actively colonize cave waters during parts of their life cycles. These species occupy subterranean waters along a continuum of residence times, spanning from only hours to days to months to seasons and possibly years. Our knowledge of them necessarily correlates with their residence times because longer residence times increase their chances of being noticed by biologists. There have been few studies of non-stygobionts at a regional scale, however, and the following summary necessarily includes much speculation and very broad treatment of these organisms, especially ones with shorter residence times.

Species with short residence times, from hours to days, are soft-bodied benthic organisms washed into cave streams due to surface storm events, because a majority of them are unlikely to survive for long the severe physical disturbance of the substratum during such events, and those that survive rapidly become prey to resident species or starve due to insufficient or inappropriate resources. These include many

W.B. White (ed.), *Caves and Karst of the Greenbrier Valley in West Virginia*, Cave and Karst Systems of the World, https://doi.org/10.1007/978-3-319-65801-8_19

species of insect groups with aquatic larvae, such as mayflies (Ephemeroptera), stoneflies (Plecoptera), caddisflies (Trichoptera), dragonflies (Odonata), and some true flies (Diptera), as well as species of aquatic invertebrate groups such as flatworms (Platyhelminthes) and segmented worms (Annelida), among others. All of them are common and abundant components of the benthic fauna in surface streams, but are rarely recorded from cave streams in the Greenbrier Karst for several reasons. First, the harsh physical condition of cave streams during and immediately after flood events that transport these species into caves also prevents most cavers and biologists from entering caves. Second, their short residence times decrease their chances of being studied by biologists. Third, most biologists ignore them anyway because they are obviously part of the surface fauna and thus are considered uninteresting to study.

Species with longer residence times, from days to weeks to possibly months, are ones with more robust bodies and more likely to survive the physical disturbance of storm events and thus persist in subterranean waters for some duration. These include species of crayfish (Cambaridae) and fish such as minnows (Cyprinidae), sculpins (Cottidae), and sunfish (Centrarchidae) as well as some frogs (Ranidae). Emaciated fish and frogs are occasionally observed by cavers deep into cave systems weeks to months after a storm event. Ultimately, they all die from starvation and their bodies benefit scavengers and predators of the scavengers. Whether their body mass is an important energy source sufficient to maintain higher population sizes or longevity or both of the permanent cave stream fauna is unclear (Fong 2011). Some soft-bodied organisms may better survive the physical process of being washed into cave streams depending on the route of entry. For example, storms flushing water from low-gradient intermittent surface streams may cause only mild disturbance to the substratum, allowing a larger proportion of soft-bodied organisms to survive the event. Some of these may persist for weeks to months if appropriate resources are available in subterranean waters. Examples are caddisflies in the family Philopotamidae, known as finger-net caddisflies because their larvae spin and live within a silken net up to 1 cm in diameter by 4 cm in length. These nets function as filters that trap fine particulate organic matter, and this organic matter and the associated microbes serve as food for the larvae. Under rocks and pebbles in the upper level streams of some caves in the Greenbrier Karst, such as Organ Cave and The Hole, dense patches of these finger-shaped nets may be found throughout the year, although most are empty and only a small fraction is occupied. It is unknown if the empty nets indicate larvae that had successfully metamorphosed into flying adults or larvae that died from starvation and had since decomposed because the quality and quantity of the available organic matter are insufficient, or both, although it is not uncommon to encounter adult caddisflies in upper level cave stream passages in the Greenbrier Karst. Interestingly, in a group of European caddisflies in the family Limnephilidae the larvae grow in temporary surface waters, but adults migrate into caves and are considered troglophiles (Salavert et al. 2008). Some snails in the family Physidae are also not uncommon in upper level cave streams in the Greenbrier Karst. These snails are grazers on the epilithic biofilm, a matrix of algae and bacteria coating rocks in surface streams, and may persist in cave streams for some duration by surviving on the biofilm but without the algae in cave streams.

A number of species continuously colonize and exhibit lengthy residence times in subterranean waters, measured in seasons and years, and most are considered stygophiles. Some have large body sizes and thus are obvious components of the aquatic fauna (Fong et al. 2012). The more charismatic ones are salamanders, such as larvae and especially adults of the Spring Salamander, *Gyrinophilus porphyriticus*, the Dusky Salamander, *Desmognathus fuscus*, and the Cave Salamander, *Eurycea lucifuga*, although adults of the Cave Salamander are mostly terrestrial. These amphibians use subterranean waters as refuge from extremes of epigean temperature and aridity and for reproduction, where they also prey on invertebrates such as annelid worms as well as amphipod and isopod crustaceans. Another obvious stygophile is the crayfish *Cambarus bartonii*, a common inhabitant of cave streams and pools, where it is probably a top predator. A number of species appear to constantly rain down into vadose passages from the epikarst through ceiling drips and other routes, and these include many copepod crustaceans (Pipan and Culver 2005). These species are minute in size and difficult to observe and collect. They colonize rimstone pools and cave streams and pools and may persist for long periods; their ultimate fate is unclear, however, as the vadose passage is not their primary habitat.

19.3 Stygobionts

Ultimately, along the same continuum, stygobionts are species with residence times measured at evolutionary rather than ecological scales. A majority of the subterranean aquatic fauna in the Greenbrier Karst is collected from cave streams, and caves streams are the main habitat of most stygobionts known in the area. Stygobionts are also collected from the epikarst (Culver and Pipan 2009), and some stygobionts collected from upper level cave streams originate from the epikarst (Fong and Culver 1994). The fauna of epikarst is understudied in the Greenbrier Karst and in North America in general, however, and more species may be discovered from the epikarst in the future. No species in the

Greenbrier Karst is known from phreatic waters, although several species occur in phreatic waters in the nearby Shenandoah Valley in Virginia (Holsinger and Culver 1988; Hutchins et al. 2010). The lack of phreatic stygobionts in the Greenbrier Karst reflects a similar pattern in karst areas of the Appalachians in general. Although stygobionts are well known from hypotelminorheic habitats (see Culver and Pipan 2014), which are shallow aquifers perched above the local water table, such habitats, however, are unknown in the Greenbrier Karst. Large springs or groundwater resurgences are common in the Greenbrier Karst (see Jones 1973, 1997), and stygobionts are occasionally collected from such habitats, probably having been passively flushed out from subterranean waters during high discharge. These resurgences harbor large populations of the amphipod crustacean *Gammarus minus* and are the origins of stygobiotic and troglomorphic populations of *G. minus* in many cave systems in the Greenbrier Karst (see Culver et al. 1995 and below).

19.4 Diversity Patterns

There are 16 species of stygobionts known from 92 caves within the Greenbrier Karst (Fong et al. 2007) (Table 19.1, Fig. 19.1). They include one flatworm (Platyhelminthes), two segmented worms (Annelida), two snails (Mollusca), and one salamander (Chordata). The other nine species are arthropods in the subphylum Crustacea (orders Malacostraca and Maxillopoda), including one copepod (Cyclopidae), one isopod (Asellidae), one crayfish (Cambaridae), one amphipod in the genus *Gammarus* (Gammaridae), and five amphipods in the genus *Stygobromus* (Crangonyctidae). The stygobiotic fauna of the Greenbrier Karst is therefore dominated by arthropods (nine of 16 species or 56%), and amphipods are dominant among the arthropods (six of nine species or 67%), while all but one of the six amphipod species (83%) belong to the genus *Stygobromus*. Thus, amphipod species in one genus, *Stygobromus*, account for almost one-third (31%) of all stygobionts in the Greenbrier Karst.

The frequency distribution of number of stygobiotic species per cave is given in Fig. 19.2. The distribution is strongly skewed to the left as is typical of species richness data. 49% of the caves (45) harbor only one species, and 23% (21 caves) are inhabited by two species, with 10% (9 caves) and 11% (10 caves) housing three and four species, respectively. Organ Cave has the highest number of species at eight, followed by The Hole with seven, then by General Davis Cave and McClung Cave at six each, while Benedicts Cave, Buckeye Creek Cave, and Fuells Fruit Cave each harbor five species. The locations of caves with five or more species are given in Fig. 19.3, and all six are located in Greenbrier County at or near the edge of the Greenbrier Karst. A map of species richness at a regional scale using 10 km × 10 km grids is given in Fig. 19.4. It mirrors the pattern of individual caves with five or more species shown in Fig. 19.3 and shows high species richness in central Greenbrier County and in southern Greenbrier County at the Organ Cave plateau and areas near General Davis Cave.

19.5 Distribution

The number of caves occupied by each of the 16 stygobionts is given in Fig. 19.5. The most widely distributed species is the amphipod *Stygobromus spinatus* (Fig. 19.6), occurring in 59 of the 92 caves. This species is ubiquitous in aquatic habitats within its range, found in both epikarst (Holsinger 1978) and small headwater cave streams (Fong and Culver 1994), but with a body size at only 5 mm it is rarely noticed. The isopod *Caecidotea holsingeri* ranks second in terms of distribution, occurring in 32 caves, but unlike *S. spinatus*, it is rarely associated with epikarstic habitats and is commonly found on the underside of rocks in cave streams (Fong et al. 2007) and frequently in hygropetric habitats, the thin film of water on flowstone (Sket 2004). The amphipod *Stygobromus emarginatus* inhabits 22 caves in the Greenbrier Karst and many more beyond. It is easily observable as a large white amphipod, at almost 10 mm in length, and is abundant in some headwater cave streams. Knapp and Fong (1999) used a mark–recapture technique to estimate its population size at 10–14 per linear meter along a 300-m-long first-order stream in Organ Cave and present evidence suggesting that, despite its relatively large body size, its primary habitat is the epikarst rather than the vadose stream.

The next three species in ranks of distribution are the snail *Fontigens tartarea*, the crayfish *Cambarus nerteius*, and the flatworm *Macrocotyla hoffmasteri*, found in 17, 16 and 15 caves, respectively. The snail, *Fontigens tartarea*, at only 1–2 mm in size is difficult to see and its basic biology is unknown. It has an extensive range outside of the Greenbrier Karst, and given its low potential for dispersal, it is likely a cryptic species complex. The other snail, *F. turritella*, is similar to *F. tartarea* in size, but has a puzzling distribution. It is found in two caves in Greenbrier County and in one other cave about 250 km to the northeast. The outlier may be a different species. The Greenbrier Cave Crayfish, *Cambarius nerterius* (Fig. 19.7), is the only stygobiotic crayfish in the Greenbrier Karst. Its range overlaps that of the related stygophile *C. bartonii*, and it is difficult to distinguish between them because of similar morphologies. Any large, white flatworm, up to 1.2 cm in length when stretched out, found gliding along rocks in cave streams in the Greenbrier Karst is probably *Macrocotyla hoffmasteri*, a predator on smaller organisms such as juvenile amphipod and isopod crustaceans. We have little knowledge of cave flatworms,

Table 19.1 List of sixteen species of stygobionts known from caves in the Greenbrier Karst arranged by higher taxonomic grouping

Class	Order	Family	Species	Abv	Sites
Trepaxonemata	Neoophora	Dendrocoelidae	*Macrocotyla hoffmasteri*	Mh	15
Oligochaeta	Lumbriculida	Lumbriculidae	***Stylodrilus beattiei***	Sb	3
			Trichodrilus culveri	Tc	1
Gastropoda	Neotaenioglossa	Hydrobiidae	***Fontigens tartarea***	Fta	17
			Fontigens turritella	Ftu	2
Maxillopoda	Cyclopoida	Cyclopidae	*Rheocyclops virginiana*	Rv	9
Malacostraca	Isopoda	Asellidae	***Caecidotea holsingeri***	Ch	32
	Amphipoda	Crangonyctidae	***Stygobromus emarginatus***	Se	22
			Stygobromus mackini	Sm	2
			Stygobromus pollostus	Sp	6
			Stygobromus redactus	Sr	1
			Stygobromus spinatus	Ss	59
		Gammaridae	*Gammarus minus*	Gm	10
	Decapoda	Cambaridae	***Cambarus nerterius***	Cn	16
Amphibia	Caudata	Plethodontidae	***Gyrinophilus subterraneus***	Gs	1

Abv is the abbreviation for each species used in Fig. 19.5. Sites indicate the number of caves where each species has been collected from within the Greenbrier Karst. Ten species with type localities within the Greenbrier Karst are identified in bold type. Species taxonomic authorities, distribution maps, and names of caves occupied are given in Fong et al. (2007)

however, because preservation and identification of flatworms are technically challenging, and systematic collection and studies of flatworms in North America have not been made for several decades.

The next species in rank in term of distribution is the amphipod crustacean *Gammarus minus* (ten caves). It is the most intensively studied stygobiont in the Greenbrier Karst, and its biology is summarized in a separate section below. Three species, the copepod *Rheocyclops virginiana* (nine caves) and two amphipods *Stygobromus pollostus* (six caves) and *S. redactus* (one cave within the Greenbrier Karst, but also occur in a few caves outside the area) are dwellers of the epikarst. All of them are minute in body size and difficult to observe, and they were collected only opportunistically rather than systematically. For example, mature individuals of *S. redactus* reach only 2 mm, and those from Parlor Cave were collected from water in caver's heel prints, the only standing water in the cave (Fong et al. 2007). A systematic survey of the epikarst fauna at the regional scale using techniques pioneered by Pipan (2005) should lead to a better understanding of their distribution. Again, we know little of their basic biology. Although the amphipod *Stygobromus mackini* occurs in only two caves in the Greenbrier Karst, these sites are located at the extreme northern edge of its large range which extends to the southwest for over 400 km. This species is mainly found in water associated with the epikarst, such as drip pools and very small first-order streams, with females being much more common than males (Hoslinger 1978).

At the opposite extreme in terms of distribution, only two species, the segmented worm *Trichodrilus culveri* and the salamander *Gyrinophilus subterraneus*, are single-site endemics. The restricted distributions of *T. culveri* and of the other segmented worm *Stilodrilus beattiei* which occurs in three caves in the Greenbrier Karst are most likely due to the lack of systematic collections since the 1970s (see Cook 1975), and further work should yield expanded ranges of extant species and additional species. The West Virginia Spring Salamander, *G. subterraneus*, is the only stygobiotic vertebrate in the Virginias and one of only four species of stygobiotic salamanders in the family Plethodontidae east of the Mississippi River. It is endemic to General Davis Cave in Greenbrier County, where it co-occurs with the closely related Spring Salamander *G. porphyriticus*. The larva is confined to the cave stream while the adult is equally adept on land and in the water (Fig. 19.8), and they act as the top predator in the cave. The status of *G. subterraneus* as a distinct species was questioned almost immediately after its description (Besharse and Holsinger 1977) because of the high degree of morphological variability in both *G. porphyriticus* and *G. subterraneus* (Blaney and Blaney 1978), and this situation remains unresolved (Niemiller et al. 2010). Interestingly, there are occasional reports of albino specimens of *G. porphyriticus* larvae deep in caves (e.g., Brandon and Rutherford 1967). It is unknown if these albino larvae will metamorphose into albino adults or if albinos are more likely than non-albinos to remain and reproduce within cave systems. These observations, however, suggest that

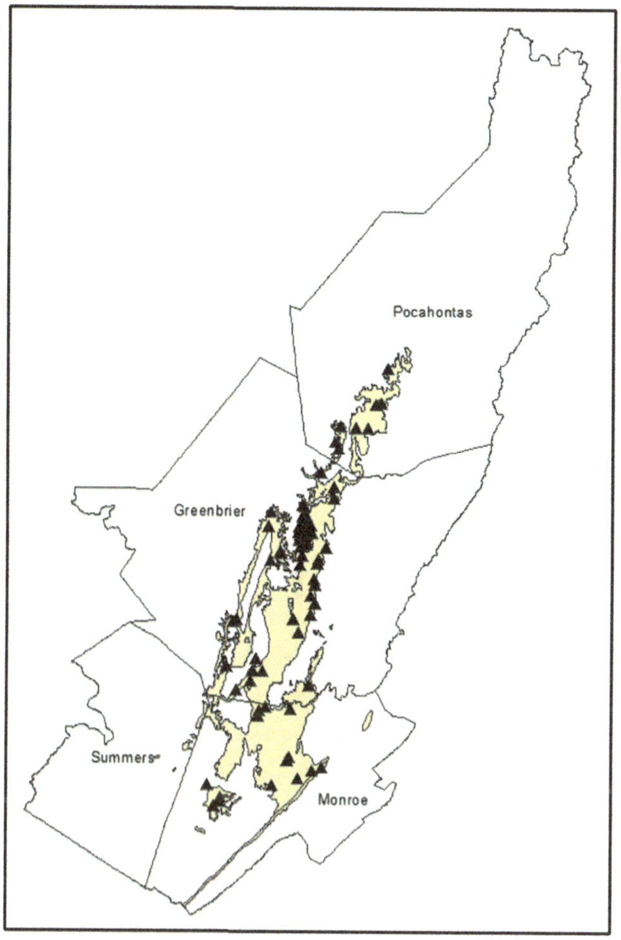

Fig. 19.1 Locations of the 92 caves harboring stygobionts in the Greenbrier Karst

stygobiotic variants are evolving in some of the stygophilic *G. phorphyriticus* populations in the Greenbrier Karst, a potential case study of speciation in action.

Little is known about the distribution of stygobionts within cave systems because most collection records associate species occurrences with a cave name only and provide no more information. Knowledge on fine-scale species distribution within cave systems is needed for a deeper understanding of the ecology of stygobiotic fauna. For example, Fong and Culver (1994) examined the fine-scale distribution of five crustacean species within the complex Organ Cave system (Stevens 1988). They found that while three crangonyctid amphipods and one asellid isopod are more abundant at first-order headwater streams and decrease in abundance at higher-order lower level streams, the opposite is true for a gammarid amphipod. They suggested that the pattern is explained by different routes of colonization of the cave system, with the crangonyctid amphipods colonizing from the epikarst and the gammarid colonizing from downstream originating from the resurgence. Furthermore, although all of the water from the Organ Cave systems exits at a single resurgence, there is a subterranean divide that partitions the water in the system into a western and an eastern drainage (Stevens 1988). While the gammarid is abundant in the eastern drainage, it is absent in the western drainage, and the crangonyctids are present in both. They explained the pattern by suggesting that the western drainage was a separate system with a different resurgence in the past and subsequently merged with the eastern drainage by a subterranean stream capture event. The gammarid is absent from the western drainage either because

Fig. 19.2 Frequency distribution of caves by number of species

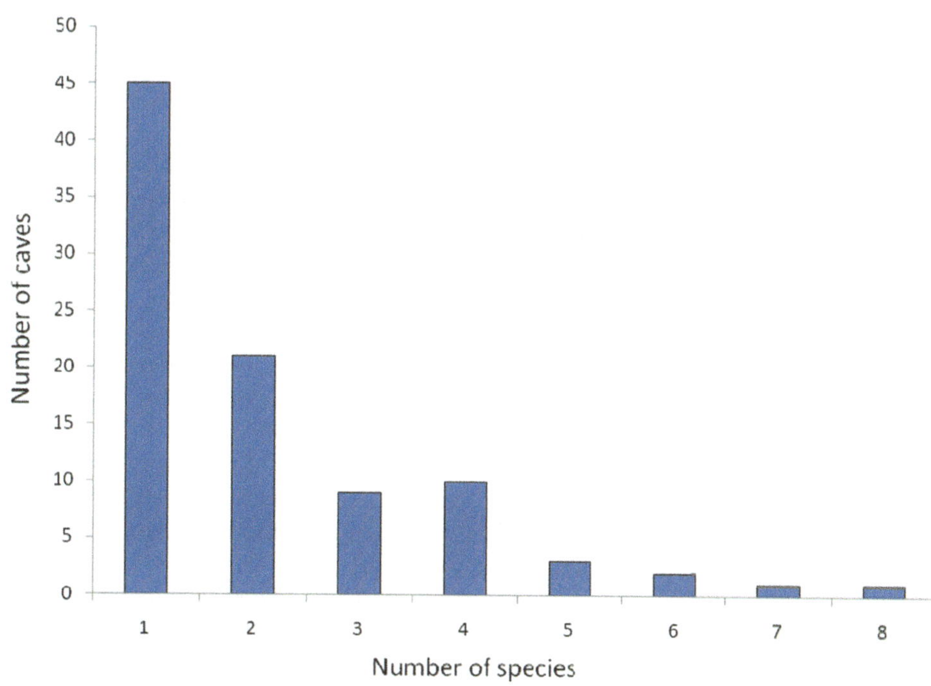

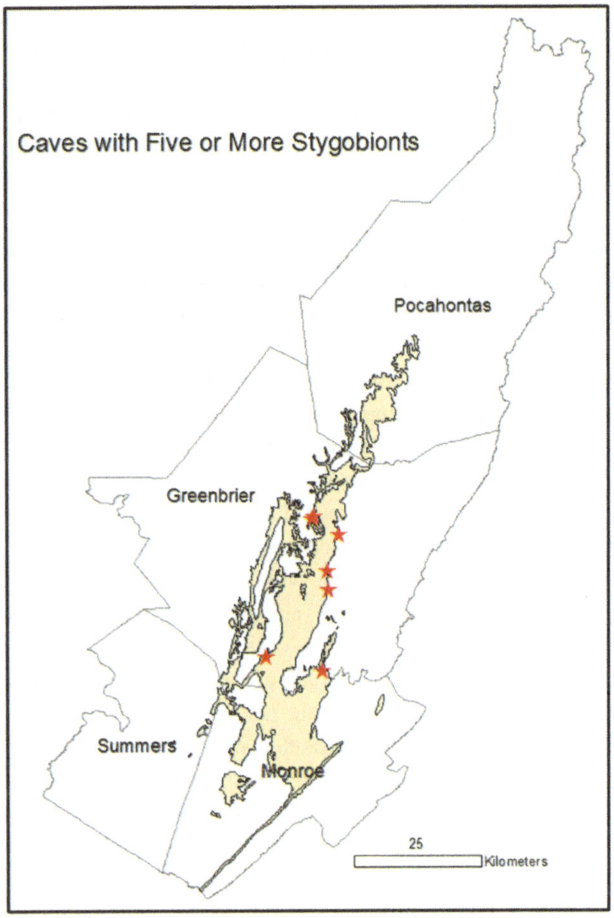

Fig. 19.3 Map of caves with five or more stygobionts each. All six caves are located at or near the contact between the Greenbrier Karst and non-carbonate rocks

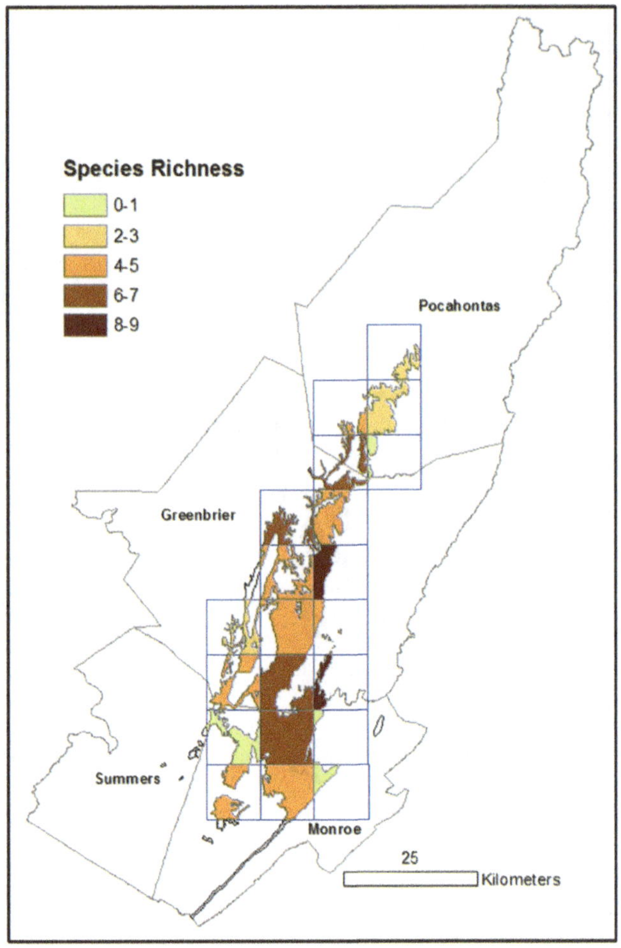

Fig. 19.4 Map of species richness at a regional scale using 10 km × 10 km grids

it colonized and adapted to the eastern drainage prior to the stream capture, or the hydrogeological setting of the stream capture physically prevented the possibility of the gammarid to travel upstream into the western drainage, or both.

19.6 The Case of *Gammarus Minus*

The most intensively studied stygobiont in the Greenbrier Karst is the amphipod *Gammarus minus* (see overview in Culver et al. 1995 and Fong 2012). Here we summarize the results of some of the studies focused on this species that contributed to a better understanding of the evolution and ecology of stygobionts. It is a common and abundant inhabitant of karst springs and cave streams, often reaching densities of 15–20 per m^2 in cave streams and an order of magnitude higher in karst springs. Populations in springs and in most cave streams are not troglomorphic. Troglomorphic populations are found only in large cave systems in Greenbrier and Monroe Counties in the Greenbrier Karst and in Tazewell County in Virginia. Troglomorphic populations show only remnants of eyes, elongated appendages, and reduced body pigmentation compared to non-troglomorphic populations (Fig. 19.9). Shoemaker (1940) considered these troglomorphic populations to be a distinct variety and named it *tenuipes*. Recent genetic evidence indicates such a designation is unwarranted (e.g., Carlini et al. 2009), and we use the term *Gammarus minus tenuipes* simply as a convenient label for the troglomorphic populations. A map of the locations of non-troglomorphic cave populations and of *G. m. tenuipes* is given in Fig. 19.10, which shows that all *G. m. tenuipes* sites are situated at or near the eastern edge of the Greenbrier Karst.

Evidence strongly indicates that *G. m. tenuipes* populations originated through colonizing cave streams at some time in the past by *G. minus* dwelling at the spring where the cave water resurges, and that *G. m. tenuipes* populations in different drainages are each independently derived from a different spring population. Populations of *G. m. tenuipes* on average have lower within population variation in the

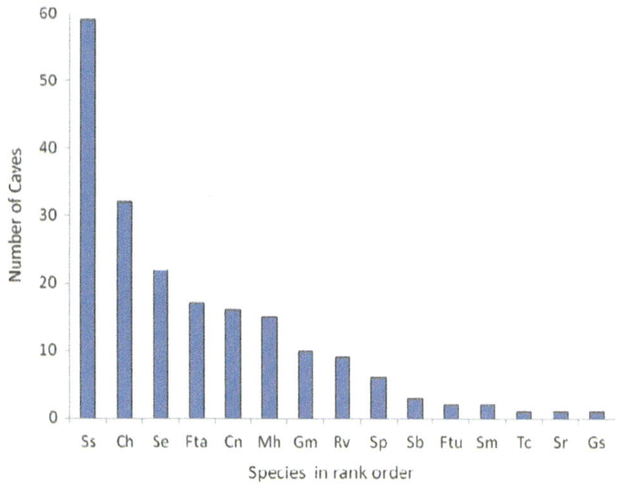

Fig. 19.5 Number of caves occupied by each species arranged by rank order. See Table 19.1 for the key to abbreviated species names

Fig. 19.7 *Cambarus nerterius*, the only stygobiotic crayfish in the Greenbrier Karst. It is only mildly troglomorphic, but has never been found outside caves. This specimen was about 8 cm in body length. Photograph by Horton H. Hobbs III. Used with permission

Fig. 19.6 *Stygobromus spinatus*, the most widely distributed stygobiont in the Greenbrier Karst. This ovigerous female, carrying eight developing embryos in a ventral marsupium, was 4 mm in length. Photograph by Michael E. Slay. Used with permission

Fig. 19.8 Larva (*top*) and adult (*bottom*) of the West Virginia Spring Salamander, *Gyrinophilus subterraneus*, the only stygobiotic vertebrate in the Greenbrier Karst. The snout-vent lengths (SVL) of larvae and adults reach up to 9 and 11 cm, respectively. Photographs by Danté Fenolio. Used with permission

sequences of the COI and ITS-1 genes than do spring populations, and are also on average more closely related to hydrologically proximate spring populations than to *G. m. tenuipes* populations in other drainages even though they differ substantially in morphology from sister spring populations (Kane et al. 1992; Carlini et al. 2009). *G. m. tenuipes* populations also show much lowered effective population sizes as measured by levels of codon bias compared to surface populations of the same drainage (Fig. 19.11; Carlini et al. 2009), indicating severe genetic bottlenecking as a result of reduced population size during colonization of the cave habitat.

Aspiras et al. (2012) compared the levels of expression of four genes, hedgehog, pax6, sine oculus and dachshund, involved in the developmental pathway of arthropod eyes in

Fig. 19.9 Specimen of *Gammarus minus* from a surface spring (*top*) and of *Gammarus minus tenuipes* from a cave (*bottom*). Both are mature males. The *tenuipes* specimen is about 10 mm in body length. The two specimens are shown at the same scale. Photographs by Michael E. Slay. Used with permission

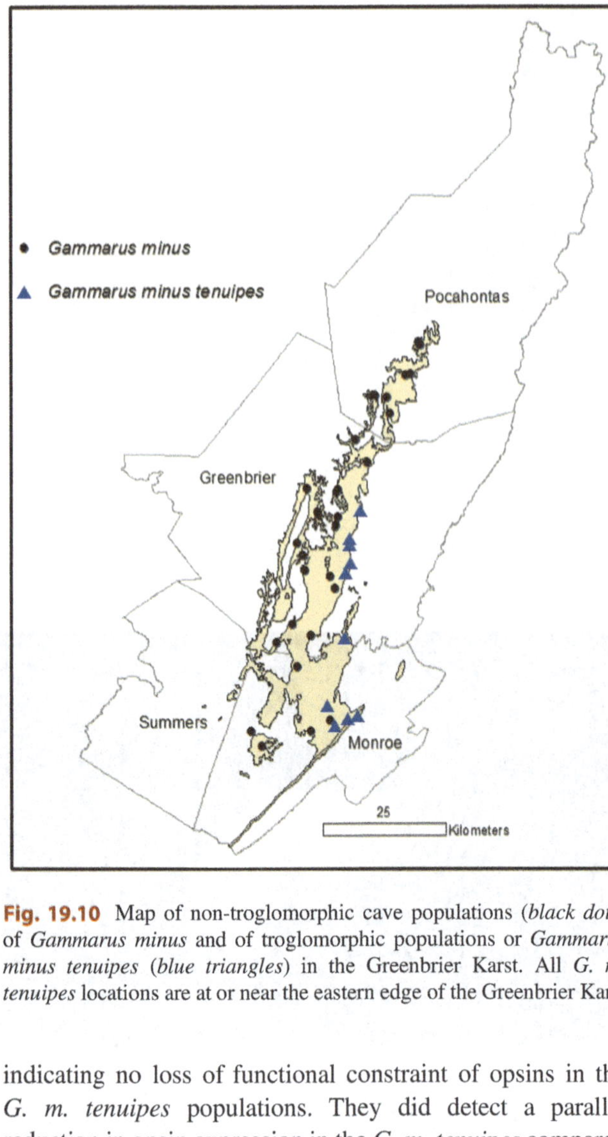

Fig. 19.10 Map of non-troglomorphic cave populations (*black dots*) of *Gammarus minus* and of troglomorphic populations or *Gammarus minus tenuipes* (*blue triangles*) in the Greenbrier Karst. All *G. m. tenuipes* locations are at or near the eastern edge of the Greenbrier Karst

three sister pairs of surface *G. minus* and *G. m. tenuipes* populations. They found a parallel reduction in expressions of only one of the genes, hedgehog, in the *G. m. tenuipes* populations compared to surface populations, but not in the other three genes upstream of hedgehog in the pathway (Fig. 19.12). Their results mirror that of Yamamoto et al. (2004), who showed that hedgehog-related genes are also involved in eye reduction in the cavefish *Astyanax mexicanus*. The conclusion is that selection may target similar genes governing eye development in a vertebrate and an invertebrate during adaptation to the subterranean environment. The implication is that the genetic mechanism behind convergent morphological adaptation among diverse species may be simpler than expected.

Carlini et al. (2013) compared DNA sequence variation and levels of expression in two paralogs of the gene for opsin, a protein that functions in phototransduction and is responsible for photosensitivity, and also in three sister pairs of surface *G. minus* and *G. m. tenuipes* populations. They discovered little sequence variation as well as little difference in synonymous to non-synonymous ratios of amino acid substitutions in the opsin genes among populations, indicating no loss of functional constraint of opsins in the *G. m. tenuipes* populations. They did detect a parallel reduction in opsin expression in the *G. m. tenuipes* compared to surface populations. The conclusion is that functions of the opsins in darkness were likely maintained by selection through pleiotropy because opsins serve an important but unknown function or functions unrelated to vision. Thus, reduction in opsin expression may be advantageous in darkness up to a point, beyond which further reduction is selected against because of the pleiotropic effects. The implication is that loss of optic structure and reduction to loss of optic function may be governed by different genetic mechanisms.

Although surface populations of *G. minus* are ecologically categorized as detrivores, consuming decaying plant material but deriving their nutrition from the associated microbes (Kostalos and Seymour 1976), the food niche of *G. m. tenuipes*is usually assumed to be similar to surface populations. MacAvoy et al. (2016) compared the food niche among surface *G. minus* and *G. m. tenuipes* populations using stable isotopes of carbon and nitrogen. They show that

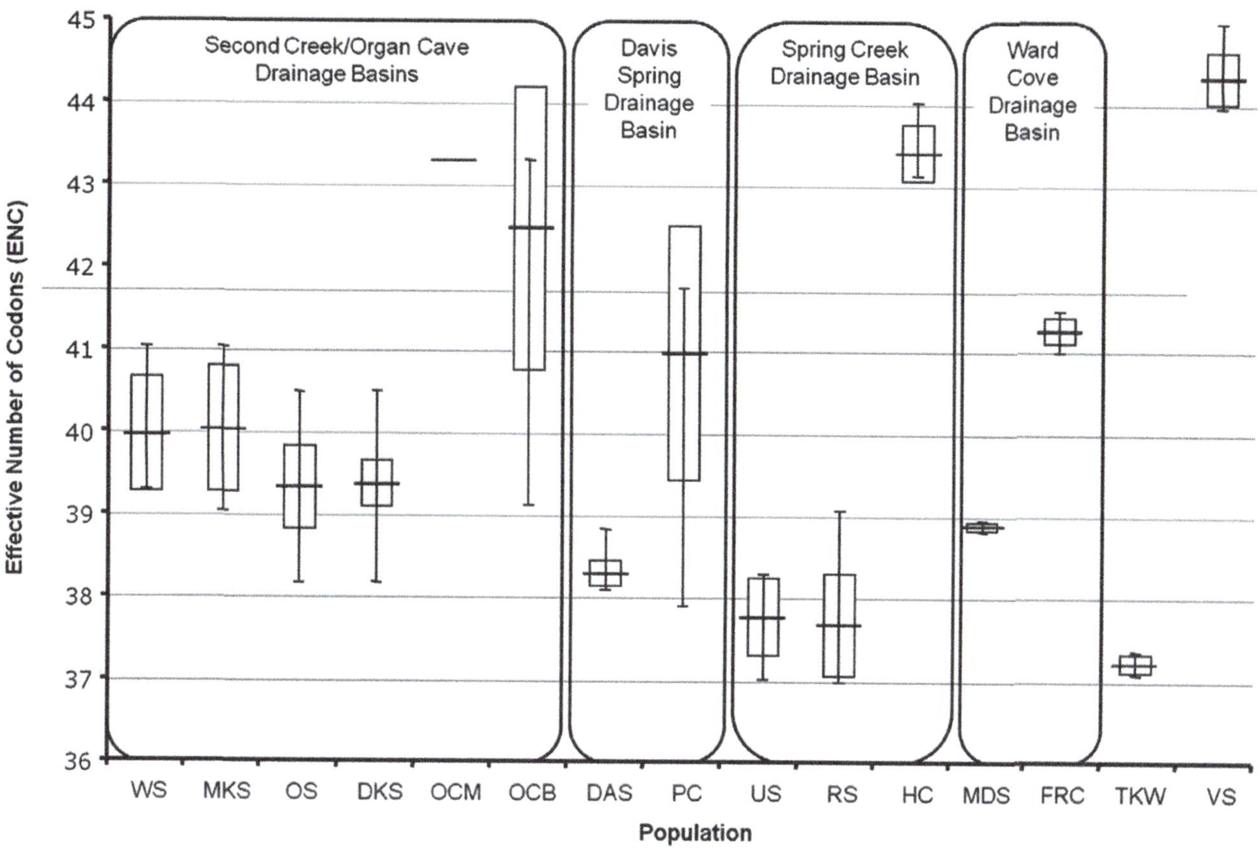

Fig. 19.11 Effective number of codons (ENC; Wright 1990) in COI sequences from 15 populations of *Gammarus minus*. High ENC indicates small effective population size, and low ENC indicates large effective population size. The mean (*horizontal line*), standard deviation (*box*), and range (*vertical bar*) of ENC are indicated for each population (except for OCM where all individuals were fixed for one sequence). Populations are grouped by drainage basin and habitat type (spring populations are designated by the letter *S*, cave populations by the letter *C*, and the single karst window population by the letters *KW*). The *TKW* and *VS* populations were the only sites sampled within their respective drainage basins and so are not grouped with any other population. Reproduced from Carlini et al. (2009). Used with permission

G. m. tenuipes populations clearly have a different nitrogen source than do surface populations, and that the change corresponds to a jump in trophic level, indicating that *G. m. tenuipes* acts as a predator on other invertebrates as well as a consumer of available detritus. Their results confirm the observations that *G. m. tenuipes* preyed on the isopod *Caecidotea holsingeri* in artificial streams in the laboratory (Culver et al. 1991) and that *G. m. tenuipes* preyed on other invertebrates in the field (Fong 2011). Broadening of food niche as an adaptation to the cave environment has also been demonstrated for subterranean amphipods in the Edwards Aquifer (Hutchins et al. 2014) and in a cave salamander (Fenolio et al. 2005) and is consistent with the general theory of expansion of the food niche in food poor environments.

19.7 Competition Among Cave Stream Invertebrates

Caves in the Greenbrier Karst have proved to be useful ecological laboratories for the study of interspecific interactions, just as they have proved to be useful evolutionary laboratories for the study of adaptation and natural selection. The cave streams in the Greenbrier Karst are dominated by four species: the amphipods *Gammarus minus, Stygobromus emarginatus, S. spinatus*, and the isopod *Caecidotea holsingeri*. This number of species is quite small compared to most surface communities and even some highly diverse cave stream communities in the Dinaric karst (Fišer et al. 2012), yet with enough species that a variety of combinations of species exist in different caves, making for natural

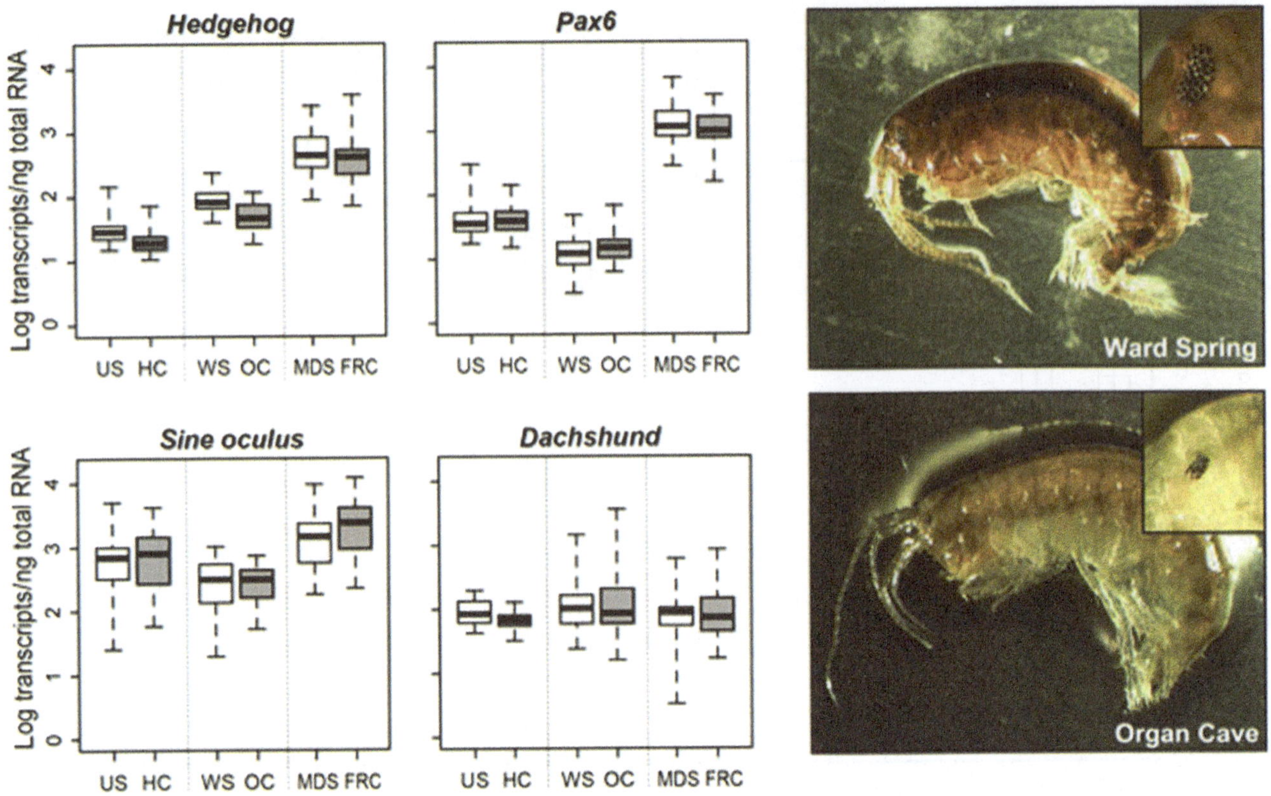

Fig. 19.12 Gene expression in the heads of adult *Gammarus minus*. **a** Hedgehog, **b** pax6, **c** sine oculus, **d** dachshund. *Bars* indicate the median value for each population, whereas boxes denote the upper and lower quartiles. Whiskers indicate the full range of sampled values. Boxes are *shaded gray* for cave populations and white for surface populations. The expression levels are significantly lower in cave compared to surface populations only in hedgehog. **e** An example individual specimen from Ward Spring with normal eye size for surface amphipods. **f** A troglomorphic specimen from Organ Cave illustrates reduced eye size. Insets in (**b**, **c**) show detail of the eyes at the same scale. Reproduced from Aspiras et al. 2012. Used with permission

experiments. In a community of four species, there are six pairs of possible interactions, whereas in a community of 10 species there are 45 such pairs. Because of the patchy nature of the habitat and the resulting patchy distribution of most species, nearby caves often have different species present. This allows these "natural experiments" to be used to study the effects of species additions and removals. The following account is adapted from Culver et al. (1991) and Culver (2012).

In many cave streams in the Greenbrier Karst, there is an alternation between deeps (pools) and shallows (riffles). The amphipods and isopods are highly concentrated in riffles as a result of the concentration of food (especially leaf detritus), increased oxygen, and the absence of salamander predators, which live in pools. In this habitat, the three obvious kinds of interactions are as follows: (1) Species may compete for food, (2) species may compete for space (the underside of gravels), and (3) species may serve as food for other species. All can and do occur in particular situations, but the most universal (and easiest to analyze) is competition for space on the underside of riffles.

The basis for competition for space on the underside of riffles is that when any two individuals meet in a riffle, space is in short supply. It is very easy to observe the behavioral response to most encounters even in a small dish in the laboratory—one or both individuals rapidly move away (Culver 1970). More realistic laboratory experiments were done in a small artificial riffle, where the washout rate of individuals put in the riffle in various combinations could be measured. For competition for space in a riffle is approximately equal to the ratio of the washout rate of species i when species j is present to the washout rate of species i when species j is not present. The resulting estimates of competition can then be compared with field data on the amount of overlap among species in caves and within riffles within a cave. The greater the amount of overlap in the field should correspond with lower intensity of competition for space, because otherwise the species would not be in proximity.

In Organ Cave, when isopod and amphipod species co-occur, they occupy different sized rocks within a riffle, different riffles, or different streams. In this study, laboratory stream studies were combined with perturbation (addition and removal) experiments in the field. Three species pairs of the six possible species pairs were studied:

Stygobromus spinatus and *Caecidotea holsingeri*
Gammarus minus and *Caecidotea holsingeri*
Gammarus minus and *Stygobromus emarginatus*

Laboratory stream competition between *C. holsingeri* and *S. spinatus* was one-sided—the isopod always seemed to dislodge the amphipod. Superficially, the *Gammarus minus–Caecidotea holsingeri* pair seemed to compete for space: Fewer individuals remained in the artificial riffle when the other species was present. However, it turned out that *G. minus* was eating, rather than dislodging, *C. holsingeri*. Thus, this was predation rather than competition. In the field, a new interaction appeared. The presence of *G. minus* had a positive effect on the abundance of *S. emarginatus*, rather than the negative effect predicted from the artificial riffle experiments. Subsequent investigation showed that *S. emarginatus* was feeding on the fecal material of *G. minus*. Thus, the two species compete at the microscale (a single rock), but have a commensal relationship at the scale of a riffle.

Even though it initially appeared that these cave streams were dominated by simple competitive interactions for space, the reality was much more complicated, with both negative and positive effects. The teasing apart of these interactions was only possible because of the small number of species involved.

19.8 Protection of Stygobionts

Strategies for the protection of stygobionts necessarily focus on the nature of threats to the fauna and methods to eliminate or mitigate such threats as well as setting criteria for prioritizing sites for protection. Potentially severe threats arise from human activity (see Humphreys 2011) that may directly destroy subterranean habitats such as construction, quarrying, and other resource extraction practices using earth moving equipment. In the Greenbrier Karst, especially in the vicinity of Lewisburg, widespread development of home sites is the most immediate threat because of the increased potential for groundwater contamination from excess household and lawn chemicals and spillage of petroleum products from increased local traffic and delivery of supplies via trucks. The expansion of impervious surfaces associated with urban development also reduces groundwater recharge, which may severely curtail the input of energy into subterranean systems, especially via the epikarst. Fong (2011) suggests that disruption of input of photosynthetically based energy and nutrients in support of the subterranean fauna is a less apparent but continuous and insidious problem, and this is especially true for the Greenbrier Karst because chemosynthesis is unknown in the area. Fortunately, organizations such as the West Virginia Cave Conservancy and The Nature Conservancy are dedicated to protecting many cave entrances and immediately surrounding areas in the Greenbrier Karst (see a list of caves in Chap. 18). However, proper management of not only the entrance areas to caves but the entire recharge area of a cave system is necessary to ensure protection of stygobionts and troglobionts.

A way to set priorities among caves for protection of stygobionts is to emphasize cave systems that are type localities or contain relatively high numbers of species or both. In the Greenbrier Karst, ten caves meet these criteria: Six are type localities for a total of ten species, and seven caves house five or more species each (Table 19.2). Some of these caves are managed by conservancies. McClung and Benedicts Caves are managed by the West Virginia Cave Conservancy and General Davis Cave is owned and managed by The Nature Conservancy. Although this affords some degree of protection, the protected areas include only

Table 19.2 Caves with high stygobiont species richness (Sp $\geq$ 5) or are type localities (type: number of type species) or both

Cave	Sp	Type
Organ	8	3
The Hole	7	–
General Davis	6	1
McClung	6	2
Benedicts	5	–
Buckeye Creek	5	–
Fuels fruit	5	–
Court street	4	1
Tub	2	2
Arbuckle	1	1

They are all located within Greenbrier County except for Tub Cave which is located within Pocahontas County

the immediate vicinities of the cave entrances and leave the recharge zones unprotected. All of the other caves are also unprotected. Organ Cave stands out because it has the highest species richness at eight and is the type locality for three species. In addition, Organ Cave is also a hot spot in terms of terrestrial biodiversity, housing 13 troglobionts (Chap. 18). This is a large cave system with multiple entrances (Stevens 1988), and although the historic, commercial entrance is gated, all other entrances and all of its recharge area are unprotected and face increased pressure for development of housing and especially intensive poultry operations. At the other extreme is Arbuckle Cave, the type locality of its only stygobiont, *Stygobromus redactus*, and the only location of this epikarstic species within the Greenbrier Karst (the species occurs in only a few caves outside the Greenbrier Karst). Although this is a small cave, it also harbors 12 troglobionts and is the only known site of the beetle, *Horologion speokoites*, the primary habitat of which is likely also the epikarst (Chap. 18). The location of Arbuckle Cave, near the Greenbrier Valley Airport and surrounded by expanding housing development, makes its fauna extremely vulnerable and urgently needs protection. All of the remainder of the unprotected caves, except for Tub Cave, are located in Greenbrier County, and all are subject to threats associated with increased development and changing land use patterns. But it is clear, however, that the Organ Cave system is biologically unique in the Greenbrier Karst and especially deserving a coordinated effort for protection.

References

Aspiras, A.C., R. Prasad, D.W. Fong, D.B. Carlini, and D.R. Angelini. 2012. Parallel reduction in expression of the eye development gene hedgehog in separately derived cave populations of the amphipod *Gammarus minus*. *Journal of Evolutionary Biology* 25: 995–1001.

Besharse, J.C., and J.R. Holsinger. 1977. *Gyrinophilu ssubterraneus*, a new troglobitic salamander from southern West Virginia. *Copeia* 1977: 624–634.

Blaney, R.M., and P.K. Blaney. 1978. Significance of extreme variation to a cave population of the salamander *Gyrinophilus porphyriticus*. *Proceedings of the West Virginia Academy of Science* 50: 23.

Brandon, R.A., and J.M. Rutherford. 1967. Albinos in a cavernicolous population of the salamander *Gyrinophilus porphyriticus* in West Virginia. *American Midland Naturalist* 78: 537–540.

Carlini, D.B., J. Manning, P.C. Sullivan, and D.W. Fong. 2009. Molecular genetic variation and population structure in cave and surface populations of the freshwater amphipod *Gammarus minus*. *Molecular Ecology* 18: 1932–1945.

Carlini, D.B., S. Satish, and D.W. Fong. 2013. Parallel reduction in expression, but no loss of functional constraint, in two opsin paralogs within cave populations of *Gammarus minus* (Crustacea: Amphipoda). *BMC Evolutionary Biology* 2013 (13): 89.

Cook, D.G. 1975. Cave-dwelling aquatic Oligochaeta (Annelida) from the eastern United States. *Transactions of the American Microscopic Society* 94: 24–37.

Culver, D.C. 1970. Analysis of simple cave communities: niche separation and species packing. *Ecology* 51: 949–958.

Culver, D.C. 2012. Species interactions. In *Encyclopedia of Caves*, ed. W.B. White, and D.C. Culver, 743–748. Amsterdam: Elsevier/Academic Press.

Culver, D.C., and T. Pipan. 2009. *The biology of caves and other subterranean habitats*. Oxford, UK: Oxford University Press.

Culver, D.C., and T. Pipan. 2014. *Shallow subterranean habits: ecology, evolution and conservation*. Oxford, UK: Oxford University Press.

Culver, D.C., D.W. Fong, and R.W. Jernigan. 1991. Species interactions in cave stream communities: experimental results and microdistribution effects. *American Midland Naturalist* 126: 364–379.

Culver, D.C., T.C. Kane, and D.W. Fong. 1995. *Adaptation and natural selection in caves: The evolution of Gammarus minus*. Cambridge, MA: Harvard University Press.

Fenolio, D.B., G.O. Graening, B.A. Collier, and J.F. Stout. 2005. Coprophagy in a cave-adapted salamander: The importance of bat guano examined through nutritional and stable isotope analyses. *Proceedings of the Royal Society of London [Biology]* 273: 439–443.

Fišer, C., A. Blejec, and P. Trontelj. 2012. Niche-based mechanisms operating within extreme habitats: A case study of subterranean amphipod communities. *Biology Letters* 8: 578–581.

Fong, D.W. 2011. Management of subterranean fauna in karst. In *Karst management*, ed. P.E. Van Beynen, 201–224. Dordrecht, Germany: Springer.

Fong, D.W. 2012. *Gammarus minus* as a model system for the study of adaptation. In *Encyclopedia of caves*, ed. W.B. White, and D.C. Culver, 341–347. Amsterdam: Elsevier/Academic Press.

Fong, D.W., and D.C. Culver. 1994. Fine-scale biogeographic differences in the crustacean fauna of a cave system in West Virginia, USA. *Hydrobiologia* 287: 29–37.

Fong, D.W., D.C. Culver, H.H. Hobbs III, and T. Pipan. 2007. *The Invertebrate Cave Fauna of West Virginia*, 2nd ed., vol. 16. Barrackville, WV: West Virginia Speleological Survey Bulletin.

Fong, D.W., M.L. Porter, and M.E. Slay. 2012. *Cave life of the Virginias. A field guide to commonly encountered species*. Huntsville, AL: Biology Section of the National Speleological Society.

Holsinger, J.R. 1978. Systematics of the subterranean amphipod genus *Stygobromus* (Crangonyctidae), Part II: Species of the eastern United States. *Smithsonian Contributions to Zoology* 266: 1–144.

Holsinger, J.R., and D.C. Culver. 1988. The invertebrate cave fauna of Virginia and a part of East Tennessee: Zoogeography and ecology. *Brimleyana* 14: 1–162.

Humphreys, W.F. 2011. Management of groundwater species in karst environments. In *Karst Management*, ed. P.E. Van Beynen, 283–318. Dordrecht, Germany: Springer.

Hutchins, B., D.W. Fong, and D.B. Carlini. 2010. Genetic population structure of the Madison Cave Isopod, *Antrolana lira* (Cymothoida: Cirolanidae) in the Shenandoah Valley of the eastern United States. *Journal of Crustacean Biology* 30 (2): 312–322.

Hutchins, B.T., B.F. Schwartz, and W.H. Nowlin. 2014. Morphological and trophic specialization in a subterranean amphipod assemblage. *Freshwater Biology* 59: 2447–2461.

Jones, W.K. 1973. *Hydrology of Limestone Karst in Greenbrier County, West Virginia*. Morgantown, WV: West Virginia Geologic and Economic Survey Bulletin 36.

Jones, W.K. 1997. *Karst hydrology Atlas of West Virginia*, vol. 4. Charles Town, WV: Karst Waters Institute Special Publication.

Kane, T.C., D.C. Culver, and R.T. Jones. 1992. Genetic structure of morphologically differentiated populations of the amphipod *Gammarus minus*. *Evolution* 46: 272–278.

Knapp, S.M., and D.W. Fong. 1999. Estimates of population size of *Stygobromus emarginatus* (Amphipoda: Crangonyctidae) in a headwater stream in Organ Cave, West Virginia. *Journal of Cave and Karst Studies* 61: 3–6.

Kostalos, M., and R.L. Seymour. 1976. Role of microbial enriched detritus in the nutrition of *Gammarus minus* (Amphipoda). *Oikos* 27: 512–516.

MacAvoy, S.E., A. Braciszewski, E. Tengi, and D.W. Fong. 2016. Trophic plasticity among spring vs. cave populations of *Gammarus minus*: examining functional niches using stable isotope and C/N ratios. *Ecological Research* 31: 589–595.

Niemiller, M.L., M.S. Osbourn, D.B. Fenolio, S.K. Pauley, B.T. Miller, and J.R. Holsinger. 2010. Conservation status and habitat use of the West Virginia Spring Salamander (*Gyrinophilus subterraneus*) and Spring Salamander (*G. porphyriticus*) in General Davis Cave, Greenbrier Co., West Virginia. *Herpetological Conservation and Biology* 5: 32–43.

Pipan, T. 2005. *Epikarst—a promising habitat. Copepod fauna, its diversity and ecology: a case study from Slovenia (Europe).* Postojna-Ljubljana, Slovenia: ZRC Publishing, Karst Research Institute at ZRC-SAZU.

Pipan, T., and D.C. Culver. 2005. Estimating biodiversity in the epikarstic zone of a West Virginia cave. *Journal of Cave and Karst Studies* 67: 103–109.

Salavert, V., C. Zamora-Muñoz, M. Ruiz-Rodríguez, A. Fernández-Cortés, and J.J. Soler. 2008. Climatic conditions, diapause and migration in a troglophilic caddisfly. *Freshwater Biology* 53: 1606–1617.

Shoemaker, C.R. 1940. Notes on the amphipod *Gammarus minus* Say and description of a new variety, *Gammarus minus* var. *tenuipes*. *Journal of the Washington Academy of Science* 30: 388–394.

Sket, B. 2004. The cave hygropetric—a little known habitat and its inhabitants. *Archiv für Hydrobiologie* 160: 413–425.

Stevens, P.J. 1988. *Caves of the Organ Cave Plateau, Greenbrier County, West Virginia*, vol. 9. Barrackville, WV: West Virginia Speleological Survey Bulletin.

Wright, F. 1990. The 'effective number of codons' used in a gene. *Gene* 87: 23–29.

Yamamoto, Y., D.W. Stock, and W.R. Jeffery. 2004. Hedgehog signaling controls eye degeneration in blind cavefish. *Nature* 431: 844–847.

Pleistocene Vertebrates from Greenbrier Valley Caves

20

E. Ray Garton and Frederick V. Grady

Abstract

Every cave has the potential to be a repository of Pleistocene and older bones and past life. The Greenbrier Valley Karst in particular has a rich repository of late Pleistocene faunal remains. Cave entrances lend themselves to be nesting places for birds of prey who deposit the bones of their meals onto cave floors. The very nature of sinkholes and surface pits creates natural traps for animals. Streams are a source for depositing animals in caves where they die and are buried in cave sediments. Caves are also used as dens for a variety of extinct and living animals that died or left the remains of their prey. The unique environment of caves contributes to the entrapment and preservation of Ice Age bones.

20.1 Introduction

West Virginia is blessed with a rich and diverse natural history. Nowhere is this more evident than in its Ice Age fauna. What is popularly thought of as the Ice Age in West Virginia began about 2 million years ago and ended about 11,700 years ago. This period, known as the Pleistocene, was marked by several dramatic climatic changes and giant animals such as the woolly mammoth, saber-toothed cat, and giant ground sloth. The Pleistocene is perhaps best known for its great ice sheets and glaciers. However, there were several periods of long, very warm, interglacial climates, much warmer than even today. In fact there were once armadillos and vampire bats in West Virginia. How do we know? Laymen and scientists have been finding the remains of long extinct Pleistocene age animals in West Virginia for over 200 years.

There is no evidence that any of the great ice sheets of the Pleistocene ever reached into West Virginia but they still had a significant and lasting impact on the state's flora, fauna, and geomorphology. For example, during the last great ice sheet known as the Wisconsinan Glacier what is now the Monongahela River flowed north across Pennsylvania and into the St. Lawrence River system. When the ice dammed the river near Beaver, Pennsylvania, a great lake formed and extended south 120 miles to Weston and was up to 50 miles wide. The lake called Glacial Lake Monongahela filled the valleys to an elevation of 1100 ft above sea level making the lake hundreds of feet deep in some places. The lake eventually filled and overflowed to the west and south along the margins of the ice sheet forming what is now the upper portion of the Ohio River. Thick layers of varved (banded) clay and fine sand are still found along the margins of the Lake Monongahela shoreline.

Because of another ice dam near Chillicothe, Ohio another great lake formed that backed up the waters of present-day Scioto, Ohio, and Kanawha rivers. This lake extended from Chillicothe, Ohio, past Ashland, Kentucky, Huntington, and Charleston and all the way to Hawks Nest, West Virginia. This lake is called Glacial Lake Teays and was probably in existence for about 25,000 years during the last (Wisconsinan) ice age. The lake level rose until it overflowed at a low point near Portsmouth, Ohio. These overflowing waters formed the lower Ohio River and followed a westerly route toward Cincinnati then continued along the southern boundary of the ice sheet across Indiana and Illinois to the Mississippi River. As is found along the margins of Glacial Lake Monongahela thick layers of varved (banded) clay and fine sand are still found along the margins of the Lake Teays shoreline.

E. R. Garton (✉)
PO Box 200Barrackville, WV 26559, USA
e-mail: garton@prehistoricplanet.com

F. V. Grady
2240 S. Walter Reed Drive, Arlington, VA 22206, USA

W.B. White (ed.), *Caves and Karst of the Greenbrier Valley in West Virginia*,
Cave and Karst Systems of the World, https://doi.org/10.1007/978-3-319-65801-8_20

On the high mountains such as Spruce Knob, Canaan Valley, and Dolly Sods the bitter cold of the ice age formed permafrost and patterned ground which remain as irregular polygons and circles of surface rocks and debris formed during intensive frost action. This relic patterned ground is still visible on some of the state's highest peaks. Even though glaciers never reached West Virginia, they still had a dramatic effect on the land and in shaping the topography of the state.

The first ice age fossils found in West Virginia were discovered by saltpeter miners who sent the bones to President Thomas Jefferson in 1797. Jefferson wrote a scientific paper titled "A Memoir on the Discovery of Certain Bones of a Quadruped of the Clawed kind in the Western Parts of Virginia" which was published in 1799. Jefferson thought the large bones and claws found in a Monroe County cave were that of a giant cat, but later they turned out to really be that of an extinct giant ground sloth. This new species was named *Megalonyx jeffersonii*, after our 3rd President. *Megalonyx jeffersonii* was designated at the "official" state fossil for West Virginia in 2008.

Most of West Virginia's ice age fossil animals have been and continue to be found in the state's 3500 plus limestone caves. Why in caves? This is mostly because caves provide a place for stream and talus sediments to accumulate. Caves also provide a stable temperature and humidity, which are important to preservation. In order for a bone to be preserved, it needs to be buried quickly because on the surface, bones decay rapidly and are often eaten by rodents. In the limestone regions of the state, there are many sinkholes, fissures, sinking streams, and caves. When an animal dies, there is a better chance that its carcass will fall or wash into one of these bone preserving natural traps. Still as important as caves are to bone preservation, a few important specimens have been found in surface streams and surface alluvial deposits. For example, none of the state's 12 mammoth specimens have been found in caves, and the state's first musk ox (*Bootherium bombifrons*) specimen was found on a farm in Brooke County.

You do not have to be a scientist to make important discoveries. In fact cave explorers, hikers, farmers, and students make many discoveries. While you may not find a new species, every bone is important because it may be the first of that species from the state or it may extend the range of that species in the state or region. For example, until recently, the only specimen of armadillo from West Virginia came from Organ Cave in Greenbrier County, and that specimen was a single scute (shell plate). In 1995, some people digging the foundation for a house discovered what would turn out to be dozens of pieces of an armadillo (*Dasypus bellus*) in Berkeley County. This extended the range of armadillo in West Virginia by 125 miles to the north and also making it the most northern record of armadillo in the eastern USA. Other warm climate indicators include a species of vampire bat (*Desmodus*), tapirs, peccaries, and ground sloths. These animals indicate that once West Virginia had a climate similar to northern Florida. At the other end of the climate extreme are animals that indicate that the climate in West Virginia was similar to northern Canada and Alaska. These animals include musk ox, arctic shrews, northern bog lemmings, and caribou.

Some of the other dramatic finds in West Virginia include extinct species such as *Smilodon* (saber-toothed cat) from Greenbrier and Pendleton Counties, a cheetah (*Micracinonyx inexpectatus*) and a jaguar (*Panthera onca*) from Pendleton County, several Dire Wolves (*Canis dirus*), the great cave bear (Arctodus pristinus), and several occurrences of woolly mammoths (*Mammuthus*) and mastodons (*Mammut americanum*). But just as important as the mega-fauna are the smaller animals such as the extinct trout vole (*Atopomys salvelinus*), Cumberland pocket gopher (*Plesiothomomys*), the Cumberland vole (*Pitymys cumberlandensis*), or the Cumberland mouse (*Peromyscus cumberlandensis*) all of which may be over 800,000 years old. Indeed big or small each species played an important role in shaping the future make-up of West Virginia's animal kingdom.

20.2 Previous Work

Thomas Jefferson's (1797) description of the sloth Megalonyx made the first contribution to the Pleistocene of West Virginia when it was then still part of Virginia. Little work was done regarding the Quaternary of West Virginia during the nineteenth century. There is one reference to deer bones in sandstone cave and Henry C. Mercer apparently checked out one cave in West Virginia in the 1890s (Mercer 1897).

In 1902, J.B. Hatcher reported a partial skull of an extinct musk ox and a scapula of a mammoth from a sand pit in Brooke County. About a decade later, bones of an extinct peccary were found in Rennick Quarry Cave Greenbrier Count (Gidley 1918).

It was not until the middle of the twentieth century when caving became increasingly popular that researchers turned their attention to the potential of Pleistocene deposits in West Virginia caves. Handley (1956) reported on finds in six caves in Greenbrier, Monroe, and Pocahontas counties including the first finds in Organ Cave. About 10 years later, field parties from Carnegie Museum of Natural History (CMNH) in Pittsburgh under the direction of John Guilday and led by Allen McCrady, and Harold Hamilton began to systematically investigate West Virginia caves for bone sites. They made significant excavations in Trout, Eagle, Hoffman School, Mandy Walters, and Smoke Hole Caves, all in Pendleton County. A group from the Charleston Grotto and elsewhere coalesced into the West Virginia Association for Cave Studies (WVACS) and in the process of exploring and mapping caves in Greenbrier County made some bone discoveries. Some of these sites were further investigated by Carnegie Museum. Also in the 1960s, a group from the Baltimore Grotto found teeth of a short-faced bear in Grapevine Cave (now Lost World Caverns). The discovery of a partial mastodon tooth in Bowden Cave, Randolph County by members of Monongahela Grotto led E. Ray Garton to develop a professional interest in Pleistocene bone sites in West Virginia. In 1976, Garton had a manuscript listing thirty-five sites in the state.

A fortuitous meeting of E. Ray Garton and Fred Grady at the Carnegie Museum of Natural History New Paris, PA field station led to a productive collaboration. Interestingly enough, Grady's interest in bone caves also had started in the 1960s with the chance find of some bones in an eastern Pennsylvania Cave. Both Garton and Grady were led to John Guilday of Carnegie Museum and both corresponded with him until his death in 1982. The discovery of the main site in New Trout Cave by a party including both Garton and Grady in 1978 led to a project lasting some 2 years during which some four tons of bone bearing cave earth usually known as matrix was removed from the cave for screening. On one trip, four individuals took out one hundred bags of matrix weighing about one ton. The New Trout Cave finds included several bones of a large vampire bat (Grady et al. 2002).

The discovery of a jaguar skeleton in Hamilton Cave by Grady in December of 1980 led to several other spectacular finds including partial skeletons of the saber-toothed cat *Smilodon* and bones of an extinct cheetah-like cat. There was a brief hiatus of about 2 years while ownership of the property containing Hamilton Cave was transferred to the National Speleological Society (NSS). Following this, more discoveries were made in Hamilton Cave including most of the remainder of the cheetah-like cat skeleton and more *Smilodon* parts. A great deal of matrix was also hauled out for screening resulting in the finding of many small vertebrates including several extinct species. Some of this work was supported by a grant from the National Geographic Society.

During the ownership change, at what is now the John Guilday Cave and Nature Preserve discoveries were made in Patton Cave by Marshall Holmes that led to the collection by Grady and others of a large number of bones of the extinct peccary *Platygonus compressus* including some virtually complete skulls (Grady 1988a). Smaller vertebrates were also collected from Patton Cave at this time. Ammunition boxes and fishing tackle boxes were used to take out the bones, which required going up a rope drop and traversing a considerable amount of cave passage. George Dasher was especially helpful with the rigging of the drop and coordinating the groups that assisted in getting so many bones out. Also in the early 1980s, the DC Grotto of the NSS was resurveying Organ Cave and Grady and Garton took the opportunity to check out earlier reported sites in this large cave system and discovered new ones found during the resurvey. Fragmentary teeth of *Smilodon* was among the discoveries at this time.

Following the purchase of the Guilday Cave and Nature Preserve Carnegie Museum parties made a large excavation in Trout Cave where a survey party from DC Grotto had found peccary teeth and preliminary work by Grady had revealed many small vertebrates.

The 1990s saw the discovery of parts of an extinct armadillo in Alaina P. Cave in Berkeley County and a virtually complete skeleton of the large extinct Peccary *Platygonus vetus* in Poorfarm Cave, Pocahontas County. Significant sites were found in Helix Cave, Greenbrier County, and Nut Cave in Pendleton County. As the millennium dawned discoveries of Mastodon and other vertebrates were made in the far reaches of Scott Hollow Cave.

20.3 Description of Pleistocene Vertebrate Faunas from the Greenbrier Valley Karst

20.3.1 Acme Quarry #5 Cave, Monroe County

A single upper molar tooth of a horse *Equus* sp. was donated to US National Museum by Bob and Chris Alderson. This tooth may or may not be a Pleistocene species.

20.3.2 Alexander Cave #2, Greenbrier County

A tooth of a gray fox *Urocyon cinereoargenteus* was found in this cave and donated to US National Museum.

20.3.3 Benedicts Cave, Greenbrier County

The partial right and left mandibles of a pine martin *Martes cf. americana* were donated to Carnegie Museum by John Rutherford and identified by John Guilday. The specimen is cataloged as number CMNH 24698. This is an extirpated species in West Virginia and may indicate a Pleistocene age.

20.3.4 Bransfords Cave, Greenbrier County

A small collection of bones and teeth was made by Tom Hay. The collection included several *Platygonus compressus* teeth along with black bear, deer, woodchuck, *Peromiscus*, beaver, and pine vole. The peccary indicates a late Pleistocene age.

20.3.5 Buckeye Creek Cave, Greenbrier County

Roger Baroody found two Pleistocene species in this cave. A lower jaw with 3 deciduous teeth was identified as *Tapirus veroensis* CMNH 24326, and several bones of a peccary *Platygonus compressus* were cataloged as CMNH 24430 (Grady and Garton 1994). Suzzane Strait from Marshall University also collected at least one mastodon molar *Mammut americanium* from this cave.

20.3.6 Cabble Cave, Greenbrier County

Handley (1956) reported bones of a mouse *Peromyscus* sp. and a weasel *Mustela cf. rixosa* that is now identified as *Mustela cf. nivalis*. These specimens are at University of Michigan.

20.3.7 Cass Cave, Pocahontas County

During the resurvey of Cass Cave Greg Springer and Mark Botkin found some bones and teeth on the upper level of the cave. A small American mastodon *Mammut americanum* tooth is probably a lower deciduous fourth premolar or a small first molar. There is also a horse tooth that may or may not be modern. A total of nearly twenty animal species were found along with small bird, snake, frog, and salamander bones. The mammals included two different shrews, a mole, two bats, several rodents, and carnivores. Two carnivores are of interest though not specifically identifiable. The shaft of a large humerus appears to be some sort of large bear probably too big to be a black bear. An anterior premolar seems to be a large wolf possibly a dire wolf. There was also a pig tooth found in the same area indicating that modern material is mixed with probably Pleistocene material. Clearly, more work is required.

20.3.8 Cassel Cave, Pocahontas County

A skull of a bat *Pipistrellus subflavus* was collected in this cave and donated to the US National Museum.

20.3.9 Cave Near Patton Cave, Monroe County

Bob Leibman donated to the US National Museum skulls of a rabbit *Sylvilagus* sp. and a skunk *Mephitis* from this cave.

20.3.10 Friars Hole Cave System, Pocahontas County

Surprisingly, for the largest cave system in West Virginia, there are only two main bone finds. A mandible of a big brown bat *Eptesicus fuscus* was found in the Snedegars area of the cave, and a woodchuck *Maromta monax* incisor was found at the south end of the system. Both specimens were donated to the US National Museum.

20.3.11 Fletchers Cave, Monroe County

George Dasher photographed bones of two species in this cave, a long-nosed peccary *Mylohyus fossilis* jaw and parts of the skull of a wolf, *Canis* sp. It is hoped that in the near future permission can be obtained to collect these specimens.

20.3.12 Helix Cave, Greenbrier County

Surveyors found several large teeth in Helix Cave in the mid-1990s. These were identified as the Peccary *Platygonus compressus*, extinct musk ox *Bootherium bombifrons,* and a horse *Equus* sp. Excavations were made along a stream bank near the entrance, and many small vertebrates were recovered. A total of about thirty mammal species have been identified from Helix Cave. There are several shrews including the Pigmy shrew *Sorex hoyi*, a mole and very few bat bones. Rodents include the northern species *Phenacomys intermedius* and *Synaptomys borealis*. There is also the thirteen-lined ground squirrel *Speromphillus tridecemlineatus* and the pocket gopher *Geomys* sp. Small carnivores from Helix include the striped skunk *Mephitis*, the least weasel *Mustel nivalis,* and the ermine *Musteli erminea*, a northern species. There is also the tooth of a coyote *Canis latrans*, a species that may have been responsible for much of the bone accumulation. A part of a coprolite from a medium-sized carnivore with bone fragments was also found. Besides the large teeth previously noted, a premolar of caribou, *Rangifer tarandus* was also found. A few bones of salamander, frog, snake, and bird were also found in Helix Cave.

20.3.13 Haynes Cave (Gromers Cave), Monroe County

Bones from a cave owned by Frederic Gromer were sent to Thomas Jefferson in 1796 who identified them as an extinct

Fig. 20.1 Left hand claws of *Megalonyx jeffersonii* from Haynes Cave, Monroe County. First described by Thomas Jefferson in 1797

genus which he named *Megalonyx* believing them to be that of an extinct lion. Before publishing, Jefferson came across an illustration in a periodical of a skeleton found in Paraguay of a large extinct species related to the sloth. Jefferson realized the bones he had might in fact also be a sloth (Jefferson 1797). The bones included an ulna, a radius, three metacarpals (Figs. 20.1 and 20.2), several phalanges and reportedly an epiphysis of a femur. The bones were re-described in more detail by Wistar (1799), and casts were sent to Couvier in France. The bones were eventually given the species name *Megalonyx jeffersonii* and are in the collection of the American Philosophical Society in the Academy of Natural Sciences in Philadelphia. The actual location of the cave has been a source of much debate and research. The owners of Organ Cave have long maintained that they were found in that cave but Frederic Gromer never owned the property that Organ Cave is on. The late Peter Hauer

Fig. 20.2 Left ulna and radius of *Megalonyx jeffersonii* from Haynes Cave, Monroe County. First described by Thomas Jefferson in 1797

determined that Organ Cave was not the cave from which the *Megalonyx* was found. Grady has spent several years working on this problem and has concluded that the cave was actually one in Monroe County which was split off from Greenbrier County in 1799. Hopefully, a more detailed account of this research will be published in the near future.

20.3.14 Higginbothan Cave #4, Greenbrier County

The mountain lion *Felis concolor* was reported from this cave by Stafford (1961, 1963).

20.3.15 Laurel Creek Cave, Monroe County

Remains of black bear *Ursus americanus* were found in this cave (Barr 1953).

20.3.16 Lightner Pit #2, Greenbrier County

Two specimens from this cave were found and donated to Carnegie Museum. A partial skull of a porcupine *Erethizon dorsatum* was cataloged as CMNH 24307 and a premolar either caribou *Rangifer tarandus* or the extinct deer *Sangamona fugitiva* was cataloged as CMNH 24306.

20.3.17 Lost World Caverns (Grapevine Cave), Greenbrier County

Members of Baltimore Grotto found a canine and a lower first molar of a lesser short-faced bear *Arctodus pristine* in this cave in 1968. A second trip resulted in the recovery of bone fragments including the upper end of a femur (Shockley 1968). The specimens were donated to the Carnegie Museum and cataloged as CMNH 12992. More recently, after the cave was commercialized as Lost World Caverns, Grady noted several partial skeletons of *Procyon lotor* and collected a mandible of a spotted skunk *Spilogale pictorials*. Also, several recent skeletons have been noted in the commercial area of the cave including an owl and an opossum. A partial skull with no teeth, femur, vertebra, and other post cranial bones of *Ursus americanus*, black bear, were found in January 2015. Some of the bones were submitted for carbon-14 dating resulting in a date of 32,200 ± 400 years BP, making this specimen the oldest known black bear from West Virginia. The bones are curated at Lost World Caverns.

20.3.18 Ludingtons Cave, Greenbrier County

A lower jaw and some vertebrae of a porcupine *Erethizon dorsatum* were found in this cave by John Rutherford in 1967 and identified by John Guilday of Carnegie Museum.

20.3.19 McClung Cave, Greenbrier County

Vertebrate remains were found in two areas of this large system. Bill Balfour reported one find and then Grady collected much of the skeleton of a bobcat, *Lynx rufus* and fragments of a second along with part of a skeleton of a woodland jumping mouse *Napaeozapus insignis* near the lightner entrance. Horse teeth, *Equus*, along with a few bird and rabbit, bones were donated to the US National Museum by Tisha Springer.

20.3.20 McFerrin Saltpeter Cave, Greenbrier County

There is an unsubstantiated report of a mastodon tooth *Mammut americanus* from this cave.

20.3.21 Organ Cave, Greenbrier County

The Organ Cave system has some thirty-eight miles of mapped passages. Vertebrate remains have been found in several different areas of the cave, some quite far from any known entrance (Grady 1988a). It had been suggested during the twentieth century that the bones Thomas Jefferson described as the *Megalonyx* came from Organ Cave though it is now believed that these bones came from a cave in Monroe County. Nevertheless, there have been several important finds in Organ Cave. A single dermal plate of the extinct armadillo *Daypus bellus* was found in the Bone Room (Guilday and McCrady 1966). Saber-toothed cat, *Smilodon*, teeth, and bone fragments were found in two different areas of the cave. The most complete bones of the long-nosed Peccary *Mylohyus fossilis* and the only find of a grizzly bear *Ursus arctos* in West Virginia were also found in Organ Cave. A total of over thirty mammal species are known from the fauna of Organ Cave. A single carbon-14 date of 21,040 ± 760 years was obtained on fragments of bone associated with *Smilodon* teeth in the crawl. Recent finds include an elk *Cervus elaphus cannadensis* antler and the identification of a femur in the US National Museum collection as badger *Taxidea taxus*, a western species. Specimens from Organ Cave are at Carnegie Museum, the US National Museum and UMMVP.

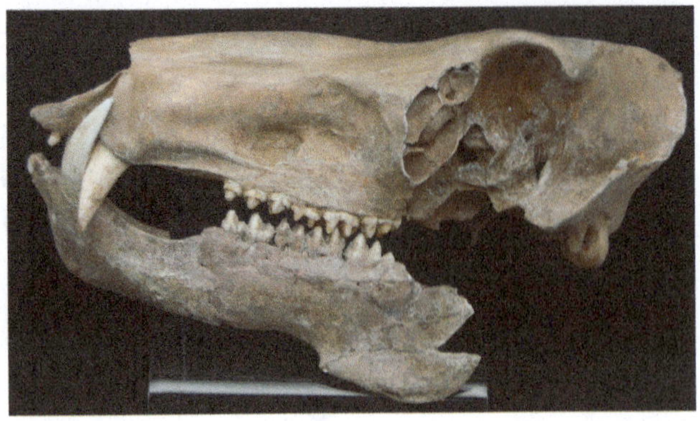

Fig. 20.3 One of several skulls of *Platygonus compressus* found in Patton Cave, Monroe County

20.3.22 Patton Cave, Monroe County

The first specimen from Patton Cave, the partial maxilla of the flat-headed peccary *Platygonus compressus*, was brought to the US National Museum by Lynn Ferguson in 1976 (Fig. 20.3). In 1988, while surveying the cave, Marshal Holmes found more bones of *Platygonus* and then about a year later, after digging through a narrow passage, substantially more bones were found at the bottom of a drop. A series of trips were made to collect this material assisted by many cavers, especially George Dasher. More than thirty species of mammals were found. Over twenty individuals of the flat-headed peccary were found at the bottom of the drop, some represented by virtually complete skulls. A second site above the drop produced about six more individuals of that species. *Platygonus* bones at the bottom of the drop carbon-14 dated at 13,350 YBP—while a jaw of the yellow-cheeked vole *Microtus xantognathus* dated at 14,450 YBP. *Platygonus* bones above the drop dated at 22,260 YBP. Most of the other mammals recovered were small, including shrews, moles, and rodents. The shrews included the northern *Sorex arcticus*, while the rodents included the northern species *Phenacomys intermedius, Synaptomys borealis*, and *Microtus xanthognathus*. The western rodent *Spermophillus tridesem lineatus* was also found along with a species that eats it, the badger *Taxidea taxus*. Other carnivores included the pine marten *Martes americana*, fox *Vulpes,* and a probably a black bear represented by a single foot bone and several tracks of a small individual. A single molar of the extinct deer *Sangamona fugitiva* was also found (Grady 1988b).

20.3.23 Piercys Mill Cave, Greenbrier County

A single molar initially identified as *Canis lupus* was donated to the US National Museum by Bob Leibman. There is now information to suggest that the wolves in the Northeastern states were in fact the red wolf *Canis rufus,* so the identity of this specimen is changed to *Canis* sp.

20.3.24 Piddling Pit Cave, Pocahontas County

The canine of a black bear *Ursus americanus* was loaned to the US National Museum for molding and casting. A few bones of *Neotoma magister* were later donated to the US National Museum from this cave.

20.3.25 Ramp Hole Cave, Pocahontas County

Handley (1956) reported a small collection of eight mammal species from this cave. The species included two shrews, a bat, several rodents including the porcupine *Erethozon dorsatum*, raccoon, and deer.

20.3.26 Roadside Pit, Pocahontas County

A collection of vertebrate material from roadside pit was donated to the US National Museum by Rock Bridges in 1970. There are twelve species of mammals, one bird bone and some snake vertebrae. The mammals include shrew, a mole, rabbit or hare, and various rodents including porcupine *Erethizon dorsatum.* Carnivores from roadside pit include a mandible of the pine marten *Martes americana*, a northern species, and a split part of a canine crown tentatively identified as mountain lion *Felis concolor.* There are two juvenile bones identified as *Bison* sp. The fauna, if it is all contemporaneous, may be late Pleistocene especially with the pine marten present.

20.3.27 Rapps Cave, Greenbrier County

Two Pleistocene mammals have been found in this cave. John Rutherford found the right mandible of a pika *Ochotona* sp. in the entrance room which was donated to Carnegie Museum and cataloged as 24290. More recently, Tom Hay found parts of two upper third molars of the mastodon *Mammut americanum* further in the cave. These teeth are on loan to the West Virginia Geological Survey and casts are at the USNational Museum. A human molar was also donated to Carnegie Museum. A large mound in the front of the cave is an important archeological site. Human and animal bones have been observed here including woodchuck *Marmota monax*, black bear *Ursus americanus*, and deer *Odocoileus virginianus.* Bird bones were also observed here.

20.3.28 Rennick Quarry Cave (Upper Bone), Greenbrier County

Bones of an extinct Peccary *Platygonus vetus* have been found beginning in 1913 and reported by Gidley (1918). The bones include a nearly complete skull, two other skull parts, parts of three left, and three right mandibles, and various post cranial elements. The skull was cataloged as USNM 8003 and figured by Gidley (1918). An upper canine collected in 1929 is also in the US National Museum collection. The West Virginia University Geology Department has both sides of a mandible of *Platygonus vetus* collected in 1913 by I.O. Smith. This peccary represents a middle Pleistocene species.

20.3.29 Rockland Cave, Monroe County

A tooth of a tapir *Tapirus* sp. was reported to have been found in this cave by RE Whittemore.

20.3.30 Scott Hollow Cave, Monroe County

A significant collection of mostly large species has been made from this cave. There are various parts of mastodon, including most of both sides of the upper jaws of a young individual, a small tusk, a juvenile humerus from an individual much removed from the first finds, and several teeth of adult individuals (Fig. 20.4). There is a jaw of a wolf probably a red wolf *Canis rufus* and some bones of raccoon. There is also a single toe bone of the ground sloth *Megalonyx* believed to be the middle Pleistocene species *Megalonyx wheatlyi.* There is one end of a radius of a peccary *Platygonus* sp. a partial tooth of a horse, and the distill end of a tibia that is probably elk *Cervus elaphus cannadensis.* There is also much of a toad skeleton. These finds were made in several parts of the cave and are likely different ages. Carbon-14 dates on the mastodon humerus and one of the teeth were 21,830 and 11,350 YBP, respectively. Most of the specimens have been retained by the owners though casts of the mastodon teeth, upper jaws and sloth phalanx are at US National Museum and West Virginia Geological and Economic Survey.

20.3.31 Steam Cave, Pocahontas County

A lower third premolar of a horse *Equus cf. fraternus* was identified by Morris Skinner and is in the collection of Carnegie Museum cataloged 29551.

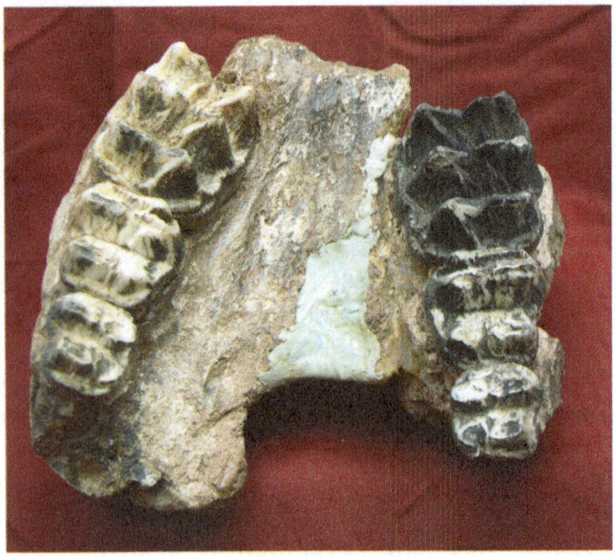

Fig. 20.4 Baby mastodon palate, *Mammut americanum*, from Scott Hollow Cave, Monroe County

20.3.32 The Hole (Cave) Greenbrier County

In the 1960s, members of WVACS found several teeth, a jawbone, and a fourth metacarpal of dire wolf *Canis dirus* in this cave. The specimens were donated to Carnegie Museum and cataloged as 24327. They represent the best find of a dire wolf in West Virginia.

20.3.33 Windy Mouth Cave, Greenbrier County

Bones have been found in several places in Windy Mouth Cave. Handley (1956) described a jaw of a gray bat *Myotis cf. griscescens*. The humerus of a porcupine *Erethizon dorsatum* was donated to Carnegie Museum by Jim Hixson. Various bones of black bear *Ursus americanus* have also been found in this cave, and a skull is at Carnegie Museum. The most significant find in Windy Mouth Cave consists of a femur and a calcaneus of a Jaguar *Panthera onca*, CMNH as 24699, found by John C. Hempel and Jim Hixon. Grady has observed bones in several places in this cave but so far permission from the landowner to make collections has not been granted.

20.4 Summary

20.4.1 Pleistocene Vertebrate Fauna and Taxa List from Greenbrier Valley Karst

Approximately, 135 Pleistocene specimens have been reported from 33 caves in the Greenbrier Valley Karst. Represented are at least 52 genera and 63 species (Table 20.1) sorted by species name.

20.4.2 Carbon-14 Dates for Greenbrier County Karst Pleistocene Vertebrate Faunas

Nine Carbon-14 dates have been obtained from 6 Greenbrier Valley Karst caves. The dates range from 35,960 to 11,350 years (Table 20.2).

20.5 Paleozoic Vertebrates from Greenbrier Valley Karst Caves

The caves of the Greenbrier Valley Karst are developed in the Mississippian Greenbrier Limestone group (Chap. 2). Since, all limestones of the group are marine and filled with many taxa of invertebrate fossils, it stands to reason that there were also vertebrates living in these seas. Fish and sharks to be sure. While no fish fossils have been reported, observant cavers over the years have spotted several caves with shark teeth and spines embedded in cave walls made visible by solution process (Fig. 20.5). Some of the specimens have only been reported and some have been molded and cast and a few have been removed. A selected table from Garton's manuscript "Paleozoic Vertebrates from West Virginia" shows the reported Paleozoic vertebrates from the Greenbrier Valley karst caves (Table 20.3).

Table 20.1 Greenbrier Valley bone caves: short list

Cave	County	Genus	Species	Name
Patton cave	Monroe	*Martes*	*americana*	Pine marten
Roadside pit	Pocahontas	*Martes*	*americana*	Pine marten
Buckeye Creek cave	Greenbrier	*Mammut*	*americanium*	Mastodon
Bransford cave	Monroe	*cf. Ursus*	*americanum*	Black bear
Cass cave	Poncahontas	*Mammut*	*americanum*	Mastodon
McFerrin Saltpeter cave	Greenbrier	*Mammut*	*americanum*	Mastodon
Organ cave	Greenbrier	*Mammut*	*americanum*	Mastodon
Rapps cave	Greenbrier	*Mammut*	*americanum*	Mastodon
Scott Hollow cave	Monroe	*Mammut*	*americanum*	Mastodon baby's pallate
Scott Hollow cave	Monroe	*Mammut*	*americanum*	Mastodon tusk fragment
Scott Hollow cave	Monroe	*Mammut*	*americanum*	Mastodon, 7 teeth, 3 indiv
Scott Hollow cave	Monroe	*Mammut*	*americanum*	Mastodon, humerus
Laurel Creek cave	Monroe	*Ursus*	*americanus*	Black bear
Organ cave	Greenbrier	*Ursus*	*americanus*	Black bear
Patton cave	Monroe	*Sorex*	*arcticus*	Arctic shrew
Organ cave	Greenbrier	*Ursus*	*arctos horribilis*	Grizzly bear
Organ cave	Greenbrier	*Dasypus*	*bellus*	Armadillo
Helix cave	Greenbrier	*Bootherium*	*bombifrons*	Harlan' muskox
Patton cave	Monroe	*Lasiurus*	*borealis*	Red bat
Patton cave	Monroe	*Synaptomys*	*borealis*	Northern bog lemming
Patton cave	Monroe	*Blarina*	*brevicauda*	Short-tailed shrew
Roadside pit	Pocahontas	*Blarina*	*brevicauda*	Short-tailed shrew
Patton cave	Monroe	*Parascalops*	*breweri*	Hairy-tailed mole
Bransford cave	Monroe	*Castor*	*canadensis*	Beaver
Benedicts cave	Greenbrier	*Martes*	*cf. americana*	Pine marten
Cass cave	Poncahontas	*Ursus*	*cf. americanus*	Black bear
Piddling pit	Pocahontas	*Ursus*	*cf. americanus*	Black bear
Ramp hole	Pocahontas	*Blarina*	*cf. brevicauda*	Short-tailed shrew
Steam cave	Pocahontas	*Equus*	*cf. fraternus*	Horse
Windy Mouth cave	Greenbrier	*Eptesicus*	*cf. grandis*	Big brown bat
Patton cave	Monroe	*Myotis*	*cf. gresescens*	Gray bat
Organ-Hedrick's cave	Greenbrier	*Myotis*	*cf. grisescens*	Gray bat
Windy Mouth cave	Greenbrier	*Myotis*	*cf. grisescens*	Gray bat
Ramp hole	Pocahontas	*Procyon*	*cf. lotor*	Raccoon
Cabble cave	Greenbrier	*Mustela*	*cf. nivalis*	Least weasel
Ramp hole	Pocahontas	*Sorex*	*cf. palustris*	Water shrew
Bransford cave	Monroe	*Microtus*	*cf. pinetorum*	Pine vole
Grapevine cave	Greenbrier	*Arctodus*	*cf. pristinus*	Short-faced bear
Patton cave	Monroe	*Pipistrellus*	*cf. subflavus*	Pipistrelle bat
Renick Quarry cave	Greenbrier	*Platygonus*	*cf. vetus*	Peccary
Organ-Hedrick's cave	Greenbrier	*Odocoileus*	*cf. virginianus*	White-footed mouse
Ramp hole	Pocahontas	*Odocoileus*	*cf. virginianus*	White-footed mouse
Patton cave	Monroe	*Sorex*	*cinereus*	Masked shrew
Beckeye Creek cave	Greenbrier	*Platygonus*	*compressus*	Flat-head paccary
Bransford cave	Monroe	*Platygonus*	*compressus*	Flat-head paccary

(continued)

Table 20.1 (continued)

Cave	County	Genus	Species	Name
Organ cave	Greenbrier	*Platygonus*	*compressus*	Flat-head paccary
Patton cave	Monroe	*Platygonus*	*compressus*	Flat-head paccary
Higginbotham's cave	Greenbrier	*Felis*	*concolor*	Mountain lion
Roadside pit	Pocahontas	*Felis*	*concolor*	Mountain lion
Patton cave	Monroe	*Condylura*	*cristata*	Star-nosed mole
Renick Quarry cave	Greenbrier	*Platygonus*	*cumberlandensis*	Peccary
Organ cave	Greenbrier	*Canis*	*dirus*	Dire wolf
Renick Quarry cave	Greenbrier	*Canis*	*dirus*	Dire wold
The hole	Monroe	*Canis*	*dirus*	Dire wolf
Lightner pit #2	Greenbrier	*Erethizon*	*dorsatum*	Porcupine
Ludington's cave	Greenbrier	*Erethizon*	*dorsatum*	Porcupine
Organ cave	Greenbrier	*Erethizon*	*dorsatum*	Porcupine
Patton cave	Monroe	*Erethizon*	*dorsatum*	Porcupine
Ramp hole	Pocahontas	*Erethizon*	*dorsatum*	Porcupine
Roadside pit	Pocahontas	*Erethizon*	*dorsatum*	Porcupine
Windy mouth cave	Greenbrier	*Erethizon*	*dorsatum*	Porcupine
Organ cave	Greenbrier	*Mylohyus*	*exortivus*	Peccary
Patton cave	Monroe	*Neotoma*	*floridana*	Woodrat
Roadside pit	Pocahontas	*Neotoma*	*floridana*	Woodrat
Cave near Patton cave	Monroe	*Sylvilagus*	*floridanus*	Eastern cottontail rabbit
Patton cave	Monroe	*Sangamona*	*fugitiva*	Fugitive deer
Friars Hole Cave system	Pocahontas	*Eptesicus*	*fuscus*	Big brown bat
Patton cave	Monroe	*Clethrionomys*	*gapperi*	Red-backed vole
Organ cave	Greenbrier	*Myotis*	*grisescens*	Gray bat
Patton cave	Monroe	*Microsores*	*hoyi*	Pigmy shrew
Patton cave	Monroe	*Tamiasciurus*	*hudsonicus*	Red squirrel
Roadside pit	Pocahontas	*Tamiasciurus*	*hudsonicus*	Red squirrel
Patton cave	Monroe	*Zapus*	*hudsonicus*	Meadow jumping mouse
McClungs cave	Greenbrier	*Napaeozapus*	*insignis*	Woodland jumping mouse
Patton cave	Monroe	*Phenacomys*	*intermedius*	Heather vole
Organ-Hedrick's cave	Greenbrier	*Myotis*	*keenii*	Keen's bat
Patton cave	Monroe	*Myotis*	*keenii*	Keen's bat
Scott Hollow cave	Monroe	*Procyon*	*lotor*	Raccoon
Piercys Mill cave	Greenbrier	*Canis*	*lupus*	Gray wolf
Ramp hole	Pocahontas	*Neotoma*	*magister*	Woodrat
Cave near Patton cave	Monroe	*Mephitis*	*mephitis*	Striped skunk
Bransford cave	Monroe	*Marmota*	*monax*	Woodchuck
Friars hole cave system	Pocahontas	*Marmota*	*monax*	Woodchuck
Organ-Hedrick's cave	Greenbrier	*Marmota*	*monax*	Woodchuck
Patton cave	Monroe	*Marmota*	*monax*	Woodchuck
Roadside pit	Pocahontas	*Marmota*	*monax*	Woodchuck
Organ cave	Greenbrier	*Mylohyus*	*nastus*	Peccary
Windy Mouth cave	Greenbrier	*Panthera*	*onca*	Jaguar
Patton cave	Monroe	*Microtus*	*pennsylvanicus*	Meadow vole
McClungs cave	Greenbrier	*Lynx*	*rufus*	Bobcat

(continued)

Table 20.1 (continued)

Cave	County	Genus	Species	Name
Rapps cave	Greenbrier	*Homo*	*sapiens*	Human
Roadside pit	Pocahontas	*Bison*	*sp.*	Bison
Organ cave	Greenbrier	*Equus*	*sp.*	Horse
Patton cave	Monroe	*Myotis*	*sp.*	Little brown bat
Ramp hole	Pocahontas	*Myotis*	*sp.*	Little brown bat
Rapps cave	Greenbrier	*Ochotona*	*sp.*	Pika
Bransford cave	Monroe	*Peromyscus*	*sp.*	Deer or white-footed mouse
Cabble cave	Greenbrier	*Peromyscus*	*sp.*	Deer or white-footed mouse
Organ-Hedrick's cave	Greenbrier	*Peromyscus*	*sp.*	Deer or white-footed mouse
Patton cave	Monroe	*Peromyscus*	*sp.*	Deer or white-footed mouse
Patton cave	Monroe	*Peromyscus*	*sp.*	Deer or white-footed mouse
Ramp hole	Pocahontas	*Peromyscus*	*sp.*	Deer or white-footed mouse
Roadside pit	Pocahontas	*Peromyscus*	*sp.*	Deer or white-footed mouse
Lightner pit #2	Greenbrier	*Rangifer?*	*sp.*	Caribou
Lightner pit #2	Greenbrier	Sangamona?	*sp.*	Deer
Organ cave	Greenbrier	*Smilodon*	*sp.*	Saber-toothed cat
Buckeye Creek cave	Greenbrier	*Tapirus*	*sp.*	Tapir
Rockland cave	Monroe	*Tapirus*	*sp.*	Tapir
Roadside pit	Pocahontas	*Tamias*	*striatus*	Eastern chipmunk
Cassel cave	Pocahontas	*Pipistrellus*	*subflavus*	Pipistrelle bat
Patton cave	Monroe	*Pipistrellus*	*subflavus*	Pipistrelle bat
Organ cave	Greenbrier	*Rangifer*	*tarandus*	Caribou
Fletchers	Monroe	*Equus*	*tau*	Horse
Patton cave	Monroe	*Taxidea*	*taxus*	Badger
Bransford cave	Monroe	*Odocoileus*	*virginianus*	White-tail deer
Patton cave	Monroe	*Odocoileus*	*virginianus*	White-tail deer
Patton cave	Monroe	*Vulpes*	*vulpes*	Red fox
Scott Hollow cave	Monroe	*Megalonyx*	*wheatleyi*	Ground sloth
Patton cave	Monroe	*Microtus*	*xanthognathus*	Yellow-cheeked vole
Patton cave	Monroe	*Bear*		Bear
Unnamed cave	Greenbrier	*Bird*		Bird
Scott Hollow cave	Monroe	*Bufo*		Toad
Carpenters-Swago cave	Pocahontas	*Cat*		Cat
Scott Hollow cave	Monroe	*Cf. bison or musk ox*		Cf. bison or musk ox
Scott Hollow cave	Monroe	*Cf. moose or elk*		Cf. moose or elk
Unnamed cave	Greenbrier	*Elk or muskox*		Elk or muskox
Bransford cave	Monroe	*Frog or toad*		Frog or toad
Unnamed cave	Greenbrier	*Horse*		HORSE
Roadside pit	Pocahontas	*Microtine*		Microtine
Fletchers	Monroe	*Mylohyus*		Peccary
Scott Hollow cave	Monroe	*Peccary*		Peccary
Unnamed cave	Greenbrier	*Peccary*		Peccary
Patton cave	Monroe	*Rabbit or hare*		Rabbit or hare
Roadside pit	Pocahontas	*Rabbit or hare*		Rabbit or hare
Fletchers	Monroe	*Wolf*		Wolf

Table 20.2 Carbon-14 dates from Greenbrier Valley bone caves

Locality/cave	County	C14 Age	±Years	Description	Lab	Date	Author
Haynes	Monroe	35960	210	*Megalonyx jeffersonii*	WVGS 160	April, 2005	Grady and Garton
Helix	Greenbrier	26300	240	*Platygonus*	TRA3	May 31, 2011	Grady and Garton
Organ	Greenbrier	21040	760	*Smilodon site*	Beta 18349	June 8, 1905	Grady and Garton
Patton	Monroe	14450	70	*Microtus xantognathus*	NSRL 2626	June 7, 1985	Graham and Stafford
Patton	Monroe	13350	240	Upper site	Beta 4309	June 4, 1985	Grady and Garton
Patton	Monroe	22260	760	Lower site	Beta 4308	June 4, 1985	Grady and Garton
Scott Hollow	Monroe	11350	360	*Mastodon* tooth	30837	March, 2004	Grady and Garton
Scott Hollow	Monroe	21830	660	*Mastodon* bone	30838	March, 2004	Grady and Garton
Lost World Caverns	Greenbrier	32200	400	*Ursum americanus*	GX-33837-AMS	June 26, 2015	Garton and Silverberg

Fig. 20.5 *Orthocanthus* sp. shark spine found in a cave wall in Organ Cave, Greenbrier County

Table 20.3 Caves with Paleozoic fossils

Cave	County	Kind of fossil
Benedicts cave	Greenbrier	Teeth
Cassell cave	Pocahontas	Shark spines
Haynes cave	Monroe	Shark teeth
Haynes cave	Monroe	Shark crusher teeth
Laurel Creek cave	Monroe	Shark spine
Organ cave	Greenbrier	Shark spine
Unnamed cave	Greenbrier	Shark tooth

References

Barr, T.C., Jr. 1953. Fossil vertebrate remains from Laurel Creek Cave. *N.S.S. News* 11 (6): 2.

Gidley, J.W. 1918. A Pleistocene cave deposit of Western Maryland. *Smithsonian Institution Annual Report* 281–288.

Grady, F. 1988a. Fossil vertebrates. In *Caves of the Organ Cave Plateau*, ed. P. Stevens, vol. 9, 51–55. West Virginia Speleological Survey Bulletin.

Grady, F. 1988b. A preliminary account of the Pleistocene mammals from Patton Cave, Monroe County, West Virginia. *National Speleological Society Bulletin* 50 (1): 9–16.

Grady, F., and R. Garton. 1994. The Pleistocene peccaries *Platygonus vetus* and *Platygonus compressus* from West Virginia caves. *Proceedings of the West Virginia Academy of Science* 66 (1): 52.

Grady, F., J. Arroyo-Cabrales, and E.R. Garton. 2002. The northernmost occurrence of the Pleistocene vampire bat *Desmodus stocki* Jones (Chiroptera: Phyllostomatidae: Desmodontinae) in eastern North America. *Smithsonian Contributions to Paleobiology* 93: 73–75.

Guilday, J.E., and A. McCrady. 1966. Armadillo remains from Tennessee and West Virginia caves. *National Speleological Society Bulletin* 28 (4): 183–184.

Handley Jr., C.O. 1956. Bones of mammals from West Virginia caves. *American Midland Naturalist* 56 (1): 250–256.

Jefferson, T. 1799. A memoir on the discovery of certain bones of a quadruped of the clawed kind in the western parts of Virginia. *American Philosophical Society Transactions* 4: 246–260.

Mercer, H.C. 1897. The finding of remains of the fossil sloth at Big Bone Cave, Tennessee in 1896. *American Philosophical Society Transactions* 36 (154): 36–70.

Shockey, R. 1968. Remains of giant bear found in Grapevine. *Baltimore Grotto News* 9 (12): 317–342.

Stafford R. 1961. Bones in a cave. *N.S.S. News* 19(7): 80.

Wistar, C. 1799. A description of the bones deposited by the President in the Museum of the Society. *American Philosophical Society Transactions* 4: 526–531.

The manufacturer's authorised representative in the EU is Springer Nature Customer Service Centre GmbH, Europaplatz 3, 69115 Heidelberg, Germany. If you have any concerns regarding our products, please contact ProductSafety@springernature.com

Printed and bound by CPI Group (UK) Ltd, Croydon, CR0 4YY
30/09/2026
02225516-0010